WDM Systems and Networks

For further volumes:
http://www.springer.com/series/6976

Optical Networks

Series Editor Biswanath Mukherjee
University of California
Davis, CA

Neophytos (Neo) Antoniades
Georgios Ellinas · Ioannis Roudas
Editors

WDM Systems and Networks

Modeling, Simulation, Design and Engineering

Springer

Prof. Neophytos (Neo) Antoniades
Department of Engineering Science
and Physics
College of Staten Island
The City University of New York
Staten Island, NY 10314
USA
e-mail: neo.antoniades@csi.cuny.edu

Georgios Ellinas
Department of Electrical
and Computer Engineering
University of Cyprus
Nicosia, Cyprus
e-mail: gellinas@ucy.ac.cy

Ioannis Roudas
Department of Electrical Engineering
University of Patras
26504 Rio, Greece
e-mail: roudas@ece.upatras.gr

Series Editor
Biswanath Mukherjee
University of California
Davis, CA
USA

ISSN 1935-3839
e-ISSN 1935-3847
ISBN 978-1-4614-1092-8
e-ISBN 978-1-4614-1093-5
DOI 10.1007/978-1-4614-1093-5
Springer New York Dordrecht Heidelberg London

Library of Congress Control Number: 2011941711

Printed on acid-free paper

Springer is part of Springer Science+Business Media (www.springer.com)

Neo Antoniades dedicates this book to his son Andreas for his love of science

Georgios Ellinas dedicates this book to his nephew Nicolas and his niece Carina

Ioannis Roudas dedicates this book to Linda, Nicholas and Adrian

Foreword

High-capacity optical networks employing wavelength-division multiplexing (WDM) form the underpinnings of our telecom backbone networks. The continued developments of WDM systems and networks and related technologies are crucial for the evolution of the future Internet. Important problems include accurate modeling, simulation, design, and engineering of components, devices, subsystems, systems, and the network as a whole. The authors of this book must be complimented for capturing these important problems across this entire ecosystem. This book presents in a clear and illustrative format the technical and scientific concepts that are needed to accomplish the design of WDM systems and networks (and hence the future Internet).

The book has been co-edited by Professor Neo Antoniades of The City University of New York, Professor George Ellinas of the University of Cyprus, and Professor Ioannis Roudas of the University of Patras, Greece. The editors bring nearly 40 years of combined experience on various aspects of optical WDM systems and networks in preparing this book. They provide very good balance between theory and practice, having worked in the telecom industry (Bellcore, Tellium, Corning) as well as in academe.

The book is divided into three parts: (1) Tools and Methods; (2) Implementations; and (3) The Logical Layer. Tools and Methods capture the state of the art on (a) Device and network element modeling; (b) Network modeling and performance evaluation; (c) Modeling of the transport systems; and (d) Commercial software simulation tools. Particularly noteworthy is the fact that modeling and simulation methods in software tools from leading vendors (such as VPI Systems, RSoft, and Optiwave) are covered.

Part II on Implementations captures the relevant problems and challenges from short range to longer range systems such as: (a) Optical interconnects; (b) In-building systems (using plastic optical fibers); (c) Radio-over-fiber (RoF) systems; (d) Passive optical networks (PON) (for Fiber-to-the-home (FTTH) applications); and (e) Optical communication systems.

Part III on Logical Layer includes networking problems such as: (a) Physical-to-logical layer interactions and (b) Multi-domain optical networks and survivability.

The editors deserve praise because chapters on the above topics have been contributed by an excellent lineup of authors, many of whom work in leading companies in the field as well as in reputable universities, and they come from many different countries from around the world.

This book is highly recommended as it offers timely, accurate, and authoritative information on WDM systems and networks, particularly on their accurate modeling, simulation, design, and engineering. The reader will enjoy the book and learn a lot.

USA, August 2011 Biswanath Mukherjee

Preface

The demand for more and better applications for telecommunications networks has always driven the need for more bandwidth from these networks. At the same time, as the bandwidth has steadily increased over the last few decades, more applications have emerged that utilize this bandwidth. And so the "cycle" continues.

Fiber-optic networks have been in the forefront of the effort to provide the users with more bandwidth, enabling them to run a number of new applications that span the whole spectrum of human interaction, including day-to-day activities, entertainment, and business, among others. Optical systems and networks have seen an incredibly rapid evolution. Only a few decades separate the invention of the laser and the low-loss fiber from the wide usage of wavelength-division-multiplexed (WDM) fiber-optic communication systems in the current telecom networks. These systems are now in every part of the telecom infrastructure including undersea, backbone, metropolitan area, as well as access networks, reaching all the way into our homes and businesses. Although some parts of the network still include networking functions such as switching and regeneration in the electronic domain, the direction is more and more toward networks that are transparent to signal rate, protocol, and format, where the signal stays in the optical domain as long as possible.

This book provides the latest developments in the ever-expanding field of optical communication system and network design and engineering. It presents the industry, as well as current research, in state-of-the-art architectures of WDM optical systems and networks and takes a vertically layered (across the network layers) approach to system/network modeling, design, and engineering. Modeling and simulation techniques, in conjunction with experimental verification and engineering, are presented for different systems and networks. This book is different from a number of other books on optical systems and networks that are either general textbooks on optical networks or focus exclusively on the technology and point-to-point transmission. This contributed volume looks at both networking and system design issues, and focuses on the latest research developments in a number of areas including ultra long haul (ULH), metro, and access networks, as well as enabling technologies. It can be a very good supplement to any of the general optical networks/technology books for those interested in probing and understanding this area further. The aim of the editors is to present a body of work in this book that can provide the research scientist, company engineer, as well as the university professor/researcher with a better understanding of how to more efficiently design and engineer optical communication systems and networks. The intention is to demonstrate the effectiveness of computer-aided design when it comes to network engineering and prototyping.

This book is divided into three parts that can appeal to different readers, who are interested in different types of networks and different applications. The first part of the book (Chaps. 2–5) presents modeling approaches and simulation tools mainly for the physical layer (including transmission effects, devices, subsystems, and systems), whereas the second part (Chaps. 6–11), presents more engineering/design issues for various types of optical systems (including ULH, access, and in-building systems). The third part of the book (Chaps. 12–13) presents mostly networking issues related to the design of provisioning and survivability algorithms for impairment-aware and multi-domain networks. Finally, Chaps. 1 and 14 provide some introductory remarks and future directions respectively.

Acknowledgments

We are indebted to our past advisors, supervisors, colleagues, students and friends, who have motivated and inspired us in the optical communications paths that we took in our careers. In particular, we express our deep gratitude to our advisors and mentors Profs. Thomas E. Stern, Gee-Kung Chang, Mohamed Ali, Alfred Levine, Syed Ahamed, Syed Rizvi, Yves Jaouen, Constantin Caroubalos, Aurel Lazar and Manolis Sangriotis, and to our former colleagues Drs. Krishna Bala, Rich Wagner, Sarry Habiby, Aly F. Elrefaie, Jean-Francois Labourdette, Eric Bouillet, Janet Jackel, Richard Vodhanel, B. Roe Hemenway, Vassilis Keramidas, K. P. Ho, and Michael Sauer. All of them offered us their invaluable advise, exceptional insight, and foresight. They provided us with valuable guidance throughout the years and helped us better understand and appreciate various aspects of optical networking. We would also be remiss if we did not extend a thank you to Professor Biswanath Mukherjee for taking the time to write the foreword for this book and for his patience and encouragement while this book was being prepared and delivered. We also wish to express our thanks to Alex Greene, Allison Michael, and the entire publishing team at Springer Science+Business Media, for their effort and patience in order to bring this project to fruition. Special thanks go to Profs. N. Madamopoulos' and N. Antoniades' Ph.D. student Sasanthi Peiris for her great editorial support on this project. Also, special thanks goes to all the authors who contributed chapters for this book. Finally, Georgios Ellinas is greatly indebted to his family for their understanding and patience during this undertaking. Neo Antoniades wishes to express his profound gratitude to his wife Beatriz for her support and help while working on this project, to his parents for their lifelong inspiration and love and lots of thanks to Andreas, Pablo, George and Allison for making it fun. Ioannis Roudas gratefully acknowledges his family, whose loving support, help, and understanding has made his involvement in this book possible.

Neophytos (Neo) Antoniades
Georgios Ellinas
Ioannis Roudas

Contents

Contributors

Marianna Angelou Athens Information Technology, Peania, Attiki, Athens, Greece, e-mail: mang@ait.edu.gr

Neophytos (Neo) Antoniades Department of Engineering Science and Physics, College of Staten Island/The City University of New York, Staten Island, NY 10314, USA, e-mail: neo.antoniades@csi.cuny.edu

Siamak Azodolmolky Athens Information Technology Peania, Attiki, Athens, Greece, e-mail: sazo@ait.edu.gr

Sergey Burtsev Xtera Communications Inc., Allen, TX, USA, e-mail: sergey.burtsev@xtera.com

Georgios Ellinas Department of Electrical and Computer Engineering, University of Cyprus, Nicosia, Cyprus, e-mail: gellinas@ucy.ac.cy

Patricia Fernández Department of Signal Theory, Communications and Telematic Engineering, University of Valladolid, Valladolid, Spain, e-mail: patfer@tel.uva.es

Nasir Ghani Department of Electrical and Computer Engineering, University of New Mexico, Albuquerque, NM, USA, e-mail: nghani@ece.unm.edu

Timothy B. Gibbon Department of Photonics Engineering, Technical University of Denmark, Lygnby, Denmark, e-mail: timothygibbon@yahoo.com

Vladimir S. Grigoryan Ciena Corporation, Linthicum, MD, USA, e-mail: vgrigory@ciena.com

Ramón Gutiérrez-Castrejón Universidad Nacional Autónoma de México (UNAM), Mexico City, Mexico, e-mail: RGutierrezC@iingen.unam.mx

Ronald Holzlöhner Laser Guide Star Department, European Southern Observatory (ESO), Garching, Germany, e-mail: rholzloe@eso.org

Alicia López Photonic Technologies Group, i3A, University of Zaragoza, Zaragoza, Spain, e-mail: aliclope@unizar.es

María Angeles Losada, Photonic Technologies Group, i3A, University of Zaragoza, Zaragoza, Spain, e-mail: alosada@unizar.es

Nicholas Madamopoulos Department of Electrical Engineering, City College of New York/The City University of New York, New York, NY 10031, USA, e-mail: nmadamopoulos@ccny.cuny.edu

Juan José Martínez Photonic Technologies Group, i3A, University of Zaragoza, Zaragoza, Spain, e-mail: jjosemar@unizar.es

Javier Mateo Photonic Technologies Group, i3A, University of Zaragoza, Zaragoza, Spain, e-mail: jmateo@unizar.es

Noemí Merayo Department of Signal Theory, Communications and Telematic Engineering, University of Valladolid, Valladolid, Spain, e-mail: noemer@tel.uva.es

Idelfonso T. Monroy Department of Photonics Engineering, Technical University of Denmark, Lygnby, Denmark, e-mail: idtm@fotonik.dtu.dk

Tania Panayiotou Department of Electrical and Computer Engineering, University of Cyprus, Nicosia, Cyprus, e-mail: panayiotou.tania@ucy.ac.cy

Min Peng Wuhan University, Wuhan, People's Republic of China, e-mail: pengm@whu.edu.cn

Kamau Prince Department of Photonics Engineering, Technical University of Denmark, Lygnby, Denmark, e-mail: kpri@fotonik.dtu.dk

Ammar Rayes Cisco Systems, Inc., San Jose, CA, USA, e-mail: rayes@cisco.com

Dwight H. Richards Department of Engineering Science, College of Staten Island/The City University of New York, Staten Island, NY 10314, USA, e-mail: drichards@csi.cuny.edu

André Richter VPIphotonics Division, VPIsystems, Berlin, Germany, e-mail: Andre.Richter@VPIphotonics.com

Ioannis Roudas Department of Electrical Engineering, University of Patras, 26504 Rio, Greece, e-mail: roudas@ece.upatras.gr

Oleg V. Sinkin System Modeling and Signal Processing Research, Tyco Electronics Subsea Communications, Eatontown, NJ, USA, e-mail: osinkin@subcom.com

Ioannis Tomkos Athens Information Technology, Peania, Attiki, Athens, Greece, e-mail: itom@ait.edu.gr

Xianbin Yu Department of Photonics Engineering, Technical University of Denmark, Lygnby, Denmark, e-mail: xiyu@fotonik.dtu.dk

Chapter 1
Modeling, Simulation, Design and Engineering of WDM Systems and Networks: An Introduction

Georgios Ellinas, Neophytos (Neo) Antoniades and Ioannis Roudas

Abstract Optical systems and networks have evolved enormously in the last three decades with the creation of next-generation optical components, subsystems, systems, and networks that are now utilized in all aspects of the network structure starting from the in-house/building and access networks, all the way up to the backbone and ultra long-haul infrastructures. This book addresses a variety of issues related to next-generation optical systems and networks, starting from modeling and simulation and then proceeding to design and engineering, enabling the reader to gain an understanding of the latest developments in the ever-expanding field of optical communication systems and networks and of all the open technological and research issues. It presents the state-of-the-art architectures of optical systems and networks and takes a vertical approach to system/network modeling and design, starting from the transmission effects and component level and moving all the way up to the control layer. Analysis is mixed with modeling/simulation and engineering approaches to present the current industry techniques, as well as the research state-of-the-art.

G. Ellinas (✉)
Department of Electrical and Computer Engineering
University of Cyprus
1678 Nicosia, Cyprus
e-mail: gellinas@ucy.ac.cy

N. (Neo) Antoniades
Department of Engineering Science and Physics
College of Staten Island/The City University of New York
Staten Island, NY 10314, USA
e-mail: neo.antoniades@csi.cuny.edu

I. Roudas
Department of Electrical Engineering
University of Patras
26504 Rio, Greece
e-mail: roudas@ece.upatras.gr

N. (Neo) Antoniades et al. (eds.), *WDM Systems and Networks*, Optical Networks,
DOI: 10.1007/978-1-4614-1093-5_1,

1.1 Introduction

In the past few decades, innovations in optical components, subsystems, systems, and networks have made it possible to satisfy the enormously increasing network traffic demand resulting from the high-bandwidth applications required by end users. Today, 40-channel or 80-channel 10 Gbps optical communications systems are barely meeting the requirements for high-capacity transmission. Under development next-generation optical communications systems exhibit single-channel data rates of 40 and 100 Gbps for long-haul and are targeting 10 Gbps for the access with 1 Gbps symmetric access to customers being the logical step. It is expected that data traffic will maintain an exponential increase in the foreseeable future due to the rapidly increasing use of high-speed online services and broadband applications, such as telepresence, high-definition TV (HDTV), teleconferencing, telemedicine, data storage, grid computing, metacomputing, and others. This trend keeps pushing the broadband communication infrastructure towards higher speeds, greater capacity, longer reach, and closer to the end user. Thus, the need to better design and engineer all types of optical networks, including core, metro, and access networks has been a high priority of telecommunication engineers and researchers in the last few decades.

With the successful commercialization of wavelength-division multiplexing (WDM) and the advent of dense WDM (DWDM) the optical transport network (OTN) has been standardized by the ITU-T as the underlying infrastructure of the network [1]. This provides for the management of services using multiple, different wavelengths of light over the same optical fibers, with the clients of the OTN being able to occupy up to a full wavelength. Furthermore, this allows the network to transport huge amounts of data that are needed for many current and future communication services and applications. WDM was first employed in point-to-point systems where fiber was used as a direct substitute for copper. In this approach all switching and processing of the data was handled by electronics and the optical traffic-carrying signal would undergo optical-electronic-optical (OEO) conversions at every switching/routing node. The scaling limitations in signal bit rate, switch matrix port-count, and network element cost were the key motivations behind the attempt to develop large port-count transparent switches to be used in opaque network architectures [2]. The key advances within the optical networking space in the second half of the 1990s and the early 2000s were mainly due to the development of new DWDM component, subsystem, and system technologies that resulted in the second generation optical networks where the provisioning and recovery functionalities moved to the optical part of the network [3].

Clearly, the ultimate goal is the design of completely transparent optical networks, as these networks offer the most economical solution with transparency in bit rate, signal format, and protocol. However, transparent optical networks face a number of challenges, with the two main ones being the difficulty in control and management of these networks and the need for end-to-end engineering due to the accumulation of physical impairments in the optical path. A middle-of-the-path solution to this problem is the creation of networks with some transparency but also some opacity, with the signal undergoing OEO conversions at specified points in the network. These types of networks are called translucent network architectures that seek a graceful balance among network design, cost, and service provisioning performance by exploiting the advantages of both transparent and opaque optical networks while eliminating their drawbacks. As the traffic grows and the bit rates increase substantially, translucent (optical-bypass) core WDM networks using reconfigurable optical add/drop multiplexers (ROADMs) and tunable lasers appear to be on the road towards widespread deployment while all-optical networks using DWDM technology appear to be the sole approach for transporting huge network traffic in future core networks [4].

Over the years the functionality of the physical layer is enhanced, with OTN providing for carrier-grade operations, administration, and maintenance (OAM) for these managed wavelength services, forward error correction (FEC) for increased reach and margin, and protection switching for high availability. Focus on manageability has resulted in advances in control and management, including taking control techniques designed for the electrical transport layers of the network and adapting them to the optical transport layer. As an example, the control protocol known as generalized multiprotocol label switching (GMPLS) has been proposed for application to control functions in the physical layer. GMPLS is a generalization of multiprotocol label switching (MPLS), which was designed as an improvement over the packet-forwarding techniques used in IP networks and to ensure automated provisioning, to maintain connections, and to manage the network resources including providing quality-of-service (QoS) and traffic engineering (TE) [5].

A typical WDM system and network spans the entire networking space consisting of the in-house/building, access, metropolitan-area, long-haul, and ultra long-haul networks. The distances covered by these networks include everything from a few meters for in-house/building networks to ultra long-haul transmission links such as transoceanic cables that can range from a few hundred kilometers in distance to several thousands of kilometers. The topologies of these networks also vary from tree structures for newly proposed passive optical network (PON)-based access networks, to ring (and stacked/interconnected ring) structures in metro and long-haul networks, to arbitrary mesh topologies for backbone networks.

Thus, optical networks have evolved from point-to-point WDM transmission systems, to star, bus, and ring networks to arbitrary connected mesh networks utilizing intelligent network elements such as reconfigurable optical cross-connects or ROADMs among others that enable them to support network functionalities such as automated provisioning of optical services, automated protection/restoration capabilities to combat failure scenarios, and others. These networking equipment provide for the seamless transport of the traffic with the QoS and availability that meets the customer's requirements. The applications that are supported in these networks range from plain old telephone service (POTS) to Internet data and business-to-business services. Clearly, future optical networks will be required to support not only point-to-point connections but multicast and groupcast connections as well, as the number of applications that require such connections is expected to rise significantly in the near future.

New trends, new technologies, and new applications are in the forefront of optical systems and optical networks. Below we describe three simple examples that highlight new applications and new design requirements, and the need for new technologies and new techniques to address them efficiently.

As an example, consider the use of single-mode fiber (SMF), multi-mode optical fiber (MMF), and plastic optical fiber (POF) for next-generation avionics platforms. Most airplanes, including commercial and military aircrafts, have silica multimode optical fiber links installed for data transfer from/to on-board surveillance equipment and for internal communications. POF with its large diameter offers a good alternative for low-bandwidth avionics applications due to cost and ease of installation and should see implementations in the near future [6]. As the bandwidth required by different on-board applications is steadily growing, there is an urgent need for technology upgrades to higher data rates. So far, MMF transmission in avionics environments has been limited to data rates on the order of several dozen to a few hundred Mbps. Recent new applications such as entertainment on board commercial airplanes require more bandwidth, so new transmission techniques need to be considered. Since MMF bandwidth is limited by modal dispersion, a state-of-the-art approach is to use advanced modulation formats, such as quadrature phase shift keying (QPSK) and orthogonal frequency division multiplexing (OFDM) [7], as described in Chap. 10, in conjunction with multiple-input multiple-output (MIMO) techniques, in order to either overcome or be able to electronically compensate modal dispersion. Multi-level modulation and orthogonal modulation have been used broadly in wireless and radio frequency (RF) applications. These advanced modulation techniques that use orthogonal carriers to transport the phase and amplitude information and parallel streams to effectively transmit and receive a high-bit-rate stream and achieve high spectrum efficiency can be applied in this application to overcome the limitations of the existing systems. SMF on the other hand is the clear favorite for military avionics platforms where the amount of information is putting high requirements on bandwidth. As this example clearly shows, a combination of technologies (types of fiber) and transmission techniques (modulation formats) is required to address this new relatively simple application.

Another example of an optical networking application that requires new design and engineering approaches is the provisioning of broadband services inside buildings. For this application there can be two possible solutions: (1) a wireline approach with the usage of MMF or POF in these systems (discussed in Chap. 7), or (2) a wireless/wireline approach with the usage of radio-over-fiber (RoF) technology (examined in Chap. 8). RoF has recently been proposed for the transport of wireless signals particularly inside a building and has been proposed for use over glass MMF in [8]. The idea is to provide shared bandwidth to wireless users using optical fiber and at the end on the wireless side, by reducing the cell size to pico-cell scale limiting the number of users and increasing their bandwidth. This is achieved since a reduced cell size maintains a high signal-to-noise ratio (SNR) at the receiver with moderate transmit power. For applications with high user density a cell may only have a radius of a few meters and as a result covering the space of a very large room will require tens of antennas. Despite the burden on the wired infrastructure that the above seems to bring, it has been demonstrated that radio-over-fiber approaches can result in a simplified overall system design for a pico-cellular network by aggregating the RF signal generation and network management at a central location [9] and using optical fiber to distribute the signal to the various antenna sites. Clearly, this is another example that shows how new systems can be designed and engineered for the delivery of services utilizing different types of fibers and millimeter waves.

A third (networking) example detailing the new trends in the design of optical networks, deals with the inclusion of the physical layer impairments in the algorithmic design necessary for the development of provisioning and survivability techniques. If the network is transparent, where the signal stays in the optical domain for the entire path, efficient routing and wavelength assignment (RWA) and protection algorithms have to account for the physical layer impairments (noise, crosstalk, dispersion, etc.) that the optical signal encounters. Clearly, the impact of the physical layer impairments depends on the design and engineering of the network components and subsystem architectures (e.g., optical node architecture) and different designs will achieve different network performance. If this example is extended to translucent networks, a new interesting problem that arises is the placement of the opaque nodes (e.g., regenerators) in conjunction with the impairment-aware RWA and protection algorithms.

As the aforementioned examples amply demonstrate, advances are being undertaken in all aspects of optical systems and networks. This book provides the latest developments in the ever expanding field of optical communication system and network design and engineering. It presents the state-of-the-art architectures of WDM optical systems and networks and takes a vertical approach to system/network modeling and design, starting from the transmission layer and moving all the way to the control layer. A detailed discussion of the main physical transport layer effects and models is followed by a discussion of current trends in WDM technologies and their corresponding modeling. The latest in semi-analytical system simulation techniques are applied to optical WDM systems and networks and a review of various current areas of optical communication systems and optical networks is presented. Modeling and simulation is mixed with experimental verification and engineering to present the industry as well as research state-of-the-art. The goal of this book is to provide the professional scientist, engineer, and industrial/academic researcher with the basic skills, concepts, and simulation/design techniques so as to be able to design and engineer optical communication systems and networks at various layers.

1.2 Organization of the Book

This book is a combination of a number of topics related to the design (from a modeling and simulation perspective) and engineering of next-generation optical systems and networks that include components, devices, subsystems, systems, and networks, but it is by no means an exhaustive list. This book identifies engineering and research areas and topics that are in the forefront of research in fiber-optic networks and that could be of interest to the reader. It looks at both networking and system design issues and focuses on the latest research developments in a number of areas including optical interconnection, in-building, access, and long-haul (LH) networks, as well as enabling technologies, and as such, it can be of interest to all telecommunication network engineers and researchers that want to probe and understand these areas further.

The book is divided into three main parts; Readers who are interested in the modeling and simulation perspective can read the first part of the book consisting of Chaps. 2–5 that deal with the modeling methodologies of transmission effects, components, devices, subsystems, systems, and networks and the design of software simulation tools for WDM systems and networks. The second part (Chaps. 6–11), apart from the modeling methodology, also includes a more engineering flavor and examines the system engineering/design of optical interconnects, in-building systems, RoF systems, access networks, and the upgrade of terrestrial and undersea systems. This part of the book will be of particular interest to those involved in WDM systems and in optical network engineering and deployment. The final part of the book comprising Chaps. 12 and 13 moves to the control layer (logical layer) and involves algorithmic designs taking into account physical-to-logical layer interactions and the case of multi-domain networks in the design of provisioning and survivability algorithms for these networks. In summary, the reading of the book can be customized as suited to each individual's interests in exploring the issues related to the modeling, design, and engineering of WDM systems and networks. Several other new areas of engineering and research for multiwavelength optical systems and networks are outlined in Chap. 14 as areas of future and promising work that can be of interest to the reader. The reader is encouraged to explore further the references provided in that chapter to gain a better understanding of the ongoing work and the important issues related to the research work on those topics.

In more detail, the rest of the chapters in this book provide a concise treatment of the following topics:

Chapter 2 gives an overview of the computer modeling for the physical (i.e., optical transport) layer effects of multiwavelength optical networks. It initially describes different physical effects causing system degradation in the WDM transport layer including fiber propagation, optical amplification, and optical signal generation effects. In terms of fiber propagation effects this chapter considers chromatic dispersion, fiber nonlinearities (including Kerr effect, self- and cross-phase modulation, four wave mixing, and scattering effects), polarization mode dispersion (PMD) and polarization-dependent loss/gain (PDL/PDG) and then describes modeling signal propagation over SMF. Optical amplification effects include amplified spontaneous emission (ASE) noise, gain and optical signal-to-noise ratio (OSNR) ripple effects, as well as transients, while optical signal generation effects consist of different types of modulation (direct and external) and various transmitter effects including transmitter-induced distortion and chirp, as well as transmitter-induced noise. Other transmission effects discussed in Chap. 2 include crosstalk and filter concatenation-induced distortion. Chapter 3 builds on Chap. 2 and focuses on the state of the art in passive device and component-level modeling (including optical fiber, multiplexers/demultiplexers, switches, and dynamic spectrum equalizers), as well as in network element-level modeling (including optical add drop multiplexers (OADMs) and optical cross-connects (OXCs)). The chapter shows how complex structures such as wavelength selective switches and optical add drop multiplexers can be built from simple structures (e.g., switches, filters, etc), utilizing a simulation tool. Different techniques are also described on how to model more sophisticated active devices such as lasers, amplifiers, modulators, and receivers.

The development of models for the network performance evaluation of multiwavelength optical networks based on the modeling described in Chaps. 2 and 3 is crucial in the deployment of these networks. Chapter 4 addresses precisely this problem by introducing semi-analytical approaches for analyzing signal statistics in optical (nonlinear) transmission systems and ascertaining the fiber network performance. The chapter describes several semi-analytical methods, including linearization methods for the accurate computation of the bit error rate (BER) through the covariance matrix analysis and momentum-based methods for computationally efficient analysis of the amplitude and timing jitter as well as signal statistics. Chapter 5 combines the ideas presented in Chaps. 2–4 by presenting and reviewing the existing commercial optical communication software simulation tools, which include modeling and simulation of optical components, devices, subsystems, systems, and networks. Different capabilities of these simulation tools are described for different layers such as the physical, systems, and network layer.

As in the last few years optical interconnect technology has become less costly and with enhanced functionality, there is increased interest in developing optical interconnect technology for shorter and shorter distances, including rack-to-rack, board-to-board, inter-chip, and intra-chip applications. Chapter 6 describes the current state-of-the-art on optical interconnects including rack-to-rack, chip-to-chip, and on-chip interconnects and delves into the current trends on data centers and on the 100 Gbps frontier. This chapter details the latest optical interconnect systems (including system requirements and design issues, photonic technologies utilized, and architectures developed) and outlines the significant challenges that must be overcome to provide significant performance enhancements compared to their electrical counterparts.

In Chap. 7 the issue of short range (in-building) systems and networks is examined, focusing on the recent emergence of optical in-building systems including the usage of plastic optical fiber (POF) in these systems. Modeling approaches for POF and for effectively simulating these systems and networks are discussed and different applications and implementations are described and analyzed.

Chapter 8 introduces the concepts of radio-over-fiber (RoF), focusing mainly on WDM phase modulated millimeter-wave fiber systems. RoF schemes facilitate wireless microwave signal coverage within a remote area; information is transported over an optical fiber by modulating the message-bearing radio frequency (RF) signal onto an optical lightwave carrier. RoF technology allows combining of the ultra-broadband capacity of optical networks with the flexibility and mobility of wireless access networks, which is required to meet the capacity demand of future broadband wireless applications. Different architectures for the RoF systems are considered in this chapter and compared in terms of transmitter and receiver designs, BER performance, and crosstalk in order to propose an RoF system that is highly promising for future high-frequency wireless fiber communications.

There is a growing perception that copper-based access networks will soon no longer be able to meet the ever-growing consumer demand for bandwidth. This, along with a combination of regulatory and competitive forces, as well as recent rapid advances and standardization of PON technology, have prompted carriers around the world to consider PON-based fiber-to-the-curb/home systems as a possible successor to current copper-based access solutions. Among the various PON-based fiber-to-the-x (FTTx) schemes (e.g., fiber-to-the-home (FTTH), fiber-to-the-curb (FTTC), or fiber-to-the-building (FTTB)), single-channel time-division multiplexed PON (TDM-PON) and multi-channel wavelength-division multiplexed PON (WDM-PON) architectures are the two most viable candidates. Chapter 9 is devoted to these types of access systems and networks, providing a concise treatment of the state of the art in PONs with an emphasis on simulation, design, and engineering. A comprehensive overview of current deployed and novel proposed protocols and subsystem architectures for passive optical networks is presented. Emphasis is given to link layer techniques that allow resource sharing without comprising bandwidth utilization or so-called dynamic bandwidth allocation techniques. Different PON architectures are presented and examined including tree and ring architectures.

As the channel data rate moved up to 40 Gbps and 100 Gbps, traditional intensity modulation and direct detection (IMDD) technologies that are broadly adapted in optical communications systems are incapable of achieving the required performance levels. Although technological advances have allowed the IMDD transmitter and receiver to operate at signaling speeds as high as 40 Gbps, at these speeds the cost of binary IMDD transmitter and receiver increases dramatically since both optical and electronic components are required to operate at such high speed. In addition, several detrimental transmission effects such as chromatic dispersion (CD), polarization mode dispersion (PMD), and fiber nonlinearities become the major limiting impairments of reliable transmission as the data rate reaches 40 Gbps and beyond. The rapid evolution of long-haul optical communications systems witnessed in the last five years is mainly due to the gradual adoption of spectrally-efficient, multilevel modulation formats, in conjunction with polarization division multiplexing (PDM) and coherent intradyne detection assisted by electronic compensation techniques using digital signal processing (DSP). The objective of Chap. 10 is to briefly review the operating principles of long-haul coherent optical communications systems using M-ary modulation formats. This chapter initially presents the digital M-PSK transmitter and receiver optimal architectures. Subsequently it reviews the operating principles of coherent detection and the different types of coherent receivers and then focuses mainly on the modeling of coherent optical systems using QPSK modulation and the evaluation of their performance. A description of high-speed DSP algorithms utilized for coherent detection is also included in this chapter.

A number of enabling technologies described in the preceding chapters (WDM, FEC, RZ modulation format, advanced fiber designs, erbium-doped fiber amplifiers (EDFAs)) have been the principal enablers of the modern undersea fiber-optic systems. Furthermore, the introduction of Raman amplification (distributed Raman amplification (DRA) and remote optically pumped amplifiers (ROPA)), improved performance fibers, and other types of modulation formats (e.g., differential phase shift keying (DPSK)) can be utilized to increase the length of the spans between repeaters and enhance the received OSNR. Building greenfield undersea and terrestrial fiber-optic DWDM systems is a very expensive proposition for the carriers. A much cheaper solution is the upgrade of existing legacy systems upgrading either only the line terminal equipment (LTE) or upgrading the LTE and adding in-line active equipment in the transmission path, while preserving the legacy fiber plant. The goal of Chap. 11 is precisely to cover the design process of such DWDM upgrades of legacy fiber-optic systems. In this chapter the current industrial state of the art in the design process for the DWDM upgrade of terrestrial and undersea long-haul optical communications systems is presented, aiming to maximize upgrade capacity and minimize cost in these systems. Specifically, this chapter examines upgrades of undersea repeatered systems, un-repeatered undersea and terrestrial systems, and terrestrial multi-span long-haul systems. The focus of this chapter is on the optical transport part of any given fiber-optic system upgrade design, including fiber plant quality (in-line fiber type, fiber loss/chromatic dispersion/PMD data), margins for repair and aging, estimation of repeater characteristics, management of multipath interference and nonlinearities, Raman amplification aspects, and capacity results based on the models developed.

Chapter 12 combines the physical (transport) and logical (control) layers and examines the physical-to-logical layer interactions, dealing specifically with impairment-aware (IA) routing and protection including different impairment-aware RWA algorithms for both unicast and multicast applications and the required control plane extensions for impairment-aware optical networking. Specifically, the transmission impairments and their modeling discussed in Chaps. 2 and 3 are considered in this chapter, and a host of IA RWA and protection techniques are developed and analyzed in terms of various performance metrics. Eventually, impairment-aware control plane extensions are compiled in this chapter in order to demonstrate how an IA approach would be implemented in the network architecture.

Chapter 13 considers only the logical layer and goes a step further by considering multi-domain network architectures, taking into account the fact that modern wireline infrastructures have morphed into a complex interconnection of distributed, decentralized domains. In this case each domain is treated as a standalone entity running its own control and management planes, as delineated by ownership (multi-carrier), location, technology type, etc [10–12]. To address this type of network architecture, issues related to provisioning and survivability in multi-domain optical networks must be investigated and analyzed. Thus, this chapter discusses multi-domain provisioning, hierarchical routing and path computation, as well as per-domain provisioning techniques. Multi-domain survivability techniques including proactive protection strategies and dynamic restoration strategies are also described.

Finally, Chap. 14 concludes the work and summarizes the state of the art in the field while presenting some thoughts on the obvious question: what the future holds.

References

1. Bonenfant P, Rodriguez–Moral A (2000) Optical data networking. IEEE Commun Mag 38(3):63–70
2. Bouillet E, Ellinas G, Labourdette J–F, Ramamurthy R (2007) Path routing in mesh optical networks, 1st edn. Wiley-Interscience, New York
3. Saradhi C, Subramaniam S (2009) Physical layer impairment aware routing (PLIAR) in WDM optical networks: issues and challenges. IEEE Commun Surv Tutor 11(4):109–130
4. Rahbar A, Yang O (2008) Contention avoidance and resolution schemes in bufferless all-optical packet-switched networks: a survey. IEEE Commun Surv Tutor 10(4):94–107
5. Stern TE, Ellinas G, Bala K (2008) Multiwavelength optical networks: architectures, design, and control, 2nd edn. Cambridge University Press, Cambridge, UK
6. Gasulla I, Capmany J (2007) High-frequency radio over fiber QPSK transmission through a 5 km multimode fiber link. In: Proceedings of European conference on optical communications (ECOC), Berlin, Germany
7. Antoniades N, Losada MA, Mateo J, Richards D, Truong TK, Jiang X, Madamopoulos N (2011) Modeling and characterization of SI-POF and connectors for use in an avionics system environment. In: Proceedings of the international conference on plastic optical fibers (ICPOF), Bilbao, Spain
8. Chang G-K, Jia Z, Yu J, Wang T, Ellinas G (2008) Super broadband optical wireless access technologies. In: Proceedings of the IEEE/OSA optical fiber communication conference (OFC/NFOEC), paper OThD1, San Diego, CA
9. Lethien C, Loyez C, Vilcot J-P, Clavier L, Bocquet M, Rolland P (2009) Indoor coverage improvement of MB-OFDM UWB signals with radio over POF system. Optics Commun 282(24):4706–4715
10. Ghani N, Liu Q, Gumaste A, Benhaddou D, Rao NSV, Lehman T (2008) Control plane design in multidomain/multilayer optical networks. IEEE Commun Mag 46(6):78–87
11. Chamania M, Jukan A (2009) A survey of inter-domain peering and provisioning solutions for the next generation optical networks. IEEE Commun Surv Tutor 11(1):33–51
12. Alanqar W, Jukan A (2004) Extending end-to-end optical service provisioning and restoration in carrier networks: opportunities, issues, and challenges. IEEE Commun Mag 42(1):52–60

Part I
Tools and Methods

Chapter 2
Computer Modeling of Transport Layer Effects

André Richter

Abstract This chapter introduces the reader into optical signal representations and the major physical layer effects causing system degradations in the WDM transport layer. Suitable modeling approaches are presented, and typical simulation results are demonstrated. Finally, the chapter focuses on performance degrading effects due to fiber propagation, optical amplification, and signal generation.

2.1 The Value of System-Level Simulation

System-level simulations are now routinely used to theoretically evaluate the performance of transmission systems based on the parameters of existing optical equipment, and to define performance requirements for new equipment to ensure robust operation for all required application scenarios. In a sense, system simulation is used as a virtual prototyping tool that reduces cost by shortening the system construction cycle. Furthermore, simulations prove valuable for investigating possible upgrade scenarios, understanding the nature and sources for system-limiting effects, determining the full effect of component limitations, and developing new technological approaches. The most critical part in designing accurate and reliable system simulation scenarios is selection of the right parameters and accurate transport effect models.

A. Richter (✉)
VPIphotonics Division, VPIsystems
Berlin, Germany
e-mail: Andre.Richter@VPIphotonics.com

N. (Neo) Antoniades et al. (eds.), *WDM Systems and Networks*, Optical Networks,
DOI: 10.1007/978-1-4614-1093-5_2, © Springer Science+Business Media, LLC 2012

2.2 Simulation Domains and Signal Representations

For computer simulation to be effective, optical signals need to be represented into the computer. Optical signal representations with different degrees of abstraction are important in providing flexibility for modeling various aspects of photonic networks. Since the initial publications of the fundamental concept as in Lowery et al. [1], many improvements have been introduced supporting the modeling requirements of the ever-evolving advances in optical technologies and network architectures [2].

Time-domain simulations that propagate individual time samples on an iteration-by-iteration basis between modeling blocks have been used to investigate, for instance, the properties of integrated photonic circuits, high-speed transmitters, and receivers, where the analysis of bidirectional interactions in the picoseconds range are of key importance [3]. For optical systems and networking applications mixed-domain simulations (i.e. including both, the time- and frequency-domains) are more suitable where arrays of signal samples are passed between modeling blocks. Applications are mainly unidirectional propagation scenarios and bidirectional scenarios where time delays are longer than the array duration (e.g., optical switching scenarios in bidirectional ring networks), or where bidirectional interactions can be averaged in time (e.g., bidirectional Raman pumping of optical fiber).

The simulation efficiency of mixed-domain simulations can be controlled by representing optical signals either by their complex, polarization-dependent sampled waveform (time-dynamic or spectral representation), or by time-averaged parameters only. Depending on the modeling scope of interest, the time-dynamic representation can be applied to a single frequency band (SFB), or to multiple (non-overlapping) frequency bands (MFBs) for which the appropriate propagation equations describe the evolution of individual frequency bands, while interacting terms are approximated using frequency decomposition (FD) techniques.

Although waveform-dependent propagation effects are ignored when using a parameterized representation of optical signals, this approach provides significant simulation speed-up while keeping sufficient modeling accuracy when investigating, for instance, the impact of 'far-away' WDM channels or optical pumps. Parameterized Signals (PS) can also be used to track important properties along the transmission path such as power, principal states of polarization (PSP), as well as accumulated amounts of noise, chromatic dispersion (CD), self-phase modulation (SPM), differential group delay (DGD), and thus easily assess signal characteristics over topology and frequency. Additive noise that is generated by lumped or distributed optical amplifiers can be tracked separately from the optical signal, disregarding nonlinear interaction effects between noise and signal, or accounting for them at the receiver using deterministic system performance estimation algorithms. In that case, optical noise can be described by its wavelength-dependent power spectral density (PSD), which could be step-wise approximated by so-called Noise Bins (NBs) of constant PSD. Further on, the parameterized representation is useful for keeping parasitic terms resulting from optical amplification, crosstalk, and scattering processes (collectively called Distortions) separate from the optical signal, and thus, allowing the investigation of their impact on the signal quality.

2.3 Device Modeling

Different levels of abstraction in device modeling help to address the tradeoffs between the model complexity, simulation accuracy and computation speed [4] thus allowing the researcher to choose the required depth of detail. Typically, detailed models require in-depth parameters, whereas behavioral models operate on measured characteristics, or information from data sheets. At one extreme, detailed physical models represent components based on material and structural parameters. These parameters may be difficult to obtain, being proprietary, or difficult to derive from external measurements of a packaged device. Detailed physical models may require intensive computation, but can be used to design new devices and predict their performance. Black-box models are based on the physics of the device, but can be described in terms of behavioral parameters (instead of in-depth parameters), which can be directly derived from external measurements. Linear devices (such as filters) or well-specified devices (such as transmitters with a digital input) can be represented by their measured performance alone. Although novel devices cannot be designed directly using these models, the system performance of a module is easily and accurately assessed using these methods. Data sheets often provide characteristics as a series of parameters fitted to measurements (such as rise-time, spectral width). Although such data are often gathered using long-term measurements, their wide availability makes it useful for systems-level simulation. Attention should be given so that long-term averages do not misrepresent the worst case.

2.4 Fiber Propagation

Transmission effects influencing signal propagation over single-mode optical fiber can be grouped into several general categories arising from the fiber properties itself. There are first optical power loss effects, which are either regarded as being equally distributed along the fiber distance (e.g., power attenuation) or occur at isolated points along the fiber due to discrete events such as splices and fiber ends. Further on, the velocity of light traveling through the fiber depends on the wavelength, which results into dispersive effects of information-carrying optical signals. The so-called group velocity dispersion (or CD) is a linear propagation effect, which could theoretically be reversed after fiber propagation by means that are independent of the transmitted signal, for example, using components with opposite dispersion such as using dispersion compensating fiber (DCF) or fiber

Bragg gratings (FBG). Additionally, power loss and dispersive propagation effects might be dependent on the local polarization state in time and frequency due to manufacturing imperfections, stress or other effects changing the properties of the perfectly circular optical single-mode fiber. Important effects are polarization mode dispersion (PMD) and polarization-dependent loss/gain (PDL/PDG).

In addition to the above-mentioned linear effects linear scattering processes such as Rayleigh scattering, spontaneous Raman- and Brillouin- scattering are also important to consider. Rayleigh scattering can be understood as distributed reflection of the light wave on microscopic in-homogeneities of the fiber medium. This effect significantly contributes to the fiber loss. Furthermore, double Rayleigh scattering leads to a reflected lightwave propagating in the same direction as the originating signal and therefore interferes with it. While the reflection strength is typically small, the effect may become significant in high amplifying media (for instance Raman amplifiers with high pump power). Spontaneous Brillouin and Raman scattering effects are caused by the scattering of the lightwave on thermally induced acoustic waves or molecular oscillations of the fiber medium (also referred to as acoustic and optical phonons, respectively). The importance of these effects results from the fact that the spontaneously scattered signal becomes a seed for further amplification by the much stronger nonlinear (stimulated) scattering effects, which are considered below.

Besides polarization-independent or dependent linear effects, signal transmission over optical fiber may suffer from effects, which result from the nonlinear interactions of the fiber material and the light traveling through it. Nonlinear propagation effects can generally be categorized as effects involving instantaneous or almost instantaneous electronic contributions—the optical Kerr effect and its various manifestations as SPM, cross-phase modulation (XPM), and four-wave mixing (FWM) as well as nonlinear effects with a noticeable time response: stimulated Brillouin and Raman scattering (SBS and SRS) respectively. Depending on the response time the spectral bandwidth of these nonlinear effects varies from almost unlimited (in telecommunications terms) for the Kerr effect, intermediate for the Raman effect (in the vicinity of 13 THz) and rather small for the Brillouin effect (linewidth of about 100 MHz and a Stokes shift of about 10 GHz).

Finally, it is important to note that the nonlinear propagation effects occur in coexistence to CD, polarization effects, and attenuation, and may produce complex degradations or beneficial interactions (depending on the system parameters under investigation). For example, the FWM efficiency depends strongly on the CD of the fiber; SPM can support very stable pulse propagation in the presence of the right amount of CD; the impact of PMD can be reduced by nonlinear fiber interactions; the polarization dependence of SRS may introduce PDG. Thus, accurate fiber modeling needs to include nonlinear and linear phenomena, so that detrimental and beneficial interactions between these characteristics are predicted correctly.

2.4.1 Linear Propagation Effects

2.4.1.1 Attenuation

Fiber attenuation is mainly caused by absorption and scattering processes. It determines the resulting exponential decay of optical input power propagating through the fiber. Absorption arises from impurities and atomic effects in the fiber glass. Scattering is mainly due to intrinsic refractive index variations with distance (Rayleigh scattering) and imperfections of the cylindrical symmetry of the fiber. The usable bandwidth ranges from approximately 800 nm (increased Rayleigh scattering scales with λ^{-4} [5, 6]) to approximately 1,620 nm (infrared absorption due to vibrational transitions). Older types of single-mode fiber show additionally an attenuation peak at approximately 1,400 nm due to the absorption of water molecules.

Typically being specified in logarithmic units as dB/km, sometimes it is useful to work with attenuation values in linear units so:

$$\alpha = \left(\frac{\log_e(10)}{10}\right)\alpha_{\mathrm{dB/km}} \approx 0.23026\alpha_{\mathrm{dB/km}} \tag{2.1}$$

For single channel applications α can be assumed to be wavelength-independent. However, for applications covering a large spectral range, such as multi-band WDM or Raman amplification, the spectral dependency of fiber attenuation should be accounted for accurately in simulations.

The effective length L_{eff} defines the equivalent fiber interaction length with respect to constant power [7] so that:

$$L_{\mathrm{eff}} = \int_0^z e^{-\alpha z_1}\,dz_1 = \frac{1 - e^{-\alpha z}}{\alpha} \tag{2.2}$$

In the above, L_{eff} is an important quality measure when rescaling the signal evolution to account for attenuation and periodic amplification, and helpful when performing quick system performance estimations.

2.4.1.2 Chromatic Dispersion

The CD or group velocity dispersion (GVD) represents a fundamental linear propagation limitation in optical fiber. It describes the effect that different spectral components propagate with different group velocities. CD alone causes spreading of chirp-less pulses along the fiber, although this statement is not true in general. For instance, in combination with nonlinear propagation effects other pulse propagation characteristics are detectable. Note that CD is a collective effect of material and waveguide dispersion [6, 8]. Depending on the manufacturing process and the radial structure of the fiber, different fiber types with various CD profiles can be designed. In standard single-mode fibers (SSMF) the GVD is positive (i.e. shorter wavelengths propagate faster) for wavelengths longer than approximately 1.3 μm and negative for shorter wavelengths. Thus, 1.3 μm is the point of zero GVD for SMFs. The fiber design allows shifting the point of zero dispersion and production of zero-dispersion fibers at 1.5 μm called dispersion-shifted fibers (DSF) or fibers with low positive or negative dispersion at that wavelength called non-zero dispersion-shifter fibers (NZ-DSF), as well as fibers with high negative dispersion values for dispersion compensation called DCF.

The impact of GVD on pulse propagation becomes clearer when expanding the modal propagation constant $\beta(\omega)$ in a Taylor series around an arbitrary frequency ω_0 as shown in Agrawal [9]:

$$\beta(\omega) = n(\omega)\frac{\omega}{c} = \beta_0 + \beta_1(\omega - \omega_0) + \frac{1}{2}\beta_2(\omega - \omega_0)^2 + \frac{1}{6}\beta_3(\omega - \omega_0)^3 \tag{2.3}$$

where $n(\omega)$ is the effective refractive index of the fiber, c is the speed of light in a vacuum, and $\beta_k = \partial^k/\partial\, \omega^k \beta(\omega)$ at $\omega = \omega_0$ for $\mathrm{k} = 0, 1, 2, 3$ (i.e. partial derivative of $\beta(\omega)$ with respect to ω evaluated at ω_0)

In the above:

- β_0 accounts for a frequency-independent phase offset during propagation corresponding to the propagation constant at a certain reference frequency.
- β_1 defines the inverse of the group velocity v_g determining the speed of energy propagated through the fiber.
- β_2 defines the GVD and describes the frequency dependence of the inverse of v_g. It is responsible for the broadening of an initially chirp-less pulse due to the fact that its Fourier components propagate with different group velocities. This also leads to pulse chirp when the leading and trailing edges of the pulse contain light with different frequencies. Equivalently, it defines the different propagation speeds of pulses in frequency-separated channels (such as in WDM systems).
- β_3 is known as the slope of the GVD or second order GVD. This term accounts for the frequency dependence of the GVD and therefore for the different broadening properties of signals or signal portions propagating at different frequencies. This term is critical for wideband transmission systems, for systems operating in frequency regions where β_2 is close to zero as well as systems utilizing DCFs to compensate dispersion for multiple WDM channels simultaneously.

In general, it is of more interest to determine the dependence of the inverse of the group velocity v_g on wavelength rather than on frequency. This dependence is described by the dispersion parameter D and its slope with respect to wavelength, S. The following relationships inter-relate the above parameters:

$$D = \frac{\mathrm{d}}{\mathrm{d}\lambda}\frac{1}{v_g} = -\frac{2\pi c}{\lambda^2}\beta_2 \tag{2.4}$$

$$S = \frac{\mathrm{d}D}{\mathrm{d}\lambda} = \frac{(2\pi c)^2}{\lambda^3}\left(\frac{1}{\lambda}\beta_3 + \frac{1}{\pi c}\beta_2\right) \tag{2.5}$$

$$\beta_2 = \frac{\mathrm{d}}{\mathrm{d}\omega}\frac{1}{v_g} = -\frac{\lambda^2}{2\pi c}D \tag{2.6}$$

$$\beta_3 = \frac{\mathrm{d}\beta_2}{\mathrm{d}\omega} = \frac{\lambda^3}{(2\pi c)^2}(\lambda S + 2D) \tag{2.7}$$

D is typically measured in ps/nm-km and can be interpreted as describing the broadening ΔT of a pulse with bandwidth $\Delta\lambda$ after propagation over a distance z, or equivalently, the time offset ΔT of two pulses after distance z that are separated in the spectral domain by $\Delta\lambda$.

ΔT is given below as:

$$\Delta T \approx \Delta\lambda\frac{\mathrm{d}T}{\mathrm{d}\lambda} = \Delta\lambda z\frac{\mathrm{d}}{\mathrm{d}\lambda}\frac{1}{v_g} = \Delta\lambda z D \tag{2.8}$$

From the above, the walk-off length L_{w} can be defined as the distance it takes for one pulse with duration T_0 traveling at frequency ω_1 to overtake another pulse traveling at frequency ω_2 and is thus given by:

$$L_{\mathrm{W}} = \frac{T_0}{|\beta_1(\omega_2) - \beta_1(\omega_1)|} \approx \frac{T_0}{|\Delta\lambda D|} \tag{2.9}$$

The dispersion length L_{D} characterizes the distance over which dispersive effects become important in the absence of other effects. Specifically, it defines the distance over which a chirp-free Gaussian pulse broadens by a factor of $\sqrt{2}$ due to GVD and is given by:

$$L_{\mathrm{D}} = \frac{T_0^2}{|\beta_2|} = \frac{2\pi c}{\lambda^2}\frac{T_0^2}{|D|} \tag{2.10}$$

However, L_{D} can also be applied as an approximation to determine the relevance of CD effects on other intensity modulated pulses. Since L_{D} is inversely proportional to the square of the signal bandwidth, as an example, the CD requirement increases by a factor of 16 when increasing the signal bitrate from 10 to 40 Gbit/s using the same modulation format and filter bandwidths that are proportional to the signal rate. In consequence, CD limits uncompensated 40 Gbit/s NRZ transmission over SSMF to approximately 4 km and over NZ-DSF to approximately 20 km. Additional dispersion variations due to changes in temperature along the link accumulate fluctuations as shown in Kato et al. [10] that have to be compensated adaptively, especially in long-haul applications.

So it would be natural to think that low- or zero-dispersion fibers are the optimum choice for WDM systems design. However, instead of using DSFs with very small local GVD values, it is more appropriate to use transmission fibers with larger local dispersion values (e.g., SSMF or NZ-DSF) and place DCFs at regular distances along the link compensating for all or part of the accumulated CD the optical signal experienced up to that point. If optical fiber were a purely linear

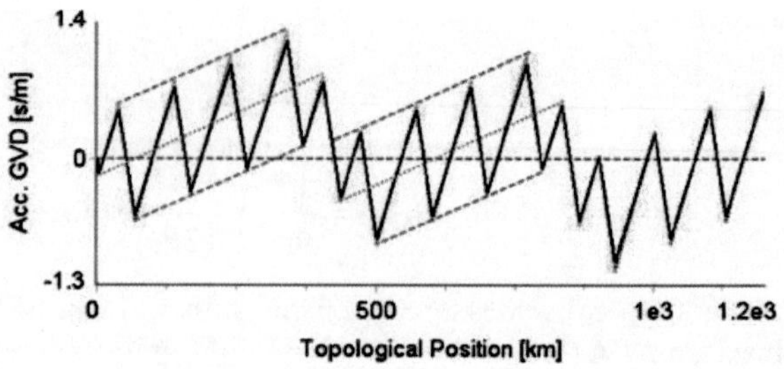

Fig. 2.1 Dispersion management for a long-haul transmission experiment where dispersion compensation is placed at regular intervals along the fiber as described in Liu et al. [11]

transmission medium it would not matter at which point along the link CD is compensated. However, as will be described in the following section, fiber propagation is slightly dependent on the intensity of the light traveling through it, thus resulting in nonlinear propagation effects. In modern WDM transmission systems with channel symbol rates of 10 Gbit/s and beyond, dispersion management taking care of the careful adjustment of local CD values (not too small) and maximum accumulated CD before dispersion compensation (not too large) helps to control these nonlinear fiber impairments (Fig. 2.1).

2.4.2 Nonlinear Propagation Effects

2.4.2.1 Kerr Effect

The Kerr effect denotes the phenomenon in which the refractive index of an optical fiber $n(\omega, t)$ is slightly dependent on the intensity of the optical signal $I(t) = |E(t)|^2$, with $E(t)$ being the electric field, passing through the fiber. As a result:

$$n(\omega, t) = n_0(\omega) + n_2 I(t) \tag{2.11}$$

where n_0 is the linear refractive index and n_2 is the nonlinear refractive index of the fiber.

When the electric field intensity of the transmitted signal bit stream varies in time it induces an intensity-dependent modulation of the refractive index, and hence, modulation of the phase of the transmitted signal. Compared to other nonlinear media, n_2 is very small (typically on the order of 10^{-20} [m^2/W]). However, even this weak fiber nonlinearity becomes fairly relevant for signal propagation as field intensities of several mWs are focused in a small fiber core (in the order of several tens of μm^2) over interaction lengths of tens to hundreds of kilometers. As a result, effects such as nonlinear interaction between signal pulses might accumulate during transmission and become of system limiting importance.

In conclusion, the Kerr nonlinearity results in several intensity-dependent propagation effects, with the most important ones being SPM, XPM, and FWM.

Self- and Cross-Phase Modulation

SPM and XPM occur when a temporal variation of the optical signal intensity induces a temporal phase shift on the originating signal (SPM) or on other co-propagating signals at different wavelengths (XPM).

For example, in the case where several optical signals at different wavelengths each with intensity $I(t)$ and initial phase φ_0 are launched into a fiber, the phase modulation of the signal corresponding to channel m depends on the local power distribution of all other channels as follows:

$$\begin{aligned}\varphi_m(t,z) - \varphi_{0,m} &= \varphi_{lin,m} + \varphi_{\mathrm{SPM},m} + \varphi_{\mathrm{XPM},m} \\ &= \frac{2\pi}{\lambda} n_{0,m} z + \frac{2\pi}{\lambda} n_2 z I_m(t) + \frac{4\pi}{\lambda} n_2 z \sum_{k \neq m} I_k(t)\end{aligned} \tag{2.12}$$

where $\varphi_m(t, z)$ is the phase modulation of channel m, $\varphi_{0,\ m}$ is the initial phase of channel m, $n_{0,\ m}$ is the linear refractive index of channel m, n_2 is the nonlinear refractive index, and k is an index denoting the neighboring WDM channels of channel m.

Furthermore:

- $\varphi_{lin,\ m}$ corresponds to the accumulated linear phase shift due to transmission.
- $\varphi_{\mathrm{SPM},\ m}$ corresponds to the accumulated nonlinear phase shift due to SPM in channel m. The SPM-induced phase shift is proportional to the local signal intensity. It induces chirp (time-varying frequency shift) and spectral broadening, so as an example, pulses behave differently in the presence of GVD and optical filtering.
- $\varphi_{\mathrm{XPM},\ m}$ corresponds to the accumulated nonlinear phase shift due to XPM in channel m describing the phase modulation that is induced by intensity fluctuations in neighboring WDM channels. XPM introduces additional nonlinear phase shifts that interact with the local dispersion as well.

It must be noted that XPM occurs only over distances where optical intensities at different frequency components co-propagate, e.g. pulses propagating in different WDM channels are overlapping. In general, the XPM effect reduces with increased CD as pulses at different frequencies propagate faster through each other, e.g. L_w (as defined in Sect. 2.4.1.2) becomes smaller. For the same reason, XPM scales inversely with the channel spacing. L_w is also called the collision length L_c as it accounts for the distance where two pulses at different frequencies collide (and thus, interact nonlinearly due to XPM) during propagation.

Four-Wave Mixing

Parametric interactions between optical field intensities at different frequencies might induce the generation of inter-modulation products at new frequencies when propagating optical signals over wide spectral ranges through the fiber.

This nonlinear effect is called FWM. FWM can occur between channels in WDM systems, between optical noise and channels, and between the tones within one channel. More generally noted, FWM occurs for instance when two photons at frequencies ω_1 and ω_2 are absorbed to produce two other photons at frequencies ω_3 and ω_4 satisfying the relation below:

$$\omega_1 + \omega_2 = \omega_3 + \omega_4 \tag{2.13}$$

It could also be understood as mixing of three waves producing a fourth one according to the electrical field equations below:

$$\begin{aligned} E_{klm} &= E_k E_l E_m^* \\ &= |E_k||E_l||E_m| e^{j(\omega_k+\omega_l-\omega_m)t} e^{-j[\beta(\omega_k)+\beta(\omega_1)-\beta(\omega_m)]z} \end{aligned} \tag{2.14}$$

where E_i is the electric field of the wave propagating at frequency ω_i, $\beta(\omega_i)$ is the modal propagation constant at ω_i, and $\omega_{klm} = \omega_k + \omega_l - \omega_m$ is the carrier frequency of E_{klm}. The energy of E_{klm} is given as the superposition of mixing products of any three waves for which $\omega_{klm} = \omega_k + \omega_l - \omega_m$ holds (with the degenerate case for $\omega_k = \omega_l$). Note that the propagation constant is frequency-dependent, so efficient interactions only occur if contributions to E_{klm} given at different times add up over distance, e.g., the phase mismatch $\Delta\beta$ between the inter-modulating fields tends to go to zero (phase matching condition):

$$\Delta\beta \to 0, \quad \Delta\beta(\omega) = \beta(\omega_k) + \beta(\omega_l) - \beta(\omega_m) - \beta(\omega_{klm}) \tag{2.15}$$

The power of the newly created wave is proportional to the power of the three interacting waves assuming no pump depletion due to FWM occurs [12] resulting in:

$$I_{klm}(z) = |E_{klm}(z)|^2 = \eta \gamma_{klm}^2 |E_k(z)|^2 |E_l(z)|^2 |E_m(z)|^2 e^{-\alpha z} L_{\text{eff}}^2 \tag{2.16}$$

where γ_{klm} is the nonlinear index of the fiber at ω_{klm}, α is the fiber attenuation, z is the propagation distance, L_{eff} is the effective length, and η is the so-called FWM efficiency given by:

$$\eta = \frac{\alpha^2}{\alpha^2 + \Delta\beta} \frac{1 + 4e^{-\alpha z}\sin^2(\Delta\beta z/2)}{(1 - e^{-\alpha z})^2} \tag{2.17}$$

FWM between the WDM channels scales inversely with channel spacing in case of non-zero GVD. Increasing the local dispersion results in an increased walk-off of the Fourier components and thus, it results in phase mismatch after shorter propagation distances, which leads to an even steeper decrease of the FWM efficiency with channel separation.

Fig. 2.2 presents the results of a case study where four 10 Gbit/s NRZ signals are propagated over 50 km of DSF and NZ-DSF fiber segments, respectively. Figure 2.2a shows the spectrum after transmission over 10 km of DSF, while Fig. 2.2b shows the corresponding spectrum after transmission over 10 km of

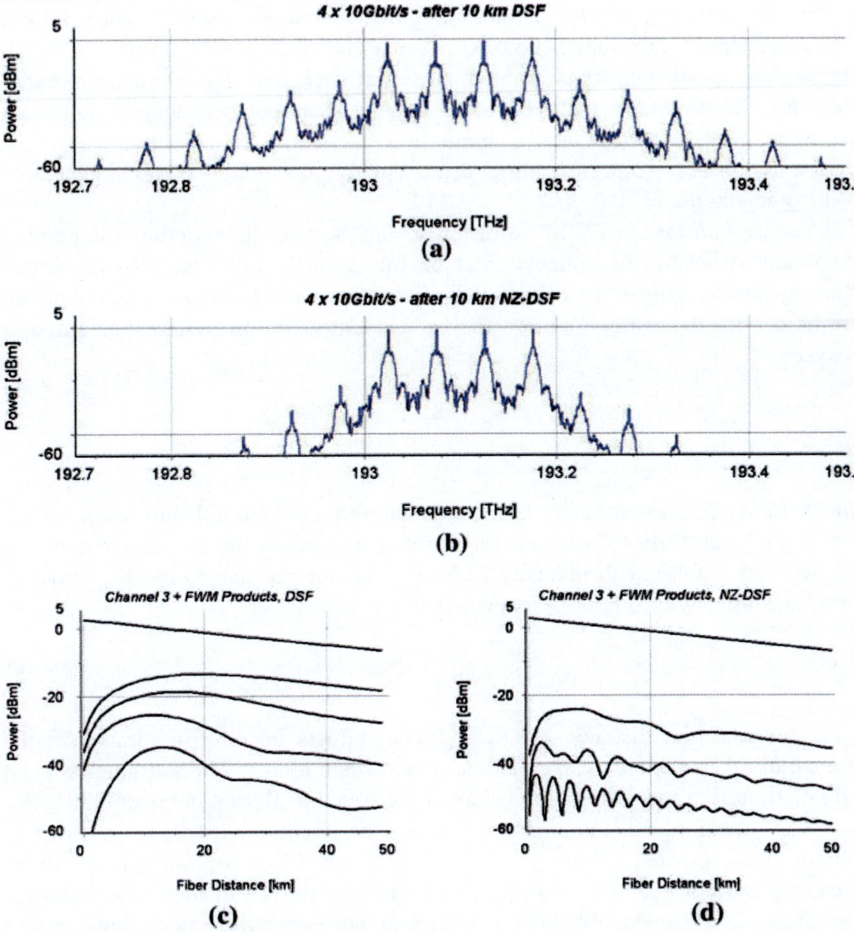

Fig. 2.2 Spectrum of four 10 Gbit/s NRZ channels after transmission over 10 km of: **a** DSF; **b** NZ-DSF segments. Power profiles of channel 3 with FWM products for propagation over: **c** DSF; **d** NZ-DSF segments

NZ-DSF. The larger local dispersion of the NZ-DSF with respect to the DSF results in an increased walk-off between the interacting tones (e.g., WDM channels and FWM distortions outside the WDM spectrum that are created during propagation). Hence, the FWM efficiency is reduced resulting in mixing products of lower power and faster power oscillations during propagation. This can be observed when comparing Fig. 2.2c showing the power levels of channel 3 and the FWM products versus DSF distance and Fig. 2.2d showing the power profiles for the corresponding NZ-DSF case.

The number of generated FWM components that are created when signals at N wavelengths mix together grows with $N^2(N-1)/2$. Using a sampled signal representation, all important components that fall into the simulation bandwidth are automatically included when solving the fiber propagation equations (see Sect. 2.4.4.1). Note that the simulation bandwidth should be chosen wide enough in order to avoid misleading results due to aliasing effects of the generated FWM components [13].

When accounting for FWM distortions using parametric modeling, the number of calculated FWM components shall be limited to avoid excessive numerical effort by considering only FWM distortions that would reach a certain power threshold, and thus, are important to be accounted for in system performance analysis.

Kerr Nonlinearities and CD

Similar to the dispersion length L_D, a scale length for Kerr nonlinearities in optical fiber could be defined: the nonlinear length L_{NL} is given as the distance over which the phase of a pulse with intensity I changes by one rad due to the Kerr nonlinearity in a fiber with a nonlinear coefficient γ

$$L_{NL} = \frac{1}{\gamma I^2} \tag{2.18}$$

L_{NL} denotes the distance where nonlinear effects become important (in the absence of other effects), also called nonlinear scale length. The ratio between L_D and L_{NL} describes the dominating behavior for signal evolution in optical fiber. For $L_D < L_{NL}$ linear effects dominate the signal propagation. In these quasi-linear systems pulse powers change rapidly in time due to dispersion-induced pulse spreading and compression over L_{NL}, and thus, the impact of nonlinear propagation effects is averaged [14]. Those nonlinear pulse-to-pulse interactions can be best described by time-equivalent terms of FWM and XPM, namely intra-channel four-wave mixing (I-FWM) and intra-channel cross-phase modulation (I-XPM) as described in Essiambre et al. [15], Mamyshev and Mamyshev [16]. Typically, I-FWM induces pulse echoes while I-XPM induces timing jitter. Pulse-to-pulse interactions in quasi-linear systems become important for bitrates of 40 Gbit/s and higher. With a careful selection of launch power, pre-chirp as well as the dispersion map, degradations due to I-XPM and I-FWM can be drastically reduced (see for instance [17–19]).

Of special interest is also the case when $L_D \sim L_{NL}$. In these nonlinearity-supporting systems cancelation of dispersive and nonlinear effects can be observed for carefully chosen system settings such as in soliton or dispersion-managed soliton (DMS) systems. Choosing system parameters such as pulse width, separation, power, as well as dispersion map and amplifier position and separation carefully, pulse chirp introduced by CD and SPM may combine and be used

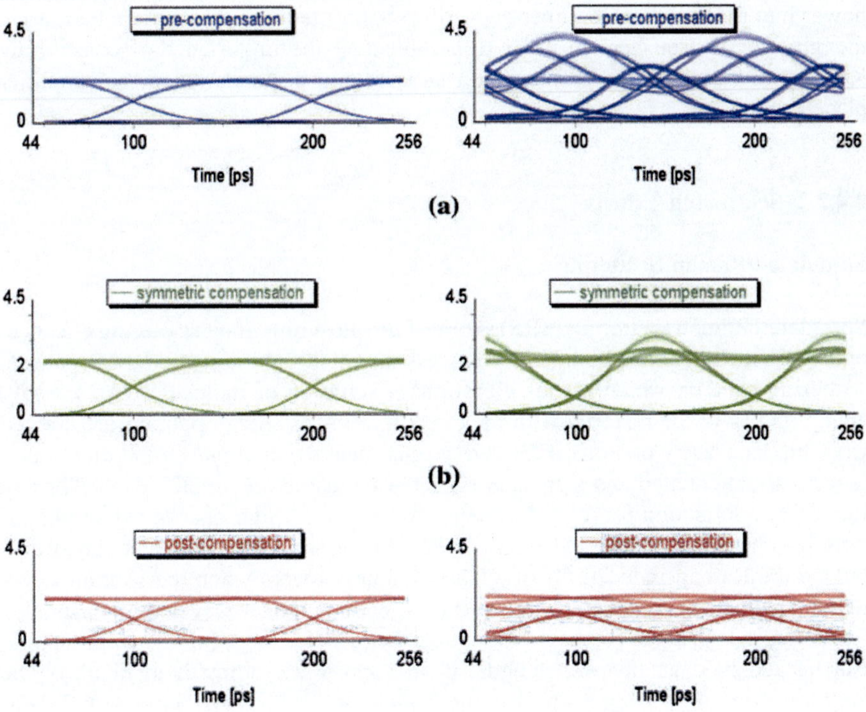

(a)

(b)

(c)

Fig. 2.3 Eye diagrams of a 10 Gbit/s NRZ signal after propagation over four spans of 80 km SSMF for three different dispersion compensation schemes: **a** pre-compensation; **b** symmetric compensation; **c** post-compensation. For each case, eye diagrams without consideration of Kerr nonlinearities are displayed on the *left* and including Kerr nonlinearities on the *right*

advantageously for stabilizing pulse propagation over very long distances. However, XPM introduces timing jitter, which in addition to timing jitter induced by added optical noise along the link is one of the ultimate limitations for such systems (see for instance [20]).

Fig. 2.3 shows results of a single channel 10 Gbit/s NRZ transmission over four spans of 80 km SSMF using different methods of dispersion compensation. Figure 2.3a shows eye diagrams (with and without considering Kerr nonlinearities) for the case that 100% of the accumulated dispersion of each SSMF span is compensated before entering the fiber (pre-compensation). Figure 2.3b shows the corresponding results when applying symmetrical compensation, e.g., equal parts of dispersion compensation are performed before and after the SSMF. Figure 2.3c shows the eye diagrams when the accumulated dispersion of each SSMF span is compensated after the fiber (post-compensation). Appropriate DCMs are placed at mid-stage access EDFAs before and/or after each span, which also set the signal power into the transmission fiber at 10 dBm. Admittedly this value is rather large, though, it is chosen here in order to demonstrate the importance of considering Kerr nonlinearities when designing for an optimum dispersion map and amplifier placement.

2.4.2.2 Scattering Effects

Stimulated Raman Scattering

Stimulated Raman scattering (SRS) is one of the most important scattering effects in optical fiber. It denotes the inelastic process in which light is scattered by material scattering centers, which causes vibrational excitation of molecules. As a result, energy is lost to the material with the effect that lower energy photons are created from higher energy photons. Effectively, this means that signals at shorter wavelengths are attenuated and signals at longer wavelength are amplified. SRS has an important application for WDM systems, as relatively high distributed amplifications may be achieved over the length of the transmission fiber. However, it may also induce the Raman-induced tilt of WDM channel powers when the shorter wavelength channels become attenuated at the expense of longer wavelength channels.

Raman scattering is present at all optical frequencies, e.g., Raman amplification can be achieved at any wavelength (if an appropriate pump is available). The amplification range using a single pump wavelength is broadband (about 6 THz at 1.5 μm) but also strongly shaped. Raman amplifiers providing flat gain over large bandwidths require multiple pumps at different wavelengths [21] and a careful adjustment of the individual pump powers [22, 23].

Besides stimulated Raman scattering, spontaneous Raman scattering occurs in the fiber as well producing broadband optical noise that is strongly wavelength- and temperature-dependent. Additionally, double Rayleigh scattering may produce significant amounts of additional noise. The pumping conditions (number and power of co- and counter-propagating pumps) play an important role on the noise characteristics, as shown clearly in Figure 2.4.

Compared with EDFAs compensating fiber loss at the ends of links, Raman amplification is low noise due to its distributed nature. Hence, Raman amplification allows lower launched powers while maintaining the OSNR, which reduces other nonlinear effects in the fiber.

Stimulated Brillouin Scattering

Stimulated Brillouin scattering (SBS) occurs when forward- and backward-propagating lightwaves induce an acoustic wave due to electrostriction with a frequency that equals the frequency difference of the optical waves, and a propagation constant that equals the sum of the propagation constants of the optical waves (phase matching). In other words, the acoustic wave forms a

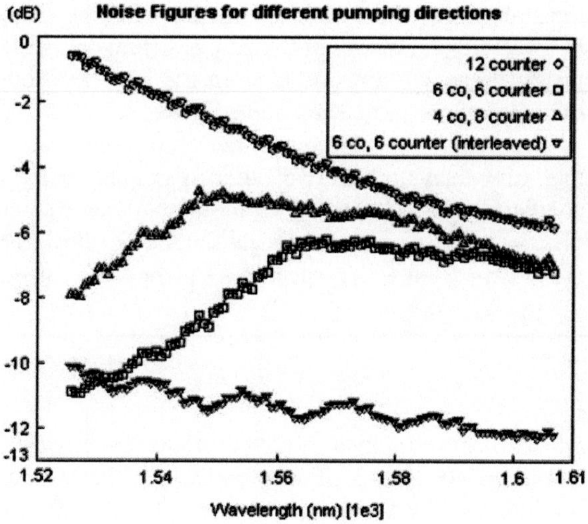

Fig. 2.4 Effective noise figures for different pumping conditions after optimization of 12 Raman pumps to generate flat gain over 80 nm signal bandwidth after transmission over 100 km NZ-DSF (From Breuer et al. [21], Fig. 9. Reproduced by permission of © 2004 The Institute of Electrical and Electronics Engineers)

traveling index grating, which reflects the incident light wave with a downward Doppler shift. The strength of SBS depends significantly on the type and modulation bandwidth of the transmitted signal. SBS is strongest for narrow-band signals, as a strong coherent grating is produced. It may be detrimental for optical communication systems, as it strongly reflects un-modulated signals and narrow-band pumps. However, the impact of SBS can be effectively lowered by employing techniques that spread the signal power more evenly in frequency (i.e., phase modulation).

2.4.3 Polarization-Dependent Propagation Effects

2.4.3.1 Polarization Mode Dispersion (PMD)

Due to manufacturing imperfections, fibers are not ideally circular and are additionally subject to temperature changes, vibrations, bending, and twisting, which make them birefringent. All of these random mechanisms cause local stress and redefine the local birefringence axes along the fiber length. Hence, light that is propagating along the single-mode fiber is split into two local polarization modes traveling at different velocities. PMD describes this dependence of the propagation velocity of optical signals on the local state of polarization.

At any point along the fiber, light travels as a combination of two polarization eigenstates, which randomly vary with fiber length. For instance, signal power originally sent into one state may partially transfer to the other along the fiber. As a final result, this means that different parts of a signal pulse travel with different velocities causing pulse spreading, which limits the bandwidth of the transmission system. With this model, PMD at a fixed frequency can be described by two PSPs representing the pair of orthogonal polarization states of slowest and fastest propagation resulting from the integrated polarization characteristics over the whole fiber at that frequency.

The maximum difference between group delays of any polarization state (e.g., the delays of the two PSPs) is defined as the DGD $\Delta\tau$. The *intrinsic* PMD coefficient $D_{\mathrm{PMD,i}}$ determines the amount of $\Delta\tau$ per unit length z for very short fiber lengths where birefringence can be assumed to be uniform [24]. Using Taylor series expansion with respect to frequency and considering the first two terms only, $D_{\mathrm{PMD},i}$ is given by

$$D_{\mathrm{PMD},i} = \frac{\Delta\tau}{z} = \frac{\mathrm{d}}{\mathrm{d}\omega}\left(\beta_{1,s} - \beta_{1,f}\right) = \frac{n_s - n_f}{c} + \frac{\omega}{c}\frac{\mathrm{d}}{\mathrm{d}\omega}\left(n_s - n_f\right) \tag{2.19}$$

where $\beta_{1,\mathrm{s}}$ is the inverse of the group velocity in the slow PSP, and $\beta_{1,\mathrm{f}}$ is the inverse of the group velocity in the fast PSP, where n_s and n_f are the effective refractive index for the slow and fast PSP, respectively. For transmission fiber lengths of interest, however, it cannot be assumed that the orientation and value of birefringence remain constant. The correlation length L_{corr} provides a measure for the strength of the accumulated random change in polarization states. It is defined as the distance where the average power in the orthogonal polarization mode, P_{orth}, is within $1/e^2$ of the power in the starting mode, P_{start}, when omitting the effect of attenuation [25]

$$\left|\langle P_{\mathrm{start}}(L_{\mathrm{corr}})\rangle - \langle P_{\mathrm{orth}}(L_{\mathrm{corr}})\rangle\right| = \frac{P_{\mathrm{total}}}{e^2} \tag{2.20}$$

For typical transmission fibers, L_{corr} is only several meters. Here, the randomly and continuously varying DGD over z represents the characteristics for a random walk problem [26], and thus, scales with the square root of z (each incremental *section* may add or subtract to the effects of previous *sections*). In consequence, the PMD coefficient of transmission fibers D_{PMD} determines the amount of average DGD per $\sqrt{z}$ [25]. The stochastic process of varying birefringence can be evaluated analytically with the result that $\Delta\tau$ follows a Maxwellian probability density function (PDF) [27]. The Maxwellian PDF originates from the quadratic addition of three Gaussian processes following an analogy with the statistical model of Brownian motion [28]. It can be used to describe the random distribution of the DGD for fiber distances that are much larger than L_{corr} as follows:

$$P(\Delta\tau) = \frac{\pi}{16}\frac{\Delta\tau^2}{\langle\Delta\tau\rangle^3}e^{-\frac{4}{\pi}\Delta\tau^2/\langle\Delta\tau\rangle^2} \tag{2.21}$$

Note that $D_{\mathrm{PMD},i}$ is not only time-dependent due to the drift of the environmental parameters but shows also a frequency dependence, which is indicated by the second term in Eq. 2.19. Using the Poincare Sphere as representation of light polarization [29] the PMD vector is, by definition, a Stokes vector pointing in the direction of the slow PSP with a length equal to the DGD [26].

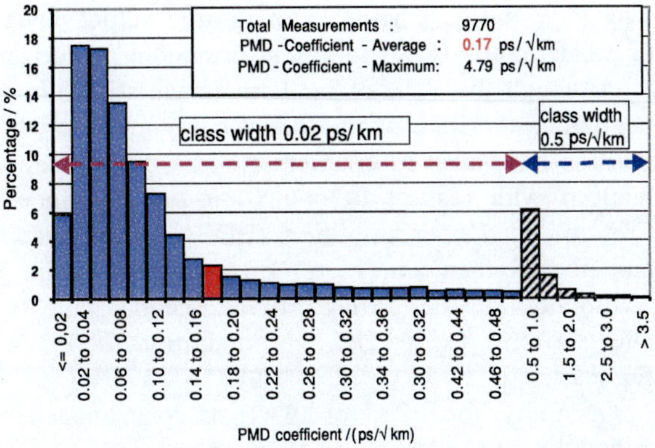

Fig. 2.5 Histogram of PMD coefficient for 9,770 evaluated fibers inside the network of Deutsche Telekom. Dark column represents PMD coefficient average. (From Freund et al. [31], Fig. 12, *left*. Reproduced by permission of © 2004 The Institute of Electrical and Electronics Engineers)

The frequency dependence of PMD can be described by introducing terms that represent the n-th order dependencies of the PMD vector on frequency. Ignoring third and higher order effects, second order PMD alone consists of two orthogonal components in the frequency domain: polarization chromatic dispersion (PCD) resulting from the linear frequency dependence of the DGD and pointing in the same direction as the PMD vector, and depolarization resulting from the linear frequency dependence of the PSPs and pointing in the orthogonal direction to the PMD vector [24, 30].

The main difference between PMD and the other fiber propagation effects is that PMD shows a strong statistical behavior that is frequency dependent, changes randomly with time, and from fiber to fiber [28, 30]. Hence, it is very time-consuming to measure the influence of PMD on implemented systems and design countermeasures. Further on, terrestrial transmission networks using fiber infrastructure that has been deployed over several years using different generations of fiber might contain fiber cascades with strongly varying PMD coefficients causing many difficulties for implementing high-speed transmission systems (see Freund et al. [31]). Figure 2.5 shows the histogram of PMD coefficients for 9,770 fibers in a Deutsche Telekom network showing the varying degrees of PMD coefficients.

2.4.3.2 Polarization-Dependent Loss/Gain (PDL/PDG)

PDL/PDG (in dB) describes the difference between the maximum and minimum loss/gain with respect to all possible states of polarization. As the polarization is randomly varying along the transmission fiber due to birefringence, fiber PDL gives a statistical value that only accounts for the maximum difference in accumulated loss that any two states of polarization at the fiber output will experience:

$$\mathrm{PDL_{dB}} = 10\log_{10}\left(\frac{P_{\max}}{P_{\min}}\right) \tag{2.22}$$

with P_{max} and P_{min} being the measured output powers. Note that those two states of polarization always represent orthogonal polarizations [32].

Although the PDL of modern transmission fibers is rather small, and mostly negligible, the PDL of individual components or discrete events along the transmission line (such as couplers, splices, filters) can cause significant power fluctuations with respect to polarization. Furthermore, optical amplifiers such as Erbium-doped fiber amplifiers (EDFA), semiconductor optical amplifiers (SOA) and fiber Raman amplifiers (FRA) or distributed Raman amplifiers (DRA) as commonly referred to may provide unequal gain for all states of polarization, an effect referred to as PDG. The definition for PDG is similar to that of PDL in Eq. 2.22.

Generally, the PDL (or PDG) of concatenated components cannot be determined by just adding the PDL of the individual components as the state of polarization of each PDL-element is different due to the random birefringence of the fiber links connecting these individual components. Consequently, this approximation would give a far too pessimistic number (as it assumes all PDL elements are polarization aligned). A statistical modeling approach is more appropriate in order to estimate the average PDL value and its variation for a given link. It has been found that for a link with many individual PDL elements, the PDF of the overall PDL at the output of that link can be approximated well by a Maxwellian distribution [33]. Additionally, attention needs to be paid to the fact that in combination with PMD, PDL (and PDG) can produce significant amounts of signal distortions, and thus, limit the system reach (see for instance [34–37]).

2.4.3.3 Polarization-Dependent Kerr

Fiber Kerr nonlinearities—resulting in inter-channel (SPM, XPM, FWM) and intra-channel (I-XPM, I-FWM) propagation effects—are actually polarization-dependent. For example, XPM causes nonlinear polarization rotation of the Stokes vector of the propagating channels as shown in Wang and Menyuk [38]. The knowledge of the polarization dependence of Kerr nonlinearities can also be used favorably in transmission system design (see for example [39, 40]). Figure 2.6 shows two eye diagrams demonstrating an effective technique for suppressing intra-channel nonlinear distortions by sending adjacent bits in alternate polarizations. Here, results are shown for a 40 Gbit/s NRZ signal after transmission over a 2,500 km dispersion-compensated fiber link. Clearly, the case using alternate polarization (see Fig. 2.6a) shows much less inter-symbol degradations (here mainly due to I-XPM) compared to the case using standard NRZ modulation (see Fig. 2.6b). Similarly, adjacent WDM channels are often launched in orthogonal polarizations to reduce inter-channel nonlinear interactions between them.

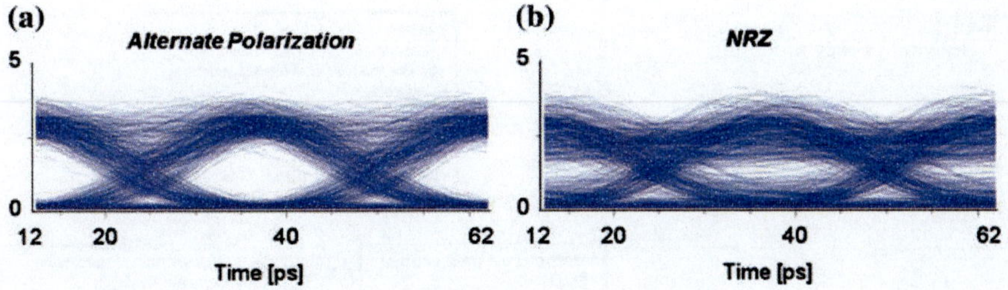

Fig. 2.6 Effective technique for suppressing intra-channel nonlinear distortions: eye diagrams of a transmitted 40 Gbit/s signal when: **a** applying alternate polarizations to adjacent bits; **b** using the standard NRZ format

2.4.4 Modeling Signal Propagation Over Single-Mode Fiber

Figure 2.7 presents a classification of modern modeling techniques for optical fiber [41]. The simplest approach is based on a linear fiber model and includes attenuation and CD effects. This approach is easy to implement and very fast as the whole fiber length can be integrated in one step. However, this approach may only be applicable for systems with very low bitrate, where the signal power does not need to be large due to the small signal bandwidth, and correspondingly, the noise power will be low within such bandwidths. In consequence, nonlinear propagation effects are negligible. Furthermore, it can be used for system designs where nonlinear effects cannot be disregarded to determine a lower bound on system performance.

2.4.4.1 Unidirectional Transmission and Scalar Field Description

With an increase of bitrate, the signal power has to be increased as well to maintain an acceptable optical signal to noise ratio (OSNR). At these powers, Kerr nonlinearities of the fiber have to be considered. As a result, signal propagation can be described by the generalized nonlinear Schrödinger (GNLS) equation containing terms responsible for linear and Kerr nonlinear propagation effects (see for instance [9, 25]). This leads to the following representation:

$$\begin{aligned} j\frac{\partial}{\partial z}A(z,t) + \sum_{k=1}^{\cdots}(-1)^{k-1}\frac{j^k}{k!}\beta_k\frac{\partial^k}{\partial t^k}A(z,t) + j\frac{\alpha}{2}A(z,t) \\ = \gamma|A(z,t)|^2A(z,t) \end{aligned} \tag{2.23}$$

where $A(z,t)$ is the complex field envelope of $E(z,t)$, with $E(z,t) = A(z,t)e^{j(\omega_0 t - \beta z)}$, ω_0 being the reference frequency and β_0 the reference phase offset. The GNLS cannot be solved analytically for the general case of arbitrarily shaped pulses propagating over the fiber. However, powerful numerical procedures have been developed with

Fig. 2.7 Evolution of Single-Mode Fiber modeling

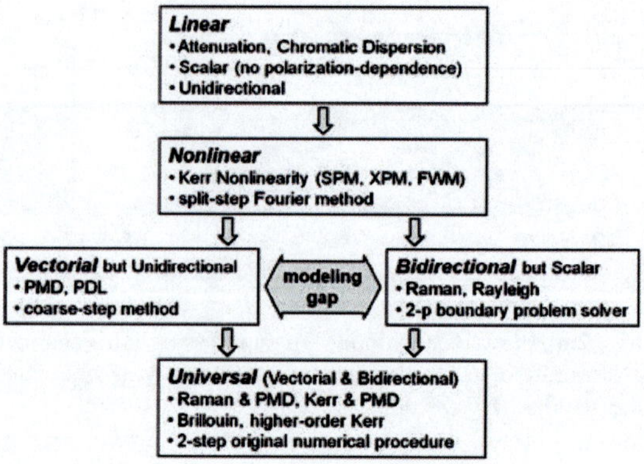

the most prominent one being the split-step method. This method is based on the assumption that linear and nonlinear propagation effects can be considered independently over a short fiber distance Δz called split step-size so that:

$$A(z+\Delta z) = \left[A(z)e^{\Delta z \underline{N}}\right]e^{\Delta z \underline{D}} \tag{2.24}$$

where $\underline{D}$ is the linear operator accounting for fiber dispersion and attenuation and $\underline{N}$ the nonlinear operator accounting for the Kerr nonlinearity.

If Δz, the so-called split-step size, becomes too large the condition for separable calculation of $\underline{D}$ and $\underline{N}$ breaks down, and the algorithm fails. Hence, careful determination of the optimum split-step size is critical in order to use minimal computational effort for a given accuracy. Typically Δz is adjusted adaptively, an example of which is shown below:

$$\Delta z = \min(\Delta\varphi_{NL} L_{NL}, \Delta z_{\max}) \tag{2.25}$$

where $\Delta\varphi_{NL}$ is the maximum acceptable phase shift due to the nonlinear operator, $\Delta z_{\max}$ is a maximum split-step size which is set to limit spurious FWM tones [42] and to avoid underestimation of XPM effects due to dispersion induced walk-off. A simple and widely applicable method for determining the minimum $\Delta z_{\max}$ is given in Rasmussen [43].

While the nonlinear operator $\underline{N}$ is favorably solved in the time domain, the linear operator $\underline{D}$ can efficiently be solved in the spectral domain (using split-step Fourier method, SSFM) or in the time-domain as well (using time-domain split-step method, TSSM). Using SSFM, the Fast Fourier Transform (FFT) can be used for converting data between time and frequency domain [13]. As the speed of the FFT is proportional to $N \log_2(N)$, where N is the number of signal samples in the time or frequency domain, careful determination of the simulation bandwidth and the simulated time duration is important for minimizing the computational effort for given accuracy constraints. Although the SSFM has proven to be the most robust technique [25], both approaches have advantages and disadvantages which depend on the field of application [44].

For small simulation bandwidths and systems with low amounts of accumulated dispersions a TSSM that uses infinite impulse response (IIR) calculations to solve $\underline{D}$ is more efficient than the SSFM (see Fig. 2.4 in Carena et al. [45]) considering some modeling approximations. Generally, this applies also to other TSSM using, for instance, overlap-and-add (OAA) to solve $\underline{D}$ [46]. The main advantage of TSSMs in an environment with an aperiodic representation of the time domain is that the data sequence can be divided easily into small blocks which are handled sequentially. However, the Singleton method [47] can be applied in an SSFM to divide the size of Fourier Transforms into smaller blocks as well. Doing this, simulation speed can be increased by parallelizing the computation effort in both cases. The walk-off between channels in a WDM system becomes significant for higher amounts of accumulated CD and a large simulation bandwidth, and thus, a relatively long bit sequence must be chosen for system performance analysis when using time-domain approaches with aperiodic representations. One further problem is associated with the buildup time of IIR filters or virtual signal delays in case of OAA methods. Simulation data generated during such buildup time must be removed before the system performance is estimated.

These problems do not exist in the case of the SSFM. Here, the problem with different group velocities is eradicated because of the periodic time-domain representation of signals. For example, channels that are delayed in the simulation will be 'wrapped-around' to the beginning of the data block: no energy is lost, and the wrap-around represents exactly the data delayed from earlier in the actual data sequence. Note that this artificial periodicity may affect the measured signal statistics, so one should carefully check potential correlations and select the simulated time duration to be long enough. Furthermore, the frequency-dependent phase shift caused by CD is exactly represented using SSFM, not only approximated by a truncated filter impulse response when using TSSM. In time domain, this can be understood as wrapping around an infinite impulse response within the periodic block of samples. This is accurate over all frequencies, whereas TSSM introduces inaccuracies close to the edges of the modeled bandwidth due to truncation effects (Gibb's phenomenon as shown in Oppenheim and Schafer [13]). SSFM also shows advantages when modeling dispersion-limited optical systems with vanishing nonlinearity. Here only a single step (equal to the total fiber length) is required. This greatly accelerates computation effort compared to time-domain techniques for solving the linear operator $\underline{D}$. Finally, the flexibility is a very important issue that should be taken into account when choosing the simulation domain. For example, it is much more difficult to include a realistic dispersion dependency on frequency or the Raman response function into the TSSM than to the SSFM. Any changes of the equations adding higher order derivatives would usually lead to a major rework of time-domain algorithms.

To make simulations of wideband systems even more effective, the SSFM can be extended for the GNLS to accept PS and NBs, and to allow individual sampling of spectrally non-overlapping signals. This approach is often referred to as frequency decomposition (FD) method or mean-field approach [48] and can be used as the main tool for multi-band WDM system simulations at moderate bitrates.

2.4.4.2 Unidirectional Transmission and Vectorial Field Description

An increase of the channel bitrate to more than 10 Gbit/s makes degradations due to PMD noticeable. Modeling PMD in the fiber requires a vectorial approach, describing the electrical field of the propagating light by its Jones vector, and not only by the SOP-averaged electrical field components. To simulate system degradations due to PMD is time-consuming, as PMD is a random phenomenon that is varying in time and frequency. Special statistical treatments such as the coarse-step model [49, 50] (possibly combined with the importance sampling technique [51]) can be applied to avoid an inefficient straightforward integration with very short steps accounting for the short scale length of PMD. Using these techniques, simulations of specific fiber conditions become possible in reasonable time. However, many of those simulation runs are still required to determine statistical information about polarization-dependent fiber characteristics.

Besides statistical modeling of all orders of PMD, the fiber model shall consider Kerr nonlinearities and CD as well as their interplay, especially when investigating high-speed transmission systems. For this purpose, a set of modified GNLS equations (similar to Eq. 2.23) for the two orthogonal polarization components of the electrical field $E(z, t)_{x,y}$ can be used to describe propagation mechanisms over birefringent optical fibers [52]. Assuming rapidly varying birefringence and that the polarization states of light average over the whole Poincare sphere, the Manakov equation has been derived from that [49], with the coarse-step method being introduced as an efficient numerical method for solving it. Later work then led to the Manakov-PMD equation [27, 50], which provides a closed-form and numerically efficiently solvable description of the interplay between CD, Kerr, and PMD effects in optical fiber.

The widely used intuitive coarse-step model, introduced by Wai et al. [49] without strict mathematical reasoning, is based on the assumption that although birefringence perturbations vary from position to position along the fiber length, their impact can be modeled by a concatenation of 'coarse' fiber sections, each longer than the correlation length L_{corr} of the fiber, with randomly rotated PSPs and step-wise constant birefringence values. For each of these fiber sections the SSFM, or alternatively, TSSM is used to numerically solve the propagation equations. The length of each polarization section should be much shorter than the total fiber distance in order to model a sufficient number of randomly oriented sections. Studies such as Wai and Menyuk [27], Marcuse et al. [50] showed the validity of the coarse-step model when applied with the correct weighting factors and scattering operations between sections for approximating solutions of the polarization-coupled GNLS equations with the precondition that L_{corr} and L_b (describing the beat length as a measure of fiber birefringence as in Kaminow and Li [24]) are much shorter than L_{NL}. This restriction ensures that the effect of fiber nonlinearity is averaged over the quick oscillation of the signal state of polarization for each polarization section. Consequently, this precondition imposes limitations on the highest possible signal intensity that can be simulated with this method. However, for typical signals used in telecommunications this condition is well satisfied.

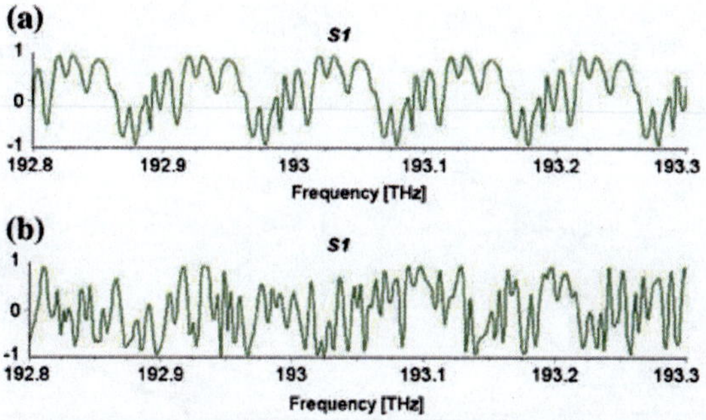

Fig. 2.8 The intrinsic frequency periodicity of the modeled polarization characteristics using the coarse-step model; **a** the above is demonstrated for the first component of the Stokes vector S_1. The above can be avoided by choosing polarization sections with random lengths; **b** suppression of frequency periodicity for 10% width deviation of the modeled steps

Table 2.1 Accuracy (Δ in %) of PDF modeling for different number of Monte-Carlo simulations

	Number of trials		
	10^3(%)	10^4(%)	10^5(%)
DGD	6.2	1.9	0.7
$\lvert PMD_2 \rvert$	13.5	4.4	1.3
PCD	5.8	1.7	0.6
Depol	13.2	4.1	1.3

The individual section lengths should be randomly selected from a Gaussian distribution to avoid a short frequency periodicity of the modeled PMD as described in Fig. 2.8. Figure 2.8a shows the first component of the Stokes vector, S_1, as a function of frequency for the case that all polarization sections have the same length, whereas Fig. 2.8b shows S_1 when choosing random section lengths (Gaussian with 10% standard deviation). In order to reproduce interesting (e.g., worst case) constellations, it is helpful to be able to store and retrieve the seeding for the random polarization scatterers between sections.

When deploying a statistical model (such as the coarse-step model described above) for emulating the random nature of PMD, it is important to use a sample set that is large enough in order to represent the underlying probability density functions correctly.

Table 2.1 shows results of a study on the modeling accuracy for the PDF of the DGD and second order PMD terms such as its absolute value— $\lvert PMD_2 \rvert$, PCD, and Depolarization. The coarse-step model without any special statistical methods such as importance sampling [51] has been used. A figure of merit Δ is chosen that is defined as the maximum absolute difference between the results from Monte-Carlo simulations (using 100 polarization sections) and analytical predictions as in Nelson and Jopson [53] for an infinite number of polarization sections.

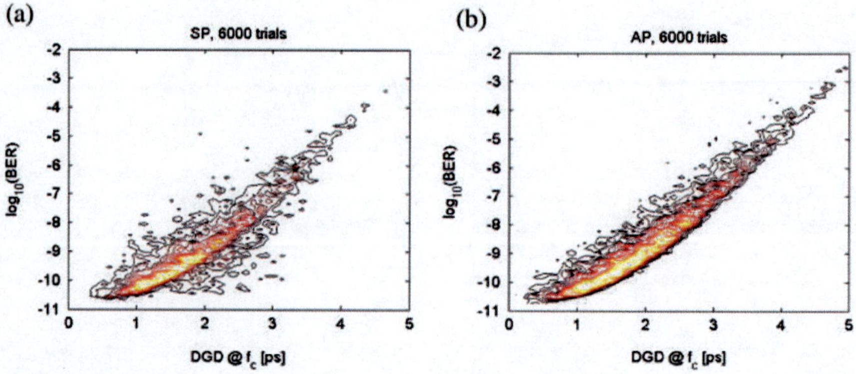

Fig. 2.9 BER versus DGD for: **a** single polarization; **b** alternate polarization. In both cases the system consists of four, 40 Gbit/s OTDM channels propagating over a noisy link with an accumulated mean DGD of 2.1 ps. (From Richter et al. [57], Fig. 4. Reproduced by permission of © 2005 The Institute of Electrical and Electronics Engineers)

Table 2.1 suggests that: (a) when increasing the number of trials by a factor of 10, the accuracy increases approximately by a factor of $\sqrt{10}$, and (b) the modeling accuracy of the PDFs of Depolarization and $|PMD_2|$ is approximately half as good as that for the PDFs of DGD and PCD using the same number of trials.

To assess the simulation accuracy of the spectral PMD modeling, the auto correlation functions with respect to frequency for the PMD vector and for DGD should be evaluated. For a given product, P, of average root-mean-squared DGD $\langle \Delta\tau_{\mathrm{rms}} \rangle$, and measurement bandwidth B_m, as defined in Shtaif and Mecozzi [54] the simulation accuracy Δ can be calculated by comparing Monte-Carlo simulations with theoretical considerations. For example, an accuracy of $\Delta < 2\%$ can be achieved for $P = \langle \Delta\tau_{\mathrm{rms}} \rangle B_m = 100$ using averaged results from 50 Monte-Carlo simulations. Note that increasing the number of simulation trials does not improve the modeling accuracy.

As an application example, the statistical signal performance for 4*40 Gbit/s OTDM transmission over 520 km dispersion-compensated SSMF with an accumulated average DGD of 2.1 ps is modeled using the method described in Richter et al. [55]. Figure 2.9 shows contour plots of bit error rate (BER) versus DGD for the 40 Gbit/s tributaries from 6,000 simulation trials with random PMD realizations using a deterministic BER estimation method accounting for arbitrary signal statistics and polarization states of optical noise and signal in front of the receiver [56]. Figure 2.9a shows results for the case that all pulses are transmitted with equal polarization, while Fig. 2.9b depicts results for the case that the 40 Gbit/s tributaries are launched in alternate polarizations.

The use of “link emulators” to investigate the impact of combined CD–PMD and PDL effects is another possible approach for modeling the transmission link. With such emulators DGD, and the second order PMD characteristics—PCD and depolarization—can be set individually and specific combinations of first and

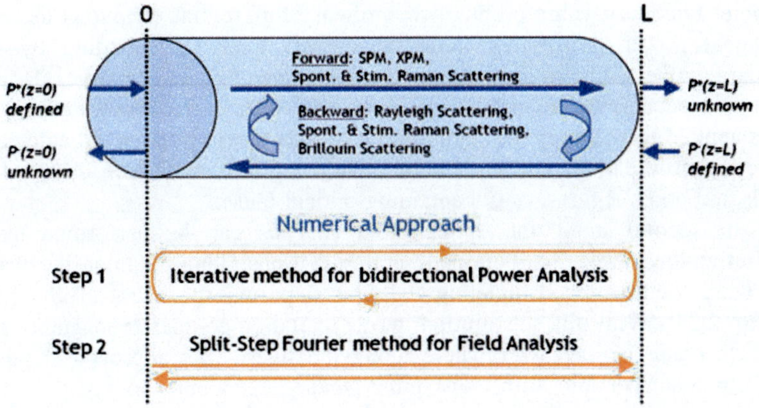

Fig. 2.10 Bidirectional signal propagation in fiber and two-point boundary problem, [63, section *UniversalFiber*] (Copyright 2011 VPIsystems. Used by permission of VPIsystems Inc)

second order PMD can be selected. Many other possible implementations have been proposed as in Francia et al. [58], Kogelnik et al. [59], Eyal et al. [60], Orlandini and Vincetti [61] which typically differ in the amount of residual third and higher order PMD terms that cannot be controlled by the emulator.

2.4.4.3 Bidirectional Transmission and Scalar Field Description

The utilization of Raman amplifiers and the increase in bitrates have raised the need for further development of fiber modeling techniques. Raman amplification requires bidirectional fiber modeling, as the pumps and signals propagate in opposite directions in many configurations of interest. Noise from spontaneous Raman scattering is always generated in both directions as well. To model bidirectional signal propagations in fibers, at least a two-point boundary value problem has to be solved taking into account the initial conditions at both fiber ends. In case of reflective and/or lossy splices (or other discrete scattering events) located along the fiber the complexity is raised to a multipoint boundary value problem. This type of problem is impossible to solve using ordinary unidirectional numerical integration methods without neglecting interactions between counter-propagating waves. To take into account bidirectional interactions an original numerical procedure consisting of two stages can be applied as presented in Fig. 2.10 and further described in Poloyko et al. [62], Photonic Modules Reference Manual [63].

In the first stage, a bidirectional power analysis is performed and so the complex amplitudes of the sampled signals at the input fiber ports are replaced by the corresponding parameterized (time-averaged power) representation. This step excludes time-dependent effects from the analysis and reduces the problem to a two-point boundary value problem for ordinary differential equations describing the signal power propagation along the fiber. The corresponding two-point boundary value problem can effectively be solved by an iterative algorithm. Simulation accuracy and speed depend strongly on the frequency discretization that is applied for this approximation process. It is useful in practical calculations to specify different resolutions for frequency regions containing sampled data signals and other regions, only containing optical noise.

In the second stage, the optical field analysis can be performed by first substituting the power distributions in both directions along the fiber distance that have been found after completing the first step into the full GNLS equation (see Eq. 2.23). With this substitution, terms including counter-propagating signal powers become defined everywhere along the fiber. This approach is possible assuming that dynamic (time and pulse shape-dependent) propagation effects induced by interactions of counter-propagating waves are negligible. As their interaction time is very short only the spatially resolved average power of counter-propagating signals is of relevance. Furthermore, it can be assumed that FWM between counter-propagating waves is negligible due to the phase mismatch. As a result, the two-point boundary problem reduces to a problem similar to the unidirectional propagation, and the SSFM can be used for effectively solving it. Depending on the specific application, the field analysis performed in the second step could be done for frequency bands propagating in the forward, in the backward, or in both directions.

When modeling bidirectional propagation effects over heterogeneous fiber links consisting of several different fiber spans it is important to consider each span by its individual fiber characteristics (e.g., attenuation, CD, Rayleigh coefficient, nonlinear index, core area, Raman gain etc.) and to add information about losses and reflections at fiber splices and fiber ends.

For an accurate modeling of wideband applications it is important to account for the wavelength-dependence of relevant fiber properties as well. For instance, the Raman gain profile of the fiber needs to be known when modeling SRS. As the Raman gain depends on the pump frequency, either many profiles measured at different pump wavelengths are to be supplied and the actual Raman gain at the frequencies of interest is obtained by interpolation, or Raman gains at different pump wavelengths should be approximated using correct frequency scaling [62, 64].

Even more care is needed when trying to adjust multiple pumps at different wavelengths in order to achieve flat amplification over a wide spectral width. Special optimization algorithms are needed for this task due to the large number of open parameters [65, 66]. To make those optimization algorithms applicable for real-world problems, it is essential that they consider not only all Raman-related propagation effects, but also other power-related effects along the fiber. For instance, signal–signal Raman interactions, Rayleigh and Brillouin scattering, and power-related variations of broadband pump sources can cause significant amplifier degradations [67].

The possibility to model the different physical effects individually or in combination, and to control their strengths is important for the assessment of sources for system degradations (and saving computation time). Further on, adjustment factors could be introduced to scale the strength of nonlinear interactions due to SPM, XPM, and SRS between co-propagating bands. These factors can for instance, account for the dependence of Raman interactions on the mutual polarization of the interacting fields, for example: when using information that was obtained by measuring the Raman gain coefficient for a short (~meters) fiber with co-polarized signal and pump, an adjustment factor of 0.5 should be used in order to treat the Raman gain correctly; when simulating long (~100 km) fiber spans, the polarizations of pump and signals vary randomly with respect to each other due to the frequency-dependence of PMD. As another example indicates, the SPM effect can be scaled by the factor 8/9 to account for the averaging of Kerr nonlinearities for applications where PMD and PDL can be neglected, if the value of nonlinear index coefficient n_2 used for the simulations has been measured without considering polarization averaging over long fibers [25]. Similarly, the XPM effect between orthogonally polarized signals propagating in different sampled bands over high-birefringence fiber can be scaled accordingly to account for the fact that XPM is much stronger for signals with aligned polarizations.

2.4.4.4 Bidirectional Transmission and Vectorial Field Description

For many years, the development of bidirectional and vectorial models was independent and went in separate directions, using completely different numerical techniques. However, high bitrate Raman-amplified systems required the consideration and inclusion of both, PMD and bidirectional Raman effects [68]. For this reason vectorial and bidirectional methods have been unified into one universal model as discussed in Photonic Modules Reference Manual [63, section *UniversalFiber*]. An advantage of this unification is that polarization-dependent effects during Raman amplification can be considered. This is for example, useful for cases of polarization-multiplexed systems, not perfectly depolarized pumps, or polarization-sensitive discrete fiber events (such as PDL at splices). Figure 2.11 shows how the Raman gain would be distributed for co-polarized pumping over fibers with different PMD coefficients if a Raman pump is used that is not depolarized. The SRS effect requires signals and pumps to be polarization-aligned, which is on average the case over half the distance in standard transmission fibers. However, depending on the actual polarization states along the fiber the alignment varies. For fibers with low PMD coefficients the statistical variation could be large, which is the reason why depolarized Raman pumps are preferred in scenarios with co-propagated pump(s).

The development of numerical methods for this challenging task makes it possible to take into account a number of additional important propagation effects; for instance, Brillouin backscattering, multiple Rayleigh scattering, and back-reflections/attenuations at fiber splices. The bidirectional vectorial consideration is very complex but represents the most powerful simulation technique that is currently available.

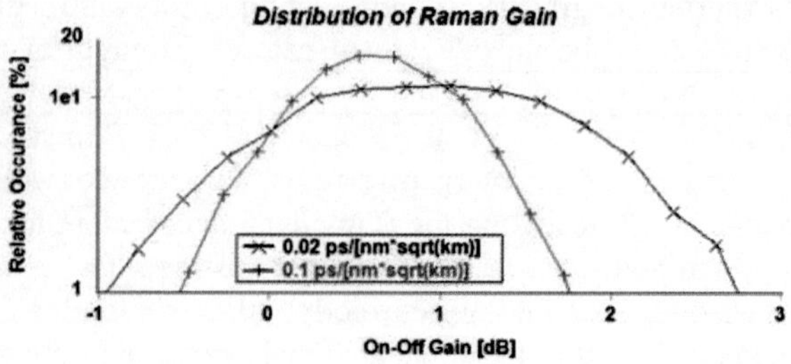

Fig. 2.11 Distributions of Raman gain for fibers with different PMD coefficients using a single co-propagated Raman pump without depolarization. (Note: 0 dB gain corresponds to perfect compensation of fiber loss for the case that a depolarized pump is used)

In summary, the ultimate (universal) model for single-mode fiber propagation supports

- the modeling of wideband bidirectional transmission and interaction of signals, pumps, and noise
- the representation of signals, pumps and noise by sampled waveforms or time-averaged parameters (e.g., to keep them separately and increase simulation efficiency), and
- the wavelength-dependent definition of all fiber parameters (attenuation, CD, effective mode area, nonlinear coefficient, Brillouin and Raman scattering responses, Rayleigh backscattering, birefringence profile, etc.)

The full vector model (Jones) for arbitrary polarization dependencies and the scalar model (SOP averaging) for a linear polarized approximation define the main modes of operation. In scalar mode, the universal model shall include attenuation, CD, Kerr nonlinearities with delayed-response terms, SRS between pumps, signals and noise, point reflections and losses, Raman-induced changes of the nonlinear refractive index, wideband noise generation due to spontaneous Raman scattering at any temperature, stimulated and spontaneous Brillouin scattering including the linewidth-dependence of the reflection coefficient, and multiple Rayleigh scattering. In vectorial mode, the model shall *additionally* be able to account for all-order PMD (with optional techniques to weight statistics and emphasize less likely events), and polarization dependencies of nonlinear Kerr and scattering effects.

Furthermore, it is very helpful to have the ability to switch individual propagation effects on/off and to define heterogeneous fibers consisting of multiple spans of different fiber types that are connected by lossy and reflective splices. Parameters that control the numerical accuracy and calculation speed of underlying algorithms as well as means for visualizing fiber characteristics versus distance or frequency and the numerical convergence of calculation algorithms are also extremely important and helpful.

2.4.4.5 Parametric Analysis

While a high level of detail is important for component and subsystem design, rapid visualization of important performance measures can aid in the overall understanding of systems, and thus, be very useful in system and network design. In conjunction with approximate design rules network engineers can access, without demonstrating very detailed photonics knowledge, whether a network (or sub-network) is likely to operate properly or not. Such a quick performance assessment can be achieved using a parametric approach that avoids time-consuming simulation runs when verifying complex network configurations and identifying potential design flaws [69].

Figure 2.12 shows a sample network configuration and typical results of the parametric system validation process. Figure 2.12a depicts the simulation setup[1] of a three-node network that consists of two parallel paths connecting N1 and N2 with H2, and a common Raman-amplified 140 km link connecting H2 with N3. A total of 16 C-band channels each carrying 10 Gbit/s are used for communication between N1 and N3, while 16 L-band channels are deployed between N2 and N3. The individual SSMF fiber spans are post-compensated by appropriate DCFs; EDFAs are placed at H1 and H2 accounting for the accumulated loss. Figure 2.12b shows optical signal powers versus topology (left) and versus frequency at N3 (right). Figure 2.12c shows a parametric plot illustrating the relationship between accumulated SPM and CD for two channels (one in C-band and one in L-band) as they propagate through the network. Without the need to run detailed simulations the results depicted in Fig. 2.12 indicate that one Raman pump per band is not sufficient to achieve flat gain and the power in both bands is very unbalanced requiring additional gain adjustments before entering into any possible links that are following. Furthermore, the designer is able to locate places where large amounts of accumulated CD and nonlinear phase shift occur simultaneously, something that can lead to significant penalties. Hence, the example discussed above emphasizes quickly the limitations of the chosen amplification and dispersion maps.

2.5 Optical Amplification

Several technologies for optical amplification have been introduced over the years addressing different system requirements and applications: semiconductor optical amplifiers (SOA), rare earth ion doped fiber (or waveguide) amplifiers (e.g., Erbium-doped fiber amplifier—EDFA), and discrete and distributed fiber Raman amplifiers (FRA). Especially with the invention of the EDFA in the late 1980s the development of fiber-optic communication systems accelerated rapidly [70]. Electro-optic repeaters could be replaced by the more robust, flexible, and cost-efficient EDFAs. The main requirements on EDFA technologies from a systems design perspective are: operation over wide wavelength ranges, high output powers, equal gain to each channel, low-noise characteristics, negligible crosstalk between channels, mechanisms for gain control, high energy efficiency, and low cost.

---

[1] using VPItransmissionMaker™Optical Systems, Version 8.6.

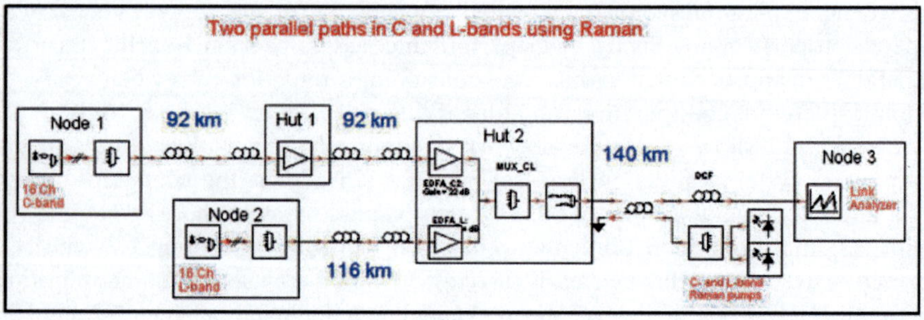

(a)

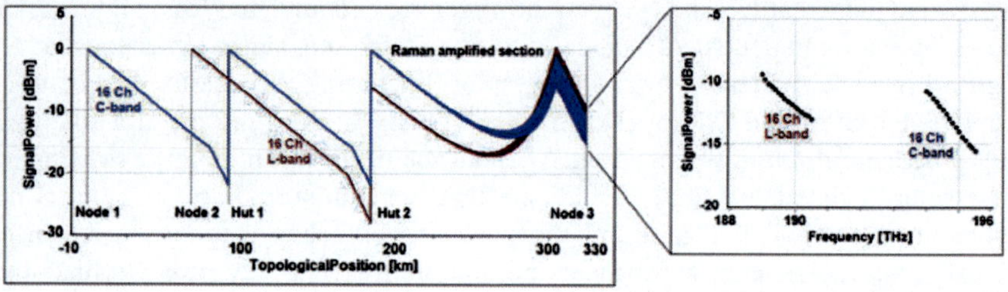

(b)

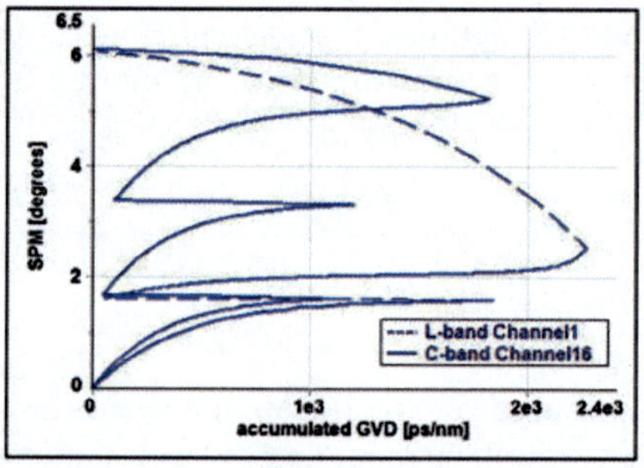

(c)

Fig. 2.12 **a** Optical network with two parallel paths, each carrying 16 × 10Gbit/s, with Raman-amplified common link; **b** optical signal powers versus topology and versus frequency; **c** parametric plot illustrating the relationship between accumulated SPM and CD for two channels as they propagate through the network

The requirements of low noise, wide bandwidth, high power, and gain flatness cannot be met with simple EDFA designs employing only a single pump, and a single amplifying fiber. Optical amplifiers operating over multiple wavelength bands with improved performance have incorporated sophisticated technological advances including multi-stage doped-fiber amplifier designs with pump reuse, mid-stage gain equalization, novel dopants and hosts, as well as topological innovations such as bidirectional or parallel amplifiers. Additionally, alternatives such as Raman amplification in non-doped fibers using multiple pump wavelengths and cascaded pumping have been introduced to increase the amplifiable system bandwidth and required repeater spacing.

Obviously there are many combinations of amplifier topology, fiber, pump and passive component technology that could produce a cost-effective, low-noise, wide-bandwidth, high-power, spectrally flat, and power-efficient amplifier. A detailed discussion on the theory and design of EDFAs is given in Desurvire [71], Becker et al. [72]. In what follows we present the main theoretical concepts and underlying fundamental operating principles for these devices that are intended to facilitate their modeling. In Chap. 3 a "black-box" approach for their modeling is also presented.

2.5.1 EDFA Theory

A single-mode Er-doped fiber can be modeled using the rate and propagation equations for a three-level laser medium or using the equations for a simplified two-level laser medium [73]. Figure 2.13 shows the energy diagram of Erbium ions in glass hosts. For 980 nm pumping, carriers are absorbed from the ground level to the third laser level. Due to phonon relaxation (in the times scale of μs) carriers transit down to the second level almost immediately where they reside with an average lifetime of approximately 10 ms and distribute due to thermalization processes. The same effect could be achieved for 1,480 nm pumping, causing absorption of carriers from the ground level directly to the second laser level. With this, and ignoring other, higher order effects in the Er-doped fiber such as excited state absorption, up-conversion, spectral-hole burning, and others, it is a reasonable assumption to model the gain and noise behavior in an EDFA using only a two-level laser medium. The third level comes into play only via the coefficients of absorption from the ground level. Note that the two-level model is applicable for silica host glasses, but not for glasses where the third level has a comparatively long lifetime (e.g. fluoride). From the above it is clear that a three-level scheme must be used.

2.5.1.1 Gain

The gain in EDFAs is strongly dependent on the carrier inversion, or in other words, the amount of carrier population in the upper state compared to the total number of carriers. It is determined by the pumping scheme, and is influenced by the existence of co-dopants such as germanium and alumina. The wavelength dependence of local gain can be written as presented in Desurvire [71]:

$$g(\lambda) = [N_2\sigma_e(\lambda) - N_1\sigma_a(\lambda)]\Gamma(\lambda) \tag{2.26}$$

where N_2, and N_1 denote the carrier populations of the upper and lower states respectively, σ_a, and σ_e denote the absorption and emission cross-sections respectively, and $\Gamma(\lambda)$ is the overlap factor, i.e., the area of overlap between

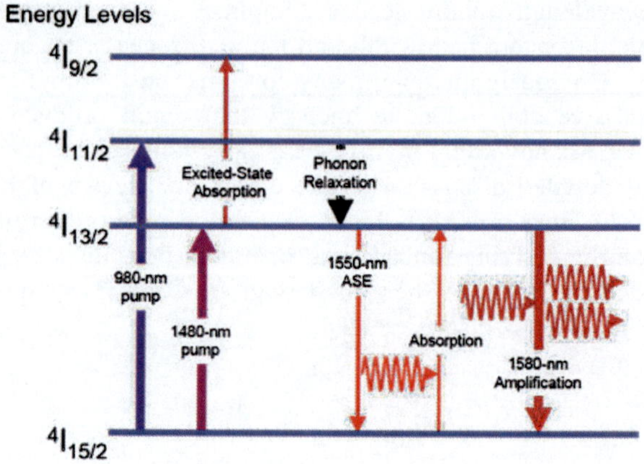

Fig. 2.13 Transitions between energy levels in an EDFA as presented in User's Manual [74, pp. 12] (Copyright 2011 VPIsystems. Used by permission of VPIsystems Inc)

Erbium ions and the optical signal mode in the doped fiber. Note that the population inversion shows a strong local dependence so gain may differ over the length of the doped fiber.

Figure 2.14 shows simulation results characterizing the performance of a single stage EDFA for different pumping schemes. Figure 2.14a shows amplifier gain as a function of population inversion around the 980 nm (pump) wavelength. Perfect inversion is possible as the emission cross-section is zero at 980 nm. This results in small noise generation for 980 nm pumped EDFA configurations. Figure 2.14b shows gain as a function of population inversion at 1,480 nm (pump) and around 1,550 nm (C-band). For the depicted inversion profile, the transparency point (g = 0 dB/m) for 1,480 nm pumping is at about 75% population inversion. So, perfect inversion is not possible, which results in increased noise generation. The figure depicts the strong wavelength dependence of the gain, which results from the wavelength dependence of population inversion.

Ignoring noise to begin with, the average signal power dependence on doped fiber length is given by

$$\frac{\mathrm{d}P}{\mathrm{d}z} = g(\lambda, P)P \tag{2.27}$$

For small signal powers, the total amplifier gain is usually independent of the incoming power. If the power launched into the doped fiber is increased over a certain level, stimulated recombination starts to affect carrier population inversion. Every photon created by stimulated emission transfers one ion from the upper state to the lower state. This results in gain reduction until absorption of pump power and stimulated plus spontaneous emissions are balancing each other. This effect is known as gain saturation. The saturation characteristic of an EDFA is very complex, as it is dependent on fiber distance. Assuming a homogeneous power distribution along the EDFA, the ratio of input power P_{in} and saturation power P_{sat} is given by Iannone et al. [25]

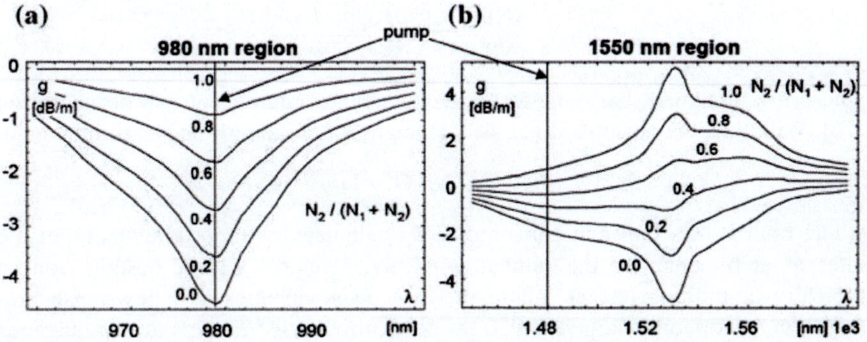

Fig. 2.14 Sample gain/loss profile for various values of population inversion in a typical EDFA. (From Richter [20]. Reproduced by permission of the author © 2002, Andre Richter)

$$\frac{P_{\text{in}}}{P_{\text{sat}}} = \frac{1}{\log_e(4)} \frac{G_0 - 2}{G - 1} \log_e\left(\frac{G_0}{G}\right) \tag{2.28}$$

where G is the amplifier gain with $G = P_{\text{out}}/P_{\text{in}}$. In the above G_0 is the small-signal (or linear) amplifier gain, and P_{sat} is defined as the input power where $G/G_0 = 0.5$.

2.5.1.2 Noise

Spontaneous transitions of photons from the upper state to the ground state accumulate along the doped fiber and stimulate other transitions which cause self-amplification, or amplified spontaneous emission (ASE). ASE noise is un-polarized and builds up in the forward and backward directions. The amount of ASE noise created at each end of the doped fiber depends on the local population inversion. For system-level applications ASE noise is well approximated by a Gaussian random process [71, 75]. Its PSD at the amplifier output can be written as:

$$S_{\text{ASE}} = \frac{P_{\text{ASE}}}{B_m} = 2n_{\text{sp}}(G-1)hf, \quad \frac{1}{n_{\text{sp}}} = 1 - \frac{N_1\sigma_a(\lambda)}{N_2\sigma_e(\lambda)} \tag{2.29}$$

where n_{sp} is the spontaneous emission factor, G is the total amplifier gain, hf is the photon energy, and P_{ASE} is the ASE noise power measured over a bandwidth of B_{m}. As the ASE noise power is proportional to the gain and n_{sp} it can be minimized by operating the EDFA at high population inversion.

The noise performance of amplifiers is usually characterized by the noise figure (NF) being defined as the degradation of the electrical signal-to-noise ratio (SNR) due to the amplifier and measured with an ideal photodetector. From Derrickson [32] the definition for the noise figure is formulated as:

$$NF = \frac{\text{SNR}_{\text{in}}}{\text{SNR}_{\text{out}}} \approx 2n_{\text{sp}}\frac{G-1}{G} + \frac{1}{G} \tag{2.30}$$

where it is assumed that the ASE–ASE beat noise component can be neglected [76]. With this, the noise PSD can be written as a function of the NF and the gain:

$$S_{\text{ASE}} \approx (NF\,G - 1)hf \tag{2.31}$$

where both values (*NF* and gain) are being calculated without knowledge of the internal architecture of the optical amplifier. Practical EDFA designs consist typically of multiple stages where the first stage operates as a low-noise pre-amplifier, preferably pumped at 980 nm to ensure small *NF*. This is advantageous as according to the chain rule, the noise figure of an amplifier cascade is mainly determined by the *NF* of the first amplifier in the chain [77]. The following stages operate as a power amplifier, preferably pumped at 1,480 nm as this concept provides higher power conversion efficiency. Isolators being placed between stages prevent saturation of the previous stages due to backward propagating ASE noise from the following stage. Also, filters are used to suppress ASE noise outside the signal bandwidth and perform gain equalization.

2.5.1.3 Optical Signal-To-Noise Ratio

One important performance parameter is the optical signal-to-noise-ratio (OSNR) which is defined as the ratio of optical signal power and optical noise power within a measurement bandwidth B_m. Using Eq. 2.31 above we can derive the following:

$$\text{OSNR} = \frac{P_{\text{out}}}{S_{\text{ASE}}B_m} \approx \frac{P_{\text{in}}}{(NF - 1/G)\,hf\,B_m}, \qquad P_{\text{out}} = GP_{\text{in}} \tag{2.32}$$

The OSNR can be directly related to system performance for applications that are limited by optical noise. For a signal with perfect extinction ratio, assuming that signal-ASE beat noise and ASE–ASE beat noise dominate the electrical noise (e.g., ignoring thermal and shot noise added by a direct-detection optical receiver) and further assuming that no other propagation effects, such as for example CD, Kerr, and others, deteriorate the eye opening, the Q factor can be easily related to the OSNR at the input of the optical receiver [24] as follows:

$$Q = 2\,\text{OSNR}\frac{\sqrt{B_o/B_e}}{1+\sqrt{1+4\text{OSNR}}} \tag{2.33}$$

In the above, B_e is the electrical bandwidth of the receiver and B_o is the bandwidth of the optical filter in front of the receiver. For $B_m \neq B_o$, the OSNR value should be scaled by the factor B_o/B_m. before being used in the Q factor calculation. In most applications, the bandwidth of optical noise sources is much wider than the signal rate, and thus it can be assumed that the OSNR is proportional to the signal rate if optical filtering effects are neglected. So when increasing the signal rate for instance from 10 Gbit/s to 40 Gbit/s using the same modulation format and filter bandwidths that are proportional to the signal rate, the OSNR requirement rises approximately by 6 dB.

2.5.2 Systems Modeling and Performance

2.5.2.1 Black-Box Modeling

A phenomenological modeling approach that requires parameters describing the behavior of EDFAs in a system context is better suited for the simulation of a WDM transmission line since the internal architecture and parameters of the amplifiers are typically not known at this level of design. For this purpose Black-Box modeling is favorable. The approach to consider amplifier gain and noise figure as fixed functions of wavelength is an approximate one since it ignores for instance dependencies of pump and signal powers on the EDFA gain and noise characteristics, yet for many system simulation applications it is fairly adequate.

More accurately, but still making a number of simplifying assumptions on the physics in the doped fiber, the spectral gain and noise behavior of the EDFA can be calculated requiring no detailed knowledge of the internal amplifier details (e.g., emission and absorption characteristics, dopant density, reflection properties) [78–82]. As a consequence of the assumption that the doped fiber saturates homogeneously, the gain characteristics (with respect to wavelength, mode of operation, possibly pump power) can be calculated from a number of data files obtained from standard optical spectrum analyzer measurements performed on an actual physical device [32] or from detailed component-level simulations of architecturally complex doped amplifier designs [83]. The noise behavior is typically estimated in a similar manner by interpolating data taken at different points of operation, even though a solid physical background for this approach does not exist. Black-box modeling is computationally fast and therefore more efficient for use in system-level simulations and provides possibilities for incorporating internal automatic gain or power control as well.

If not compensated carefully, when designing high-capacity WDM systems utilizing a cascade of EDFAs, the individual gain and noise characteristics of each EDFA may result in significant fluctuations of signal power and OSNR between WDM channels. Figure 2.15 shows results of a related case study. The optical signal powers and OSNR values of 32 WDM channels in the C-Band are plotted versus propagation distance. The WDM band has been propagated over ten 80 km dispersion compensated fiber spans with the loss of each span being compensated by an EDFA. All amplifiers are set to provide the same average gain and have the

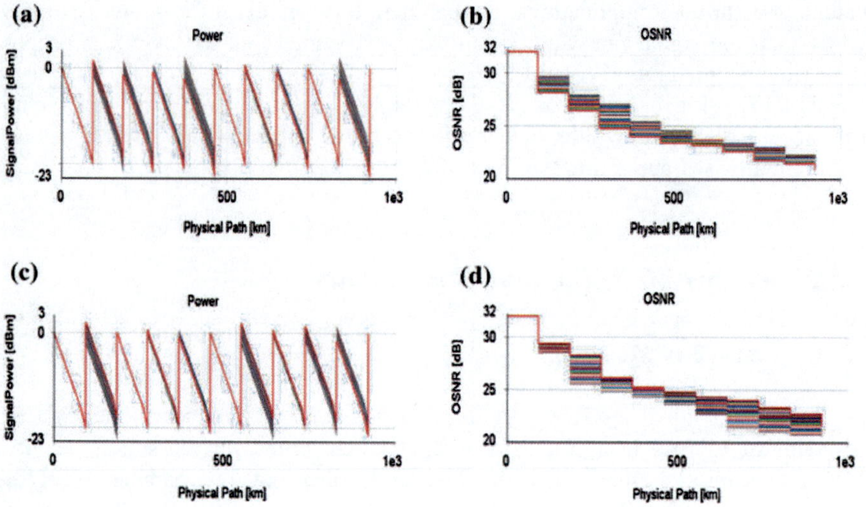

Fig. 2.15 Signal Power shown in (**a**) and (**c**) and OSNR shown in (**b**) and (**d**) for 32 channels in C-Band for propagation over cascades of 10 EDFA with different gain and noise tilts. Note that (**c**) and (**d**) present an inverted order of EDFAs compared to (**a**) and (**b**)

same average noise figure, however, they have varying gains and noise tilts. The upper graphs (Fig. 2.15a, b) are obtained using an arbitrary order of EDFAs; the lower graphs (Fig. 2.15c, d) are obtained using the inverted order of amplifiers. Overall, even though the same ten EDFAs are used in both cases, a significant spread in WDM channel signal power and OSNR is obvious in Fig. 2.15a, c for the former and Fig. 2.15b, d for the latter.

2.5.2.2 Power Transients

EDFAs, including their gain stabilization circuits, do not react instantly to abrupt changes in input power which might be caused by events such as automatic network re-configuration, failure of components, or switching of optical packets. The resulting power transients are especially problematic in networks with a complex topology due to numerous feedback mechanisms. Modeling dynamic transmission phenomena in transparent networks differs significantly from modeling stationary processes. Of particular interest is the analysis of dynamic power perturbations that propagate through the optical network and thus change the average power level at components along the optical path and at the receiver (see for example Fig. 2.16b and descriptions thereto further down). In a worst-case scenario, channel switching effects can create chaotic oscillations of signal powers that propagate through the network as described in Yoo et al. [84].

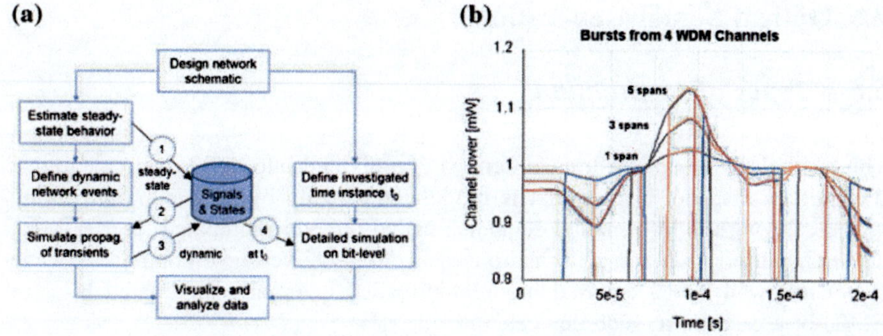

Fig. 2.16 **a** Workflow for emulating dynamic events in optical networks and investigating the resulting performance impacts; **b** Dynamics of random traffic bursts from 4 WDM channels: channel powers after one, three and five amplified spans using gain-clamped EDFAs; (From Richter et al. [85], Figs. 2 and 3. Reproduced by permission of © 2010 The Institute of Electrical and Electronics Engineers)

To emulate the characteristics of dynamic networks, it is necessary to combine investigations on bit-level (ns to sub-ps time scales) with analysis of effects with much larger response times (up to a few ms). The evolution of signals in such applications is often affected by varying noise levels and transient processes in control loops with various characteristic times. Figure 2.16a shows a general workflow for emulating dynamic events in optical networks and analyzing their performance impact [85]. First, slowly varying network events are emulated using parametric signal representations that track information about important signal attributes versus time, frequency, and topology. When starting the network emulation, only changes to these attributes are calculated in the context of a global evolution analysis considering all signal paths. The origin of the limiting effects can be detected by recording and analyzing important link characteristics at pre-defined topology points as described in Richter et al. [69]. With the gathered parametric information, estimates of the system quality can be obtained using functional relationships that are derived from measurements or experience (design rules). For example, Fig. 2.16b shows the growths of power fluctuations of random traffic bursts from four WDM channels after passing one, three, and five amplified spans using gain-clamped EDFAs.

In an optional second step, detailed simulations down to the bit-level can be performed to reveal the physical limitations at a certain, identifiable time instance of the network (e.g., during turn on, stable operation, channel switching, component failure). For this purpose, the time-dynamic information about the network (e.g., signal and equipment state data at discrete time instances) from the first step is used to define initial conditions for a detailed simulation on a bit-level.

2.6 Optical Signal Generation

2.6.1 Types of Modulation

The technology of optical transmitters can be divided into two broad categories: transmitters made of a laser that is modulated directly by the electrical driving signal carrying the information sequence called directly modulated laser (DML), and transmitters that are combinations of a laser delivering a continuous wave (CW) followed by an external modulator (called externally modulated laser or EML) or a cascade of such devices.

2.6.1.1 Direct Modulation

The applicability of directly modulated lasers in modern transmission systems is limited due to the narrow modulation bandwidth of lasers and the generation of frequency fluctuations with time (chirp) that are hard to control or eliminate. Chirp combined with CD critically affects the channel transmission performance over fiber. Recent work on chirp-managed lasers (CMLs) has proven that high dispersion tolerance could be achieved when using the directly modulated laser in combination with a carefully tailored optical filter applied to counteract the adiabatic chirp from a laser [86].

The behavior of DMLs is most easily predicted using laser rate equation models requiring physical parameters describing the internal cavity mechanisms of the laser. These physical parameters can be easily estimated by fitting theoretical curves to measured data for intensity modulation response, light–current characteristic, and time-resolved chirp [87]. Simple rate equation models ignore the details of the optical spectrum (assuming a single-mode, both in the longitudinal and transverse directions). Specifically, distributed feedback (DFB) lasers are well suited for a description by rate equations because the DFB design allows suppressing side modes efficiently. However, in order to evaluate such suppression and design optimal DFB structures, laser models that include the longitudinal structure of the laser such as the transmission-line laser model and its variants [88, 89] are required. If the variation of the carrier density is represented longitudinally these detailed models can predict chirp due to longitudinal spatial hole burning and even instabilities caused by this effect.

2.6.1.2 External Modulation

WDM transmission systems employing channel bitrates of 10 Gbit/s and higher typically deploy externally modulated transmitters. These provide a high wavelength-stability, small amount of distortions, high extinction ratio, and a defined frequency chirp characteristic that can be controlled to counteract for fiber propagation degradations due to CD and SPM [90]. The laser in an externally modulated system is operated under CW conditions with a well-defined linewidth and intensity noise. For 10 Gbit/s systems and above the frequency spectrum of the intensity noise and the detailed lineshape may be important. Also, the correlation between the instantaneous frequency and the intensity could be relevant and must be considered (these characteristics can be predicted using rate-equation laser models or even transmission-line laser modeling).

Two types of modulators are commonly used, namely Mach–Zehnder modulators (MZM) and electro-absorption modulators (EAM) [91]. MZMs provide usually a better defined transfer characteristic. They can be designed to have zero frequency chirp, or a chirp that can be controlled by the electrical drive voltage. EAMs are easier to integrate to the device, but come with an intrinsic chirp which is proportional to the gradient of the emitted optical power with respect to applied voltage. They require careful characterization over a wide range of wavelengths, bias, and temperature operating conditions. Furthermore, EAMs have a nonlinear transmission-voltage characteristic and a voltage-dependent chirp characteristic. They can be represented in simulation models by using polynomial fits, or multidimensional lookup tables [92]. More advanced models combine the spectral and dynamic characteristics (e.g., important for wideband applications) and polarization sensitivity, as well as specific electric properties of the driving signal [93]. Figure 2.17 shows static and dynamic characterization and validation results of a typical EAM. Figure 2.17a compares measured and simulated static power transfer functions of the investigated EAM using a single spectral band of about 10 THz to allow a simultaneous modulation of 13 optical carriers between 1,540 and 1,588 nm. Figure 2.17b shows comparisons of measured and simulated normalized dynamic EAM power transfer function and alpha factor characteristics for single channel 40 Gbit/s NRZ modulation.

The Mach–Zehnder modulator is based on the interference principle which is well represented by a few parameters in a physical model. The electric field of the incident optical signal is split to propagate over two branches, over which the field experiences different amounts of phase change due to the electro-optic effect. The optical signals of the two branches are then recombined which results in an interference pattern that is directly related to the phase difference between the two branches. The amount of phase change over each branch is controlled by electrical voltages. With the lumped model approximation the transfer function of the MZM can be written as:

$$\frac{E_{\text{out}}(t)}{E_{in}(t)} = \cos[\Delta\Phi(t)]e^{-j\alpha\tan(\Delta\Phi_{\text{bias}})\Delta\Phi(t)} \tag{2.34}$$

where $E_{\text{in, out}}(t)$ is the electric field at the input and output of the modulator respectively, $\Delta\Phi(t)$ is the phase difference of the electric fields in the two branches of the MZM, $\Delta\Phi_{\text{bias}}$ is the bias point of operation, and α is the so-called α-factor defining the chirping behavior of the MZM. This α-factor is further defined as the ratio of the phase change to the intensity change at the output of the MZM [92]:

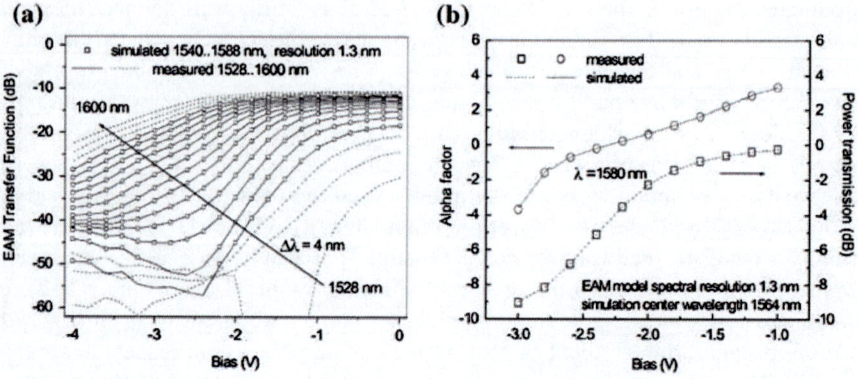

Fig. 2.17 Static and dynamic EAM modeling. **a** measured (*lines*) and simulated (*squares*) power transfer function using single ~10 THz band to allow a simultaneous modulation of 13 optical carriers in 1,540–1,588 nm; **b** dynamic power transfer function (normalized to 0 dB at 0 V) and alpha factor extracted from the modulated signal. (From Koltchanov et al. [93], Figs. 1 and 3 *right*. Reproduced by permission of © 2005 The Institute of Electrical and Electronics Engineers)

$$\alpha = 2I\frac{\mathrm{d}\phi(t)/\mathrm{d}t}{\mathrm{d}I(t)/\mathrm{d}t} \tag{2.35}$$

When operating the MZM with $\Delta\Phi_{\mathrm{bias}} = \pi/4$ ideal intensity modulation is achieved for $\alpha = 0$. Traditionally MZMs are built on $LiNbO_3$ making use of the Pockels effect (linear dependence of the refractive index on the applied bias voltage). The progress in monolithic integration is supporting the development of InP-based integrated transmitters providing benefits such as lower voltage drive requirements, shorter device dimensions and the possibility to integrate with laser sources or optical amplifiers. InP modulators are based on the Kerr and Frank–Keldysh effect (bulk structures) or QCSE (multi-quantum well structures) so that the refractive index change is quadratic with voltage and the absorption change is linear with voltage [94]. The electro-refractive effect enables the development of MZMs with low voltage drive requirements and short electrode lengths. Using a phenomenological modeling approach for the electro-refraction affects the change of the group effective index of the guided optical mode Δn as described in Arellano et al. [89] as:

$$\Delta n = r_1 V + r_2 V^2 + \frac{\mathrm{d}n}{\mathrm{d}N}(N - N_0) \tag{2.36}$$

The first two terms on the right-hand side represent the electro-refractive index change that occurs due to electro-absorption. This may be caused by the Pockels and Kerr effects and it is described by the linear r_1 and quadratic r_2 change of refractive index with voltage V. The last term represents the carrier-dependent refractive index change. Here, $\mathrm{d}n/\mathrm{d}N$ is the differential refractive index, N is the carrier density, and N_0 is the reference carrier density for which the chirp-induced frequency offset is zero.

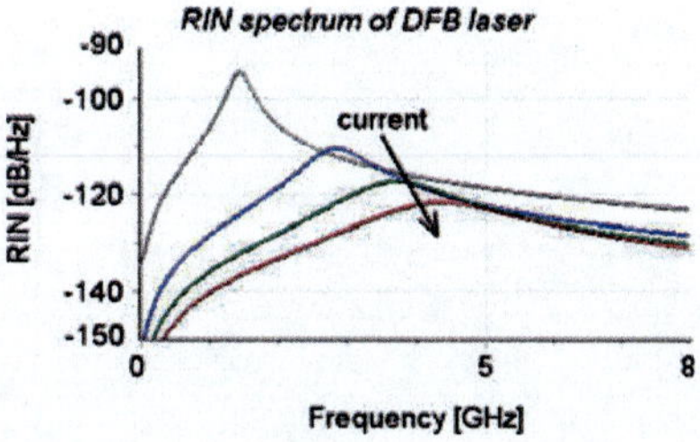

Fig. 2.18 RIN spectrum of DFB laser for different drive currents

2.6.2 Transmitter-Induced Signal Degradation Effects

2.6.2.1 Laser Noise

Due to spontaneous emission taking place inside semiconductor lasers used for direct modulation or as optical light sources for external modulators, photons with random phase are added to the coherent output field which is generated by stimulated emission inside the laser, creating small perturbations of the amplitude and phase. Random variations of the laser intensity limit the maximum achievable OSNR while random variations of the phase lead to an increased spectral linewidth of the laser [8].

The relative intensity noise (RIN) describes the ratio of the mean-squared fluctuations in the observed PSD and the average output power of the laser [32]. The amount of RIN is independent of attenuation along a link as both the nominal signal power and the noise are attenuated equally. RIN represents a fundamental limit to the transmission capacity of high-speed links and the carrier-to-noise performance of analog links. For lasers in multimode operation mode partition noise (MPN) defines the RIN in a single mode relative to the RIN of all the modes. MPN represents an important measure for lasers emitting more than one mode (and this includes standard DFB lasers), since even if the overall RIN is small, individual modes may show large intensity fluctuations.

Using a simple system-level modeling approach, intensity noise could be approximated by a Gaussian random noise process and specified with respect to a given reference power. If the laser output power is different from that number then the RIN has to be scaled inversely to emulate the effect that the laser RIN scales inverse proportionally to the drive current (assuming the ratio of stimulated and spontaneous emission scales with drive current and the laser is not saturating). This model however, produces a flat electrical noise spectrum. Although this is rather unrealistic since RIN follows the intrinsic frequency response of the laser, this assumption can be used to emulate the worst-case impact of RIN. The RIN spectrum could be calculated more accurately using rate equations for modeling the laser characteristics [95]. The RIN spectrum of a DFB laser for four different drive currents is depicted in Fig. 2.18 showing clearly the dependence of the spectral RIN peak due to the carrier-photon resonance of the laser.

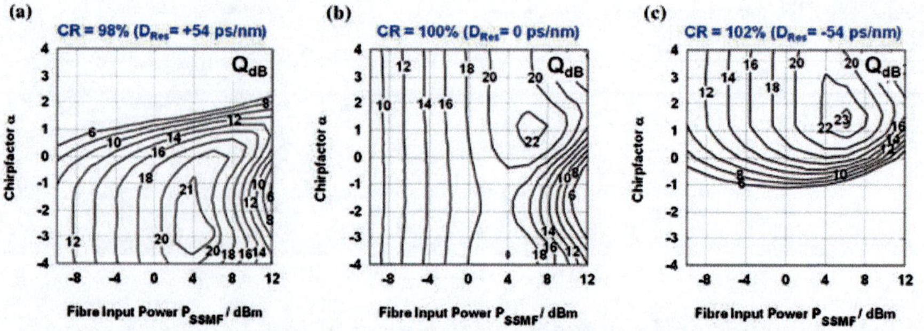

Fig. 2.19 Q factor versus transmitter pre-chirp and fiber input power for **a** under-compensated; **b** completely compensated; **c** over-compensated transmission sections for 40 Gbit/s NRZ over 2 × 80 km SSMF with one EDFA per section. (From Freund et al. [97], Fig. 4. Reproduced by permission of © 2004 The Institute of Electrical and Electronics Engineers)

The spectral linewidth of a laser is defined as the width (full-width at half-maximum, FWHM) of the CW spectrum which could be approximated by a Lorentzian shape (if correlation between intensity and phase fluctuations is neglected and relaxation oscillations are ignored) [8]. Typically the linewidth of DFB laser sources is below 20 MHz, and thus only of small importance for externally modulated systems utilizing direct-detection receivers since the modulation bandwidth is much wider (e.g., on the order of the driving symbol rate). However, phase fluctuations of optical sources with a linewidth in the MHz range could represent limitations for direct modulated systems as well as high-speed systems utilizing coherent reception. For the latter type of systems a more accurate modeling than the Lorentzian approximation of the spectral line shape is advisable (for details, see for instance [96]).

2.6.2.2 Chirp and Fiber Effects

As discussed in the previous sections, the interactions among transmitter chirp, fiber CD, and Kerr nonlinearities can have a strong impact on transmission system performance. Figure 2.19 shows sample simulation results of such interdependency. Contour plots of the Q factor are shown against fiber input power and transmitter chirp for a 40 Gbit/s NRZ transmission over two spans of 80 km SSMF for different degrees of dispersion compensation (compensation ratio, CR). Similar performance optima can be achieved for all dispersion maps for carefully selected values of the transmitter chirp. However, the system tolerance against chirp fluctuations varies significantly between completely compensated links (Fig. 2.19b) and links with under- or over-compensation (Fig. 2.19a, c). Consequently, when designing 40 Gbit/s WDM systems without exact compensation of the CD slope, each channel requires individual CD compensation depending on initial chirp and residual dispersion per channel.

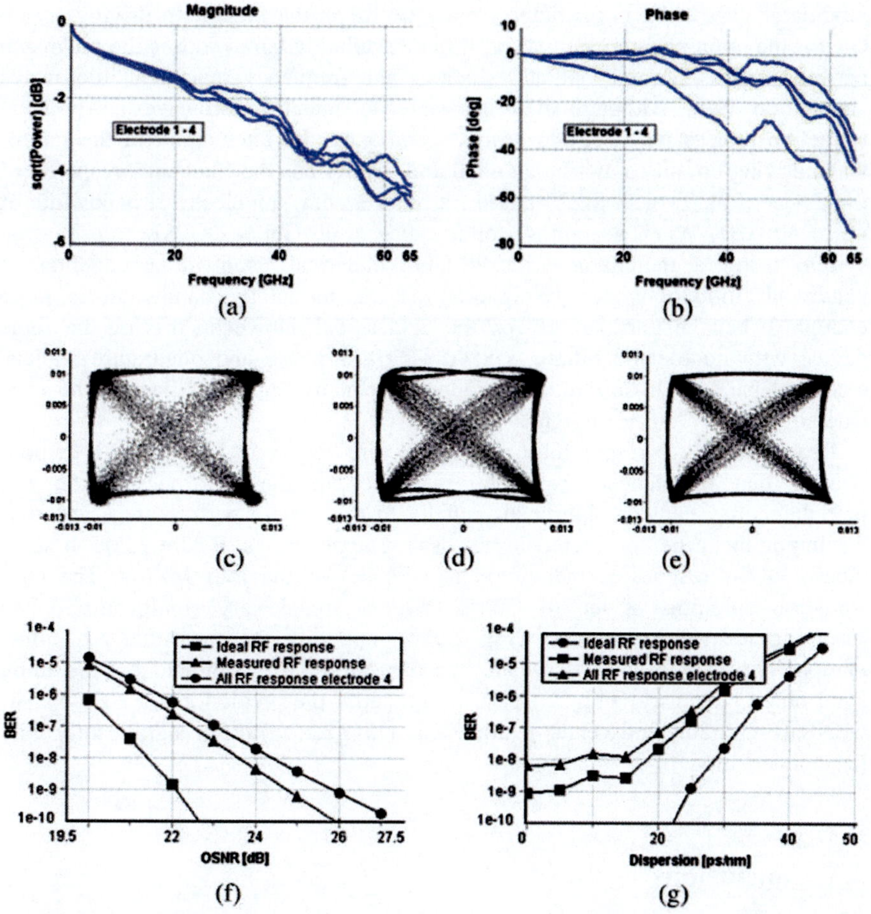

Fig. 2.20 **a** and **b** show S_{21} responses measured at input ports after smoothing; **c** Constellation diagrams of 80 Gbit/s DQPSK for an ideal S_{21} response; **d** measured magnitude and phase; **e** measured magnitude only; **f** BER of 80 Gbit/s DQPSK modulated signal versus OSNR; **g** BER of 80 Gbit/s DQPSK modulated signal versus residual CD for OSNR = 25 dB. (From Arellano et al. [100], Figs. 6,7, and 8. Reproduced by permission of © 2009 The Institute of Electrical and Electronics Engineers)

2.6.2.3 Drivers and E/O-Contact Limitations

High-speed applications that need to manage large signal bandwidths show increased requirements on electrical components and electro-optical interconnects. At first, electrical drive circuits produce signals with an intrinsic amplitude noise, timing jitter, and imbalance between rise and fall times introducing restrictions to the ideal driving scenario of optical modulators. Furthermore, the optical modulator bandwidth is restricted setting limits on the maximum driving speed. One of the main phenomena that limit the modulation bandwidth is the mismatch between microwave and optical velocities and frequency-dependent microwave attenuation [98]. Although it is possible to match microwave and optical refractive indices by electrode design techniques, the attenuation in the microwave driving circuit cannot be avoided and determines the frequency response of the device [99]. Typical MZM-based transmitters have an electrical bandwidth of up to 40 GHz. When designing, for instance, a 40 Gbit/s DQPSK transmission system using a modulator with 20 GHz electrical bandwidth, modeling of bandwidth limitations can be ignored as the modulator bandwidth is large enough to behave transparently on the data signal. However, driving the same device with double the bitrate, i.e. 40 Gbit/s in-phase and quadrature signals, electrical bandwidth limitations introduce a penalty, and thus have to be considered carefully in the design.

Results of a corresponding case study are depicted in Fig. 2.20 demonstrating that imbalanced frequency responses of the electrodes of an I-Q modulator may cause a significant penalty to the system performance, and thus are important to be considered in high-speed applications. Figure 2.20a, b show measured S_{21} responses (magnitude and phase) of the four MZMs. The constellation diagrams of an 80 Gbit/s DQPSK signal vary significantly when assuming ideal S_{21} responses (Fig. 2.20c), applying the measured S_{21} information (Fig. 2.20d), and when using the measured S_{21} magnitudes but assuming ideal phase responses (Fig. 2.20e). Consequently, the predictions of BER versus OSNR (Fig. 2.20f) and versus residual CD (Fig. 2.20 g) differ significantly with the applied model.

2.7 Conclusions

The above work introduced the reader into the concept of multiple optical signal representations and the major physical layer effects causing system degradations in the WDM transport layer. The focus was on deriving computer models for the various effects and for that matter these were grouped into categories such as fiber propagation, optical amplification, and transmitter-induced signal degradation effects. Within the fiber propagation category, effects were divided into linear, nonlinear, and polarization-dependent.

Suitable modeling approaches were presented, and typical simulation results were demonstrated.

Acknowledgment The author thanks Igor Koltchanov, Hadrien Louchet, Jim Farina, Cristina Arellano, and other (current and former) members of the team at VPIphotonics for sharing their invaluable knowledge and numerous contributions referenced in this work.

References

1. Lowery AJ, Lenzmann O, Koltchanov I, Moosburger R, Freund R, Richter A, Georgi S, Breuer D, Hamster H (2000) Multiple signal representation simulation of photonic devices, systems, and networks. IEEE J Sel Top Quant Electron 6(2):282–296
2. User's Manual (2010) VPItransmissionMaker^{TM} Optical Systems, Version 8.5
3. Piprek J (ed) (2003) Optoelectronic devices: advanced simulation and analysis. Springer, New York
4. Richter A, Louchet H, Koltchanov I (2007) Pitfalls when modeling high-speed optical transmission systems. In: Proceedings of IEEE/LEOS Summer Topicals, paper TuE1.1, Portland, OR, pp 238–239
5. Gowar J (1984) Optical communication systems. Prentice Hall, Englewood Cliffs, pp 71–80
6. Keiser G (2000) Optical fiber communications, 3rd edn. McGraw-Hill, Boston
7. Chraplyvy A (1990) Limitations on lightwave communications imposed by optical-fiber nonlinearities. IEEE/OSA J Lightwave Technol 8(10):1548–1557
8. Agrawal GP (2002) Fiber-optic communication systems, 3rd edn. Wiley Inter-science, New York
9. Agrawal GP (1995) Nonlinear fiber optics, 3rd edn. Academic Press, New York
10. Kato T, Koyano Y, Nishimura M (2000) Temperature dependence of chromatic dispersion in various types of optical fiber. Opt Lett 25(16):1156–1158
11. Liu F et al. (2002) 1.6Tbit/s (40x42.7 Gbit/s) transmission over 3600 km ultra wave fiber with all-Raman amplified 100 km terrestrial spans using ETDM transmitter and receiver. In: Proceedings of IEEE/OSA optical fiber communication conference (OFC), post-deadline paper FC7, Anaheim, CA
12. Inoue K (1992) Four-wave mixing in an optical fiber in the zero-dispersion wavelength region. IEEE/OSA J Lightwave Technol 10(11):1553–1561
13. Oppenheim AV, Schafer RW (1989) Discrete-time signal processing. Prentice Hall, Englewood Cliffs, pp 587–622
14. Mecozzi A, Clausen CB, Shtaif M (2000) Analysis of intrachannel nonlinear effects in highly dispersed optical pulse transmission. IEEE Photonics Technol Lett 12(4):392–394
15. Essiambre RJ, Mikkelsen B, Raybon G (1999) Intra-channel cross-phase modulation and four-wave mixing in high-speed TDM systems. Electron Lett 35(18):1576–1578
16. Mamyshev PV, Mamysheva NA (1999) Pulse-overlapped dispersion-managed data transmission and intrachannel four-wave mixing. Opt Lett 24(21):1454–1456
17. Clausen CB, Mecozzi A, Shtaif M (2000) Nonlinear intra-channel effects: system impairments and their remedy. In: Proceedings of European conference on optical communications (ECOC), vol 3, Munich, Germany, pp 33–34
18. Martensson J, Westlund M, Berntson A (2000) Intra-channel pulse interactions in 40 Gbit/s dispersion-managed RZ transmission. Electron Lett 36(3):244–246
19. Killey RI, Thiele HJ, Mikhailov V, Bayvel P (2000) Reduction of intrachannel nonlinear distortion in 40 Gb/s-based WDM transmission over standard fiber. IEEE Photonics Technol Lett 12(12):1624–1626
20. Richter A (2002) Timing jitter in long-haul WDM return-to-zero systems. Dissertation, Technische Universität Berlin, Germany
21. Breuer D, Schneider M, Vorbeck S, Freund R, Richter A (2004) Design analysis of upgrade strategies from single to double and triple-wavelength-band WDM transmission. In: Proceedings of IEEE Asia-Pacific optical communications conference (APOC), vol 5625, Beijing, China, pp 601–613
22. Kidorf H, Rottwitt K, Nissov M, Ma M, Rabarijaona E (1999) Pump interactions in a 100-nm bandwidth Raman amplifier. IEEE Photonics Technol Lett 11(5):530–532
23. Namiki S, Emori Y (2001) Ultrabroad-band Raman amplifiers pumped and gain-equalized by wavelength-division-multiplexed high-power laser diodes. IEEE J Sel Top Quant Electron 7(1):3–16

24. Kaminow IP, Li T (eds) (2002) Optical fiber telecommunications IVB: systems and impairments. Academic Press, San Diego, CA
25. Iannone E, Matera F, Mecozzi A, Settembre M (1998) Nonlinear optical communication networks. Wiley, New York
26. Gordon JP, Kogelnik H (2000) PMD fundamentals: polarization mode dispersion in optical fibers. PNAS 97(9):4541–4550
27. Wai PKA, Menyuk CR (1996) Polarization mode dispersion, decorrelation, and diffusion in optical fibers with randomly varying birefringence. IEEE/OSA J Lightwave Technol 14(2):148–157
28. Curti F, Daino B, de Marchis G, Matera F (1990) Statistical treatment of the evolution of the principal states of polarization in single-mode fibers. IEEE/OSA J Lightwave Technol 8(8):1162–1166
29. Collett E (1993) Polarized light: fundamentals and applications. Marcel Dekker, New York
30. Foschini GJ, Pole CD (1991) Statistical theory of polarization dispersion in single-mode fibers. IEEE/OSA J Lightwave Technol 9(11):1439–1456
31. Freund R, Molle L, Hanik N, Richter A (2004) Design issues of 40-Gbit/s WDM systems for metro and core network application. In: Proceedings of IEEE Asia-Pacific optical communications conference (APOC), vol 5625, Beijing, China, pp 532–543
32. Derrickson D (1998) Fiber optic test and measurement. Prentice Hall, Upper Saddle River
33. Mecozzi A, Shtaif A (2002) The statistics of polarization-dependent loss in optical communication systems. IEEE Photonics Technol Lett 14(3):313–315
34. Gisin N, Huttner B (1997) Combined effects of polarization mode dispersion and polarization dependent losses in optical fibers. Opt Commun 142(1–3):119–125
35. Steinkamp A, Vorbeck S, Voges E (2004) Polarization mode dispersion and polarization dependent loss in optical fiber systems. In: Proceedings of SPIE Optics East, vol 5596, Philadelphia, PA, pp 243–254
36. Antoniades N, Reichmann KC, Iannone PP, Frigo NJ, Levine AM, Roudas I (2006) The impact of polarization-dependent gain on the design of cascaded semiconductor optical amplifier CWDM systems. IEEE Photonics Technol Lett 18(20):2099–2101
37. Lee M, Antoniades N, Boskovic A (2002) PDL-induced channel power divergence in a metro WDM network. IEEE Photonics Technol Lett 14(4):561–563
38. Wang D, Menyuk CR (1999) Polarization evolution due to the Kerr nonlinearity and chromatic dispersion. IEEE/OSA J Lightwave Technol 17(12):2520–2529
39. Hodzic A, Konrad B, Petermann K (2003) Improvement of system performance in Nx40-Gb/s WDM transmission using alternate polarizations. IEEE Photonics Technol Lett 15(1):153–155
40. Forzati M, Berntson A, Martensson J (2004) IFWM suppression using APRZ with optimized phase-modulation parameters. IEEE Photonics Technol Lett 16(10):2368–2370
41. Richter A, Koltchanov I, Lowery A (2004) Photonic design automation of optical communication systems. In: Proceedings of IEEE Asia-Pacific optical communications conference (APOC), vol 5625, Beijing, China, pp 316–327
42. Bosco G, Carena A, Curri V, Gaudino R, Poggiolini P, Benedetto S (2000) Suppression of spurious tones induced by the split-step method in fiber systems simulation. IEEE Photonics Technol Lett 12(5):489–491
43. Rasmussen CJ (2001) Simple and fast method for step size determination in computations of signal propagation through nonlinear fibers. In: Proceedings of IEEE/OSA optical fiber communication conference (OFC), paper WDD29-1, Anaheim, CA
44. Taha TR, Ablowitz MJ (1984) Analytical and numerical aspects of certain nonlinear evolution equations. II. numerical, nonlinear Schroedinger equation. J Comput Phys 55(2):203–230
45. Carena A, Curri V, Gaudino R, Poggiolini P, Benedetto S (1997) A time-domain optical transmission system simulation package accounting for nonlinear and polarization-related effects in fibers. IEEE J Sel Areas Commun 15(4):751–765
46. Brigham EO (1974) The fast fourier-transform. Prentice-Hall, Englewood Cliffs

47. Singleton RC (1967) A method for computing the fast Fourier transform with auxiliary memory and limited high-speed storage. IEEE Trans Audio Electroacoustics 15(2):91–98
48. Yu T, Reimer WM, Grigoryan VS, Menyuk CR (2000) A mean field approach for simulating wavelength-division multiplexed systems. IEEE Photonics Technol Lett 12(4):443–445
49. Wai PKA, Menyuk CR, Chen HH (1991) Stability of solitons in randomly varying birefringent fibers. Opt Lett 16(16):1231–1233
50. Marcuse D, Menyuk CR, Wai PKA (1997) Application of the Manakov-PMD equation to studies of signal propagation in optical fibers with randomly varying birefringence. IEEE/OSA J Lightwave Technol 15(9):1735–1746
51. Biondini G, Kath WL, Menyuk CR (2002) Importance sampling for polarization-mode dispersion. IEEE Photonics Technol Lett 14(3):310–312
52. Menyuk CR (1987) Nonlinear pulse propagation in birefringence optical fibers. IEEE J Quant Electron 23(2):174–176
53. Nelson LE, Jopson R (2002) Introduction to polarization mode dispersion in lightwave systems Venice Summer School on PMD, Venice, Italy, pp 24–26
54. Shtaif M, Mecozzi A (2000) Study of the frequency autocorrelation of the differential group delay in fibers with polarization mode dispersion. Opt Lett 25(10):707–709
55. Richter A, Dazert M, Koltchanov I, Myslivets E, Lowery A (2002) Performance degradations in high-speed (40 Gbit/s +) transmission systems due to polarization mode dispersion. In: Proceedings of IEEE/OSA national fiber optics engineers conference (NFOEC), paper 073, Anaheim, CA
56. Richter A, Koltchanov I, Kuzmin K, Myslivets E, Freund R (2005) Issues on bit-error rate estimation for fiber-optic communication systems. In: Proceedings of IEEE/OSA optical fiber communication conference (OFC/NFOEC), paper NTuH3, Anaheim, CA
57. Richter A, Koltchanov I, Kuzmin K, Rukhlenko D (2005) Bit-error rate estimation for applications using 40 Gbit/s and higher. In: Proceedings of IEEE workshop on optical transmission and equalization (WOTE), paper B3, Shanghai, China, pp 23–24
58. Francia C, Bruyere F, Penninckx D, Chbat M (1998) PMD second-order effects on pulse propagation in single-mode optical fibers. IEEE Photonics Technol Lett 10(12):1739–1741
59. Kogelnik H, Nelson LE, Gordon JP, Jopson RM (2000) Jones matrix for second-order polarization mode dispersion. Opt Lett 25(1):19–21
60. Eyal A, Marshall WK, Tur M, Yariv A (1999) Representation of second-order polarization mode dispersion. IEEE Electron Lett 35(19):1658–1659
61. Orlandini A, Vincetti L (2001) A simple and useful model for Jones matrix to evaluate higher order polarization-mode dispersion effects. IEEE Photonics Technol Lett 13(11):1176–1178
62. Poloyko I, Khilo A, Myslivets E, Volkov V, Koltchanov I, Richter A, Lowery A (2003) Photonic design automation of Raman amplified systems. In: Proceedings of IEEE/OSA national fiber optics engineers conference (NFOEC), paper P29, Atlanta, GA
63. Photonic Modules Reference Manual (2011) VPItransmissionMaker ™Optical Systems, Version 8.6
64. Rottwitt K, Bromage J, Leng L (2002) Scaling the Raman gain coefficient of optical fibers. In: Proceedings of European conference on optical communications (ECOC), paper S3.03, Copenhagen, Denmark
65. Grant AR (2002) Calculating the Raman pump distribution to achieve minimum gain ripple. IEEE J Quant Electron 38(11):1503–1509
66. Perlin VE, Wintful HG (2002) On distributed Raman amplification for ultrabroad-band long-haul WDM systems. IEEE/OSA J Lightwave Technol 20(3):409–416
67. Richter A, Koltchanov I, Myslivets E, Khilo A, Shkred G, Freund R (2005) Optimization of multi-pump Raman amplifiers. In: Proceedings of IEEE/OSA optical fiber communication conference (OFC/NFOEC), paper NTuB4, Anaheim, CA
68. Lin Q, Agrawal G (2003) PMD effects in fiber-based Raman amplifiers. In: Proceedings of IEEE/OSA optical fiber communication conference (OFC), paper TuC4, Atlanta, GA

69. Richter A, Louchet H, Poloyko I, Karelin N, Farina J, Koltchanov I (2009) A parametric approach to optical systems design, optimization and validation. In: Proceedings of European conference on networks and optical communications (NOC), Valladolid, Spain, pp 361–367
70. Mears RJ, Reekie L, Jauncey IM, Payne DN (1987) Low noise erbium doped fiber amplifier operating at 1.54 μm. IEEE Electron Lett 23(19):1026–1028
71. Desurvire E (1994) Erbium-doped fiber amplifiers, principles and applications. Wiley, New York
72. Becker PC, Olsson NA, Simpson JR (1999) Erbium-doped fiber amplifiers, fundamentals and technology. Academic Press, New York
73. Giles R, Desurvire E (1991) Modeling erbium-doped fiber amplifiers. IEEE/OSA J Lightwave Technol 9(2):271–283
74. User's Manual (2011) VPIcomponentMaker ™Optical Amplifiers, Version 8.6
75. Abramovich F, Bayvel P (1997) Some statistical remarks on the derivation of BER in amplified optical communication systems. IEEE Trans Commun 45(9):1032–1034
76. Keiser G (2000) Optical fiber communications, 3rd edn. McGraw-Hill, Boston
77. Collin RE (1992) Foundations for microwave engineering, 2nd edn. McGraw-Hill, New York
78. Bonnedal D (1996) EDFA gain described with a black box model. In: Proceedings of OSA trends in optics and photonics, optical amplifiers and their applications, vol 5, Washington, D.C., pp 53–56
79. Burgmeier J, Cords A, März R, Schäffer C, Stummer B (1998) A black box model of EDFA's operating in WDM systems. IEEE/OSA J Lightwave Technol 16(7):1271–1275
80. Zhang X, Mitchell A (2000) A simple black box model for erbium-doped fiber amplifiers. IEEE Photonics Technol Lett 12(1):28–30
81. Jacobsen G, Persson U, Gillner L, Vanin E, Wingstrand S (2000) Pump power dependent black box EDFA model. J Opt Commun 21(5):171–177
82. Villa JAL, Bo FB, Querol VP, Teixeira AL, Prat Goma J (2008) Extended black box model for fiber length variation of erbium-doped fiber amplifiers. IEEE Photonics Technol Lett 20(24):2063–2065
83. Richter A, Devatine R, Koltchanov I, Lowery A, Khomchenko D, Yevseyenko D, Moar P (2002) Virtual product prototyping of erbium doped fiber amplifiers for applications in dense WDM systems. In: Proceedings of IEEE/OSA national fiber optics engineers conference (NFOEC), paper 153, Anaheim, CA
84. Yoo SJB, Xin W, Garratt LD, Young JC, Ellinas G, Chiao JC, Rauch M, Baran JE, Meagher B, Leblanc H, Chang G-K (1998) Observation of prolonged power transients in a reconfigurable multiwavelength network and their suppression by gain-clamping of optical amplifiers. IEEE Photonics Technol Lett 10(11):1659–1661
85. Richter A, Karelin N, Louchet H, Koltchanov I, Farina J (2010) Dynamic events in optical networks—emulation and performance impact analysis. In: Proceedings of IEEE military communication conference (MILCOM), San Jose, CA, pp 1083–1087
86. Matsui Y, Mahgerefteh D, Zheng X, Liao C, Fan ZF, McCallion K, Tayebati P (2006) Chirp-managed directly modulated laser (CML). IEEE Photonics Technol Lett 18(2):385–387
87. Cartledge JC, Srinivasan RC (1997) Extraction of DFB laser rate equation parameters for system simulation purposes. IEEE/OSA J Lightwave Technol 15(5):852–860
88. Lowery AJ (1992) A two-port bilateral model for semiconductor lasers. IEEE J Quant Electron 28(1):82–92
89. Arellano C, Mingaleev S, Novitsky A, Koltchanov I, Richter A (2009) Design of complex semiconductor integrated structures. In: Proceedings of IEEE Asia communication and photon conference (ACP) 7631, paper 7631-106, Shanghai, China
90. Gnauck AH, Korotky SK, Veselka J, Nagal J, Kemmener CT, Mindford WJ, Moser DT (1991) Dispersion penalty reduction using an optical modulator with adjustable chirp. IEEE Photonics Technol Lett 3(10):916–918

91. Hashimoto JI, Nakano Y, Tada K (1992) Influence of facet reflection on the performance of a DFB laser integrated with an optical amplifier/modulator. IEEE J Quant Electron 28(3):594–603
92. Koyama F, Iga K (1998) Frequency chirping in external modulators. IEEE/OSA J Lightwave Technol 6(1):87–93
93. Koltchanov I, Richter A, Myslivets E, Kazmierski C (2005) Complete time and frequency-dependent modeling of electro-absorption modulators. In: Proceedings of IEEE/OSA optical fiber communication conference (OFC/NFOEC), paper OME42, Anaheim, CA
94. Mendoza-Alvarez JG, Coldren LA, Alping A, Yan RH, Hausken T, Lee K, Pedrottin K (1988) Analysis of depletion edge translation lightwave modulators. IEEE/OSA J Lightwave Technol 6(6):793–808
95. Ahmed M, Yamada M, Saito M (2001) Numerical modeling of intensity and phase noise in semiconductor lasers. IEEE J Quant Electron 37(12):1600–1610
96. Betti S, De Marchis G, Iannone E (1995) Coherent optical communication systems. Wiley Inter-science, New York
97. Freund R, Molle L, Hanik N, Richter A (2004) Design issues of 40 Gbit/s WDM-systems for metro and core network application. In: Proceedings of European conference on optical communications (ECOC), paper Tu351, Stockholm, Sweden
98. Kaminow IP, Koch TL (eds) (1997) Optical fiber telecommunications IIIB. Academic Press, San Diego, CA
99. Dutta AK, Dutta NK, Fujiwara M (eds) (2002) WDM technologies: active optical components. Academic Press, New York
100. Arellano C, Louchet H, Koltchanov I, Richter A (2009) Important device limitations of transmitter and receiver concepts when designing 100 G transmission systems. In: Proceedings of IEEE international conference on transparent optical networks (ICTON), paper Tu.C2.3, Azores, Portugal

Chapter 3
State-of-the-Art in Device and Network Element Level Modeling

Ramón Gutiérrez-Castrejón

Abstract This chapter presents an overview of the most recent modeling and simulation techniques for the analysis and engineering of all the major devices and network elements that comprise a state-of-the-art optical communications system and network. The subject is presented by creating different abstraction levels in device modeling and building on the simulation fundamentals through computer modeling paradigms.

3.1 Abstraction Levels in Device Modeling

The increasing demand for optical telecommunications infrastructure, driven by the constantly growing bandwidth-need to support applications such as high–performance computing, video-on-demand and massive data-centers, asks for advanced software tools to efficiently design and analyze current and future photonic networks [1]. Modeling and simulation at the physical (or optical) layer level deal mainly with the propagation and transfer characteristics of optical data channels through a linear or nonlinear transport medium such as optical fibers, optical amplifiers, optical cross-connects (OXC) and other devices or network elements. In general, modeling of physical devices is carried out at two different abstraction levels (or somewhere in-between). Depending on the amount of detail required from the simulation results, system or detailed device models can be developed. System-level models are usually based on assumptions about the device's microscopic physical characteristics, following a phenomenological

R. Gutiérrez-Castrejón (✉)
Universidad Nacional Autónoma de México (UNAM)
Mexico City, Mexico
e-mail: RGutierrezC@iingen.unam.mx

N. (Neo) Antoniades et al. (eds.), *WDM Systems and Networks*, Optical Networks,
DOI: 10.1007/978-1-4614-1093-5_3,

approach. Normally, each component or device is described with a simple but efficient model in order to keep the computational times of the entire system within reasonable limits. Computation efficiency is therefore a must. These models are usually employed on engineering projects to design a transmission or transport system that is composed of tens to hundreds of optical components. The corresponding simulations deal with the entirety of the system and represent an aid to choose certain device parameter settings, find the optimum dispersion map, set transmission limits and so on [2]. On the other hand, device-level models are usually developed to study certain behavior of only one particular optical component or transmission link under various operating conditions, which cannot be easily realized today in the laboratory or that are very costly. They are fairly complex and normally deal with the solution of multi-dimensional partial differential equations to find the electrical field and temperature distribution within a particular structure. This kind of model sometimes also includes quantum electronic interactions described for instance through a density matrix approach. Their solution involves advanced mathematics (finite differences or finite element methods), especially if nonlinear effects are taken into account [3].

The design and engineering of wavelength division multiplexed (WDM) networks is nowadays based on the use of powerful simulation software. Such computational engines are attractive because they have the potential to replace hardware prototyping, thus reducing costs. They have evolved from relatively simple lightwave link simulators [4–6], to photonics design software spanning a wide range of abstraction scales [7]. Today, these sophisticated automatic design systems represent complete, integrated, commercially successful, lightwave simulation tools, which are discussed elsewhere in this book. In this chapter we will briefly review some advances on simulation at the photonics device level.

3.2 Main Elements of a Photonics Simulation Tool

A basic software tool for sub-systems, systems and optical networks simulation is usually composed of three main elements, namely:

1. A library of modules for each component and network element.
2. Data structures to transfer information among the modules.
3. Graphic displays.

The first element represents the core of the simulation tool and it encompasses the continuously expanding simulation units that are interconnected to build components, network elements or even the network topology. In turn, every module in the optical network is composed of three constituents that are closely interrelated: the physical model, the solver or computational algorithm to calculate the numerical solution of the mathematical model and a list of parameters belonging to that particular model. These parameters uniquely characterize the component or device under study. Some models can be fairly simple. For instance, a multi-mode interference coupler (MMI) can be modeled as a four-port device that re-scales, phase-shifts and combines the received signals according to their destination output ports. Some other models, such as those describing optical amplifiers or optical receivers, require a considerable amount of modeling effort since most of them should be able to efficiently capture large-signal nonlinear effects. Furthermore, all these models must be compatible with each other; otherwise interconnection among them would be impossible [8].

There is always more than just one way to numerically tackle the system of algebraic or differential equations that define a physical model. The algorithm of choice depends on the speed versus accuracy trade off, peculiarities of the initial conditions or geometry of the structure and the available information about the solution sought, such as periodicity. Numerical analysis is a field on its own and the reader is therefore referred to the vast available literature (e.g. [9]).

A general model can only represent a specific component when the list of associated physical parameters is properly set. A parameter can be as general as the network bit rate, and as specific as the carrier–carrier scattering time of an individual laser diode. Each model has its own list of parameters, which can usually be varied along well-determined ranges. This provides flexibility to the simulation tool because it allows modeling of similar devices that share the same model but exhibit different characteristics. Nonetheless, finding an adequate set of parameter values to characterize a particular device is sometimes one of the most daunting tasks in device modeling. It requires acute knowledge of the particular device physics and/or access to experimental characterization curves to be used as benchmark.

As mentioned before, the second main element of an optical network simulation tool is the data structures used to transfer information among the simulator modules. According to the modeling abstraction level, these structures may contain complex signals consisting of one or various sampled versions of an optical waveform (in time or frequency-domain), sampled electrical signals, statistical signals such as parameterized signals and noise bins (broad noise spectra efficiently represented as mean power spectral density within a defined frequency range), logical information (e.g. simulation central wavelength and simulation mode) and others [7]. The former structures are employed to carry out waveform-level simulations. Here, band-limited continuous-time signals are represented by discrete-time signals sampled with a frequency higher than the Nyquist rate. In WDM systems, high frequencies and interactions over bandwidths of possibly thousands of GHz lead to computationally intensive simulations [10]. Parameterized signals [7] or wavelength-domain representations [10] are continuous wave-like signals defined by their carrier wavelength, average power and, occasionally, their polarization information. They are useful for signal-to-noise calculations, where it is assumed that the optical network components do not linearly or nonlinearly alter the shape of the signal waveform. Some optical network simulation tools rely on a wise combination of sampled and parameterized signals to reduce execution time and memory consumption [11]. In this chapter we will mainly focus on the waveform-level simulation approach.

Although of great importance, the third of the three elements mentioned above is relatively simple. It is related to the inspection of the simulated signal or some of its characteristics that are presented in a graphical manner. Its relevance stems from the fact that it can be considered "the result" of the overall simulation process, or "a measurement" of the propagating signal at an intermediate simulation point. Graphic displays, visualizers or graphic-output interfaces are helpful in providing information about the analyzed signal. Such information can be presented in the form of the signal itself, its spectrum, its eye diagram, its autocorrelation and graphs of performance metrics such as bit-error rate (BER), average power, quality-factor (Q), eye-opening, etc. Their implementation is strongly aligned to the programing language used to code the simulation tool. We will therefore make no further reference to them within this chapter.

Optical network simulators have traditionally been coded in text-based programing languages such as FORTRAN or C. More recently, software environments that offer a considerable library of mathematical and other signal processing tools, such as Matlab® or Python have also been employed. An attractive alternative are graphical programing languages, such as G or LabVIEW that incorporate their own graphical user interface (GUI) and have the ability to easily create, integrate, re-utilize and manage complex libraries in a modular environment [12]. Due to its efficiency, LabVIEW has been utilized to code an optical systems simulation tool. This simulation engine has been employed, for instance, to analyze the system performance and feasibility of the just released 100 Gb/s Ethernet 40-km PHY standard [13].

We have divided this chapter into two main sections. In the first part, modeling and simulation of passive elements, such as fibers, filters, multiplexers, switches and equalizers, will be discussed. The second part is devoted to active elements. It then deals with simulation aspects of transmitters and optical amplifiers, while receivers are only briefly reviewed.

3.3 Passive WDM Network Building Blocks

3.3.1 Optical Fibers

Lightwave propagation in optical fibers is governed by Helmholtz equation [14]. Aided by the slowly varying envelope approximation and the method of separation of variables, a finite number of guided modes, which represent a solution that satisfies the optical fiber boundary conditions, can be found. Each mode has a specific propagation velocity that is determined by the mode propagation constant and the wavelength of the propagating light. However, when dealing with the simulation of optical telecommunications systems, one is concerned with the propagation of optical pulses rather than the actual modal structure of the field. In this chapter we will therefore assume that the set of transverse modes for the fiber under study is already known. Its derivation in the case of cylindrically symmetric optical waveguides can be found in several text books.

3.3.1.1 Silica and Polymer Multi-mode Optical Fibers

Pulse broadening in graded-index multi-mode fibers (GRIN MMFs) was first theoretically analyzed in 1976 [15]. The authors dealt with silica MMFs having an exponential refractive index profile (with exponent $\alpha > 0$) described by the following expression:

$$n(r,\lambda) = \begin{cases} n_1(\lambda)\left[1 - 2\Delta(\lambda)(r/a)^{\alpha}\right]^{1/2}, & \text{for } 0 \le r \le a \\ n_2(\lambda)\left[1 - 2\Delta(\lambda)\right]^{1/2}, & \text{for } r \ge a \end{cases} \tag{3.1}$$

where r is the distance from the axis located at the core center, a is the core radius, $n_1(\lambda)$ is the wavelength-dependent refractive index at the center of the core, $n_2(\lambda)$ is the refractive index at the cladding and $\Delta = (n_1^2 - n_2^2)/2n_1^2$. Based on a Wentzel-Kramer-Brillouin (WKB) analysis, the propagation constant $\beta(m,\lambda)$ for each mode can be approximately derived. The corresponding expression can be found elsewhere ([15], Eq. 36), where $m = 2\mu + \nu + 1$ stands for the principal number, and μ and ν are referred to as radial and azimuthal mode numbers. From a system simulation standpoint, however, what is more interesting is the dispersion behavior of the modes as they propagate along the fiber. In fact, it can be demonstrated that chromatic and modal dispersions can be considered to be independent effects leading to a MMF transfer function made up of two separate terms [16]. Under a linear input–output power relationship, the transfer function describing the distortion of a propagating pulse in a MMF due to chromatic dispersion on one hand, and mode delay, mode loss and mode coupling on the other hand, can be expressed as:

$$H(z,\omega) = H_{\text{chromatic}}(z,\omega)H_{\text{modal}}(z,\omega) \tag{3.2}$$

where z is the fiber length and ω is the optical frequency. Each term can then be written as [16]:

$$H_{\text{chromatic}}(z,\omega) = \int_{-\infty}^{\infty} P(\lambda)\exp\left(-i\omega z\left[D_0(\lambda_0)(\lambda-\lambda_0) + \frac{1}{2}S_0(\lambda_0)(\lambda-\lambda_0)^2\right]\right)d\lambda \tag{3.3}$$

$$H_{\text{modal}}(z,\omega) = \int_{1}^{M0} 2mC_{eff}(m,\lambda_0)L(m,\lambda_0,z)\exp(-i\omega\tau(m,\lambda_0)z)dm \tag{3.4}$$

where $P(\lambda)$ is the power spectrum density of the laser source, $D_0(\lambda_0)$ and $S_0(\lambda_0)$ are the averaged chromatic dispersion and slope over all guided modes, C_{eff} is the light coupling efficiency into each mode, L accounts for the mode-dependent loss at position z, τ represents the propagation delay per unit length of mode m and the integral upper limit stands for the total number of mode groups. Both, $H_{\text{chromatic}}$ and H_{modal} can be modeled as independent filters, and thus the propagation characteristics of the waveform can be calculated numerically. Actually, an analytical expression for the former transfer function can be derived provided that a Gaussian lineshape for the laser source is assumed. This is not the case with the latter transfer function, where an appropriate finite-difference or similar method has to be employed to numerically solve the equation [16].

Silica MMFs are dominant in Local Area Networks (www.fols.org), mainly because their larger core size imposes less severe requirements on cabling, connecting, splicing and handling than their single-mode counterparts, but also due to the lower cost of the associated active equipment. Although glass fibers have traditionally been deployed in optical communication networks, polymer optical fibers (POFs), in particular those made of polymethylmetacrilate (PMMA), represent strong candidates for short-reach communication links with lengths not exceeding 300 meters [17–20]. The reasons for this are that they offer less-expensive connectorization and installation, they have a large core diameter ($\sim$1 mm), they exhibit complete immunity to electromagnetic interference and, especially, they present greater resistance to mechanical damage and smaller weight as compared to silica fibers. As a result of this, POFs have found application in the automotive and aerospace sector, supercomputing, and as extremely low-cost optical interconnects [21] and in the ever developing in-house networks described in Chaps. 7 and 8. They have also been dubbed the "do-it-yourself" or "consumer" optical fibers. POFs are mainly limited by high multimodal dispersion (propagation of 2 million modes are estimated) and attenuation (about 0.08 dB/m in the green wavelength region). However, they present the advantage of a negligible Kerr effect. Light-emitting diodes (LEDs) and inexpensive Vertical cavity surface emitting lasers (VCSELs) are typically used as light sources operating at $\lambda = 650$ nm (visible red light).

Perfluorinated amorphous polymer-based GRIN fibers represent another class of POFs that exhibit a low-loss wavelength region from 500 to 1300 nm. This characteristic has favored its possible use in WDM transmission systems. In fact, a 2.5 Gb/s three-channel experiment over 200 m and a 2.5 Gb/s two-channel trial over 328 m have been successfully demonstrated with a BER $< 10^{-10}$ [22]. Moreover, 3-channel demultiplexers specifically designed for POFs have been recently proposed [23, 24]. Both devices were engineered using advanced computer aided design (CAD) tools to simulate the operation of the diffraction gratings that split the signals. These milestones show the feasibility of high-capacity transmission over POFs using WDM technology. Modeling of GRIN POFs can be carried out with the formalism presented in Eqs. 3.2–3.4. However, some simplifications, such as a negligible contribution from mode-coupling phenomena, simplify the analysis [25]. Advances in POFs modeling and simulation are continuously providing a deeper insight into the features of light propagation along these novel waveguide structures.

3.3.1.2 Single-mode Fibers

Single-mode fibers (SMFs) support only the HE_{11} fundamental mode. This behavior can be obtained in optical fibers by restricting their normalized frequency to be less than 2.4 [14]. The transverse structure of the fundamental mode can then be approximated by a Gaussian distribution. The aim of most optical transmission simulation software is thus to calculate the spatial and temporal evolution of the slowly varying envelope (SVE) of the optical field (i.e. lightwave pulses) as it propagates along the SMF, disregarding the aforementioned transverse structure. Assuming linear polarization and a frame of reference moving at the group velocity, the nonlinear Schrödinger equation (NLSE) that describes this phenomenon is given by [14]:

$$\frac{\partial A}{\partial z}+\frac{i\beta_2}{2}\frac{\partial^2 A}{\partial t^2}-\frac{\beta_3}{6}\frac{\partial^3 A}{\partial t^3}+\frac{\alpha}{2}A-i\gamma|A|^2A=0 \tag{3.5}$$

where A is the SVE of the propagating field, β_2 and β_3 are the group velocity and third-order dispersion parameters, α accounts for fiber losses and $\gamma = k_0\upsilon/A_{\mathrm{eff}}$. In contrast to MMFs, simulation of SMFs, whose core diameter is shorter than 10 microns, involves modeling of the nonlinear Kerr effect. This is accounted for in Eq. 3.5 through parameter γ, where υ is the nonlinear refractive index and A_{eff} is the fiber effective area. Some authors also incorporate the Raman response by adding extra terms to the NLSE [14].

Equation 3.5 represents one of the most significant device models in the design and engineering of WDM networks, since practically all the network nodes are connected through SMFs. Accurate network analyses at the physical layer level invariably involve simulation of single or multi-channel pulse propagation by solving the NLSE. Effects such as dispersion-induced distortions, attenuation, cross- (and self-) phase modulation and four-wave mixing are all considered within this powerful model. Solving Eq. 3.5 is therefore of utmost importance. This can be accomplished using finite-difference methods, especially when dealing with bidirectional beam propagation. An alternative approach is the split-step Fourier method (SSFM). Due to its efficiency, simplicity and ease of implementation, this numerical method can be found in most commercial software tools and some free simulators available on the web (SSPROP, NLSE Solver at Univ. of Rochester, photonics.iv.org.br).

In general terms, the SSFM solves the NLSE by assuming that in propagating the optical field over a small distance h, the dispersive and nonlinear effects act independently on the fiber [14]. This is schematically shown in Fig. 3.1. Under this assumption, Eq. 3.5 can be written as:

$$\frac{\partial A}{\partial z}=(\hat{D}+\hat{N})A \tag{3.6}$$

where the dispersion and nonlinear operators, respectively, are given by:

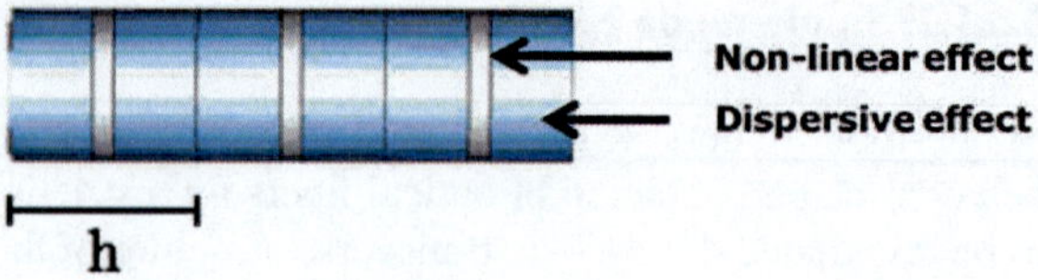

Fig. 3.1 Segmentation of an optical fiber in pieces of length h to independently calculate the non linear and dispersive effects using the SSFM

$$\hat{D} = \frac{-i\beta_2}{2}\frac{\partial^2}{\partial t^2} + \frac{\beta_3}{6}\frac{\partial^3}{\partial t^3} - \frac{\alpha}{2} \tag{3.7}$$

$$\hat{N} = i\gamma|A|^2 \tag{3.8}$$

By ignoring the non-commuting nature of $\hat{D}$ and $\hat{N}$ and assuming that the former operator is z-independent for short segments, the solution to Eq. 3.6 is then found to be:

$$A(z+h) \approx \left\{\exp\left(\frac{h}{2}\hat{D}\right)\exp\left(\frac{h}{2}\left[\hat{N}(z)+\hat{N}(z+h)\right]\right)\exp\left(\frac{h}{2}\hat{D}\right)\right\}A(z) \tag{3.9}$$

where the trapezoidal rule was utilized for the nonlinear operator. According to Fig. 3.1, the electric field at each segment can be calculated in three steps [14]. First, assuming dispersive effects only, the optical field is propagated for a distance $h/2$; this can be more easily accomplished in Fourier domain, where the n-derivatives in Eq. 3.7 become $(-i\omega)^n$. Back in time-domain, the field is affected by the fiber nonlinear effect, which is assumed to be lumped at the midplane of each segment. Finally, the field is propagated the remaining distance $h/2$ in Fourier domain. This algorithm can be implemented by means of an iterative procedure through the use of an FFT subroutine [9]. It must be noted that if no nonlinear effect is taken into account, Eq. 3.6 can be solved using a single step only. Under these circumstances, the fiber acts as a filter [5], where the transfer function corresponds to that already given in Eq. 3.3. In other words, (no surprisingly) a dispersive only SMF can be modeled as a MMF with $H_{\mathrm{modal}}(z,\omega) = 0$. Moreover, it can be demonstrated that Eq. 3.3 can be derived by solving Eq. 3.5 in Fourier domain with $\gamma = 0$.

Considering its importance in the design and analysis of WDM systems, several methods derived from the original SSFM have been proposed during the last decade to solve the NLSE. Some of them incorporate the simulation of further physical effects, while others look for optimized procedures to reduce computational time without loss of accuracy. An example can be found in [26], where different criteria to select the optimum step-size h, and thus speed up the calculation of the field along the fiber, are analyzed. This work demonstrates the importance of selecting the right value of h. This is also confirmed in [27], where an adaptive-step SSFM with fourth-order accuracy [28] is shown to increase computational efficiency by one order of magnitude in comparison to the traditional scheme. A different approach consists in eliminating the use of time-consuming FFTs through calculation of both the linear and nonlinear operators in time-domain. A simulator based on a combination of FFTs and infinite impulse response filters is then proposed in [29]. The split-step wavelet collocation method represents another alternative. It is based on the approximation of the field by a sum of several translated wavelets, transforming the NLSE into a system of ordinary differential equations that can efficiently be solved using standard methods [30]. Other schemes, such as predictor–corrector methods [31], have also been put forward. Accurate and efficient solution of the NLSE for WDM systems simulation can still be considered a topic of ongoing research.

Finally, an alternative approach to the use of the SSFM to estimate WDM system performance in the presence of nonlinear penalties consists in dividing the key transmission system impairments into three groups, and then calculates the Quality-factor (Q) for each of them. The three components of the Q-factor are then added (in dB), resulting in an accurate estimate of the system performance [32]. This approach results particularly useful in the design of long-haul systems, where simulations exhibiting short execution times are well valued. In the first design stage, a linear system is assumed and signals are modeled by a continuous wave having a certain power. The simulation tool then accounts for noise accumulation (channel crosstalk, receiver eye closure penalty and filter effects) and back-to-back system performance. In the second stage, optical pulse distortions due to the combined action of dispersion, SPM, XPM and FWM, which translates into electrical eye closure at the receiver, are estimated via a simplified split-step algorithm. Finally, the nonlinear interaction between signal and noise that affects signal variance is calculated through a transfer-matrix technique [33].

3.3.2 Optical Multiplexer/Demultiplexer: Filters

A WDM system and network realization would be impossible without multiplexers/demultiplexers. These are passive devices that combine/separate optical channels into/from one output/input. Since any demultiplexer can be used in reverse because of reciprocity, the same type of device is commonly used for both functions. However, demultiplexers present more stringent design margins. An optical demultiplexer is made up of a bank of optical filters that are expected to exhibit low-loss, flat-topped pass-band with low ripple, very steep skirts, linear spectral phase characteristics and minimum inter-channel crosstalk (<-30 dB). These specifications are becoming tighter as the network tolerance in terms of WDM channel spacing and spectral efficiency becomes more critical. Powerful software tools are then needed for optimum filter design at the device level, and filter impairment (concatenation, in-band and out-of-band cross-talk) evaluation at the system level [34]. In typical photonic systems simulators, optical filters are normally modeled via a complex transfer function using either measured characteristics or ideal filter forms. Filtering of the sampled signal is thus carried out in the frequency-domain. Typical filter shapes include Gaussian, Super-Gaussian,

Bessel and Butterworth functions. Well-known transfer functions of Fabry–Perot, Mach–Zehnder and other interferometers can also be used, especially in connection to interleavers [35].

Two main technologies dominate the demultiplexer market today. Multilayer thin-film dielectric filter (TFF) demultiplexers for low-channel counts of up to 16 channels, and arrayed waveguide gratings demultiplexers (AWGs) for large channel counts [36]. Both are commercially available at 50, 100 and 200 GHz bandwidths. TFFs are made of thin layers of film materials (SiO_2 and Ta_2O_5) with alternating high- and low-index of refraction. Band-pass filtering takes place through optical interference between the layers. Some attributes of TFF technology are environmental and thermal stability, good optical properties and scalability. TFF characteristics are well approximated via third-order Butterworth filters [34]. Moreover, since TFFs can be considered minimum phase filters, the filter phase response can be derived from its (calculated or measured) amplitude response using the Hilbert transform [37]. TFFs can also be accurately modeled in the time-domain using infinite impulse response (IIR) filters, which can be designed using standard digital signal processing techniques [37]. An AWG is a planar integrated device, mainly built using silica-on-silicon or InP technology. They consist of two free-space star couplers joined by a closely packed array of waveguides arranged along an arc. The waveguides are designed such that there is a constant optical path length difference between adjacent waveguides, thus forming a grating [36]. Some characteristics of AWGs (also known as phased arrays and waveguide grating routers) are temperature sensitivity, low-propagation loss, compactness and the possibility to fabricate multiple devices in a single wafer, which makes them robust and fabrication-tolerant. The operation principles of AWGs are well-known [38]. A good explanation on their modeling characteristics can be found in [39] and [40]; we therefore refer the reader to these articles.

3.3.3 Optical Switches

Depending on the abstraction level, modeling of passive optical switches can be as simple as directing a data structure (e.g. a sampled waveform) to one of two ports using a Boolean operator, or as complicated as modeling a free-space optics device using a ray-tracing or beam-propagation-based diffraction software. The former approach is commonly found in the design of fiber-optic telecommunications systems, whereas the latter is used in the design of a specific standalone component such as a micro-electromechanical system (MEMS). Since this book is rather concerned with WDM systems and networks, we will focus on the former approach that implies a higher abstraction level. Modeling of a switch at a low-abstraction level requires knowledge of the physical phenomena and technology behind the operation of the specific device. Such a description, even for today's market most prominent contenders, falls beyond the scope of this chapter and we hence refer the reader to [41].

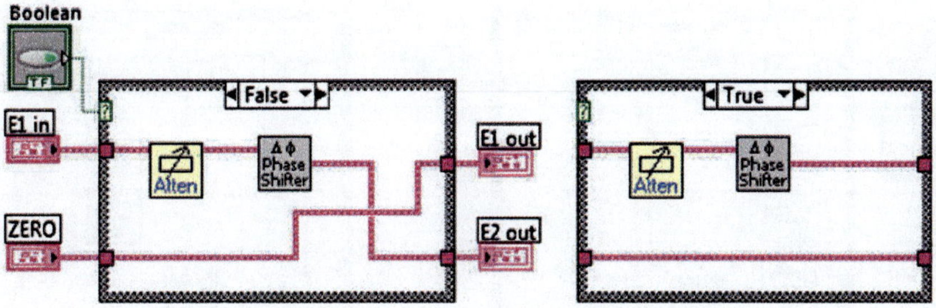

Fig. 3.2 LabVIEW block diagram of a 1 × 2 switch implementation

The simplest switching element is a 1 × 2 switch. A straightforward implementation written in LabVIEW graphical programing language is shown in Fig. 3.2. The Boolean operator determines whether E1_in bypasses the switch to E1_out (True), or is directed to E2_out (False). Attenuation and phase shifting of the input optical field is taken into account. This simple element can be scaled up to become a 1 × N switch by cascading N-1 1 × 2 switches. Alternatively, a 1 × N switch following a parallel approach can also be built using a "case" structure. Two 1 × 2 elements can be coupled to build a 2 × 2 switch. This can be more easily implemented by substituting the ZERO dummy variable in the 1 × 2 switch (Fig. 3.2) with a second input optical field. An additional pair of attenuation and phase shifter modules, unnecessary for the dummy variable, can also be included. An implementation of the 2 × 2 switch that includes the effect of cross-talk is presented in Fig. 3.3. By combining multiple 2 × 2 elements, an N × M switch can be built. However, complexity increases with port count. For instance, an N × N switch requires 2N 1 × N elements, using N 1 × N switches on the N input fibers, and N 1 × N switches on the output fibers [36]. Parallel architectures are thus preferred for large size cross-connects.

Serial and parallel switch fabric architectures have been implemented using MEMS. The former schemes resort to 2D MEMS arranged in an integrated array of N^2 tilting micro-mirrors, which can either be in a reflecting (up) or non-reflecting (down) state. Its operation follows a typical crossbar configuration [42]. The parallel scheme uses two arrays of N 3D MEMS, which are micro-mirrors that could be tilted in two orthogonal directions. The switch operation consists in directing a set of collimated beams, generated at an array of N optical fibers, into the first array of MEMS. Each of these mirrors is tilted to direct its beam to the element of the second array of MEMS associated with a desired output fiber. The second mirror then couples the beam into the corresponding element of the second array of fibers [36]. Although only a few years ago the opportunities to turn large size optical cross-connects into commercial products looked promising [42] that place is now occupied by wavelength selective switches. These next-generation components dynamically select individual wavelengths from multiple input fibers

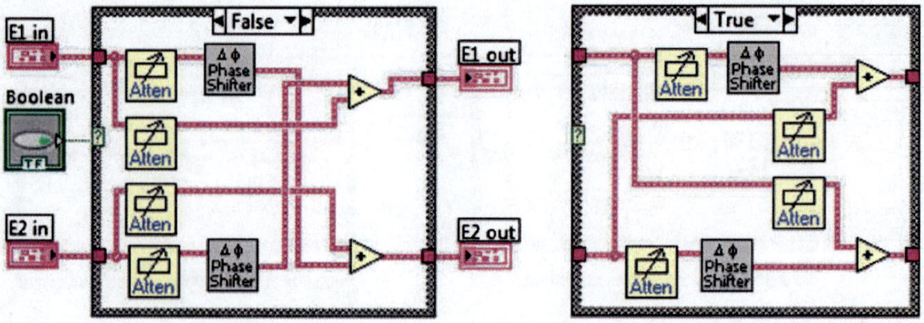

Fig. 3.3 LabVIEW block diagram of a 2 × 2 switch implementation considering crosstalk

and switch them to a common output fiber [43]. Their operation principles are outlined later in this chapter.

3.3.4 Dynamic Spectrum Equalizers

Dynamic spectrum equalizers (DSE) allow compensating the power tilt occurring in reconfigurable WDM networks due to channel add/dropping, signal power fluctuations in EDFAs, stimulated Raman scattering and the effect of gain dispersion in amplifiers, among other phenomena. A DSE can be understood as a linear array of dynamically controlled variable optical attenuators (VOAs) placed between a DEMUX/MUX pair. The filters separate the incoming WDM signal into spectral bands. Each band is passed through its corresponding VOA for equalization. Then, all bands are recombined at the multiplexer, producing the output signal. The DEMUX filters are designed to permit the necessary crosstalk to obtain a smooth loss spectrum. Among possible technological approaches, free-space and integrated solutions dominate a small market. Free-space DSEs are based on MEMS, liquid crystal (LC) or polymer dispersed LC technology. All of them use the same basic platform that combines a diffraction grating and an array of VOAs in a reflective architecture [44]. In the planar waveguide approach, AWGs are used to separate/combine the spectral bands, and electronic VOAs, controlled through in-line power monitors, are used to equalize the WDM signal [45]. The DSE can be monolithically fabricated on a single wafer, and it is thus compact and stable.

Simulation of DSEs for both, integrated and free-space approaches, can be carried out with the same model if a sufficient level of abstraction is admitted. Provided that adequate simulation modules are available, a DSE simulator can be built according to the LabVIEW block diagram shown in Fig. 3.4. The simulation modules (such as attenuators, power monitors and filters) were developed as part of an optical networks simulation tool written in this graphical programing language [46]. They can be interconnected in an intuitive manner since the program

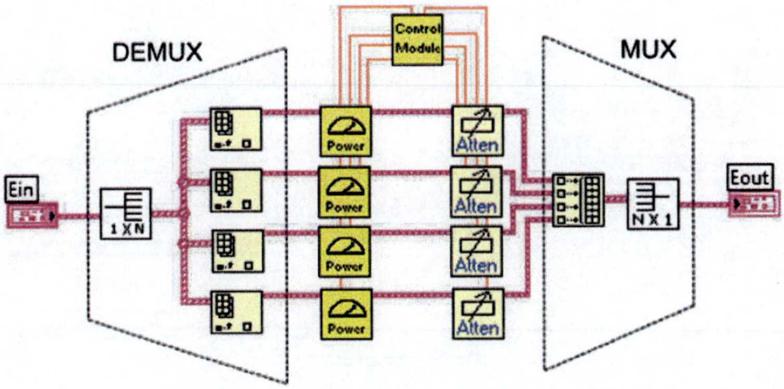

Fig. 3.4 LabVIEW block diagram for the implementation of a four band-DSE

code practically mimics the structure of the subsystem to be simulated. In Fig. 3.4, the control module calculates the necessary attenuation level for each band. Dynamic equalization takes place by dynamically varying the attenuators' characteristics in response to the power level measured by the monitor at each band.

Dynamic channel equalizers (DCE) follow the same operational principle as DSEs. However, they operate in a per-channel rather than in a per-band basis. When the DCE attenuation is high enough so as to actually block a demultiplexed channel, the equipment is called wavelength blocker. DSEs are mainly employed in long-haul systems to prevent signal distortion produced by optical amplifier chains, whereas DCEs and wavelength blockers are more common within metropolitan networks to implement reconfigurable optical add/drop multiplexers (ROADMs).

3.3.5 Wavelength Selective Switches

Nowadays DSEs are being substituted by wavelength selective switches (WSS). These components can do the same tasks as DCEs and wavelength blockers, but with the added functionality of dynamically route wavelength channels. They represent the cornerstone of modern reconfigurable agile optical networks. In a few words, a WSS is an optical spectrometer with a switching engine back end [47]. Its function is to dynamically distribute the channels of a WDM signal arriving to a network node into N multi-wavelength signals, with the possibility of independently attenuating each channel to allow for equalization. This is schematically illustrated in Fig. 3.5 for N = 4. The device consists of a single input port and N output ports, such that any input channel (or set of them) can be routed to any of the output ports. Due to the reciprocity of the optical elements involved in its construction, WSSs operate in the opposite direction as well. In this case,

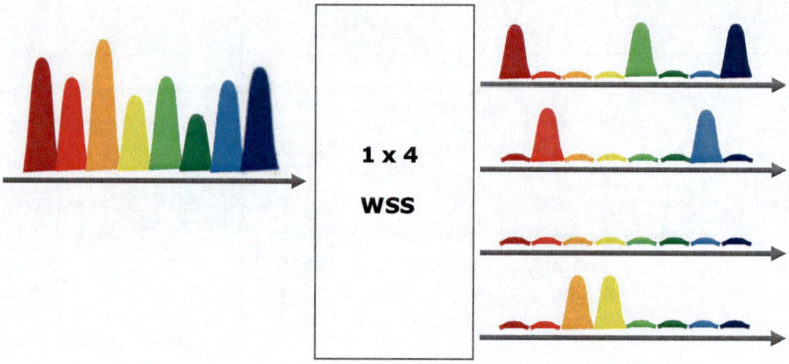

Fig. 3.5 A WDM signal decomposed into four, multi-wavelength signals using a WSS

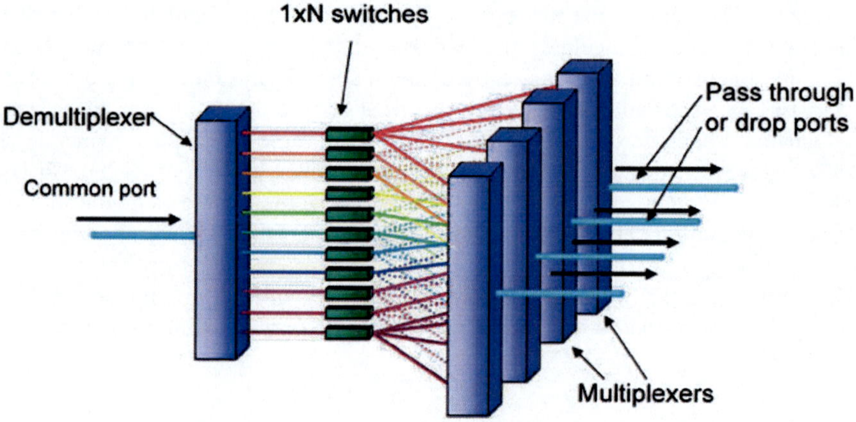

Fig. 3.6 Schematic optical functionality of a wavelength selective switch (From [36], Fig. 9. Reproduced by permission of © 2008. The Institute of Electrical and Electronics Engineers.)

N multi-wavelength signals are injected into the N (now) input ports. Each signal can share the same wavelength (or set of them) with the other N-1 signals. However, only one lambda from one of the input ports can be selected to be routed to the output port, thus blocking the rest of the input channels having the same lambda.

Besides the complementary optics, a WSS is mainly composed of a dispersion element and a switching engine. The dispersion element acts as a DEMU/MUX pair, and therefore a WSS can schematically be conceived as shown in Fig. 3.6 [36]. The input WDM signal is demultiplexed. Each channel is routed to a $1 \times N$ switch. M of these switches compose the switching engine, where M is the number of WDM channels. Each switch, controlled by means of a digital communications interface, routes the incoming signal to the appropriate MUX, where the routed

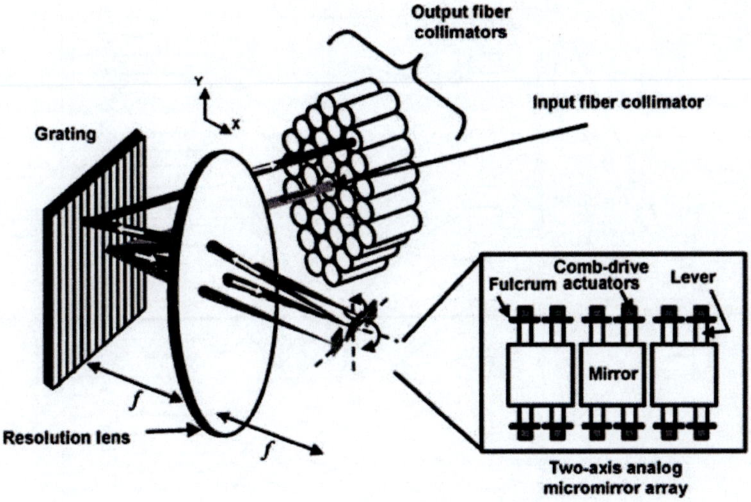

Fig. 3.7 Schematic of a WSS using a MEMS micro-mirror array as switching engine (From [48], Fig. 1. Reproduced by permission of © 2006. The Institute of Electrical and Electronics Engineers.)

lambdas are combined and directed to the corresponding output port. There are N multi-wavelength ports, one for each MUX.

Similar to DSEs, WSSs can be implemented either using integrated optical elements or free-space optics. Current commercial products follow the latter approach. A single diffraction grating is used as dispersion element, which, thanks to its 3D nature, simultaneously acts as MUX and DEMUX. The switching engine consists of an array of M switches that employ some physical feature of the incoming single-wavelength signal, such as angle, phase, polarization or displacement, to carry out the switching process. The choice of switching engine technology determines the overall optical design, cost and size of the WSS. Among the current competing technologies, the following lead: MEMS in single- and two-dimensional arrays, LC-based polarization beam deflection, and LC on silicon phased array beam steering. Details about each implementation can be found elsewhere [43]. In general terms, WSS designs follow a parallel architecture, where the input/output ports are arranged as a 1D array of fibers. The linear separation between ports is mapped to an angular separation at the switching engine such that each of the M 1 × N switches directs the input beam to the selected output port [47]. A schematic of a MEMS-based WSS is presented in Fig. 3.7.

Simulation of a WSS module can be accomplished aided by the conceptual schematic presented in Fig. 3.8 [36]. Considering that proper simulation modules for MUX, DEMUX, VOAs and 1 × N switches are available (see previous sections), a WSS module can be coded in a modular manner following the aforementioned schematic. Naturally, the use of a graphical programing language,

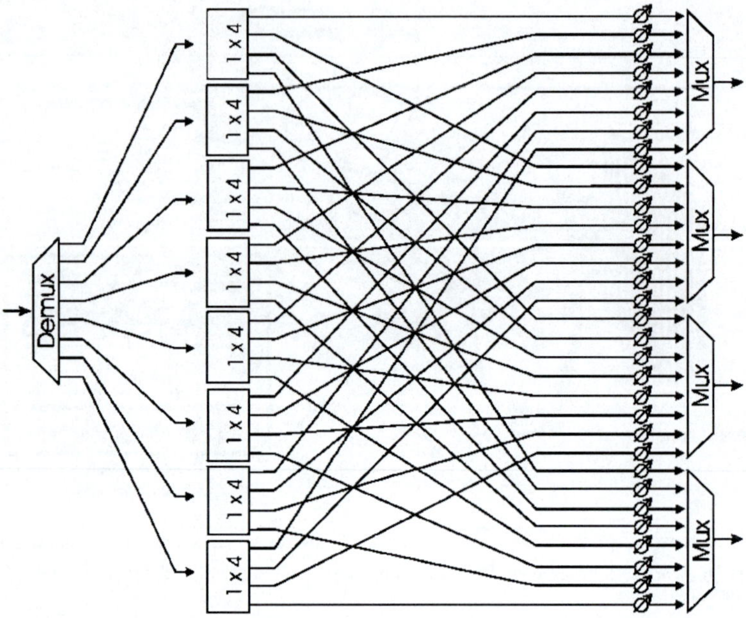

Fig. 3.8 Block diagram of a 1 × 4 wavelength selective switch for eight channels. Eight 1 × 4 switches and 32 attenuators are required

such as LabVIEW, eases the task. Actually, Fig. 3.8 resembles the interconnection diagram of an integrated WSS, where channel (de)multiplexing is carried out using N + 1 AWGs. Due to its complexity as compared to the 3D approach, planar WSSs are not expected to become commercial in the near future, even for systems with modest port counts. Simulation of free-space systems at the abstraction level shown in Fig. 3.8 requires adequate modeling of the physical phenomena that affect the WSS performance. Channel bandwidth and port isolation are two important performance parameters. The former can be taken into account within the (DE)MUX implementation since it is related to the effect of filter concatenation already mentioned in connection to modeling of demultiplexers. The latter has to do with the thoroughness with which the WSS blocks the non-provisioned channels. This effect must be considered during switch modeling by adequately taking into account the effect of cross-talk (see discussion related to Fig. 3.3).

3.3.6 Add/Drop Multiplexers

Optical cross-connects (OXCs) and optical add/drop multiplexers (OADMs) are two of the most prominent elements in current WDM networks [49]. Their function is to quickly and cost-efficiently route optical connections to their

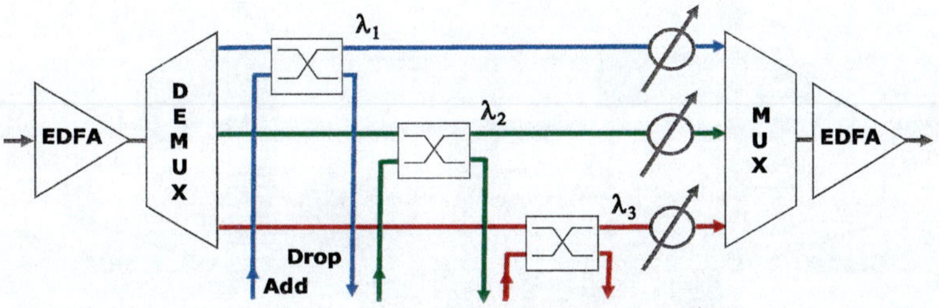

Fig. 3.9 Parallel ROADM architecture for a 3-channel WDM network

destination They are the key elements that enable the required lightpaths across a wavelength routing network. Although the deployment of large port count OXCs boomed at the beginning of the century, the current trend is to use smaller elements called reconfigurable OADMs (or ROADMs) and scale them according to the network needs. These elements distribute the WDM channels at the network nodes. Some wavelengths are allowed to pass through (called express channels), while the rest are dropped from the information main stream. The wavelengths of the dropped channels are then used to add traffic that is generated at the node. Reconfigurability refers to the ability to select the desired wavelengths to be dropped and added dynamically (as opposed to fixed) without having to plan ahead the logical topology of the network. ROADMs provide the ability to dynamically optimize the network for changing patterns of externally offered traffic, subject to availability of wavelengths and node equipment [50]. These network elements have become key building blocks for metro and regional networks due to the flexible network provisioning and transmission management benefits they provide [43].

Figure 3.9 presents a simple implementation of a ROADM that follows a parallel architecture [43]. All incoming channels are demultiplexed. Depending on the state of the switches, which can be dynamically reconfigured, some channels are dropped and the remaining pass through. Parallel designs are cost-effective only when a large fraction of the total number of channels is expected to be added/ dropped.

Since the aforementioned ROADM architecture resembles that of a DSE, its simulation can be carried out with a similar LabVIEW implementation as that shown in Fig. 3.4. This is shown in Fig. 3.10. The program clearly resembles the actual design, where simple simulation units (switches, attenuators, multiplexers, etc.) described in previous sections of this chapter, are simply interconnected through adequate data structures. A modular text-based language could have also been utilized for the same purpose.

Figure 3.11 shows another variant of ROADM architectures. It consists of cascaded single-wavelength OADMs, and thus represents a serial approach [51]. In contrast to parallel designs, in this architecture only the individual dropped/ added channels are demultiplexed. This structure, however, requires prior

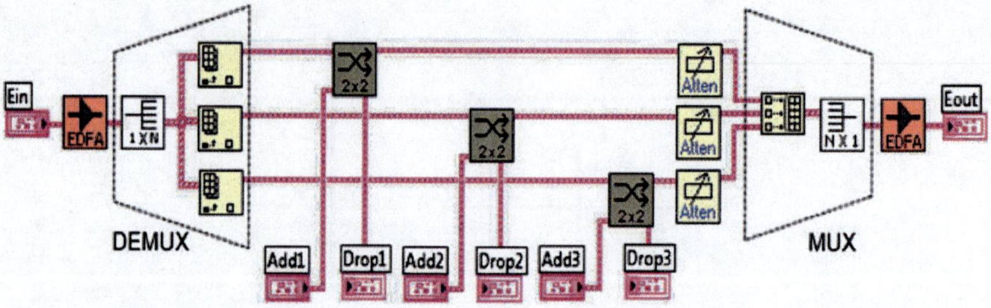

Fig. 3.10 LabVIEW block diagram of the parallel implementation of a 3-channel ROADM

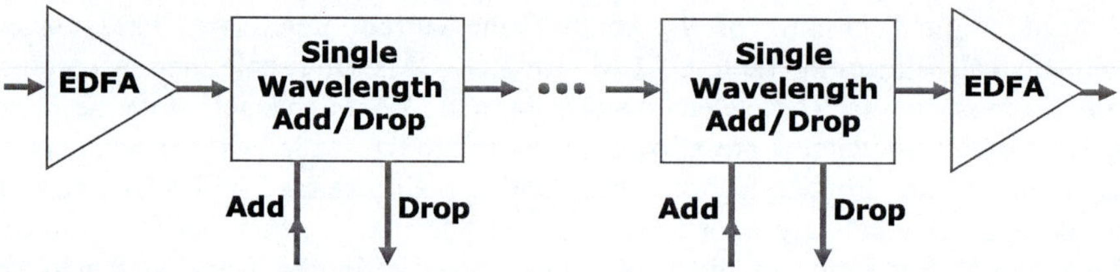

Fig. 3.11 Serial ROADM architecture

knowledge of the channels that eventually will be dropped/added before installation, although individual modules can always be added on a per-needed basis. This design is thus preferred when only a small number of channels are to be added/dropped. It also exhibits high losses.

So far, only two-degree ROADMs have been discussed. However, simulation of N-degree ROADMs using a modular software tool should present no problem. N-degree ROADMs accept traffic from more than two fibers, rearranging express channels and adding/dropping local wavelength channels. They make extensive use of WSSs, and therefore adequate design and analysis of current optical networks requires proper modeling of such elements. Figure 3.12a shows a two-degree ROADM based on a broadcast-and-select architecture [43]. First, the fiber power is split making all the wavelengths available for a local drop through a demultiplexer. Then, the 2×1 WSS passes through the express channels and blocks those that are locally dropped (and that are about to be added). The locally added wavelengths are combined at a multiplexer and fed into the other available input port of the WSS. Finally, the express and locally added channels are routed at the WSS to the output port, individually equalizing their optical power.

The extension of this architecture to an N-degree ROADM is presented in Fig. 3.12b [43]. Only one Nx1 WSS is shown. It does not simply route locally added and express channels, but traffic from the other WDM fibers as well. This ROADM implementation also requires extra splitters to route a copy of the

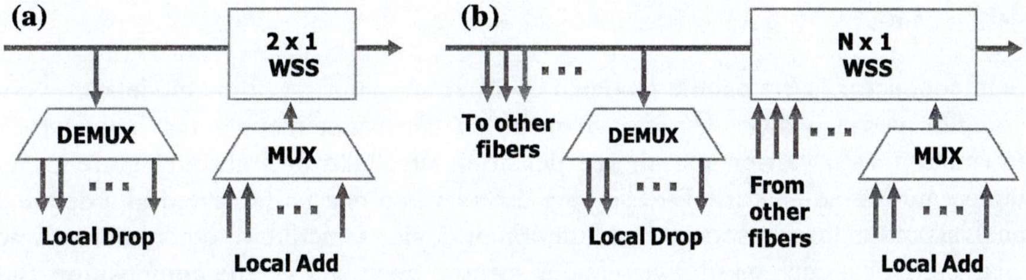

Fig. 3.12 Two-degree **a** and multi-degree **b** implementations of ROADMs. (From [43], Fig. 20.10. Used with permission from the authors.)

incoming information to other WSSs in the node. Simulation of these complex elements requires an efficent and accurate software tool, since in these integrated structures small numerical inaccuracies might quickly yield unrealistic results.

3.4 Active WDM Network Building Blocks

The field of optical transceivers or optical amplifiers modeling and simulation is so vast that at least one book can be written about this subject. Since a thorough explanation of these topics is far beyond the goals and scope of this chapter, we will focus on passing to the reader key information that will help him/her to become introduced in this fascinating subject, especially by pointing out some of the most recent advances in the field.

3.4.1 Optical Transmitters

Among several different classifications, transmitters for WDM systems can be divided into directly modulated lasers (DMLs) and externally modulated lasers (EMLs). The former class of transmitters is usually less expensive; however, their adoption has been limited by their inherent frequency chirp, which interacts with fiber dispersion, reducing in some cases, but extending in others, the reach of actual communication systems [52]. External modulation involves the use of an optical modulator, and it is not only the most popular On-Off Keying (OOK) modulation approach, but since the introduction of advanced modulation formats in optical communications (discussed in Chap. 10), it practically has become the standard modulation technique. In this section, we will first refer to the laser and then to the optical modulator modeling. Irrespective of the modulation scheme, both lasers and modulators have to be electronically driven. For brevity, modeling of wave-function, pulse or data generators [5] will not be dealt with in this chapter.

3.4.1.1 Lasers

Semiconductor lasers can be modeled in different manners. Some models are very complex because they characterize in a self-consistent manner the optical field interaction with carrier and charge densities, and take into account current and temperature gradients in 3D. They are useful when one is interested in a detailed analysis of the laser response as a function of device structure or other microscopic characteristics, such as the waveguide specific geometry or the composition and shape of quantum dots in size-quantized hetero-structures [3]. On the other hand, there are other models that are very simple, such as that of the continuous wave (CW) laser that is used to feed a modulator. This model only requires a good characterization of its emission wavelength, linewidth and the noise produced by amplified spontaneous emission, probably modeled through a white noise source. For the simulation of WDM systems, normally a rate equation formalism suffices. These are $(1 + 1)$-dimensional models, where only time and longitudinal or axial coordinate (z) variations are taken into account. The transverse character of the optical waves is accounted for via a confinement factor model and carrier diffusion is neglected. The rate equation model normally consists of three coupled differential equations: one for the photon density, one for the phase change of the optical field and one for the carrier density [53, 54]. These equations follow a semi-classical approach, and are derived from Maxwell's equations and density matrix theory assuming a two-level system. Intra-band nonlinear effects are included from a phenomenological standpoint. The first equation always contains at least two terms that account for spontaneous and stimulated emission. Phase dynamics are modeled through the inclusion of a linewidth enhancement (or Henry) factor that couples the refractive index variations to those of the modal gain. This equation is important because it describes the chirp dynamics of the laser. It might also include a term to account for cavity losses via the photon lifetime parameter, which essentially describes how fast the photon density in the cavity would decay in the absence of stimulated emission. Finally, the third equation has terms to consider carrier density increase due to carrier injection into the active region (this term is responsible for capturing the carrier density variations produced by the effect of the driving current waveform), carrier density decrease due to spontaneous emission and non-radiative recombination and, of course, a stimulated emission term. This equation usually also has a fourth term that accounts for coupling of a fraction of the spontaneous emission noise to the actual mode lasing in the cavity.

Single-rate equation models can be extended to multi-mode laser models by adding extra photon density rate equations, one for each mode [55]. They can also account for intermodulation-induced interactions between modes by including the corresponding cross-coupling terms. Quantum well structures can be modeled in a similar manner by simply adding (sometimes involved) extra carrier density equations to account for local carrier capture, recombination and other phenomena [56]. The system of coupled rate equations mentioned above can be solved using Runge–Kutta or similar analyses [9]. One drawback of rate equation (and many other) models is that their parameters do not coincide with those that can be measured. Although there have been multi-institutional efforts to standardize laser model parameters [57], there is still not a straightforward manner to interrelate experimentally extracted with model parameters [58].

Another equivalent approach for diode lasers modeling is to use electrical circuits [59]. This is realized by representing the laser diode as a circuit of electrical components having their own impedance [60]. The current injected into the laser is actually represented by an impulse to the circuit, whereas the voltage across a component represents the light output. Since in this abstraction level an inductor represents the optical element of the laser and a capacitor represents the carriers, multi-mode lasers can be modeled by simply adding additional series inductor–resistor combinations in parallel. Although this formalism might look more complicated and less physically intuitive than the rate equation approach, it has the advantage that software packages for simulation of complex circuits are readily available, and therefore it is a straightforward matter to solve the laser equations. Circuit modeling using commercial software has been applied, for instance, to study bistability of multi-mode laser diodes [61].

Somewhat related to laser modeling using circuit elements is the concept of transmission line modeling [62]. This technique is based on slicing the laser in small longitudinal elements whose length is determined by the sampling period of the simulated waveform. Every slice can be considered a small laser element where the local carrier density, calculated from one or more rate equations, is assumed to be homogeneous. Knowing the local carrier density at all the slices allows reconstruction of a discrete version of the whole distribution along the longitudinal axis of the laser. In this way, spatial in-homogeneities such as spatial hole-burning can be taken into account. This contrasts with the rate equation approach, where only a spatially averaged carrier density can be known. In the transmission line model, the local carrier density in each slice is employed to calculate the local optical field through the use of scattering matrices, which are thought of as if located in scattering nodes. All nodes are connected through transmission lines (hence the name) representing the field propagation delay along the waveguide. Scattering matrices take incident waves and scatter them to produce forward and backward reflected waves, thus taking into account the bidirectional propagation nature of the optical field within the cavity. After propagation through adjacent transmission lines, the reflected waves will impinge on neighboring nodes and corresponding scattering matrices, becoming incident waves [8]. The scattering matrices account for the three fundamental light-mater interactions: spontaneous emission, stimulated emission and absorption, and they can even include the effect of gain and index gratings. Laser dynamics are simulated by an iterative algorithm that calculates the carrier density at every section and scatters the optical field at every node. Since all modes share the same carrier density within a section, the transmission line approach also implicitly includes power transfer between modes by intermodulation due to coupling through gain saturation and the carrier population [63]. Laser simulation modules based on the transmission line formalism are useful in the analysis and design of multi-section lasers by simply connecting similar modules [8]. It also includes modeling of multi-mode operation. The transmission line laser modeling approach has been used, for instance, to analyze the effect of carrier transport in high-speed quantum well lasers [64].

3.4.1.2 External Modulators

There are basically two commercially available types of external modulators: electroabsorption modulators (EAMs) and Mach–Zehnder modulators (MZMs). The former relies on varying the light absorption level of a semiconductor device to modulate the intensity of an incident laser beam. The latter modulates the phase and/or intensity of a continuous optical waveform by controlling the optical path length in one or both arms of a Mach–Zehnder interferometer. In both cases modulation is achieved by varying the applied electric field.

EAMs are characterized by low-driving voltages (about 2 V), relatively high speed, and the possibility to be integrated with a CW laser (or other components) on a single chip, which may lead to low-production costs. On the other hand, their absorption characteristics are wavelength- and temperature-dependent, and they show high insertion losses and limited optical power-handling capabilities. Modulation in bulk semiconductors occurs via the Franz–Keldysh effect. When a reverse bias voltage is applied to the absorption layer of a p-i-n waveguide, the effective bandgap becomes reduced and this, in turn, induces absorption [65]. Multi-quantum well structures generally present higher absorption levels. This is the result of the quantum-confined Stark effect. In WDM simulation systems, EAMs can be efficiently modeled using a time-varying transmission function of the driving voltage. The model consists of the product of an absorptive term and an α-dependent expression, which accounts for the associated change in refractive index [66]. The α-parameter decreases as the EAM voltage increases. A low α-parameter, more often found in multi-quantum well EAMs, produces an output signal with small chirp. Although the transmission function modeling approach suffices for most WDM system analyses, more accurate device-level simulations require consideration of carrier dynamics within the EAM [67]. More sophisticated 3D simulations can be carried out using the effective index method and the beam-propagation method. Dynamic modeling, however, is attractive because it allows investigation of time-varying phenomena within the semiconductor structure [68].

MZMs rely on the use of phase modulators generally manufactured on lithium niobate ($LiNbO_3$). This material shows temperature and wavelength insensitivity and a large electro-optic (Pockels) effect. $LiNbO_3$ MZMs present broad optical bandwidths, large extinction ratios (up to 20 dB) and low insertion loss [65]. Their main advantage, notwithstanding, is their capability to independently modulate the phase and intensity of the electromagnetic field. This is not only valuable to control or even suppress chirp when used as intensity modulators, but it also represents the basis to generate advanced optical modulation formats (explained in Chap. 10 of this book).

MZMs can be found as single-arm or as dual-drive modulators. They are depicted in Fig. 3.13. The most common approach to model MZMs, especially in

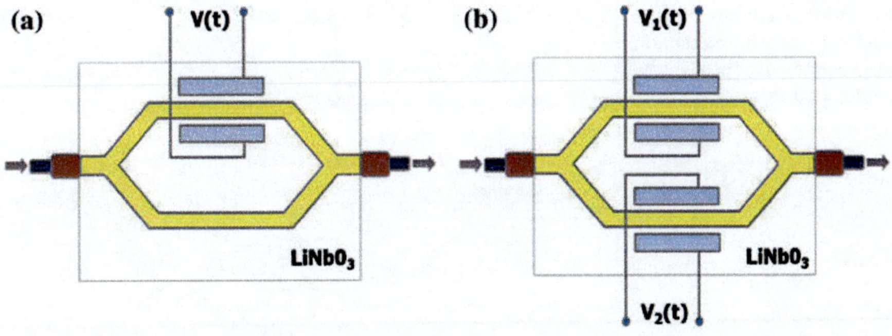

Fig. 3.13 Single-arm **a** and dual-drive **b** Mach–Zehnder external modulators

the context of transmission systems simulation, is the use of a transmission function. If coupling and propagation losses are ignored, the optical field transmission function of a balanced-arms MZM in the case of single-arm modulation can be written as [65]:

$$T(V) = \sqrt{2}\cos\left(\frac{\pi}{2}\frac{V}{V_\pi} + \frac{\varphi_0}{2}\right)\exp\left[j\left(\frac{\pi}{2}\frac{V}{V_\pi} + \frac{\varphi_0}{2}\right)\right] \tag{3.10}$$

where linear dependence of phase modulation in applied voltage (V) is assumed. V_π is the modulation voltage that is required to induce a phase change of π, and φ_0 is the modulator bias. In single-arm modulators, a voltage-induced change in amplitude is accompanied by a phase shift. Dual-drive modulators provide an extra degree of freedom because by adequately choosing the values of V_1 and V_2, one can determine the amount of phase modulation to be imprinted on the signal. In the so-called pull-pull configuration, V_1 equals V_2, and therefore an identical phase shift is induced in both arms. Under this condition, pure phase modulation is achieved. On the other hand, zero-chirp at all voltages can be realized under push–pull operation. For this, the two modulator arms are driven by the same amount, but in different directions; that is, $V_1 = -V_2$. The resulting transmission function is similar to Eq. 3.10, but without the phase term. By carefully selecting the differential voltage, V_1–V_2, and the bias voltage, optical signals encoded in different advanced modulation formats can be generated [69]. This can also be realized using an optical IQ modulator, which is also depicted in Fig. 3.14. As explained in [70], the IQ modulator consists of two MZMs integrated into a Mach–Zehnder interferometer structure. The MZMs are operated in push–pull mode to individually amplitude-modulate the split signals. A relative phase shift of $\pi/2$ is adjusted in one arm using an extra phase modulator. Coherent recombination of the signals traveling along the in-phase (I) and the quadrature (Q) arms allows generation of an output signal with arbitrary amplitude and phase. Any constellation point can hence be reached in the complex IQ-plane. Mach–Zehnder and IQ modulators can be conveniently modeled using transmission simulation software

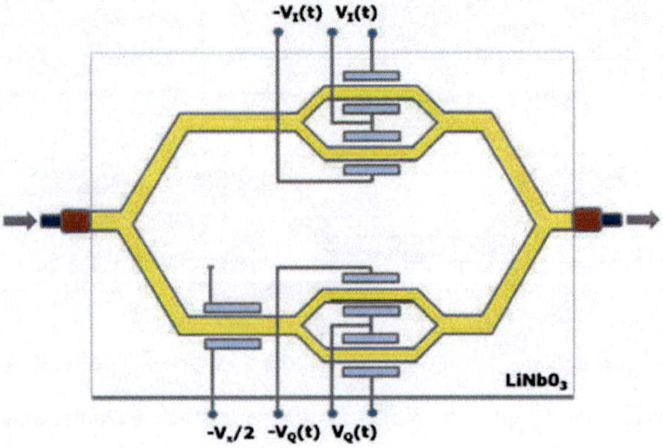

Fig. 3.14 Schematic diagram of an IQ modulator to generate advanced modulation formats

by properly integrating phase modulators within the interferometric geometry. In this case, phase modulators can be modeled as a phase shift given by $\pi V/V_\pi + \varphi_0$.

3.4.2 Optical Receivers

Digital optical receivers usually comprise a photodiode followed by the electronics that amplify and shape the signal. Normally, pin diodes or avalanche photodiodes (exhibiting built-in gain) are used to convert the incident optical field into an electrical signal. The receiver electronics include a trans-impedance amplifier (TIA) and a linear-frequency shaping filter that can be modeled, for instance, as an RC low-pass or fifth-order Bessel filter [71]. Clock-recovery and decision circuits, respectively, synchronize and determine whether the detected bit is a ONE or a ZERO. The bit-error ratio is then calculated by means of Gaussian probability density functions of the sampled electrical currents of ONEs and ZEROs. The mean and variance of the functions can be measured from the received eye diagrams or can be numerically calculated for arbitrary optical and electrical filters [71]. In general, modeling of optical receivers represents a more complex task than modeling of optical transmitters because it deals with processing of a noisy and distorted signal. Relatively simple models for direct detection schemes can be found in well-known text books [49] and review articles [71]. They are sufficiently accurate for most optical fiber systems simulations.

Although direct detection receivers have traditionally been employed in connection to intensity modulation formats, it has been shown that they are capable of detecting arbitrary modulation formats with differentially encoded phases such as DQPSK. For this, an optical interferometer (usually a delay-line one) is placed in front of the photodiode. Its function is to transform the phase difference information of two consecutive symbols into intensity variations, thus allowing the square-law receiver to access the encoded data [70]. Future trends in the improvement of detector-receiver subsystems include enhanced avalanche photodetectors, whose free-carrier multiplication processes resembles that of a multistage photo-multiplier tube, the use of coherent communications systems to reduce the optical energy per pulse required at the receiver for a constant spectral efficiency, and the research on nano-antennas [71].

3.4.3 Optical Amplifiers

Two types of amplifiers can be found in current optical networks: fiber and waveguide amplifiers. Er-doped fiber amplifiers (EDFAs) and fiber Raman amplifiers (FRAs) belong to the former type. They are more frequently employed in today's WDM systems than waveguide amplifiers because of the high gain, high-output saturation power, low-coupling losses and relatively low noise figure (NF) that they exhibit. Semiconductor optical amplifiers (SOAs), also known as semiconductor laser amplifiers, and Er-doped waveguide amplifiers (EDWAs) belong to the latter type. They offer potentially low production costs and monolithic integration with other components in a compact structure. The characteristics, operational principle and application of these devices are appealing research topics. However, even a simple introduction on the subject falls beyond the extent of this chapter. We therefore refer the reader to the vast bibliography. A classical reference for EDFAs is [72]. Distributed and discrete FRAs, their applications and related concepts are introduced in [73] and [74]. An overview of SOAs are more difficult to find, but there are worth reading references [75–77]. A discussion of noise in linear amplifiers can be found in [78].

From a modeling standpoint, EDFAs, EDWAs and FRAs can be described with a similar mathematical formalism despite the different nature of their amplification processes. This formalism consists of a system of coupled differential equations that governs the forward and backward power propagation of pumps, signals and ASE. Of course, each amplifier model contains its own physical effects and corresponding parameters, but the same numerical method can be applied to solve the equations that characterize the behavior of any of these three amplifiers. In contrast to SOA models, EDFAs, EDWAs and FRAs models do not include time-dependent equations. This description is possible in the case of EDFAs and EDWAs (which operate mainly on the same principles as EDFAs but with lower gain) because their first excited level has a long lifetime of about 10 ms. Therefore, these active devices are unaffected by signal modulation. In the case of FRAs, since Raman scattering is a sub-picosecond non-resonant process with practically instantaneous upper-state lifetime, the Raman Stokes interaction between the pump and signal photons converts the pump into a replica of the signal photon (plus an optical phonon) in an adiabatic manner. As a consequence of this, there is no need to set the corresponding time-dependent differential equation. The situation with SOA modeling is different because the carrier lifetime in SOAs is in the order of picoseconds [79], and they exhibit a nonlinearity that is orders of magnitude higher as compared to optical fibers. Moreover, in contrast to EDFAs and FRAs, SOAs are not only employed as amplifiers [80], but they are also utilized to carry out high-speed optical signal processing tasks, such as wavelength conversion, packet switching and optical gating [81, 82]. Therefore, similar to laser diodes, modeling of SOAs not only includes a propagation equation for the amplifying signals and possibly the ASE, but it also requires a system of nonlinear coupled rate equations that takes into account the variation of the total carrier density and, if ultra-fast intra-band nonlinear effects are taken into account, the local carrier densities [83].

Another difference between fiber amplifiers and SOAs pertains to the numerical method that is used to solve the system of equations, which rarely accepts closed-form analytical solutions. Since in fiber amplifiers there are usually bidirectional pump and signal flows, the mathematical model corresponds, most of the times, to a two-point boundary value problem. Shooting or similar methods then have to be used. In SOAs, anti-reflection coatings are applied to avoid facet reflections, thus turning a laser diode into a traveling-wave amplifier. Under these circumstances, all beams co-propagate, and hence a $(1 + 1)$-dimensional initial value problem that follows an input/output approach has to normally be solved.

Thanks to their pumping efficiency and their gain spectral characteristic, which coincides with the lowest-loss transmission window of silica fibers, EDFAs have traditionally been the amplifier of choice for most telecommunication applications. Notwithstanding, the increasing demand for wider gain bandwidths and lower noise figure has fueled the recognition of FRAs as a more attractive amplification technology. There are plenty of amplifier models to aid in the design of amplified communications systems. Propagation of multi-channel signals (including pump beams) along an EDFA in arbitrary directions is discussed in [84]. The authors' solution is based on turning a system of differential equations into a transcendental equation. It thus represents a simple and efficient approach. This description, however, is only valid as long as the ASE has negligible effect on the amplifier gain. This deficiency can be circumvented at the expense of a non-analytical solution [85], which can also be efficiently found using the transmission line method [86], already discussed in connection to laser modeling. The average power analysis is another well-known approach to model EDFAs [87] and FRAs [88]. It divides the amplifier into a series of concatenated elemental amplifying sections having uniform upper-level population density. Based on this approximation, small-signal traveling-wave solutions can be invoked in each elemental amplifier section.

More general computer models can be used to solve both EDFAs and FRAs, where signals, pumps and ASE propagation in forward and backward directions are taken into account [89]. They also include the interaction effects between those beams. This is important in the design of ultra-broad bandwidth ($\sim$110 nm) low-ripple FRAs, where multiple pumps propagate along considerable fiber lengths. The short wavelength pumps amplify the longer wavelengths, and hence, for proper gain-flattening, powerful pumps are needed at the shortest wavelength side of the spectrum. Naturally, this kind of detailed simulations increases the calculation loads. Therefore, most of the amplifier computational models are designed to take advantage of the exponential dependence of light amplification on the propagation distance. This allows achieving a faster convergence to a solution. Computational efficiency can be gained through the use of an adaptive integration step-size and by disregarding some phenomena such as spontaneous Raman emission, multiple-reflection Rayleigh scattering and high-order Stokes generation [90]. For a thorough understanding of these effects, a more complete model is necessary [91]. On the other hand, if only amplification of a signal by a monochromatic pump is analyzed, a simplified model can be used [73]. Furthermore, in the small-signal regime, where the un-depleted pump approximation holds, an analytical solution exists. In this case, the FRA on–off gain can thus be calculated as $G_{on\text{–}off} = \exp(C_R L_{eff}[P^+(0) + P^-(L)])$, where C_R is the Raman gain efficiency of a fiber of length L, and P^+ and P^- represent forward and backward propagating pumps. This equation clearly uncovers the exponential nature of gain evolution, typical of all amplifiers.

Advanced physical modeling and efficient simulation techniques will be critical for the enhancement of current fiber amplifier designs. Some forward-looking considerations are: pumping schemes, magnitude and bandwidth of noise transfer from pump to signal, severity of non-linear effects, etc. Among current research lines, those worthwhile to mention are cost-effective time-dependent pumping of distributed FRAs, and optical signal amplification in silicon [92].

During the last twenty years, several dynamic and static SOA models have been proposed [93]. For telecommunications applications, scalar longitudinal dynamical models have normally prevailed. However, due to the recent impact of polarization-related topics such as advanced modulation formats, polarization mode dispersion and cross-polarization modulation, a renewed interest in models dealing with the anisotropic nature of semiconductor waveguides and the vectorial character of the electromagnetic field have arisen [94]. Although more elaborated, these models are required when birefringence, either produced by the intrinsic amplifier structure, or by the energy exchange mechanisms that take place within the device, is relevant to analyze or explain the application process under study. For the design and engineering of WDM systems, SOA simulators that exhibit very fast execution time and accurate modeling of nonlinear carrier dynamics are normally preferred. This is because, in this context, fast SOA simulators can be more efficiently interconnected with other amplifiers and network elements without markedly stressing the computational resources.

SOA models are usually formulated in either time- [95] or frequency-domain [79]. Time–domain (TD) models deal with a simple, single–beam equation, where all interacting beams are multiplexed together. These models are normally flexible in terms of applications, and are appealing because, in general, the number of multiplexed channels can be increased at no extra cost in memory or computational time. An efficient balance between mathematics and computer power can be achieved through the use of frequency–domain (FD) models, leading sometimes to shorter computational times as compared to TD models. In the FD formalism, each

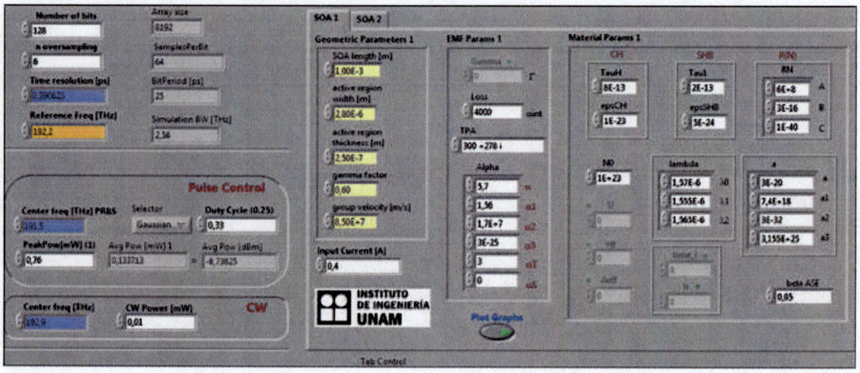

Fig. 3.15 Front panel of an SOA simulator written in LabVIEW programing language

WDM channel is associated to one propagation equation. Nonlinear cross-channel interactions are analytically calculated using coupled–wave or similar theory, and hence, modeling of large channel counts becomes impractical. FD models are rather useful in the analysis of specific amplifier tasks or when accurate modeling of frequency-dependent parameters (e.g. gain) is a must [96]. TD models, being more flexible, are preferred for system simulations.

The regime of applicability of an SOA model is defined by the set of physical effects it considers. Some of the most relevant phenomena to be taken into account to develop a general SOA model operating at 10–40 Gb/s are [83]: the light-matter interaction responsible for the (slow) inter-band dynamics and associated nonlinearities, including the corresponding refractive index changes [97]; gain dispersion; carrier heating (CH) and spectral hole-burning (SHB) gain compression [75]; ASE propagation and the carrier density depletion it produces [98], especially for long (>0.5 mm) SOAs, and material losses. A clarification of these terms and indications on whether to include a particular effect into an SOA model can be found elsewhere [99]. For instance, CH and SHB effects are only relevant if pulse widths shorter than about 10 ps are involved [100]. A uni-directional time-domain model that captures most of the aforementioned effects has been implemented in LabVIEW [83]. It is part of the photonics system simulator mentioned throughout this chapter. Among other applications, this model has been useful in the analysis of an active Mach–Zehnder interferometer operated as 40 Gb/s tunable wavelength converter [46]. Figure 3.15 shows the front panel of the SOA module.

Photonic systems simulators usually include more than one SOA implementation in its menu. Very often, laser diode implementations can also be used to simulate SOAs by reducing the laser facets reflectivity. Two examples are the equivalent electrical circuit type of models [101] and the transmission line formalism [102]. Rate equation models, especially conceived to simulate SOAs, represent more efficient alternatives. One that has received considerable attention [95] has been implemented in Matlab [103]. It is based on the analytic integration of the photon density propagation equation along the longitudinal coordinate, resulting in a system of time-varying differential equations that can be solved using well-known techniques [9]. Due to its efficiency, this theoretical description is adequate for integration into a photonics system simulator. It therefore has been utilized, for instance, to analyze the feasibility of the recently released 100 Gb/s Ethernet 40-km PHY standard [13], where an SOA is employed as pre-amplifier. A good trade off between accuracy and calculation speed can also be found in the so-called state-variable model [104], which resembles the fiber amplifier models mentioned above. Although SOA formalisms that include a propagation equation are less common, an SOA model following this approach has successfully been used to match some measured characterization curves [105]. The physical equations were numerically solved using the finite-differences beam propagation method. This model can be applied, for instance, to the design of ultrafast SOA-based schemes that require adequate characterization of amplifier noise features.

SOAs are not only appealing as all-optical signal processing devices, but also as optical amplifiers. This is mainly due to their potential low cost and their wider and more flexible amplification bandwidth as compared to EDFAs. Despite these advantages, the application of SOAs as in-line amplifiers has been hindered due to their fast and strong nonlinear response, which produces cross-gain modulation (channel crosstalk) and pattern effects. Because of these drawbacks, research on SOAs as amplifiers has been limited to metro/access networks with moderate span lengths and low channel counts [106], where cost-effective technologies are required to extend power budget margins. In this respect, a single SOA has been shown to amplify up to eight 2.5 Gb/s coarse WDM channels over a bandwidth of 140 nm for 130 km [107]. Longer spans can be reached using more advanced technologies. Among them, we can mention the use of more sophisticated internal nanometric SOA structures referred to as quantum dots [108]; the use of compensating nonlinear effects in bulk amplifiers [109, 110]; advanced modulation formats [111, 112] or the inclusion of a wavelength modulated signal per channel to maintain a constant amplitude [113]. Most of these solutions are rather cumbersome, and therefore the design and fabrication of enhanced SOAs, tailored for in-line amplification, is a more straightforward approach. Among the SOA characteristics that are necessary for this application, the most relevant are low NF, high-output saturation power, low-coupling losses and broad gain bandwidth with low ripple and preferably low-polarization dependency. High-performance devices show an output saturation power above 19 dBm and a NF of 6 dB or lower, and a gain bandwidth of at least 100 nm [114].

3.5 Conclusion

An overview of recent modeling and simulation techniques for the analysis and engineering of the most relevant passive and active elements found in current optical telecommunications networks was presented. The chapter structure followed a modular approach, similar to that suggested for the development of a photonic network simulator. In particular, aided by a simulation tool written in LabVIEW, we showed how complex structures such as wavelength selective switches can be built from simple structures, namely, filters, attenuators and 1×2 switches. Different approaches to model more sophisticated structures such as lasers and amplifiers were also discussed and a significant list of references were provided. We hope this work represents the starting point in this fascinating topic for newcomers, and a useful reference for experienced researchers and engineers.

References

1. O'Mahony MJ, Politi C, Klonidis D, Nejabati R, Simeonidou D (2006) Future optical networks. IEEE/OSA J Lightwave Technol 24(12):4684–4696
2. Agrawal GP (2005) Lightwave technology: telecommunication systems. Wiley, New York
3. Piprek J (2009) Optoelectronic devices: advanced simulation and analysis. Springer, Heidelberg
4. Bellato L, Polacco B, Pupolin SG, Tamburello M (1979) Problem in the computer simulation of a fiber optic digital transmission system. In: Proceedings of IEEE international conference on communications (ICC) 2, Boston
5. Elrefaie AF, Townsend JK, Romeiser MB, Shanmugan KS (1988) Computer simulation of digital lightwave links. IEEE J Select Areas Commun 6(1):94–105
6. Elrefaie AF, Wagner RE, Atlas DA, Daut DG (1988) Chromatic dispersion limitations in coherent lightwave transmission systems. IEEE/OSA J Lightwave Technol 6(5):704–709
7. Lowery A, Lenzman O, Koltchanov I, Moosburger R, Feund R, Richter A, Breuer D, Hamster H (2000) Multiple signal representation simulation of photonic devices, systems and networks. IEEE J Select Topics Quantum Electron 6(2):282–296
8. Lowery AJ (1997) Computer-aided photonics design. IEEE Spectrum 34(4):26–31
9. Press WH, Teukolsky SA, Vetterling WT, Flannery BF (2007) Numerical recipes. The art of scientific computing, 3rd edn. Cambridge University Press, Cambridge
10. Roudas I, Antoniades N, Richards DH, Wagner RE, Jackel JL, Habiby SF, Stern TE, Elrefaie AF (2000) Wavelength-domain simulation of multiwavelength optical networks. IEEE J Select Topics Quantum Electron 6(2):348–362
11. Yu T, Reimer WM, Grigoryan VS, Menyuk CR (2000) A mean field approach for simulating wavelength-division multiplexed systems. IEEE Photonics Technol Lett 12(4):443–445
12. Gutiérrez-Castrejón R, Duelk M (2006) Using LabVIEW for advanced nonlinear optoelectronic device simulations in high-speed optical communications. Comp Phys Commun 174(6):431–440
13. Duelk M, Gutiérrez-Castrejón R (2008) 4 × 25-Gb/s 40-km PHY at 1310 nm for 100 GbE using SOA-based preamplifier. IEEE/OSA J Lightwave Technol 26(12):1681–1689
14. Agrawal GP (2001) Nonlinear fiber optics, 3rd edn. Academic Press, New York
15. Olshanky R, Keck DB (1976) Pulse-broadening in graded-index optical fibers. Appl Opt 15(2):483–491
16. Yabre G (2000) Comprehensive theory of dispersion in graded-index optical fibers. IEEE/OSA J Lightwave Technol 18(2):166–177
17. Cárdenas D, Nespola A, Camatel S, Abrate S, Gaudino R (2008) The rebirth of large-core plastic optical fibers: some recent results from the EU project "POF-ALL". In: Proceedings of IEEE/OSA optical fiber communication conference (OFC/NFOEC), San Diego, CA

18. Breyer F, Lee SCJ, Randel S, Hanik N. (2007) 1.25 Gbit/s transmission using FFE or DFE equalisation schemes over up to 100 m standard 1 mm step-index polymer optical fibre. In: Proceedings of European conference on optical communications (ECOC), Berlin, Germany
19. Visani D, Okonkwo CM, Loquai S, Yang H, Shi Y, van den Boom HPA, Ditewig AMH, Tartarini G, Schmauss B, Randel S, Koonen AMJ, Tangdiongga E. (2010) Record 5.3 Gbit/s transmission over 50 m 1 mm core diameter graded-index plastic optical fiber. In: Proceedings of IEEE/OSA optical fiber communication conference (OFC/NFOEC), San Diego, CA
20. Yu J, Jia Z, Huang M, Haris M, Ji PN, Wang T, Chang GK (2009) Applications of 40–Gb/s chirp-managed laser in access and metro networks. IEEE/OSA J Lightwave Technol 27(3):253–265
21. Loquai S, Kruglov R, Ziemann O, Vinogradov J, Bunge C-A (2010) 10 Gbit/s over 25 m Plastic optical fiber as a way for extremely low-cost optical interconnection. In: Proceedings of IEEE/OSA optical fiber communication conference (OFC/NFOEC), San Diego, CA
22. Van den Boom HPA, Li W, van Bennekom PK, Tafur Monroy I, Khoe G-D (2001) High-capacity transmission over polymer optical fiber. IEEE J Select Topics Quantum Electron 7(3):461–470
23. Haupt M, Fischer UHP (2009) Advanced integrated WDM system for POF communication. In Proc of the SPIE 7234:72340C
24. Bartkiv LV, Bobitski YV, Poisel H (2005) Optical demultiplexer using a holographic concave grating for PON-WDM systems. Optica Applicata 35(1):59–66
25. Yabre G (2000) Theoretical investigation on the dispersion of graded-index polymer optical fibers. IEEE/OSA J Lightwave Technol 18(6):869–877
26. Sinkin OV, Holzlöhner R, Zweck J, Menyuk CR (2003) Optimization of the split-step Fourier method in modeling optical fiber communications systems. IEEE/OSA J Lightwave Technol 21(1):61–68
27. Liu X (2009) Adaptive higher-order split-step Fourier algorithm for simulating lightwave propagation in optical fiber. Opt Commun 282(7):1435–1439
28. Hult J (2007) A fourth-order Runge–Kutta in the interaction picture method for simulating supercontinuum generation in optical fibers. IEEE/OSA J Lightwave Technol 25(12):3770–3775
29. Plura M, Kissing J, Gunkel M, Lenge J, Elbers JP, Glingener C, Schulz D, Voges E (2001) Improved split-step method for efficient fibre simulations. Electron Lett 37(5):286–287
30. Kremp T, Freude W (2005) Fast split-step wavelet collocation method for WDM system parameter optimization. IEEE/OSA J Lightwave Technol 23(3):1491–1502
31. Liu X, Lee B (2003) A fast method for nonlinear Schrödinger equation. IEEE Photonics Technol Lett 15(11):1549–1551
32. Golovchenko EA, Pilipetskii AN, Bergano NS, Davidson CR, Khatri FI, Kimball RM, Mazurczyk VJ (2000) Modeling of transoceanic fiber-optic WDM communication systems. IEEE J Select Topics Quantum Electron 6(2):337–346
33. Hui R, Chowdhury D, Newhouse M, O'Sullivan M, Poettcker M (1997) Nonlinear amplification of noise in fibers with dispersion and its impact in optically amplified systems. IEEE Photonics Technol Lett 9(3):392–394
34. Downie JD, Ruffin AB (2003) Analysis of signal distortion and crosstalk penalties induced by optical filters in optical networks. IEEE/OSA J Lightwave Technol 21(9):1876–1886
35. Cao S, Chen J, Damask JN, Doerr CR, Guiziou L, Harvey G, Hibino Y, Li H, Suzuki S, Wu K-Y, Xie P (2004) Interleaver technology: comparisons and applications requirements. IEEE/OSA J Lightwave Technol 22(1):281–289
36. Tomlinson WJ (2008) Evolution of passive optical component technologies for fiber-optic communication systems. IEEE/OSA J Lightwave Technol 26(9):1046–1063
37. Lenz G, Eggleton BJ, Giles CR, Madsen CK, Slusher RE (1998) Dispersive properties of optical filters for WDM systems. IEEE J Select Topics in Quantum Electron 34(8):1390–1402

38. Smit MK (2005) Progress in AWG design and technology. In: Proceedings of IEEE/LEOS workshop on fibers and optical passive components, Mondello, Italy, pp 26–31
39. Smit MK, Van Dam C (1996) PHASAR-based WDM-devices: principles, design and applications. IEEE J Select Topics Quantum Electron 2(2):236–250
40. Takahashi H, Suzuki S, Nishi I (1994) Wavelength multiplexer based on SiO_2-Ta_2O_5 arrayed-waveguide grating. IEEE/OSA J Lightwave Technol 12(6):989–995
41. El-Bawab TS (2006) Optical switching. Springer, New York
42. Lin LY, Goldstein EL (2002) Opportunities and challenges for MEMS in lightwave communications. IEEE J Select Topics Quantum Electron 8(1):163–172
43. Antoniades N, Ellinas G, Homa J, Bala K (2010) ROADM architectures and WSS implementation technologies. In: Krzysztof I (ed) Convergence of mobile and stationary next-generation networks. Wiley, New York
44. Afonso A, de la Tocnaye JL, Barge M (2006) Dynamic gain and channel equalizers or wavelength blocker using free space: common elements and differences. IEEE/OSA J Lightwave Technol 24(3):1534–1542
45. Liu Y, Chow CW, Kwok CH, Tsang HK, Lin C (2007) Dynamic-channel-equalizer using in-line channel power monitor and electronic variable optical attenuator. Opt Commun 272(1):87–91
46. Gutiérrez-Castrejón R, Duelk M, Bernasconi P (2006) A versatile modular computational tool for complex optoelectronic integrated circuits simulation. Opt Quantum Electron 38(12–14):1125–1134
47. Wall P, Colbourne P, Reimer C, McLaughlin S (2008) WSS switching engine technologies. In: Proceedings of IEEE/OSA optical fiber communication conference (OFC/NFOEC), San Diego, CA
48. Tsai J-C, Wu MC (2006) A high port-count wavelength-selective switch using a large scan-angle, high fill-factor, two-axis MEMS scanner array. IEEE Photonics Technol Lett 18(13):1439–1441
49. Ramaswami R, Sivarajan KN (2002) Optical networks, a practical perspective. Morgan Kaufmann Publishers, Massachusetts
50. Chan WS, Hall KL, Modiano E, Rauschenbach KA (1998) Architectures and technologies for high-speed optical data networks. IEEE/OSA J Lightwave Technol 16(12):2146–2168
51. Stern T, Ellinas G, Bala K (2008) Multiwavelength optical networks: architectures, design and control, 2nd edn. Cambridge University Press, Cambridge, UK
52. Tomkos I, Chowdhury D, Conradi J, Culverhouse D, Ennser K, Giroux C, Hallock B, Kennedy T, Kruse A, Kumar S, Lascar N, Roudas I, Sharma M, Vodhanel RS, Wang C-C (2001) Demonstration of negative dispersion fibers for DWDM metropolitan area networks. IEEE J Select Topics Quantum Electronics 7(3):439–460
53. Corvini PJ, Koch TL (1987) Computer simulation of high-bit-rate optical fiber transmission using single-frequency lasers. IEEE/OSA J Lightwave Technol 5(11):1591–1595
54. Habibullah F, Huang W-P (2006) A self-consistent analysis of semiconductor laser rate equations for system simulation purpose. Opt Commun 258(2):230–242
55. Ahmed M, Yamada M (2009) Inducing single-mode oscillation in Fabry–Perot InGaAsP lasers by applying external optical feedback. IET Optoelectron 4(3):133–141
56. McDonald D, O'Dowd RF (1995) Comparison of two- and three-level rate equations in the modeling of quantum-well lasers. IEEE J Select Topics Quantum Electron 31(11): 1927–1934
57. Morthier G, Baets R (1999) Introductory physics. In: Guekos G (ed) Photonic devices for telecommunications. Springer, Germany, p 153
58. Tomkos I, Roudas I, Hesse R, Antoniades N, Boskovic A, Vodhanel R (2001) Extraction of laser rate equation parameters for representative simulations of metropolitan-area transmission systems and networks. Opt Commun 194(1–3):109–129
59. Kibar O, Van Blerkom D, Fan C, Marchand PJ, Esener SC (1998) Small-signal-equivalent circuits for a semiconductor laser. Appl Opt 37(26):6136–6139

60. Chen W, Liu S (1996) Circuit model for multi-longitudinal-mode semiconductor laser. IEEE J Select Topics Quantum Electron 32(12):2128–2132
61. Ganesh Madhan M, Vaya PR, Gunasekaran N (1999) Circuit modeling of multimode bistable laser diodes. IEEE Photonics Technol Lett 11(1):27–29
62. Morthier G, Lowery A (1999) Modelling of DFB laser diodes. In: Guekos G (ed) Photonic devices for telecommunications. Springer-Verlag, Germany, p 183
63. Nguyen LVT, Lowery AJ, Gurney PCR, Novak D (1995) A time-domain model for high-speed quantum-well lasers including carrier transport effects. IEEE J Select Topics Quantum Electron 1(2):494–504
64. Li GL, Yu PKL (2003) Optical intensity modulators for digital and analog applications. IEEE/OSA J Lightwave Technol 21(9):2010–2030
65. Cartledge JC, Chrsitensen B (1998) Optimum operating points for electroabsorption modulators in 10 Gb/s transmission systems using nondispersion shifted fiber. IEEE/OSA J Lightwave Technol 16(3):349–356
66. Højfeldt S, Mørk J (2002) Modeling of carrier dynamics in quantum-well electroabsorption modulators. IEEE J Select Topics Quantum Electron 8(6):1265–1276
67. Vukovic A, Bekker EV, Sewell P, Benson TM (2006) Efficient time domain modeling of rib waveguide RF modulators. IEEE/OSA J Lightwave Technol 24(12):5044–5053
68. Winzer PJ, Essiambre R-J (2006) Advanced optical modulation formats. Proceedings of the IEEE 94(5):952–985
69. Seimetz M (2009) High-Order modulation for optical fiber transmission. Springer, Germany
70. Winzer PJ, Pfennigbauer M, Strasser MM, Leeb WR (2001) Optimum filter bandwidths for optically preamplified NRZ receivers. IEEE/OSA J Lightwave Technol 19(9):1263–1273
71. Personick SD (2008) Optical detectors and receivers. IEEE/OSA J Lightwave Technol 26(9):1005–1020
72. Desurvire E (1994) Erbium-doped fiber amplifiers: principles and applications. Wiley, New York
73. Bromage J (2004) Raman amplification for fiber communications systems. IEEE/OSA J Lightwave Technol 22(1):79–93
74. Islam MN (2002) Raman amplifiers for telecommunications. IEEE J Select Topics Quantum Electron 8(3):548–559
75. Wiesenfeld JM (1996) Gain dynamics and associated nonlinearities in semiconductor optical amplifiers. J High Speed Electron Syst 7(1):179–222
76. Cotter D, Manning RJ, Blow KJ, Ellis AD, Kelly AE, Nesset D, Phillips ID, Poustie AJ, Rogers DC (1999) Nonlinear optics for high-speed digital information processing. Science 286(5444):1523–1528
77. Connelly MJ (2010) Semiconductor optical amplifiers. Springer, Germany
78. Yamamoto Y, Inoue K (2003) Noise in amplifiers. IEEE/OSA J Lightwave Technol 21(11):2895–2915
79. Gutiérrez-Castrejón R, Schares L, Occhi L, Guekos G (2000) Modeling and measurement of longitudinal gain dynamics in saturated semiconductor optical amplifiers of different length. IEEE J Quantum Electron 36(12):1476–1484
80. Spiekman L, Piehler D, Iannone P, Reichmann K, Lee H (2007) Semiconductor optical amplifiers for FTTx. In: Proceedings of IEEE international conference on transparent optical networks (ICTON), Rome, Italy, pp 48–50
81. Pleros N, Zakynthinos P, Poustie A, Tsiokos D, Bakopoulos P, Petrantonakis D, Kanellos GT, Maxwell G, Avramopoulos H (2007) Optical signal processing using integrated multi-element SOA–MZI switch arrays for packet switching. IET Optoelectron 1(3):120–126
82. White I, Penty R, Webster M, Chai YJ, Wonfor A, Shahkooh S (2002) Wavelength switching components for future photonic networks. IEEE Commun Mag 40(9):74–81
83. Gutiérrez-Castrejón R, Duelk M (2006) Uni-directional time-domain bulk SOA simulator considering carrier depletion by amplified spontaneous emission. IEEE J Quantum Electron 42(6):581–588

84. Saleh AAM, Jopson RM, Evankow JD, Aspell J (1990) Modeling of gain in erbium-doped fiber amplifiers. IEEE Photonics Technol Lett 2(10):714–717
85. Giles CR, Desurvire E (1991) Modeling erbium-doped fiber amplifiers. IEEE/OSA J Lightwave Technol 9(2):271–283
86. Wu AWT, Lowery AJ (1998) Efficient multiwavelength dynamic model for erbium-doped fiber amplifier. IEEE J Quantum Electron 34(8):1325–1331
87. Hodgkinson TG (1992) Improved average power analysis technique for erbium-doped fiber amplifiers. IEEE Photonics Technol Lett 4(11):1273–1275
88. Min B, Lee WJ, Park N (2000) Efficient formulation of Raman amplifier propagation equations with average power analysis. IEEE Photonics Technol Lett 12(11):1486–1488
89. Zhu L, Ma Y, Wang G, Xia L, Xie S (2002) General computer model for both erbium-doped fiber amplifier and fiber Raman amplifier. Opt Eng 41(8):1805–1808
90. Liu X (2005) Effective numerical algorithm for fiber amplifiers. Opt Eng 44(3):1–7
91. Kidorf H, Rottwitt K, Nissov M, Ma M, Rabarijaona E (1999) Pump interactions in a 100-nm bandwidth Raman amplifier. IEEE Photonics Technol Lett 11(5):530–532
92. Kalavally V, Premaratne M, Rukhlenko I, Win T, Shum P, Tang M (2009) Novel directions in Raman amplifier research. In: Proceedings of international conference on information, communications and signal processing (ICICSP), Macau, China
93. Mork J, Nielsen ML, Berg TW (2003) The dynamics of semiconductor optical amplifiers: modeling and applications. Opt Photonics News 14(7):42–48
94. Maldonado R, Soto Ortiz H, Solís K (2008) Simplified model for estimating the cross-polarization modulation in a bulk semiconductor optical amplifier. IEEE J Quantum Electron 44(9):850–857
95. Mecozzi A, Mork J (1997) Saturation effects in nondegenerate four-wave mixing between pulses in semiconductor laser amplifiers. IEEE J Select Topics Quantum Electron 3(5):1190–1207
96. Kim Y, Lee H, Kim S, Ko J, Jeong J (1999) Analysis of frequency chirping and extinction ratio of optical phase conjugate signals by four-wave mixing in SOAs. IEEE J Select Topics Quantum Electron 5(3):873–879
97. Occhi L, Schares L, Guekos G (2003) Phase modeling based on the alpha-factor in bulk semiconductor optical amplifiers. IEEE J Select Topics Quantum Electron 9(3):788–797
98. Talli G, Adams MJ (2003) Amplified spontaneous emission in semiconductor optical amplifiers: modelling and experiments. Opt Commun 218(1–3):161–166
99. Gutiérrez-Castrejón R, Duelk M (2007) Modeling and simulation of semiconductor optical amplifier dynamics for telecommunication applications. In: Bianco SJ (ed) Computer physics research trends. Nova Science Publishers, New York, pp 89–124
100. Borri P, Scaffetii S, Mork J, Langbein W, Mecozzi A, Martelli F (1999) Measurement and calculation of the critical pulsewidth for gain saturation in semiconductor optical amplifiers. Opt Commun 164(1–3):51–55
101. Morel P, Sharaiha A (2009) Wideband time-domain transfer matrix model equivalent circuit for short pulse propagation in semiconductor optical amplifiers. IEEE J Quantum Electron 45(2):103–116
102. Lowery AJ (1988) New inline wideband dynamic semiconductor laser amplifier model. IEE J Optoelectron 135(3):242–250
103. Cassioli D, Scotti S, Mecozzi A (2000) A time-domain computer simulator of the nonlinear response of semiconductor optical amplifiers. IEEE J Quantum Electron 36(9):1072–1080
104. Mathlouthi W, Lemieux P, Salsi M, Vannucci A, Bononi A, Rusch LA (2006) Fast and efficient dynamic WDM semiconductor optical amplifier model. IEEE/OSA J Lightwave Technol 24(11):4353–4436
105. Bogoni A, Potì L, Porzi C, Scaffardi M, Ghelfi P, Ponzini F (2004) Modeling and measurement of noisy SOA dynamics for ultrafast applications. IEEE J Select Topics Quantum Electron 10(1):197–205

106. Rogowski T, Faralli S, Bolognini G, Di Pasquale F, Di Muro R, Nayar B (2007) SOA-based WDM metro ring networks with link control technologies. IEEE Photonics Technol Lett 19(20):1670–1672
107. Iannone P, Reichmann K, Spiekman L (2003) In-service upgrade of an amplified 130-km metro CWDM transmission system using a single LOA with 140-nm bandwidth. In: Proceedings of IEEE/OSA optical fiber communication conference (OFC), vol 2. Dallas, TX, pp 548–550
108. Uskov A, Berg TW, Mørk J (2004) Theory of pulse-train amplification without patterning effects in quantum-dot semiconductor optical amplifiers. IEEE J Quantum Electron 40(3):306–320
109. Xu Q, Yao M, Dong Y, Cai W, Zhang J (2001) Experimental demonstration of pattern effect compensation using an asymmetrical Mach-Zehnder. IEEE Photonics Technol Lett 13(12):1325–1327
110. Gutiérrez-Castrejón R, Filios A (2006) Pattern-effect reduction using a cross-gain modulated holding beam in semiconductor optical in-line amplifier. IEEE/OSA J Lightwave Technol 24(12):4912–4917
111. Li Z, Dong Y, Mo J, Wang Y, Lu C (2004) 1050-km WDM transmission of 8 × 10.709 Gb/s DPSK signal using cascaded in-line semiconductor optical amplifier. IEEE Photonics Technol Lett 16(7):1760–1762
112. Downie JD, Hurley J, Mauro Y (2008) 10.7 Gb/s uncompensated transmission over a 470 km hybrid fiber link with in-line SOAs using MLSE and duobinary signals. Opt Express 16(20):15759–15764
113. Kim HK, Chandrasekhar S, Srivastava A, Burrus CA, Buhl L (2001) 10 Gbit/s based WDM signal transmission over 500 km of NZDSF using semiconductor optical amplifier as the in-line amplifier. Electron Lett 37(3):185–187
114. Tanaka S, Tomabechi S, Uetake A, Ekawa M, Morito K (2006) Record high saturation output power (+20 dBm) and low NF (6.0 dB) polarisation-insensitive MQW-SOA module. Electron Lett 42(18):1059–1060

Chapter 4
Semianalytical Models for Network Performance Evaluation

Ronald Holzlöhner, Oleg V. Sinkin and Vladimir S. Grigoryan

Abstract We discuss the linearization and the momentum methods as two complementary approaches for analyzing signal statistics in optical communications systems governed by the nonlinear Schrödinger equation. Based on the linearization, we derive the covariance matrix method that allows us to accurately compute the bit error rates. The momentum method represents an alternative approach for computationally efficient analysis of the amplitude and timing jitter, as well as signal statistics.

4.1 Preface

This chapter presents a review of the semianalytical methods and techniques developed by the authors for computationally efficient and accurate evaluation of fiber network performance. These semianalytical models can be divided into two categories: (1) linearization-based models and (2) momenta-based models.

R. Holzlöhner
Telescope Laser Systems Department, European Southern Observatory (ESO)
Karl-Schwarzschild-Str. 2, 85748 Garching b.
München, Germany
e-mail: rholzloe@eso.org

O. V. Sinkin (✉)
Tyco Electronics Subsea Communications
250 Industrial Way West
Eatontown, NJ 07740, USA
e-mail: osinkin@subcom.com

V. S. Grigoryan
Ciena, 1201 Winterson Road
Linthicum, MD 21090, USA
e-mail: vgrigory@ciena.com

N. (Neo) Antoniades et al. (eds.), *WDM Systems and Networks*, Optical Networks,
DOI: 10.1007/978-1-4614-1093-5_4,

Although both categories are based on a common foundation, their capabilities and numerical cost are quite different. Both methods can be applied to highly nonlinear transmission systems, which is an important and fortunate result. However, while the linearization approach directly leads to the bit error rate (BER) computation through the covariance matrix analysis, the momenta-based methods only couple the calculated momenta indirectly to the BER. On the other hand, the momenta method is much faster and easier to implement than the numerical propagation of the covariance matrix with numerous complex elements. Consequently, the linearization-based and the momenta-based models complement each other.

Over the last years, the authors noticed a significant interest in semianalytical methods for BER analysis using various modulation formats, e.g., Refs. [37, 39, 58] to mention a few. The purpose of this chapter is twofold: first, to highlight once again the fundamental aspects of how the semianalytical models can be employed for the accurate and computationally efficient calculation of the BER by presenting the analytical material for doing that in a detailed fashion, applied to different realistic transmission systems. Second, to help the reader analyze these methods by combining them in one chapter. Some supplemental mathematical details can be found in [62, 124].

4.2 Linearization Methods for the Accurate Computation of the BER and the Covariance Matrix Method

4.2.1 Linearization of the NLS

4.2.1.1 Introduction

Computing accurate BERs is one of the most important tasks when designing optical fiber transmission systems. The covariance matrix method that we derive in this section yields a speedup in simulation time of orders of magnitude over the commonly used Monte Carlo simulation technique. Unlike some ad hoc approaches, the covariance matrix method is based on a sound mathematical foundation that makes use of the result that the optical noise in a transmission system is multivariate-Gaussian distributed after phase and timing jitter are separated. This result is derived from the assumption that the signal evolution is linearizable, which means that the interactions of the noise with itself in the fiber are negligible once phase and timing jitter are separated. We show that this assumption is valid for currently used systems. Unlike Monte Carlo methods, the covariance matrix method is completely deterministic; it does not require random number generators. In the remainder of this introduction, we will outline the problem of computing BERs and provide a guide to the following subsections.

The task of computing BERs essentially consists of computing the evolution of the amplified spontaneous emission (ASE) noise that is caused by optical amplifiers, taking into account its interaction with the noise-free signal due to the fiber nonlinearity. Typical BERs are in the range of 10^{-15}–10^{-9} and below, although most systems now employ forward-error correction (FEC) and can tolerate raw BERs of 10^{-3} and more [19, 25]. A commonly used method to simulate these rare events is still to run Monte Carlo simulations of the noise and extrapolate the results [17]. This approach relies on the speed of the signal propagation equation solver, which usually employs a split-step fast Fourier transform method [7] that requires a large number of discrete Fourier transforms. Running a Monte Carlo simulation is like repeatedly rolling dice or flipping a coin, and it is easy to implement on a computer and very robust [101]. However, when evaluating the probability of extremely rare events, such as that of the coin standing on its rim after being flipped, standard Monte Carlo simulations are too inefficient. One can increase the Monte Carlo simulation efficiency by biasing the perturbations toward regions of the phase space where bit errors are believed to be more likely to occur [20, 59], but the success of this method relies on detailed knowledge of the system, and the biasing must be optimized in order to achieve a significant performance gain over the unbiased simulation. On the other hand, this detailed system knowledge is sometimes the result of a simulation, and hence it is not known at the time the simulation is started.

Monte Carlo methods are very expensive computationally. For that reason, they are often replaced in practice with simplified models. The most common of these models is to run a noise-free simulation and then to simply add white Gaussian-distributed noise at the receiver that corresponds to the noise that would arrive at the receiver in the absence of any noise-signal interaction due to fiber nonlinearity. It is known, however, that parametric pumping of noise by the signal is a significant issue in some systems. For this reason, Hui et al. [68, 70] introduced a model, later extended by Pilipetskii et al. [115] that takes into account parametric pumping by treating the signal as a cw wave. Due to difficulties in carrying out extensive Monte Carlo simulations, this simplified approach is not well-validated in general.

Systems that suffer from noise are well-known in many areas of science. In all communications systems there is a certain amount of noise that distorts signals during transmission or storage of information, and therefore limits the capacity [42, 108]. Unless a given physical system is completely dominated by nonlinearities, perturbation theory based on the assumption that the noise interaction with itself is negligible is always a powerful technique for computing error rates. The linearization assumption has been successfully applied to optical fiber communications and the known results can be divided into several groups:

- Soliton perturbation theory was applied to transmission with gain and loss, ASE noise, soliton–soliton interactions, filtering, and synchronous modulation [47, 54, 61, 77]. However, the results are only strictly valid for classical solitons.
- Standard perturbation theory was applied to an unmodulated (cw) signal [22, 27, 68, 70].
- Computation of the evolution of noise moments such as timing and amplitude jitter has been carried out for an arbitrary signal [56].

However, none of the approaches mentioned above are capable of computing the full probability density function (pdf) of the optical noise in the presence of an arbitrary signal modulation and the pdf of the electrical receiver current. The goal is hence to develop a method that:

1. linearizes the optical noise propagation in a wide range of transmission systems,
2. yields the accurate pdf of the optical noise over the entire transmission distance,
3. yields the pdf of the electrical receiver current after passing a narrow-band filter,
4. can be easily implemented and is computationally efficient.

We derive such an approach in this section and call it the covariance matrix method. The approach is capable of computing accurate probability density functions (pdfs) of the receiver current after narrow-band filtering. With this information, the BER and contour eye diagrams can then be obtained. The most complex system that we treat in this section is a 10 Gb/s five-channel WDM chirped return-to-zero (CRZ) system, in which we simulate the transmission of 32 bits. Demir [39] summarizes different approaches to noise linearization and extends the covariance matrix method to a larger number of channels and to coherent detection.

We outline different previous linearization approaches in Sect. 4.2.2 and formulate the theory of the covariance matrix method in Sect. 4.2.3. We present the fully deterministic covariance matrix method in Sect. 4.2.4. Section 4.2.5 contains some mathematical details and a description of the numerical algorithms.

4.2.1.2 The Nonlinear Schrödinger Equation

In order to study the light evolution in optical fibers, we begin by writing the electric field in the fiber as [100]

$$\mathscr{E}(\mathbf{r},\tau_p) = \sqrt{\frac{\omega_0}{2\varepsilon_0 c^2 \beta(\omega_0, z_p)}} e(z_p,\tau_p)\mathbf{R}(x,y)\exp\left[i\int_0^{z_p} \beta(\omega_0, z_p')\mathrm{d}z_p' - i\omega_0\tau_p \right], \quad (4.1)$$

where $\mathscr{E}$ is the electric field vector in MKS units, while $\mathbf{r} = (x, y, z_p)$ and τ_p are position in the fiber and physical time, $\omega_0/2\pi \approx 10^{14}$ Hz is the optical carrier frequency, $\varepsilon_0 = 8.85 \times 10^{-12}$ F/m is the permittivity of the vacuum, $c = 2.99 \times 10^8$ m/s is the speed of light in a vacuum, and $\beta(\omega_0, z_p)$ is the wavenumber at the carrier frequency ω_0, which may vary slowly relative to the wavelength along the fiber. The coordinate z_p measures distance along the fiber. The transverse modal field is normalized so that

$$\int \mathrm{d}x \int \mathrm{d}y \left|\mathbf{R}_{\perp(x,y)}\right|^2 = 1, \quad (4.2)$$

where $\mathbf{R}_{\perp(x,y)}$ is a vector obtained by projecting $\mathbf{R}$ onto the plane perpendicular to the propagation direction. With this normalization, $|e(z_p,t)|^2$ equals the local power. We note that we are assuming that only one polarization is propagating. From Eq. 4.1 for the electric field, it is possible to derive the modified nonlinear Schrödinger equation (NLS) in the following dimensionless form [7, 100],

$$i\frac{\partial u}{\partial z}+\frac{1}{2}D(z)\frac{\partial^2 u}{\partial t^2}+|u|^2u=ig(z)u+\hat{F}(z,t), \tag{4.3}$$

where $u(z,t)=e(z,t)\sqrt{\gamma(z)L_D}$, while $\gamma(z)=\gamma_0 A_{\mathrm{eff},0}/A_{\mathrm{eff}}(z)$ is the local nonlinear coefficient with its longitudinal average $\gamma_0=n_2\omega_0/(A_{\mathrm{eff},0}\,c)$, and $A_{\mathrm{eff}}(z)$ and $A_{\mathrm{eff},0}$ are, respectively, the local and the average effective fiber cross sections. The symbol L_D denotes a characteristic length and distance z is normalized as $z=z_p/L_D$. We have transformed from physical time τ_p to retarded time $t=\tau_p-\int_0^z\beta'(z')\mathrm{d}z'$. The quantity $n_2=2.6\times 10^{-16}\,\mathrm{cm}^2/\mathrm{W}$ is the Kerr coefficient, and $A_{\mathrm{eff}}=30\text{–}200\,\mu\mathrm{m}^2$ is the effective fiber core area. The quantity $D(z)=-\beta''(z)/\beta''_0$ is the normalized dispersion parameter with the local dispersion $\beta''(z)$ and a scaling dispersion β''_0. The dispersion β'' is measured in units of $\mathrm{ps}^2/\mathrm{nm}$. The characteristic length L_D equals $T_0^2/|\beta''_0|$, where T_0 is a characteristic time scale. When modeling soliton systems, it is customary to set T_0 equal to the FWHM soliton duration and β''_0 equal to the path average dispersion, in which case L_D is the *dispersive scale length* [7]. The normalized field gain coefficient $g(z)$ is

$$g(z)=\begin{cases} g_m(z), & z_m<z<z_m+L_{\mathrm{amp}}/L_D,\\ -\Gamma_l & \text{elsewhere,}\end{cases} \tag{4.4}$$

where g_m represents the gain coefficient inside the mth amplifier normalized by L_D, which we assume to begin at $z=z_m$ and to be of length L_{amp}, and Γ_l is the normalized fiber loss coefficient. Since the typical length of an EDFA is only a few meters, the dispersion and nonlinearity in the EDFA can be neglected and hence amplification and ASE noise input can be considered as lumped. For later use, we define the *nonlinear scale length* [7]

$$L_{\mathrm{nl}}=\frac{1}{\gamma_0 P_{\mathrm{peak}}}, \tag{4.5}$$

which is the distance over which the nonlinear phase rotation is approximately 2π.

Summarizing the meaning of the terms that appear in (4.3), the second term on the left-hand side describes chromatic dispersion, the third term describes the Kerr nonlinearity, and the first term on the right-hand side describes fiber attenuation and gain within the amplifiers. The last term is a *Langevin* noise term and the quantity $\hat{F}$ represents the ASE white noise contribution with zero mean

$$\langle\hat{F}(z,t)\rangle=0, \tag{4.6}$$

and the autocorrelation

$$\langle\hat{F}(z,t)\hat{F}^*(z',t')\rangle=2\eta\delta(z-z')\delta(t-t'), \tag{4.7}$$

where

$$\eta = \frac{n_{\text{sp}} g_m \hbar \omega_0 L_D \gamma(z)}{T_0}. \tag{4.8}$$

In the test systems that we consider in this section, $n_{\text{sp}} = 1.4 - 2.0$ inside the fiber amplifiers and $n_{\text{sp}} = 0$ in the transmission fiber. The angular brackets $\langle \cdot \rangle$ denote the noise ensemble average, and the asterisk denotes complex conjugation. Equation 4.3 can be extended to include effects such as higher-order dispersion, saturated absorption, and Raman and Brillouin scattering [100]. However, saturable absorbers are not used in the test systems that we study, while Raman and Brillouin scattering effects are small and can be neglected. Mu et al. [103, 106] showed that third-order dispersion does not have a strong impact on the pulse propagation in the dispersion-managed soliton (DMS) system and the CRZ system that we simulate in this section. We solve (4.3) using a standard split-step approach [7].

4.2.2 Transmission Linearization

4.2.2.1 Introduction

The covariance matrix method that we describe in this section can be divided into two steps:

1. the computation of the covariance matrix that describes the accumulation of ASE noise from the optical amplifiers and its evolution due to optical fiber nonlinearity, and
2. the computation of the pdf of the electrical current in the receiver after a photodiode and an electrical filter.

As we will show in the next subsections, the first step, in which we treat transmission through an optical fiber with a series of erbium-doped fiber amplifiers along the path is amenable to a linearization approach that assumes that the nonlinear interaction of the optical noise with itself in the fiber is negligible once phase jitter and, in some cases, timing jitter are separated. By contrast, the noise interaction with itself in the receiver cannot be ignored. The aim of this subsection is to describe some of the previous work that was done on transmission linearization and to outline the limitations of the linearization approach.

The starting point of the linearization approach is the NLS (4.3). One can decompose the noisy optical signal $u = u_0 + \delta u$ as a sum of a noise-free signal, $u_0 = \langle u \rangle$, and accumulated transmitted noise, δu. The accumulated noise is the sum of all the individual contributions and experiences nonlinear interaction with the signal, as well as dispersion, attenuation, and amplification. The angular brackets $\langle \cdot \rangle$ denote the noise ensemble average. The difference of (4.3) and the statistical average of (4.3) then yields the evolution equation for δu

$$i\frac{\partial \delta u}{\partial z} + \frac{D}{2}\frac{\partial^2 \delta u}{\partial t^2} + 2|u_0|^2\delta u + u_0^2(\delta u)^* = ig\delta u + \hat{F}, \tag{4.9}$$

where the terms that are quadratic and third order in δu, describing the optical noise–noise interactions, are omitted. The third and fourth terms on the left-hand side stem from the nonlinear term $|u|^2 u$ in (4.3) and describe the interactions of the noise-free signal u_0 with the noise δu. The term $2|u_0|^2\delta u$ leads to a phase rotation that can be regarded as a cross-phase modulation (XPM) between u_0 and δu. The term $u_0^2(\delta u)^*$ describes an energy exchange between u_0 and δu. Since $|u_0^2| \gg |\delta u|^2$, energy mainly flows from the signal to the noise and leads to *parametric growth* of the noise, an effect that we will describe in Sect. 4.2.4. Parametric gain can be viewed as a four-wave mixing (FWM) in which the annihilation of two signal photons at ω_1 create two noise photons at ω_3 and ω_4, so that $\omega_4 = 2\omega_1 - \omega_3$ [7]. The energy conversion process is most efficient if the frequency mismatch $|\omega_1 - \omega_3| = |\omega_1 - \omega_4|$ is small. We note that the linearized NLS (4.9) depends on both δu as well as on its complex conjugate $(\delta u)^*$. As a consequence, (4.9) is not Hermitian, which substantially complicates the solution, as we will show in the next section. By contrast, the original nonlinear equation (4.3) is Hermitian.

From soliton perturbation analysis, one finds that a perturbation to the soliton, such as ASE noise, causes the four soliton parameters amplitude, frequency, central time, and phase to fluctuate [47, 54, 61, 72, 77]. The rest of the noise produces a radiation continuum and disperses. Noise can be expanded into the four eigenfunctions and the continuum, and the parameter perturbations are given by the projections of the noise on the eigenfunctions that pertain to these four parameters [56]. The reader may refer to [62] for mathematical details. Although this subsection does not deal with standard solitons, these qualitative observations will be of importance in the following.

4.2.2.2 Noise Moments

Considering a single pulse within a pulse sequence, the most important moments are the pulse energy U, the central time t_p, and the central angular frequency Ω_p, defined as

$$U = \int_0^{T_{\text{bit}}} |u|^2 \mathrm{d}t, \tag{4.10a}$$

$$t_p = \frac{1}{U}\int_0^{T_{\text{bit}}} t|u|^2 \mathrm{d}t, \tag{4.10b}$$

$$\Omega = \frac{1}{U} \int_{-\infty}^{\infty} \omega |\tilde{u}(\omega)|^2 d\omega = \frac{1}{U} \mathrm{Im} \int_{0}^{T_{\mathrm{bit}}} u_t^* u \mathrm{d}t, \tag{4.10c}$$

where T_{bit} is the inverse of the data rate, $\tilde{u}(\omega)$ is the Fourier transform of $u = u(t)$, and $u_t = \partial u(t)/\partial t$.

As we discussed in the previous subsections, one would like to linearize around a computationally determined solution u_0 of the signal. In 1999, Grigoryan et al. [56] generalized the Gordon–Haus result [54] to arbitrary signal shapes and extended the analysis to different kinds of jitter. They found that the deviation of t_p and Ω_p with distance is given by projections of the noise on the field

$$\frac{\partial t_p}{\partial z} = D\Omega + \frac{i}{U} \int_{0}^{T_{\mathrm{bit}}} (t - t_p)\left(u\hat{F}^* - u^*\hat{F}\right) \mathrm{d}t, \tag{4.11}$$

$$\frac{\mathrm{d}\Omega}{\mathrm{d}z} = -\frac{i\Omega}{U} \int_{0}^{T_{\mathrm{bit}}} \left(u\hat{F}^* - u^*\hat{F}\right) \mathrm{d}t - \frac{1}{U} \int_{0}^{T_{\mathrm{bit}}} \left(u_t\hat{F}^* - u_t^*\hat{F}\right) \mathrm{d}t, \tag{4.12}$$

where D is the local dispersion. Linearizing the propagation equation around an arbitrary field u, one can now numerically compute the evolution of the timing jitter

$$\sigma_p = \left(\left\langle t_p^2 \right\rangle - \left\langle t_p \right\rangle^2\right)^{1/2} \tag{4.13}$$

and the frequency jitter $\sigma_\Omega = \left\langle (\delta\Omega)^2 \right\rangle^{1/2}$. Grigoryan et al. [56] successfully applied this method to DMS, return-to-zero (RZ), and non-return-to-zero (NRZ) systems. We will discuss it in detail in Sect. 4.3.2.3.

4.2.2.3 Limits of Linearization

What lies beyond linearization? Is it possible and is it even necessary to derive results for transmission systems in which noise–noise interactions are relevant? Mecozzi [98] has taken an interesting approach by neglecting the dispersion terms in the NLS and, using Itô's formalism [11], derives exact expressions for arbitrary field averages of the form $\langle u(t,z)^n (u^*(t,z))^m \rangle$. We note that neglecting the dispersion is a very strong simplification, and hence the resulting equation describes a system which is significantly different from a realistic optical communications system. Mecozzi is able to define characteristic distances that define two regimes of propagation, a linear regime where the noise is additive, and a nonlinear one where significant signal–noise interactions occur. Moreover, he shows that the signal spectrum broadens due to SPM and the signal–noise interactions add a noise background to the signal spectrum. However, the results are strongly limited by the constraint that the dispersion equals zero.

A traditional method for obtaining the full pdf of the optical noise at the receiver is to run Monte Carlo simulations that pick random noise realizations and hence sample the noise ensemble. From this noise distribution, one can then calculate the distribution of the receiver current. This method is robust and works in principle in all settings, including situations with strong amplifier gain saturation, frequency-dependent dispersion, and strong nonlinearity. This method can handle arbitrarily strong pulse distortion, including complete pulse breakdown. In addition, it is easy to program. If a system is to be investigated in a significantly nonlinear regime, then Monte Carlo simulations might be one of the few possible options. However, we will show in the following subsections that the linearization assumption is valid in a soliton system, whose error-free ($\text{BER} < 10^{-9}$) transmission distance is hundreds of times larger than the nonlinear scale length, once phase and timing jitter are properly treated. Consequently, we claim that linearization is capable of describing the noise evolution in a wide range of transmission systems—probably any practical system.

4.2.3 Statistics of the Noise Evolution

4.2.3.1 Theoretical Foundation

In this subsection, we describe a covariance matrix method that yields a complete and accurate statistical description of the optical noise at the end of the transmission line. In Sect. 4.2.3.6, we develop exact equations to compute the pdf of the received current from this information. The only limitation of the procedure is that the noise–noise interactions *in the fiber* are neglected in an appropriate basis set in which phase, and in some cases, timing jitters are separated. The receiver contains a square-law detector and hence it is substantially nonlinear. Therefore, we retain noise–noise terms in the receiver.

The utility of the linearization assumption stems from two key mathematical results. The first is the Karhunen–Loève theorem [113], which states that a combination of signal and noise over any finite time can be expanded in an orthonormal basis whose coefficients are independent random variables. When the noise is white, any orthonormal basis will satisfy the Karhunen–Loève theorem. In optical fiber communications systems, the ASE noise is effectively white when it is contributed by the amplifiers, but it only remains white for short distances over which the nonlinear interactions between the signal and the noise can be neglected. Over longer distances, the noise becomes correlated, and the Karhunen–Loève basis becomes unique and distance-dependent. The second mathematical result is Doob's theorem [41], which states that when the system is linearizable and driven by Gaussian-distributed noise, each of these independent random variables is

Gaussian. Thus, it suffices in principle to determine the Karhunen–Loève modes, as well as the mean and variance of its coefficients, to calculate the effective noise pdf! We emphasize that this powerful result allows the signal to interact nonlinearly with itself and with the noise; it only requires that the noise does not interact with itself. In practice, one must use an approximate static basis from which to compute the Karhunen–Loève modes. The choice of this basis set is important. The linearization assumption may hold in one basis set and not in another. As we will see, it is necessary to use a basis set in which the phase jitter, and in some systems also the timing jitter, are explicitly separated. Otherwise, the linearization assumption only holds for short distances.

The study of the DMS system that we introduce in the next subsection and in [64] shows that it is possible to use the linearization assumption to calculate the effects of accumulated noise. However, this assumption breaks down after a short propagation distance, unless the phase and timing jitter are handled separately from the other noise components, as we show in Sect. 4.2.3.2. This separation is necessary because the nonlinear equation that governs the fiber transmission (4.3) implies that small amounts of amplitude and frequency noise can lead to large amounts of phase and timing jitter, respectively, which invalidates the linearization assumption. We have discussed the relationship between the parameter perturbations in [62] for a standard soliton, but the qualitative behavior remains the same for arbitrary pulse shapes. As we will explain below, the nonlinear propagation equation is also phase and time invariant, which implies that phase and timing jitter can be separated without affecting the subsequent evolution. We will find that it is necessary to separate the phase and timing jitter separately for each pulse, in which case the phase invariance no longer strictly holds. However, it holds sufficiently well for the linearization to remain valid over the distances of interest. Once phase, and in some cases, timing jitter are separated, the coefficients of the modified Fourier basis, along with the phase and timing jitter, remain multivariate-Gaussian distributed far longer than the original Fourier coefficients [64]. We note that the phase is only Gaussian distributed if one tracks the phase change in the interval $(-\infty, \infty)$. If one only tracks it in the range $[0, 2\pi]$, then it is Jacobi-Θ function distributed. This function is the periodic analog of a Gaussian distribution [6]. Therefore, soliton perturbation theory is a special case where the expansion of optical noise into discrete modes and a continuum is appropriate, rather than an expansion in the usual Fourier basis [47, 61, 77], as shown in [62]. We validate this assumption using extensive Monte Carlo simulations.

We note again that the approach assumes that noise–noise interactions in an appropriate basis set are negligible during the transmission, but we take them into account in the receiver, which we also assume has a realistic, narrow-band electrical Bessel filter (we note that some existing receiver systems employ other electrical filters than Bessel). Thus, the work presented in this subsection is a generalization of [96], in that the noise that enters the receiver is not white, but is determined by the actual transmission, and we apply realistic electrical filtering.

The optical noise in a wide range of optical transmission systems is multivariate-Gaussian distributed with zero mean, after the phase and timing jitter are separated. This pdf only depends on the covariance matrix of the noise. The entire discussion in this subsection applies to one optical polarization only, which is appropriate for the example transmission systems that we are using [103]. Moreover, this choice somewhat simplifies the theoretical development. There is no reason to doubt that the covariance matrix method can be extended to take polarization effects into account.

The remainder of this subsection is organized as follows: We derive the theory of the linearization approach and define the covariance matrix in Sect. 4.2.3.2. A deterministic computation based on the ODE is described in Sect. 4.2.3.4. In Sect. 4.2.3.6, we describe how the pdfs of the receiver current and the accurate BERs are computed. Finally, in Sect. 4.2.3.8, we consider the computation of the BER in systems with multiple bits.

4.2.3.2 Noise Covariance Matrix

The starting point is the system of equations for u_0 and δu, as shown in the previous subsection

$$i\frac{\partial u_0}{\partial z}+\frac{D(z)}{2}\frac{\partial^2 u_0}{\partial t^2}+|u_0|^2 u_0 = ig(z)u_0, \tag{4.14a}$$

$$i\frac{\partial \delta u}{\partial z}+\frac{D(z)}{2}\frac{\partial^2 \delta u}{\partial t^2}+2|u_0|^2\delta u+u_0^2(\delta u)^* = ig(z)\delta u+\hat{F}, \tag{4.14b}$$

where (4.14a) describes the evolution of the noise-free solution u_0 and (4.14b) describes the accumulated noise if $\delta u \ll u_0$. The signal u_0 must be known when solving (4.14b) and in practice it is most convenient to solve both equations in parallel. Note that we neglect any influence of δu on u_0. Considering all the terms in the Kerr nonlinearity of the NLS with $u = u_0 + \delta u$,

$$|u|^2 u = |u_0|^2 u_0^* + 2|u_0|^2\delta u + u_0^2\delta u^* + u_0^*(\delta u)^2 + 2u_0|\delta u|^2 + |\delta u|^2\delta u, \tag{4.15}$$

where one finds the second and third terms on the right-hand side of (4.15) that appear in (4.14b). The fourth and fifth terms on the right-hand side are linear in u_0 and represent the next order beyond noise linearization. The term $|\delta u|^2\delta u$ is cubic in δu and can be expected to contribute least. The term $u_0^*(\delta u)^2$ describes the depletion of u_0 due to the four-wave mixing with the noise, and the term $2u_0|\delta u|^2$ describes a cross-phase modulation of the signal due to the noise. These two terms can in principle be included into (4.14a), which is the statistical average of (4.3) for $u = u_0 - \delta u$, yielding

$$i\frac{\partial u_0}{\partial z}+\frac{D(z)}{2}\frac{\partial^2 u_0}{\partial t^2}+|u_0|^2 u_0 = ig(z)u_0 - 2u_0\left\langle|\delta u|^2\right\rangle - u_0^*\left\langle(\delta u)^2\right\rangle. \tag{4.16}$$

The averages $\langle |\delta u|^2 \rangle$ and $\left\langle (\delta u)^2 \right\rangle$ could be obtained by solving (4.14b), in which case the Eqs. 4.16 and 4.14b would form a mutually dependent equation system. However, if second-order terms become relevant, the linearization assumption is invalid and optical noise will not be Gaussian distributed anymore. We neglect the influence of δu on u_0 by solving the system (4.14a, b), and we show that the linearization assumption holds in an appropriate basis set for highly nonlinear optical fiber transmission systems.

One can expand $u_0(t)$ and $\delta u(t)$ in Fourier series,[1]

$$u_0(t) = \sum_{n=-N_{\mathrm{FFT}}/2}^{N_{\mathrm{FFT}}/2-1} A_n(z) \exp(i\omega_n t), \tag{4.17a}$$

$$\delta u(t) = \sum_{n=-N_{\mathrm{FFT}}/2}^{N_{\mathrm{FFT}}/2-1} a_n(z) \exp(i\omega_n t), \tag{4.17b}$$

where N_{FFT} is the number of sample points in the Fourier transform, and $\omega_n \equiv 2\pi n T_0/T$, where T is the period. Typical values of N_{FFT} are powers of 2 such as 1,024, 2,048, and 4,096. The values of T and N_{FFT} are chosen so that the simulation result does not change when either one is increased. After substituting (4.17a and b) into (4.14b), one finds that

$$\frac{\mathrm{d}a_k}{\mathrm{d}z} = \left(g - i\frac{D}{2}\omega_k^2\right) a_k + i \sum_{n,l,m=-N_{\mathrm{FFT}}/2}^{N_{\mathrm{FFT}}/2-1} \left[2A_n A_l^* a_m \delta_{n-l,k-m} + A_n A_l a_m^* \delta_{n+l,k+m}\right] - i\Gamma_k(z), \tag{4.18}$$

where the Γ_k are the Fourier coefficients of the white noise input $\hat{F}$, and δ is Kronecker's delta. Using (4.7), the correlation of the Γ_k is $\langle \Gamma_k(z)\Gamma_m^*(z')\rangle = (2\eta T_0/T)\delta(z-z')\delta_{k,m}$, where $\eta = n_{\mathrm{sp}} g_m \hbar\omega_0 L_D \gamma / T_0$ is again zero outside of the amplifiers.

We now introduce a frequency cutoff and will only consider $N = 2Tf_{\max}$ frequencies with $|f| < f_{\max} = (N/2)\Delta f_{\mathrm{FFT}} = N/(2T)$ in the following. We choose the frequency range $[-f_{\max}, f_{\max}]$ so that the signal power outside of this range is smaller than 1% of its total value. Hence, we neglect frequencies from the linearization at which the signal power is small, as justified by the discussion in [62]. Typical values of N lie in the range 50–150. The large ratio of $N_{\mathrm{FFT}}/N \gtrsim 10$ demonstrates the advantage of working in the frequency domain: the relevant modes are all concentrated at low frequencies, and high frequency modes can be neglected. The covariance matrix can be expressed in any basis, for example in the time domain, but then reducing the number of modes might not be as simple.

---

[1] Note that by choosing this sign convention in the exponential functions, the optical frequency that corresponds to ω_n lies at $\omega_{\mathrm{total}} = \omega_0 - \omega_n$, where ω_0 is the carrier frequency.

We define the complex column vectors $\boldsymbol{\alpha} = (a_{-N/2}, \ldots, a_{N/2-1})^T$ and $\boldsymbol{\alpha}^* = (a^*_{-N/2}, \ldots, a^*_{N/2-1})^T$, as well as $\boldsymbol{\Gamma} = (\boldsymbol{\Gamma}_{-N/2}, \ldots, \boldsymbol{\Gamma}_{N/2-1})^T$, where the superscript T indicates the transpose operation. Then we can rewrite (4.18) in matrix form as

$$\frac{\mathrm{d}\boldsymbol{\alpha}}{\mathrm{d}z} = \mathsf{B}\boldsymbol{\alpha} + \mathsf{E}\boldsymbol{\alpha}^* - i\boldsymbol{\Gamma}, \tag{4.19}$$

where the complex matrices B and E are defined as[2]

$$\mathsf{B}_{km} = \left(g - i\frac{D}{2}\omega_k^2\right)\delta_{km} + 2i\sum_{n,l} A_n A_l^* \delta_{n-l,k-m}, \tag{4.20a}$$

$$\mathsf{E}_{km} = i\sum_{n,l} A_n A_l \delta_{n+l,k+m}. \tag{4.20b}$$

The matrix E is symmetric ($\mathsf{E}_{km} = \mathsf{E}_{mk}$), and, if g is zero, the matrix B is anti-Hermitian ($\mathsf{B}^*_{km} = -\mathsf{B}_{mk}$). The matrix B is circulant and thus corresponds to a convolution in the time domain, while E can be termed anti-circulant. (A matrix M is circulant if there is a vector x with $M_{km} = x_{k-m}$). The number of operations required to evaluate B and E grows like N^3. Equation 4.18 depends on both a_k and a_k^* since the linearized noise propagation is not Hermitian [68, 70].

The probability space of the optical noise in the frequency domain is spanned by the $2N$ real variables $a_{k,R}$ and $a_{k,I}$. It is therefore convenient to split (4.18) into its real and imaginary parts and consider the resulting system of equations. We define the real-partitioned vector $\boldsymbol{a} = (\boldsymbol{\alpha}_R, \boldsymbol{\alpha}_I) = (\boldsymbol{a}_{-N/2,R}, \boldsymbol{a}_{-N/2+1,R} \cdots, \boldsymbol{a}_{N/2-1,R}, \boldsymbol{a}_{-N/2,I}, a_{-N/2+1,I} \cdots, a_{N/2-1,I})^T$ of length $2N$, consisting of the real and imaginary parts of α at the N lowest frequencies $-N/2, -N/2+1, \ldots N/2-1$. Similarly, we define $w = (\Gamma_I, -\Gamma_R)^T$. One can express (4.19) as

$$\frac{\mathrm{da}}{\mathrm{d}z} = \mathscr{R}(z)a + w(z), \tag{4.21a}$$

$$\mathscr{R} = \begin{bmatrix} \mathsf{B}_R + \mathsf{E}_R & -\mathsf{B}_I + \mathsf{E}_I \\ \mathsf{B}_I + \mathsf{E}_I & \mathsf{B}_R - \mathsf{E}_R \end{bmatrix}, \tag{4.21b}$$

where $\mathscr{R}$ is a real $2N \times 2N$ block matrix, and we have used the notation $\mathsf{B} = \mathsf{B}_R + i\mathsf{B}_I$ and $\mathsf{E} = \mathsf{E}_R + i\mathsf{E}_I$. One may formally write the solution to (4.21a) as [40]

[2] In the rest of this subsection, we will use a sans serif font to denote complex $N \times N$ matrices like B, a script font for real $2N \times 2N$ matrices like $\mathscr{R}$, and a bold face font for real $2N$-vectors such as a. The only exception will be α, which is a complex N-vector.

$$a(z) = \Psi(z, z_0)a(z_0) + \int_{z_0}^{z} \Psi(z, z')w(z')\mathrm{d}z', \tag{4.22}$$

where $\Psi(z, \zeta)$ is a propagator matrix that obeys the equation

$$\frac{\mathrm{d}}{\mathrm{d}z}\Psi(z, \zeta) = \mathscr{R}(z)\Psi(z, \zeta), \qquad \Psi(\zeta, \zeta) = \mathscr{I} \quad \forall z, \zeta, \tag{4.23}$$

and $\mathscr{I}$ is the identity matrix. Equation 4.21b describes the spatial evolution of the noise Fourier modes. We now assume that the $a(z)$ satisfies a multivariate-Gaussian distribution, which is completely described by its first two moments. Making this assumption is equivalent to assuming that the linearization assumption holds in a Fourier basis set. For the system that we consider in this subsection, this assumption only remains valid for short distances. However, it is a useful starting point for our subsequent discussion. From (4.6), it follows that the mean of $a(z)$ is zero, while the second moments are given by the covariance matrix $\mathscr{K}(z)$. The pdf p_a of a may be written as [114]

$$p_a(a, z) = (2\pi)^{-N}\sqrt{\det \mathscr{K}^{-1}(z)}\exp\left[-\frac{1}{2}a^T\mathscr{K}^{-1}(z)a\right], \tag{4.24}$$

where the real symmetric $2N \times 2N$ covariance matrix $\mathscr{K}$ is defined by,

$$\begin{aligned} \mathscr{K} &= \left\langle \left(a - \langle a\rangle\right)\left(a^T - \langle a\rangle^T\right)\right\rangle \\ &= \left\langle \begin{bmatrix} \alpha_R\alpha_R^T & \alpha_R\alpha_I^T \\ \alpha_I\alpha_R^T & \alpha_I\alpha_I^T \end{bmatrix}\right\rangle \equiv \begin{bmatrix} \mathrm{RR} & \mathrm{RI} \\ \mathrm{IR} & \mathrm{II} \end{bmatrix}, \end{aligned} \tag{4.25}$$

where RR, RI, IR, and II are four real $N \times N$ block matrices. In (4.25) all vector products above are outer products. This definition of $\mathscr{K}$ embodies the full covariance information in $4N^2$ real numbers of which $N(2N + 1)$ are independent. The alternative complex $N \times N$ matrix $\langle a_k a_l^* \rangle = \langle \alpha\alpha^\dagger \rangle_{kl}$, where the † denotes the conjugate transpose, contains only $2N^2$ real numbers of which N^2 are independent, and thus $\langle a_k a_l^* \rangle$ does not contain complete information. From (4.21b) and (4.22), one now finds that $\mathscr{K}$ evolves over distance according to

$$\frac{d}{\mathrm{d}z}\mathscr{K} = \mathscr{R}\mathscr{K} + \mathscr{K}\mathscr{R}^T + \frac{\eta T_0}{T}\mathscr{I}, \tag{4.26}$$

where η is defined after (4.7). Equation 4.26 is a Lyapunov equation [40] and is the linear evolution equation for the covariance matrix. The right-hand side of (4.26) is symmetric since $(\mathscr{R}\mathscr{K})^T = \mathscr{K}^T\mathscr{R}^T = \mathscr{K}\mathscr{R}^T$, so that $\mathscr{K}$ remains symmetric as it evolves overz. Initially, $\mathscr{K}$ is zero since the launched signal is assumed to be noise-free. The matrix $\mathscr{R}$ is distance-dependent and includes amplification/attenuation as well as the nonlinear interactions of the signal with the noise. The last term describes the white noise input and is only nonzero inside the optical amplifiers. Newly-added noise only contributes to the diagonal elements $\mathcal{K}_{kk}$. In addition to being symmetric, $\mathcal{K}$ is also positive definite, so that its determinant is positive.

We note that the direct derivation of (4.26) from (4.21a) is only one possible way of obtaining (4.26). One can show that the pdf in (4.24), where the covariance matrix $\mathcal{K}(z)$ is described by (4.26), represents the exact solution of the Fokker–Planck equation [117] corresponding to the Langevin equation (4.21a). Yet another approach to derive (4.26) uses Itô's method [11]. All of these methods are, of course, equivalent.

4.2.3.3 Separation of Phase and Timing Jitter

Let $u_0(z,t)$ denote a signal pulse shape, e.g., an RZ pulse or a soliton. The part of the noise that is responsible for a phase perturbation of $u_0(z,t)$ is to first order proportional to $f_\Phi = iu_0$, while the part of the noise that shifts its central time is proportional to $f_\tau = (\partial u_0/\partial t)$.

The expansion $u_0\exp(i\varphi) = u_0[1 + i\varphi + O(\varphi^2)]$ shows that the phase of u_0 is rotated by φ when adding $i\varphi u_0$. Analogously, consider $\partial u_0(t)/\partial t \approx [u_0(t+\tau) - u_0(t)]/\tau$ for small τ, implying $u_0(t) + \tau u_0'(t) = u_0(t+\tau)$. It is possible to integrate arbitrarily large perturbations: Consider the initial value problem

$$\frac{\partial u}{\partial z} = i\varphi u + q\frac{\partial u}{\partial t}, \tag{4.27a}$$

$$u(t, z = 0) = u_0(t), \tag{4.27b}$$

which is exactly solved by $u(t,z) = u_0(t + zq)\exp(i\varphi z)$. Equation 4.27a describes the optical field in an idealized transmission line without nonlinearity, dispersion, or loss, driven by noise that produces phase and timing jitter. Moreover, the functions

$$f_\varphi(t) \equiv iu_0(t), \tag{4.28a}$$

$$f_\tau(t) \equiv \frac{\partial u_0(t)}{\partial t}. \tag{4.28b}$$

are solutions of the linearized NLS (4.14b) [56], showing that both perturbations are stable to first order when propagating with an arbitrary signal u_0.

Based on these observations, we define the *pulse energy U*, the *average phase* φ, the *central time* τ, and the *central frequency* Ω of a pulse *u*(*t, z*) that is confined to the bit slot $[t_0, t_0 + T_{\text{bit}}]$ as

$$U = \int\limits_{t_0}^{t_0+T_{\text{bit}}} |u|^2 \mathrm{d}t \tag{4.29a}$$

$$\tau = \frac{1}{U} \int_{t_0}^{t_0+T_{\mathrm{bit}}} t|u|^2 \mathrm{d}t, \tag{4.29b}$$

$$\varphi = \frac{1}{U} \int_{t_0}^{t_0+T_{\mathrm{bit}}} \arg(u)|u|^2 \mathrm{d}t, \tag{4.29c}$$

$$\Omega = \frac{1}{U} \mathrm{Im} \int_{t_0}^{t_0+T_{\mathrm{bit}}} u_t^* u \mathrm{d}t, \tag{4.29d}$$

respectively, where $\arg(x) = \arctan[\mathrm{Im}(x)/\mathrm{Re}(x)]$. Moreover, the set $U/2$, τ, φ, Ω is the appropriate perturbation expansion parameter set for arbitrary pulse shapes because perturbations in those are independent of each other to first order at the point in the fiber where the perturbation occurs [56]. The pulse energy can be varied without influencing the other parameters by scaling $u(t)$, the central time can be shifted in isolation, and so forth.

Soliton perturbation theory shows that phase and timing jitter grow proportional to $z^{3/2}$ [54]. In order for the linearization assumption to remain valid over large propagation distances, one must separate the phase and timing jitter from the covariance matrix, while keeping track of their magnitude. If these contributions are not separated, they distort the distribution functions so that the noise a is no longer Gaussian distributed. Physically, small amounts of amplitude and frequency noise can lead to large amounts of phase and timing jitter respectively, which can invalidate the linearization. The nonlinear propagation equation is phase and time invariant, which implies that phase and timing jitter can be removed from the accumulated noise surrounding a pulse without affecting the pulse's subsequent evolution. While this statement is strictly valid only for isolated pulses, not for systems with multiple pulses and/or multiple channels, we have found it works well enough in practice for the linearization assumption to hold, once phase and timing jitter are properly separated from the individual pulses.

Figure 4.1 shows the distortion of the marginal distributions due to phase jitter. The impact of timing jitter on a system is similar in concept but not as easy to depict. In the left diagram, we show the vectors of one Fourier coefficient A_k of the noise-free signal u_0 in the complex plane and one of the Fourier coefficients a_k of the accumulated noise δu. The contour of constant noise probability density around A_k in the presence of weak phase jitter is in general an ellipse and the projection of the pdf onto the axes yields the marginal pdfs. If the accumulated noise δu is multivariate-Gaussian distributed, then all the marginals must be one-dimensional Gaussian pdfs [113]. However, if the phase jitter becomes large and the linearization breaks down, the jitter tends to rotate the signal around the origin. The contour of constant probability density spreads out along a circle around the origin, and it is slightly skewed since noise realizations with large phase deviations

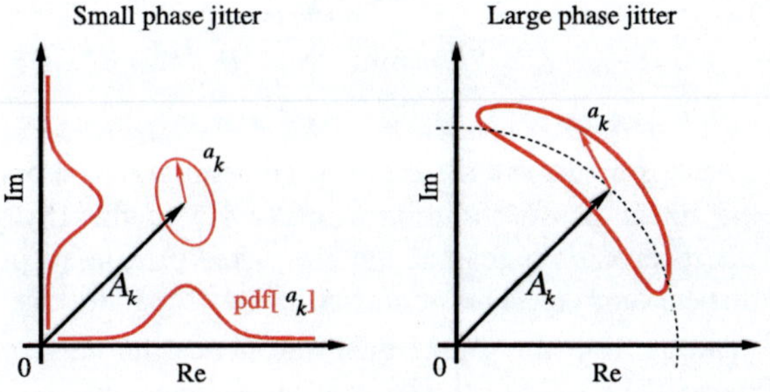

Fig. 4.1 Distortion of the marginal noise distributions due to phase jitter. The vectors are shown in the complex plane, A_k noise-free vector, a_k accumulated noise vector at frequency component ω_k

are usually caused by large amplitude deviations, resulting in a tilted, banana-shaped contour as shown in the right diagram. The projections of this contour are not Gaussian. A similar discussion applies to timing jitter.

In the DMS system that we consider in the next subsection, the timing jitter grows only linearly with distance, rather than proportionally to $z^{3/2}$, due to inline filtering in the loop [56]. Nevertheless, timing jitter can become on the order of the pulse duration once it reaches the receiver (24,000 km propagation distance) so that it is not a small perturbation anymore. Phase jitter is mainly driven by amplitude jitter, so that it is not reduced by the inline filter. We find that phase jitter has to be separated in the DMS system that we introduce in the next subsection, as well as in the much less nonlinear CRZ system that we will introduce in Sect. 4.2.4.2. Timing jitter only has to be separated in the DMS system.

In the following, we show how to separate large phase and timing jitter from the calculation of the covariance matrix by using Monte Carlo simulations. We will deal with the deterministic covariance matrix method and the incremental separation of small phase and timing jitter in the next section.

In Monte Carlo simulations, one focuses on the received optical signal $u(t) = u_0(t) + \delta u(t)$, where $u_0 = \langle u \rangle$ is the signal averaged over all noise realizations, and δu is one particular realization of the accumulated noise at the receiver. The Fourier expansion of u is $u(t) = \sum_{k=-N/2}^{N/2-1} B_k \exp(i\omega_k t)$, where $B_k = A_k + a_k$ and $\omega_k = 2\pi k T_0/T$, conforming to (4.17a and b). The phase and timing offsets φ and τ cannot be considered small anymore at the receiver, corresponding to the situation depicted in Fig. 4.1b. In fact, the jitter can become so strong that the marginal pdfs become Gaussians around zero and then the averages $A_k = \langle B_k \rangle$ tend to zero!

The approach that we use is to apply the nonlinear transformation from B_k to $\overline{B}_k$

$$\overline{B}_k \equiv B_k \exp[-i(\varphi + \omega_k \tau)] = \overline{A}_k + r_k, \tag{4.30}$$

where $\overline{A}_k \equiv \langle \overline{B}_k \rangle$ is the new signal average and $\langle r_k \rangle = 0$ is a residual noise. For single pulse transmission and for each noise realization, we determine φ and τ by fitting the linear function $\alpha + \beta\omega_k$ to the phase of the B_k using the least-squares criterion

$$H = \min_{\alpha,\beta} \sum_{k=-N/2}^{N/2-1} |B_k|^2 \left[\arg B_k - (\alpha + \beta\omega_k) \right]^2, \tag{4.31}$$

where $B_k = B_{k,R} + iB_{k,I}$, and then setting $\varphi = \alpha$ and $\tau = \beta$. We have found that the linear phase assumption of (4.31) in the DMS system is good as long as the receiver is placed at the chirp-free maximum pulse compression point of the dispersion map. This renormalization separates the large phase and timing fluctuations from the total signal and hence the new average power $|\overline{A}_k|^2 \gg |A_k|^2$ is larger than without the separation. Conversely, the average power of the residual noise $\langle \sum_k |r_k|^2 \rangle \ll \langle \sum_k |a_k|^2 \rangle$ is reduced.

One can compute the pdfs of φ and τ by averaging over all Monte Carlo noise realizations. Moreover, with the real partitioned $2N$-vector $r = (r_{-N/2,R}, \ldots, r_{N/2-1,R}, r_{-N/2,I}, \ldots, r_{N/2-1,I})^T$ and, analogously, $\overline{B} = (\overline{B}_{-N/2,R}, \ldots)^T$, one may define a reduced covariance matrix $\mathcal{K}^{(r)}$as

$$\mathcal{K}^{(r)} = \left\langle \overline{B}\overline{B}^T \right\rangle - \langle \overline{B} \rangle \langle \overline{B} \rangle^T = \langle rr^T \rangle. \tag{4.32}$$

We will use (4.32) in the next subsection. The quantity r obeys a multivariate-Gaussian distribution, and one may replace a by r and $\mathcal{K}$ by $\mathcal{K}^{(r)}$ in (4.24).

Figure 4.2 shows simulation results of the DMS system that we will treat in the next subsection. We transmit a single soliton with a peak power of 7.9 mW and a FWHM duration of 9 ps. The dots show 750 Monte Carlo noise realizations of the 10 GHz Fourier mode B_8 for $n_{sp} = 1.2$ in the complex plane. Figure 4.2a–c show B_8 without jitter separation, while Fig. 4.2d–f show B_8 after employing phase and timing jitter separation. The solid curve is a probability density contour under the assumption that the magnitude and the angles of the dots are independently Gaussian distributed. The radius of the dash–dotted circle equals $\langle |B_8| \rangle \approx |A_8|$, the average magnitude of the dots. Next, we consider the set of the angles $\{\arg(B_8)\}$. The quantity r is the ratio of the standard deviation of these angles over the standard deviation of the magnitude of the samples. When r is large, the contour assumes a banana-like or even doughnut-like shape and wraps around the unit circle. The upper row of figures shows that r grows quickly with the transmission distance, indicating strong phase jitter. However, r can be kept close to unity if we separate the phase and timing jitter. We note also that if n_{sp} or the degree of fiber nonlinearity is reduced, the value of r converges toward unity.

Figure 4.2g–i show $Ue^{i\varphi}$ for the same 750 noise realizations, using (4.29a and c). The quantity r here is the ratio of the standard deviation of φ to that of U. Again, the solid sickle-shaped contours are drawn at a constant probability density. The contours are visibly skewed, indicating a coupling between U and φ that is analogous to a similar coupling that occurs in soliton perturbation theory [72]. This skew is not visible in the upper six subfigures, since the B_k

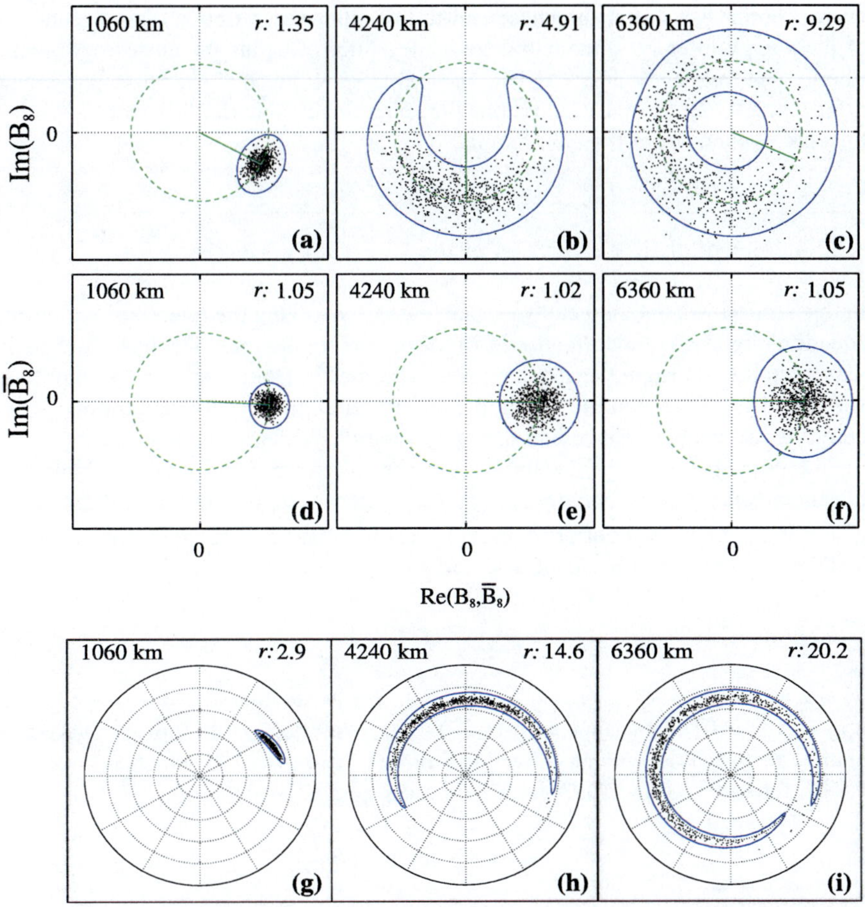

Fig. 4.2 **a** Simulation results in the DMS system introduced in the next subsection with ASE noise. The dots show 750 Monte Carlo realizations of the 10 GHz noisy Fourier mode B_8 for $n_{\mathrm{sp}} = 1.2$. **a–c** B_8 without jitter separation, **d–f** $\overline{B}_8$ with jitter separation. **g–h** $Ue^{i\varphi}$ for the same 750 noise realizations, using (4.29a and c)

only represents a small fraction of the signal energy, and hence their individual fluctuations overwhelm the skew.

After the jitter separation, the coefficients of the Fourier basis remain multivariate-Gaussian distributed far longer than the original Fourier coefficients. In principle, one should take into account the off-diagonal matrix elements of the covariance matrix that result from the interactions of the modified Fourier coefficients $\overline{B}_k$ and the phase and timing jitter. However, these elements can be ignored for reasons that we will explain later. If one assumes that soliton perturbation theory can be approximately generalized to arbitrary pulse shapes, then the noise in the Fourier basis after the phase and timing jitter separation is a representation of the energy jitter δU and central frequency jitter $\delta\Omega$, plus the noise continuum. Energy and frequency jitter can be expected to grow much slower than phase and timing jitter [72], and the simulations presented here show that it is not necessary to separate them from the continuum.

4.2.3.4 Deterministic Calculation of the Covariance Matrix

In this section, we obtain the covariance matrix by solving the linearized evolution equation directly, without the use of Monte Carlo simulations. The basic approach that we follow is to propagate the covariance matrix from amplifier to amplifier, while projecting out and separating the contribution to the phase and timing jitter, before the fluctuations have had the opportunity to accumulate significantly.

The propagation of the accumulated noise is governed by (4.14b), which is a linear equation and is homogeneous everywhere except at the amplifiers since $\hat{F} = 0$ in the fiber. Its Fourier transform must be linear and homogeneous as well, and we can write it in terms of a as in (4.21b),

$$\frac{\mathrm{da}}{\mathrm{dz}} = \mathscr{R}(z)a, \tag{4.33}$$

where we set the ASE term $w(z)$ to zero. We write the solution of (4.33) as $a(z) = \Psi(z)a(0)$, consistent with (4.22). The evolution of the noise covariance matrix $\mathscr{K} = \langle aa^T \rangle$ over one fiber leg from $z = 0$ to $z = L$, in which no noise is added, followed by an EDFA with the power gain G is given by

$$\mathscr{K}(L) = G\Psi\mathscr{K}(0)\Psi^T + \eta\frac{T_0}{T}\mathscr{I}, \tag{4.34}$$

where Ψ is a propagator matrix, $\mathscr{I}$ is the identity matrix, and η is defined in (4.8). By successive application of (4.34), one can propagate the covariance matrix from amplifier to amplifier.

We choose a perturbative method to compute Ψ. Let $u_0(t,0)$ and $u_0(t,L)$ be the noise-free optical field at the beginning and end of the fiber span, respectively. Then we perturb $u_0(t,0)$ in a single frequency mode m by a small amount ε and launch the perturbed signal

$$u^{(m)}(t,0) = u_0(t,0) + \varepsilon\exp(i\omega_m t). \tag{4.35}$$

At $z = L$, we obtain $u^{(m)}(t,L)$ by solving the nonlinear transmission equation (4.14a) and calculate the deviation $\delta u^{(m)}(t) = u^{(m)}(t,L) - u_0(t,L)$ and its Fourier space vector $a^{(m)}$. The elements of Ψ are given by

$$\Psi_{km} = \frac{a_k^{(m)}}{\varepsilon}. \tag{4.36}$$

This perturbative approach corresponds to the Lyapunov method described by Bennetin et al. [14]. Note that ϵ must be a positive real number if $1 \le k \le N$ and a purely imaginary positive number if $N+1 \le k \le 2N$, since Ψ is defined in the $2N$-dimensional space of partitioned vectors. The real and imaginary perturbations must be computed separately since the propagation equation of the noise 17 is not linear when δu is scaled by an imaginary number due to the non-Hermitian FWM term $u_0^2(\delta u)^*$.

When carrying out Monte Carlo simulations as described in the previous section, the phase and timing jitter are large by the time they reach the receiver. By contrast, when propagating $\mathscr{K}$, one can separate the jitter incrementally, on a scale that is small compared to the nonlinear scale length of the system. We do that at every amplifier. Following the discussion of the previous section, we consider the functions $f_\varphi(t) = iu_0(t)$ and $f_\tau(t) = \partial u_0(t)/\partial t$, defined in (4.28a and b). Each pulse in the signal has a different average phase φ (4.29c) and central time τ (4.29b). In a system in which pulses do not overlap, such as a soliton system, φ and τ evolve independently for each pulse and must be removed separately from each other. To deal with non-overlapping pulses, we decompose the functions $f_\varphi(t)$, $f_\tau(t)$, and $\delta u^{(m)}(t)$ into sums of mutually orthogonal pulse functions, i.e., $h(t) = \sum_l h_l(t)$, where $h_l(t) = h(t)$ for $(l-1)T_{\text{bit}} \le t < lT_{\text{bit}}$, and $h_l(t) = 0$ otherwise for an arbitrary function h. We later discuss how to generalize this decomposition to the case where pulses overlap during the transmission.

The next step is to orthogonalize the functional basis $\left[f_{\varphi,l}(t), f_{\tau,l}(t), \delta u_l^{(m)}(t)\right]$ analogous to [56]. The goal is to obtain a new set of basis vectors $\left[f_{\varphi,l}(t), \widetilde{f}_{\tau,l}(t), \widetilde{\delta u}_l^{(m)}(t)\right]$ so that $(f_{\varphi,l}, \widetilde{f}_{\tau,l}) = (\widetilde{f}_{\tau,l}, \widetilde{\delta u}_l^{(m)}) = (f_{\varphi,l}, \widetilde{\delta u}_l^{(m)}) = 0$, where $(g,h) = \operatorname{Re} \int_{-\infty}^{\infty} g(t)h^*(t)\mathrm{d}t$ denotes the scalar product of two functions g and h. We apply the Gram–Schmidt orthogonalization procedure [133]

$$\widetilde{f}_{\tau,l} = f_{\tau,l} - \frac{(f_{\tau,l}, f_{\varphi,l})}{(f_{\varphi,l}, f_{\varphi,l})} f_{\varphi,l}, \tag{4.37a}$$

$$\widetilde{\delta u}_l^{(m)} = \delta u_l^{(m)} - \frac{(\delta u_l^{(m)}, \widetilde{f}_{\tau,l})}{(\widetilde{f}_{\tau,l}, \widetilde{f}_{\tau,l})} \widetilde{f}_{\tau,l} - \frac{(\delta u_l^{(m)}, f_{\varphi,l})}{(f_{\varphi,l}, f_{\varphi,l})} f_{\varphi,l}. \tag{4.37b}$$

The Fourier decomposition vector $r^{(m)}$ of the reduced noise $\widetilde{\delta u}^{(m)} = \sum_l \widetilde{\delta u}_l^{(m)}$ then replaces $a^{(m)}$ in (4.36), and the covariance matrix $\mathscr{K}$ in (4.34) is reduced to $\mathscr{K}^{(r)}$ analogous to (4.32). Note that the functions $f_{\varphi,l}$ and $f_{\tau,l}$ are not orthogonal for asymmetric pulses; therefore, (4.37a) is not redundant in general. Even though $f_{\varphi,l}(t)$ and $\widetilde{\delta u}_l^{(m)}$ are orthogonal, the quantities φ and $\mathscr{K}^{(r)}$ are not statistically independent because φ depends on the pulse power due to the nonlinear phase rotation. A noise realization in which the noise increases the pulse power will tend to have both large $|\widetilde{\delta u}^{(m)}|$ and large $|\varphi|$, leading to a correlation. However, these correlations, such as the phase jitter itself, have no effect on a receiver with a square-law detector and can be neglected. There is a similar correlation between τ and $|\widetilde{\delta u}^{(m)}|$, and, in contrast to the phase jitter, the timing jitter cannot simply be ignored. However, we will show in Covariance Matrix Computation that the correlations between τ and the modified Fourier components has a negligible effect on the receiver current, so that these correlations do not influence the pdf of the receiver current.

We briefly return to the problem of overlapping pulses. In most modern transmission systems, optical pulses overlap during the transmission even though they are well-separated when they are launched and detected. In the CRZ system introduced in Sect. 4.2.4.2, the maximum FWHM pulse duration is 210 ps, leading to a significant overlap of adjacent pulses. In this case, the phase jitter can still be removed separately for each pulse after applying artificial dispersion compensation. One passes the functions $\{f_{\varphi,l}(t), f_{\tau,l}(t), \delta u_l^{(m)}(t)\}$ through an ideal linear and lossless fiber whose total dispersion is $-D(L)$, where $D(L)$ is the total accumulated dispersion at the point L in the transmission system. This procedure separates the pulses, Since it is linear, it is fully reversible [104]. Then, one applies the orthogonalization (4.37a and b) to the separated pulses and sends the signal corresponding to $\widetilde{\delta u}_l^{(m)}$ back to the point L through an ideal fiber with total dispersion $+D(L)$. One might argue that the phases of the overlapped pulses will not evolve independently and hence might become correlated; however, we find by comparison to Monte Carlo simulations that the procedure just described leads to accurate BERs in the CRZ system, indicating that the coupling of the signal in one pulse to the phase jitter in an overlapping pulse is negligible.

We will summarize the algorithm for propagating $\mathcal{K}$ in Sect. 4.2.5.2. Propagating the covariance matrix requires the propagation of u_0, plus the $2N$ perturbed fields. The total computational cost of this algorithm is thus roughly $2N + 1$ times that of a single Monte Carlo noise realization, plus the time required for the matrix multiplications in (4.34). Since matrix multiplications scale with the cube of the dimension N^3, there is a practical limit to the size of N. We will demonstrate in Sect. 4.2.4.2 that the solution of a problem with $N = 140$ is computationally feasible on a Pentium P4 workstation. We find that the perturbation method is numerically stable and its result is independent of the value of ε over several orders of magnitude.

4.2.3.5 Alternative Numerical Solution Methods

As mentioned in the previous subsection, the perturbative solution approach requires about the same CPU time as a Monte Carlo simulation with $2N$ noise realizations, while providing an accuracy that is in principle equivalent to running an infinite number of realizations. However, the question arises whether other, more efficient numerical solution methods exist.

We have attempted a direct solution of (4.26) using the general purpose matrix ODE solver package CVODE [38], but the solution was numerically inefficient.

When confronted with the matrix propagation equation (4.26), derived from the vector equation (4.21a), one may be tempted to consider a solution based on the matrix exponential of $\mathscr{R}$, which is known as a stable and efficient method. However, the matrix $\mathscr{R}(z)$ depends on the propagation distance due to the usually rapid signal evolution along each dispersion span, and, to a lesser extent, due to attenuation. Moreover, the most efficient exponential matrix package Expokit [123] does not compute the matrix $\exp(L\mathscr{R})$ explicitly, but only the vector $a(L) = \exp(L\mathscr{R})a(0)$, implying that the solver routine must be called $2N$ times, which likely results in poor efficiency. Finally, evaluating the convolution sums in (4.20a and b) is computationally costly in the frequency domain.

Coelho et al. [37] have simplified (4.21a) for the case of a cw signal that only varies due to attenuation, in which case (4.21a) and therefore (4.26) simplify significantly. In the receiver, the actual modulated noise-free transmission signal can still be introduced. This method is very fast and Coelho reports good agreement with a laboratory experiment of a mildly nonlinear RZ-DPSK (differential phase shift keying) signal, but obviously it cannot model the nonlinear coupling between the finite bandwidth signal spectrum and the ASE noise (Coelho does not separate jitter momenta, but his method could probably be extended without much difficulty).

Demir [39] has provided a brilliant summary of developments in calculating nonlinear optical phase noise and BERs until the year 2007. Moreover, he has devised an efficient solution of (4.21a) using the method of integrating factors (Eq. (4.25) in [39]) that removes the numerical stiffness of the ODE by transforming the dispersion terms. However, no CPU time requirement is given, so we cannot tell if Demir's method beats the perturbative approach.

4.2.3.6 Derivation of the Eye Diagram

In this section, we derive the pdf of the filtered output current of a square-law detector. Similar pdfs have already been derived by Marcuse [96], Lee et al. [85], Bosco et al. [22], and Forestieri [43]. One can use this pdf to compute an electrical eye diagram that displays a continuous probability density rather than, as is traditional in simulations, overlaying a finite number of traces of marks and spaces with different noise realizations. The following derivation is valid for both the Monte Carlo method of Sect. 4.2.3.3 and the ODE method of Sect. 4.2.3.4. The inputs are the pdf of the timing jitter p_τ, the Fourier modes $\bar{A}_k$ of the renormalized noise-free signal, and the reduced covariance matrix $\mathscr{K}^{(r)}$. Below, we will drop the superscript and just write $\mathscr{K}$ for convenience.

The photodetector in the receiver converts the optical input signal plus noise to an electrical current $I(t)$. One may assume that the photodetector is an ideal square-law detector with $I = \kappa|u|^2$, where κ is the receiver responsivity. We apply the transformation (4.30) and hence start by describing $I(t)$ in the absence of timing jitter. The electrical current is

$$
\begin{aligned}
I(t) &= \kappa|u_0(t)+\delta u(t)|^2 \\
&= \kappa\sum_{k=-N/2}^{N/2-1}\sum_{l=-N/2}^{N/2-1}(\overline{A}_k+r_k)^*(\overline{A}_l+r_l)\exp[it(\omega_l-\omega_k)] \\
&= \kappa\sum_{n=-N+1}^{N-1}\exp(it\omega_n)\sum_{k=k_1(n)}^{k_2(n)}(\overline{A}_k+r_k)^*(\overline{A}_{n+k}+r_{n+k}),
\end{aligned}
\tag{4.38}
$$

where r_k, and r_l are the residual noise coefficients, $n = l - k$, $k_1(n) = \max(-N/2, -N/2 - n)$, and $k_2(n) = \min(N/2 - 1, N/2 - 1 - n)$. The current $I(t)$ then passes through a low-pass electrical filter. The last line in (4.38) represents the Fourier decomposition of $I(t)$ in the *electrical* domain, whose $2N - 1$ coefficients are given by the sum over k. Optical and electrical filtering can be introduced by multiplying the respective Fourier components by filter functions H^{opt} and H^{el}, respectively. The combined filtering operation yields the filtered voltage $y(t)$

$$
\begin{aligned}
y(t) &= \kappa\sum_{k=-N/2}^{N/2-1}\sum_{l=-N/2}^{N/2-1}H_k^{\text{opt},*}(\overline{A}_k+r_k)^*H_l^{\text{opt}}(\overline{A}_l+r_l)H_{l-k}^{\text{el}}\exp[it(\omega_l-\omega_k)] \\
&= \kappa\sum_{n=-N+1}^{N-1}H_n^{\text{el}}\exp(it\omega_n)\sum_{k=k_1(n)}^{k_2(n)}H_k^{\text{opt},*}(\overline{A}_k+r_k)^*H_{n+k}^{\text{opt}}(\overline{A}_{n+k}+r_{n+k}).
\end{aligned}
\tag{4.39}
$$

The frequency term and all three filter terms can be combined into a complex $N \times N$ matrix

$$
\mathsf{W}_{kl}(t) = \kappa H_k^{\text{opt},*}H_l^{\text{opt}}H_{l-k}^{\text{el}}\exp(it\omega_{l-k}). \tag{4.40}
$$

One can write the filtered current $y(t)$ more compactly as

$$
y(t) = \sum_{k,l=-N/2}^{N/2-1}(\overline{A}_k+r_k)^*\mathsf{W}_{kl}(t)(\overline{A}_l+r_l). \tag{4.41}
$$

Since y is a real quantity, W must be Hermitian. We introduce the real partitioned vector $\overline{A} = (\overline{A}_{-N/2,R}, \ldots, \overline{A}_{N/2-1,R}, \overline{A}_{-N/2,I}, \ldots, \overline{A}_{N/2-1,I})^T$ to rewrite $y(t)$ as

$$
y(t) = (\overline{A}+r)^T\mathscr{W}(t)(\overline{A}+r), \tag{4.42a}
$$

$$
\mathscr{W} = \begin{bmatrix}\mathsf{W}_R & -\mathsf{W}_I \\ \mathsf{W}_I & \mathsf{W}_R\end{bmatrix} = \mathscr{W}^T. \tag{4.42b}
$$

The minus sign in the matrix in (4.42b) appears because W_I is anti-symmetric. The right-hand side of (4.42a) is a symmetric bilinear form, but, due to filtering, it is not necessarily positive. The receiver current $y_{\text{nf}}(t)$ in the absence of noise at time t is

$$y_{\mathrm{nf}}(t) \equiv \bar{A}^T \mathscr{W}(t) \bar{A}. \tag{4.43}$$

In order to obtain an eye diagram, we must derive $p_y(y,t)$, the pdf of the current y at time t. The derivation of p_y is a generalization of Marcuse's [96] in that we consider all noise correlations and allow for arbitrary optical and electrical filtering. By contrast, Marcuse assumes optical white noise at the receiver and an integrate-and-dump circuit. The procedure of computing p_y starts with computing the Karhunen–Loève modes, the basis in which (4.42a) can be decomposed into a sum of independent random variables. In a second step, we calculate the pdf $p_{y,\tau=0}(y,t)$, where the subscript $\tau = 0$ means the exclusion of the effect of the timing jitter, as the convolution of the individual pdfs of these random variables. We solve the convolution by multiplying characteristic functions. In a third step, we add the effect of the timing jitter by convolving $p_{y,\tau=0}$ with the pdf of the timing jitter p_τ.

One must first find the functional basis that diagonalizes both the square-law detection followed by the filtering, $\mathscr{W}(t)$, and the inverse covariance matrix $\mathscr{K}^{-1}$. Since both $\mathscr{W}(t)$ and $\mathscr{K}^{-1}$ are symmetric matrices and in addition $\mathscr{K}^{-1}$ is positive definite, one can apply the theorem of simultaneous diagonalization [45] which states that there is a real-square matrix $\mathscr{C}$ satisfying

$$\mathscr{K}^{-1} = \mathscr{C}^T\mathscr{C}, \qquad \text{and} \tag{4.44a}$$

$$\mathscr{W} = \mathscr{C}^T \Lambda \mathscr{C}. \tag{4.44b}$$

One procedure to obtain $\Lambda(t)$ and $\mathscr{C}(t)$ is to solve the generalized eigenproblem $\mathscr{W}\mathscr{C}^{-1} = \mathscr{K}^{-1}\mathscr{C}^{-1}\Lambda$. The solution can be performed on the computer, using a generalized eigenvalue routine such as the routine `eig()` in Matlab™, or the procedure outlined in Simultaneous Diagonalization. The matrix $\Lambda(t)$ is diagonal, and we write it as $\Lambda = \mathrm{diag}(\lambda_1, \ldots \lambda_{2N})$, where the $\lambda_k(t)$ is real [45]. Note that if the impulse response of the filter can become negative, as in the case of a Bessel filter, some of the λ_k are negative. The transformation $\mathscr{C}$ yields the Karhunen–Loève modes of y, which are the noise-free signal modes $Q_k(t) \equiv \mathscr{C}_{kl}(t)\bar{A}_l$ and the independent noise modes $q_k(t) \equiv \mathscr{C}_{kl}(t) r_l$. We simplify (4.42a) to

$$\begin{aligned} y(t) &= \left(\bar{A} + r\right)^T \mathscr{C}^T \Lambda \mathscr{C} \left(\bar{A} + r\right) \\ &= \left(Q + q\right)^T \Lambda \left(Q + q\right) \\ &= \sum_{k=1}^{2N} \lambda_k \left(Q_k^2 + 2Q_k q_k + q_k^2\right) = \sum_{k=1}^{2N} g_k, \end{aligned} \tag{4.45}$$

where the $g_k(q_k,t) \equiv \lambda_k\left(Q_k^2 + 2Q_k q_k + q_k^2\right)$ represent a new set of independent random variables.

The noise pdf p_r (4.24), where we replaced a by r, can be factored into independent Gaussian pdfs, using (4.44a)

$$
\begin{aligned}
p_r(r) &= (2\pi)^{-N}\sqrt{\det(\mathscr{K}^{-1})}\exp\left[-\frac{1}{2}r^T\mathscr{C}^T\mathscr{C}r\right] \\
&= (2\pi)^{-N}\sqrt{\det(\mathscr{K}^{-1})}\exp\left[-\frac{1}{2}\sum_{k=1}^{2N}q_k^2\right] \\
&= \sqrt{\det(\mathscr{K}^{-1})}\prod_{k=1}^{2N}f_{q_k}(q_k),
\end{aligned} \tag{4.46}
$$

where the marginal pdfs $p_{q_k}(q_k) \equiv \exp(-q_k^2/2)/\sqrt{2\pi}$ are Gaussians with zero mean and unit variance. The factor $\sqrt{\det(\mathscr{K}^{-1})} = \det\mathscr{C}$ does not appear in the definition of the p_{q_k} since $\det\mathscr{C}\mathrm{d}r = \mathrm{d}q_k$. Equation 4.45 is a sum of $2N$ random variables. The pdf $p_{y,\tau=0}(y,t)$ therefore equals the (2N-1)-dimensional convolution [113]

$$
\begin{aligned}
p_{y,\tau=0}(y,t) = &\int_{-\infty}^{\infty} p_{g_1}(y - g_2 - g_3 - \cdots - g_{2N}, t) \\
&\times p_{g_2}(g_2,t)p_{g_3}(g_3,t)\ldots p_{g_{2N}}(g_{2N},t)\mathrm{d}g_2\mathrm{d}g_3\ldots\mathrm{d}g_{2N},
\end{aligned} \tag{4.47}
$$

but this convolution can be transformed into simple multiplications using characteristic functions, taking advantage of the convolution theorem [119]. The characteristic function $\Phi_h(\zeta)$ of a random variable h is defined as the expectation $\Phi_h(\zeta) \equiv E[\exp(i\zeta h)] = \int_{-\infty}^{\infty}\exp(i\zeta h)p_h \mathrm{d}h$ [113]. With the help of the derived distribution identity $p_{g_k}\mathrm{d}g_k = p_{q_k}\mathrm{d}q_k$ [113], one can write

$$
\begin{aligned}
\Phi_{g_k}(\zeta) &= \int_{-\infty}^{\infty} \exp\left[i\zeta g_k(q_k)\right]p_{q_k}\mathrm{d}q_k \\
&= \frac{1}{\sqrt{2\pi}}\int_{-\infty}^{\infty} \exp\left[-\frac{q_k^2}{2} + i\zeta\lambda_k(Q_k^2 + 2Q_kq_k + q_k^2)\right]\mathrm{d}q_k \\
&= \frac{1}{\sqrt{1-2i\lambda_k\zeta}}\exp\left(\frac{i\lambda_kQ_k^2\zeta}{1-2i\lambda_k\zeta}\right).
\end{aligned} \tag{4.48}
$$

Again, note that the integration variables are all real, and that the entire analysis so far neglects the timing jitter. The characteristic function $\Phi_{y,\tau=0}(\zeta,t)$ of $p_{t,\tau=0}(y,t)$ equals the product of the Φ_{g_k}

$$
\Phi_{y,\tau=0} = \prod_{k=1}^{2N}\Phi_{g_k} = \left(\prod_{k=1}^{2N}\frac{1}{\sqrt{1-2i\lambda_k\zeta}}\right)\exp\left(i\zeta\sum_{k=1}^{2N}\frac{\lambda_kQ_k^2}{1-2i\lambda_k\zeta}\right). \tag{4.49}
$$

Note the identity $y_{\mathrm{nf}} = \sum_{k=1}^{2N}\lambda_k(t)Q_k^2(t)$. Additive noise sources such as electrical noise can in principle be accounted for by multiplying $\Phi_{y,\tau=0}$ with the appropriate characteristic functions [85]. From $\Phi_{y,\tau=0}$ one obtains $p_{y,\tau=0}$ [113]

$$p_{y,\tau=0}(y,t) = \frac{1}{2\pi} \int_{-\infty}^{\infty} \Phi_{y,\tau=0}(\zeta,t) \exp(-iy\zeta) d\zeta. \tag{4.50}$$

Due to the complicated dependence of $\Phi_{y,\tau=0}(\zeta,t)$ on ζ, it is not possible to evaluate the Fourier transform in (4.50) analytically. However, $p_{y,\tau=0}(y,t)$ can be computed numerically using a discrete Fourier transform.

If one sets all λ_k to the same positive value $\lambda_k = \sigma^2$, then $\Phi_{y,\tau=0}$ equals the characteristic function of a noncentral chi-square distribution [95, Eq. 20], [116, Eq. 2-1-119], and its inverse Fourier transform, the corresponding pdf, is

$$p_{\chi^2}(y) = \frac{1}{2\sigma^2} \left(\frac{y}{y_{\text{nf}}}\right)^{(n-2)/4} \exp\left(-\frac{y+y_{\text{nf}}}{2\sigma^2}\right) I_{n/2-1}\left(\frac{\sqrt{y y_{\text{nf}}}}{\sigma^2}\right), \qquad y \geq 0, \tag{4.51}$$

where I_l is the lth order modified Bessel function of the first kind [6]. This function depends on three parameters: the noncentrality parameter y_{nf}, σ^2, and the degree of freedom $n = 2N$. The pdf $p_{\chi^2}(y)$ is the pdf of the squared sum y of n independent Gaussian-distributed random variables with the means $m_k = \sigma Q_k$ and identical variances $\sigma^2 = \lambda_k$. In this sense, (4.49) is the characteristic function of a chi-square distribution generalized to unequal and possibly negative "variances" λ_k. The reason why the λ_k also appears in the means $m_k = \sigma Q_k = \sqrt{\lambda_k} Q_k$ is due to our choice of diagonalizing $\mathcal{K}^{-1}$ and $\mathcal{W}$ according to (4.44a and b). An alternative would be the diagonalization $\mathcal{K}^{-1} = \mathcal{C}^T \Lambda \mathcal{C}$ and $\mathcal{W} = \mathcal{C}^T \mathcal{C}$, but this choice requires a matrix $\mathcal{W}$ that is positive definite, implying electrical filters whose impulse response is positive.

We note that the mean of $p_{\chi^2}(y)$ is $\langle y \rangle = y_{\text{nf}} + n\sigma^2$, where the term $n\sigma^2$ is the expected current from the noise–noise terms in the receiver, its variance is $\sigma_y^2 = 4\sigma^2 y_{\text{nf}} + 2n\sigma^4$, where the terms are due to signal–noise and noise–noise beating in the receiver, respectively [116], and the third-order central moment $\left\langle (y - \langle y \rangle y)^3 \right\rangle$ equals $24\sigma^4 y_{\text{nf}} + 8n\sigma^6$. The limit of (4.51) for $y_{\text{nf}} \longrightarrow 0$ is the central chi-square pdf [116, Eq. 2-1-112]

$$p_{\chi^2}(y)\Big|_{y_{\text{nf}}=0} = \frac{1}{\sigma^2 2^{n/2} \Gamma(n/2)} y^{(n/2-1)} \exp\left(-\frac{y}{2\sigma^2}\right). \tag{4.52}$$

The relationship of the final pdf p_y that includes timing jitter with $p_{y,\tau=0}$ is

$$p_y(y,t) = \int_{-\infty}^{\infty} p_{y,\tau=0}(y, t-\tau') p_\tau(\tau') \mathrm{d}\tau', \tag{4.53}$$

where p_τ is the pdf of the timing jitter. We will show examples of p_τ and $p_y(y,t)$ in the next two subsections. Because y is phase-independent, the phase variation does not contribute to $\Phi_{y,\tau=0}$. Note that the integral in (4.53) is a convolution with respect to τ and could be expressed in terms of characteristic functions; however,

$\Phi_{y,\tau=0}$ is a characteristic function with respect to y, and one cannot use it to simplify (4.53). Since the timing jitter is Gaussian distributed, we use a numerical Gauss–Hermite integration [6] to solve (4.53). Although the λ_k is not identical in the simulations, $p_y(y)$ can sometimes be approximated by the chi-square pdf $p_{\chi^2}(y)$ with the parameters $\sigma^2 = \sigma_y^2/2(\langle y\rangle + y_{\mathrm{nf}})$ and $n = 2(\langle y\rangle^2 - y_{\mathrm{nf}}^2)/\sigma_y^2$.

Equation 4.53 is only valid if τ and r are uncorrelated. For example, if the average pulse shape in a transmission system depends on τ, then τ would be correlated with the residual noise r, and τ would have to be incorporated into an extended covariance matrix and undergo the diagonalization procedure in order to decouple it from the noise continuum. We discuss this issue further in Covariance Matrix Computation. We have found that these correlations can be safely neglected in optical transmission systems.

4.2.3.7 Coherent Detection

When the covariance matrix method was derived in the years 1999–2003, virtually all optical transmission systems used direct detection and thus neglected the optical phase. In the meantime, industry has largely embraced phase modulation formats. First, the differential detection techniques based on the differential binary phase shift-keying (DBPSK) and differential quadrature phase shift-keying (DQPSK) formats have been employed as they enable better signal-to-noise ratios, higher local dispersion, and PMD tolerance [24, 50]. Second, thanks to high-speed analog-to-digital conversion and signal processing, truly coherent polarization-multiplexed systems using BPSK, QPSK, and quadrature amplitude modulation (QAM) formats found their way into commercial systems enabling significantly higher spectral efficiencies and dispersion, as well as PMD tolerance [136].

Coelho et al. [37] apply the covariance matrix method to DBPSK detection and find that the resulting formalism for the statistic of the decision variable is very similar to that described above for direct detection, although the definition of the matrix **W** is of course completely different (Coelho's eqation (38) is equivalent to (4.42a) and his Eq. (39) to (4.44b); note that Coelho's matrices **K** and **W** correspond to our $\mathscr{W}$ and $\mathscr{C}$, respectively).

We therefore conclude that the accurate covariance matrix method is fully applicable to coherent detection. As stated before, phase jitter must be separated from the covariance matrix, but in contrast to direct detection receivers, it must be re-introduced at the coherent receiver.

4.2.3.8 Pattern Dependences

So far, the entire treatment of noise linearization in this subsection has dealt with the signal–noise interactions in the optical fiber. The previous sections show how an accurate pdf can be computed for a given bit sequence. The function $p_y(y,t)$ yields the probability density of receiving the filtered current y at time t in the presence of the noise-free signal $u_0(t)$. However, there are other physical effects in optical fiber systems besides ASE noise that lead to fluctuations in the current y. One example is polarization mode dispersion (PMD), which we do not discuss in this subsection. A second example is phase noise that is introduced by the transmitter. Another example is the interaction of neighboring bits. In single-channel systems, neighboring pulses can overlap and interact nonlinearly [95, 5]. This interaction leads to a distortion of the pulse chirp, which in turn causes the pulses to walk off, inducing timing jitter and amplitude jitter [63]. We will show an example of signal distortion in a quasilinear CRZ system in Sect. 4.2.4. Signal-signal interactions are deterministic and can easily be computed for a given signal using a numerical simulation. The bit pattern of $u_0(t)$ has an important influence on the signal-signal interactions and we elaborate on this issue in the following. We restrict the discussion to single-channel systems. In WDM systems, all channels interact, leading to more complicated interactions. However, these interactions are independent of ASE noise and can in principle be treated separately.

The existence of pattern dependences affects the noise distributions. There is no longer a single pdf for the marks and a single pdf for the spaces. Instead, there is a different conditional pdf for each possible mark and each possible space in the system. The total pdfs for the marks and the spaces are determined by summing the conditional probabilities. As the number of channel in a WDM system grows, the number of possible patterns grows explosively, making a calculation of the complete pdfs impractical. A complete resolution of this problem is beyond the scope of this subsection. However, work to date indicates that it is sufficient to focus on some number of the worst patterns.

A computational eye diagram is traditionally produced by overlaying the signal traces in all the bit slots in a given channel. We introduce the *average pdf* $p_{y,\text{eye}}(y, t_{\text{eye}})$ for the current y in the eye diagram in the time range $0 \le t_{\text{eye}} \le T_{\text{bit}}$. This pdf depends on $u_0(t)$ and on $p_y(y,t)$ at all points in time $t_k = t_{\text{eye}} + kT_{\text{bit}}$ that are overlaid in the eye diagram. One then obtains

$$p_{y,\text{eye}}(y, t_{\text{eye}}) = \frac{1}{n}\sum_{k=1}^{n} p_y(y, t_{\text{eye}} + kT_{\text{bit}}), \tag{4.54}$$

where $n = T/T_{\text{bit}}$ is the number of bits in the eye diagram. We will call the $p_y(y, t_{\text{eye}} + kT_{\text{bit}})$ *partial pdfs*. Equation 4.54 can be refined by dividing $p_{y,\text{eye}}(y, t_{\text{eye}})$ into two pdfs $p_{y,\text{eye}}(y, t_{\text{eye}}) = \left[p_{y,\text{eye},0}(y, t_{\text{eye}}) + p_{y,\text{eye},1}(y, t_{\text{eye}})\right]/2$, one containing all the spaces (0's) and the other containing the marks (1's). Each of the pdfs $p_{y,\text{eye},0}(y, t_{\text{eye}})$ and $p_{y,\text{eye},1}(y, t_{\text{eye}})$ at $t_{\text{eye}} \approx T_{\text{bit}}/2$ near the center of the eye usually consists of very similar partial pdfs. In other words, $p_y(y,t)$ for a given $u_0(t)$ mainly depends on the noise-free current y_{nf} at time t, defined in (4.43). To make this more concrete, we conjecture that if $a = y_{\text{nf}}(t_2)/y_{\text{nf}}(t_1)$ is the ratio of the noise-free currents at the times t_1 and t_2, the approximation

$$p_y(y, t_2) \approx a p_y(ay, t_1) \tag{4.55}$$

holds in the case of $a \approx 1$. Note that $\int_{-\infty}^{\infty} p_y(y', t_2)dy' = \int_{-\infty}^{\infty} a p_y(ay', t_1)dy' = 1$.

4.2.4 Covariance Matrix Method: Results

In this subsection, we apply the deterministic method described in Sect. 4.2.3.4 to propagate the noise covariance matrix, rather than running Monte Carlo simulations. We consider two different systems: a DMS system with the reduced path average dispersion of 0.02 ps/nm-km and a CRZ system with a total transmission distance of 6,100 km.

4.2.4.1 Dispersion-Managed Soliton System

System Setup

We simulated a DMS system with a transmission distance of 24,000 km. This system is very well-characterized both in simulation and in experiment [31, 51, 103, 147]. The system does not operate in the quasilinear DMS regime [84], but at an optical peak power of about 8 mW, which renders the transmission significantly nonlinear. The evolution of the pulse shape is approximately periodic with the period given by the dispersion map. We verified that the phase jitter obeys a Jacobi-Θ distribution, which is the periodic analog of a Gaussian distribution, and we also verified that the timing jitter is Gaussian distributed. Finally, we verified that the real and imaginary parts of the residual noise Fourier coefficients, after the jitter is separated, are Gaussian distributed.

The simulated transmission line is shown in Fig. 4.3 and consists of 225 periods of a dispersion map of length 106.7 km [103]. Each map contains a fiber span of length 4×25 km long with a normal dispersion of -1.03 ps/nm-km and a span of length 6.7 km with an anomalous dispersion of 16.7 ps/nm-km, denoted by the circles labeled N and A, respectively. The path average dispersion equals 0.08 ps/nm-km, which is larger than in [103]. Third-order dispersion is not relevant in this system [103] and is set to zero. The carrier wavelength is 1551.49 nm, matching the experimental value. The fiber loss is compensated by five EDFAs. One EDFA follows each of the four 25-km segments of normal-dispersion fiber, and the fifth follows the segment of anomalous-dispersion fiber. There is a 2.8 nm (350 GHz) optical bandpass filter (OBF) in each map period to reduce the amount of noise.

The receiver is modeled as an ideal square-law detector followed by an electrical low-pass fifth-order Bessel filter with a one-sided 3-dB bandwidth of 4.3 GHz. This bandwidth is much smaller than the commonly used bandwidth of 70–80% of the data rate, but it was shown to be advantageous in this experiment to suppress the effects of timing jitter [103] (Fig. 4.4).

We use the split-step Fourier method to solve the scalar NLS, which only takes into account one optical polarization. In the recirculating loop that we are modeling, the polarization-dependent loss (PDL) is large and the polarization controllers are optimized to pass the signal with minimum loss. Consequently, the signal is dominated by one polarization, and the orthogonal polarization can be neglected. The nonlinear propagation equation in all simulations in this section is solved using a third-order split-step algorithm [129] (Fig. 4.5).

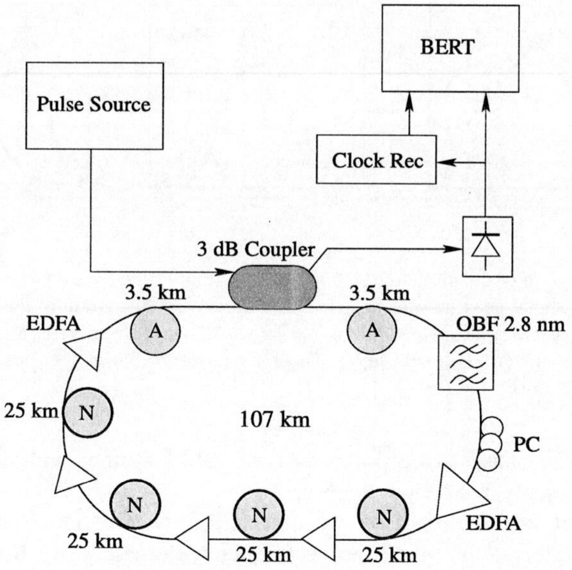

Fig. 4.3 Cartoon of the DMS system. PC denotes the manual polarization controller, BERT the bit error rate tester, ClockRec the clock recovery circuit, and OBF the optical bandpass filter

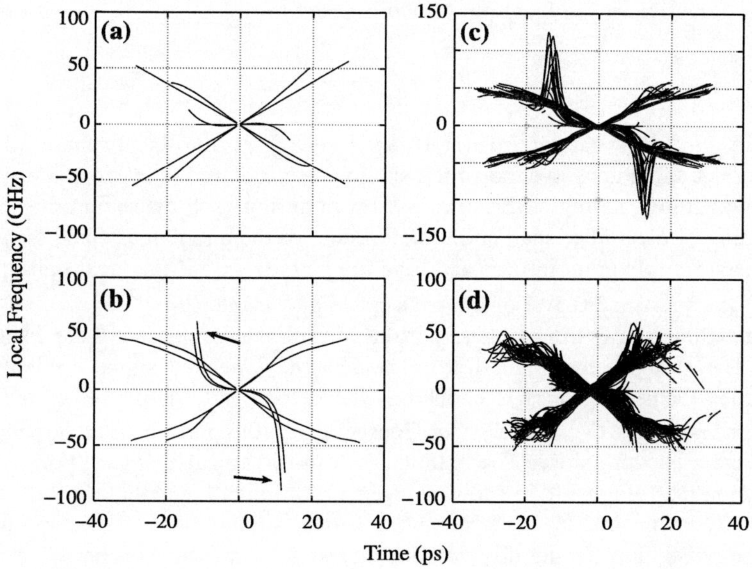

Fig. 4.4 Phase space portraits for the DMS system at the beginning of each fiber span. **a** First map period; **b** Last map period, *arrows* middle of the anomalous span; **c** All 225 periods; **d** Same as (**b**), but with added ASE noise

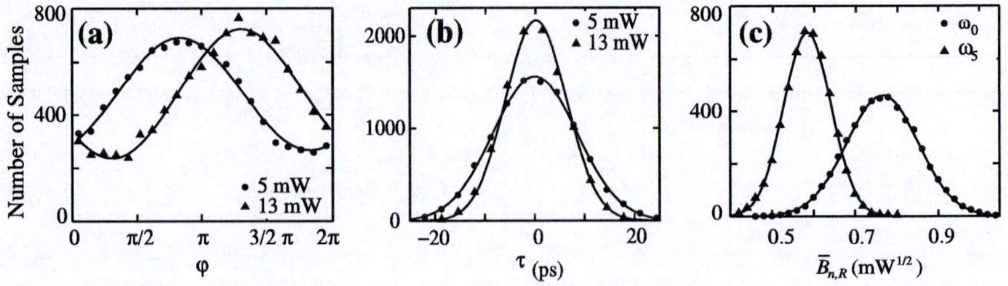

Fig. 4.5 **a** Histogram of the phase offset φ and **b** histogram of the time offset τ for two different simulations with signal peak powers $P_{\text{peak}} = 5$ and 13 mW. **c** Histogram of the two real Fourier coefficients $\bar{B}_{0,R}$ and $\bar{B}_{5,R}$ at $\omega_0 = 0$ and $\omega_5 = 2\pi \times 25$ GHz, respectively, after the phase and time offsets are removed ($P_{\text{peak}} = 5\,\text{mW}$). The *solid lines* are fits of the Jacobi-Θ function in (**a**) and Gaussians in (**b**) and (**c**)

We present a phase space portrait of the DMS system with the purpose of demonstrating some characteristic properties of a periodically stationary DMS system and for later comparison with the CRZ system. Phase space portraits contain more information than time- or frequency domain plots; for instance they are a good way of visualizing chirp. Fig. 4.4 shows four different phase portraits, by which we mean trajectories in the space of local frequency versus time. Related plots have been produced experimentally for various systems, using a technique called Frequency Resolved Optical Gating (FROG) [90]. The following discussion applies to noise-free signals. The local frequency of an optical signal $u(t)$ is defined as the derivative of the local phase with respect to time

$$f_{\text{loc}}(t) = \frac{1}{2\pi} \frac{d\arg[u(t)]}{dt},$$

where $\arg[u(t)] = \arctan[\operatorname{Im} u(t)/\operatorname{Re} u(t)] = \varphi(t)$ is the local phase of $u(t)$. To produce the graphs, we transmit a single pulse centered at t=0 in a 400 ps time window without adding ASE noise. Six times during each dispersion map, at the beginning of each fiber span and in the middle of the anomalous span, we iterate over each sample point in time and save the pairs (t, f_{loc}) if the local pulse power $|u(t)|^2$ is larger than 0.5\% of the peak pulse power $\max(|u(t)|^2)$.

Figure 4.4a shows the six phase portraits during the first period of the dispersion map. The straight horizontal line represents the launched signal, which is an unchirped Gaussian. Chirp is defined as the second time derivative of the local phase, $d^2 \arg[u(t)]/dt^2 = 2\pi\, df_{\text{loc}}/dt$ Consequently, the portrait of a chirped pulse will have a nonzero slope. The oblique lines show the pulse at later points in the dispersion map. Figure 4.4b shows the same portrait as Fig. 4.4a, but for the final map period (period 225). The difference between Figs. 4.4a and b demonstrates that the pulse chirp of the dispersion-managed soliton evolves. The curve that is marked by the arrows shows the pulse portrait at the middle of the anomalous span. The chirp is small for $5\,\text{ps} < t < +5\,\text{ps}$; however, the pulse tails are strongly chirped because the chirp-free point in a map is only located in the middle of the anomalous span if the fiber is assumed to be lossless, but it moves closer to the beginning of the anomalous span in a lossy system [97].

Figure 4.4c shows the pulse portrait at the beginning of each fiber span for the entire transmission over 225 map periods, and Fig. 4.4d shows the same, but with added ASE noise and a power cutoff of 3\%. In all four graphs, one can see that the solitons never extend beyond $-40\,\mathrm{ps} < t < +40\,\mathrm{ps}$ with a significant pulse power, and hence the overlap of the pulse tails is negligible. Note that phase portraits are not directly related to the least-squares phase fit of (Eq. 4.31) since the phase portraits show phase in the time domain, not in the frequency domain.

Figures 4.5a and b show histograms of φ and τ for the two different signal peak powers P_{peak} of 5 ~ mW and 13 ~ mW\@. The two histograms are approximations to $p_\varphi(\varphi)$ and $p_\tau(\tau)$ converges to a Gaussian distribution, and $p_\varphi(\varphi)$ converges to the Jacobi-Θ function, the periodic analog of a Gaussian pdf [6]. The phase and timing jitter can be separated and mathematically reintroduced at the receiver when computing the BERs.

Covariance Matrix Computation

In the following, we refer to the reduced covariance matrix $\mathscr{K}^{(r)}$ as $\mathscr{K}$, dropping the superscript. The calculation that we present here is completely deterministic. Compared to Monte Carlo simulations, this approach requires substantially less CPU time while producing a much higher degree of accuracy. However, for the 4-bit sequence 1,000 the covariance matrix method required still 5 h of CPU time on a 400 MHz Pentium III PC in the year 2003. The nonlinear propagation equation is solved by a third-order split-step algorithm [129]. One transmits the 8-bit deBruijn sequence 11010001 in a time window of 800 ps in a single channel; all pulses are co-polarized. The Fourier vector length N_{FFT} is 2,048.

Figure 4.6a shows the optical power $|u_0(t,L)|^2$ of the noise-free 8-bit signal at the end of the transmission in the time domain. The tic marks indicate the boundaries of the bit slots. Fig. 4.6b shows the optical power spectrum at the end of the transmission as the circles. The 10 GHz tones and their harmonics are clearly visible. The dots show the average power spectrum of the noise as obtained from $|a_k|^2 + |a_{k+N}|^2 = \mathscr{K}_{k,k} + \mathscr{K}_{k+N,k+N}$. The number of modes in $\mathscr{K}$ is $N = 120$. The optical signal-to-noise ratio (OSNR), which we define as the ratio of total signal power to total noise power in the shown bandwidth, is 6.98, or 8.44 dB. The bandwidth of 160 GHz in this definition of the OSNR seems very large, but the OSNR would not change very much by reducing the bandwidth because of the inline filter that attenuates both the signal and the noise at high frequencies.

Figure 4.7a shows three slices through the RR part of the reduced covariance matrix $\mathscr{K}$, where the inset gives a pictorial representation of a matrix whose elements are located at the grid points. The diagonal lines show the location of the three slices. The solid line hence runs along the principal diagonal, the circles run on the secondary diagonal, and the asterisks show elements on the nearest parallel

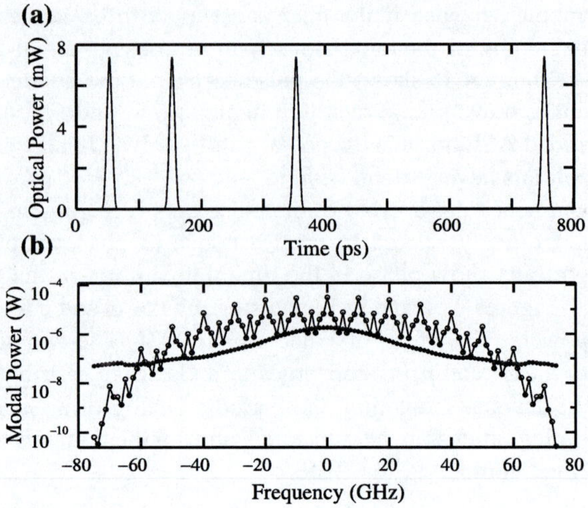

Fig. 4.6 DMS system. **a** Noise-free optical power in the time domain $|u_0(t,L)|^2$ of the noise-free 8-bit signal at the end of the transmission. **b** *Circles* optical power spectrum $|A_k|^2$ of the signal in logarithmic scale. The 10 GHz tones and their harmonics are clearly visible. *Dots* average power spectrum of the noise from $\mathscr{K}$. The OSNR is 8.44 dB and $N = 120$

to the principal diagonal. The large ratio of the noise power at frequency $f = 0$ to the power at $f = 75$ GHz is primarily due to the inline optical filter in the recirculating loop and only secondarily to parametric gain.[3]

Figure 4.7b shows the eigenvalue spectrum of $\mathscr{K}$, where the the eigenvalues are sorted by decreasing magnitude. The eigenfunctions connected to the spectrum are related to the *optical* Karhunen–Loève modes, namely the modes in which the accumulated noise can be expanded as independent Gaussian noise neglecting the receiver. If most of the eigenvalues were very small, one could make the matrix propagation more efficient by transforming the noise into the optical Karhunen–Loève basis and focusing on the modes with the largest noise power, but as Fig. 4.7b shows, the magnitude of the eigenvalues does not fall off very steeply.

Figure 4.8 compares the average pdf in the marks and spaces, as defined in (4.54), of the narrow-band filtered receiver current from the linearization method with the histogram of a standard Monte Carlo simulation. Figure 4.8b shows the corresponding eye diagram as a contour plot of the logarithm of the pdf as a function of time.

4.2.4.2 Submarine CRZ System

System Setup

The simulated transmission line of the CRZ system is shown in Fig. 4.9. It consists of 34 dispersion map periods each of length 180 km, for a total distance of 6,120 km. Each map period contains a 160 km span of normal dispersion fiber with

[3] See Sect. 4.2.2.1 for a discussion of parametric gain.

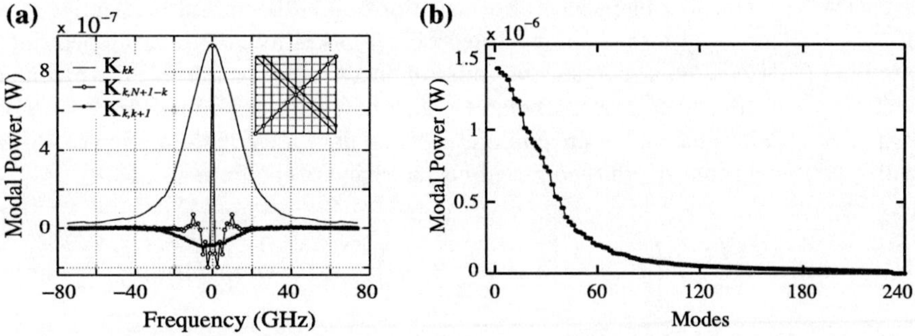

Fig. 4.7 DMS system. **a** Three slices through the RR part of the reduced covariance matrix $\mathscr{K}$, indicated by the inset cartoon; **b** Eigenvalue spectrum of $\mathscr{K}$. The eigenvalues are ordered by magnitude

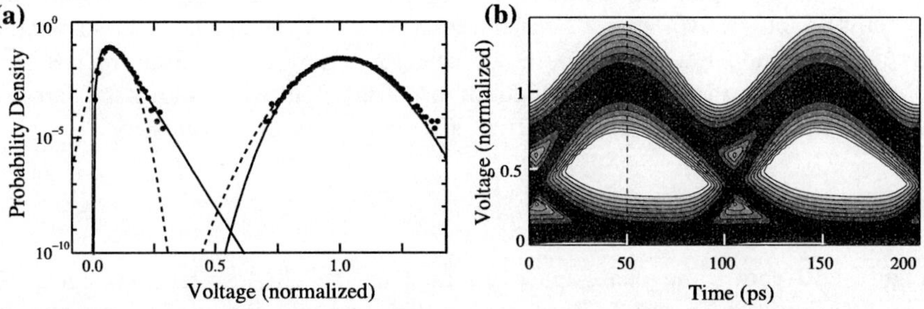

Fig. 4.8 DMS system. **a** Histogram from a traditional Monte Carlo simulation (*dots*) with Gaussian fits of the data points in the marks and spaces based on their mean and variance (*dashed lines*) and the result of the covariance matrix method (*solid line*). **b** Eye diagram as a contour plot of the logarithm of the pdf as a function of time. The pdfs in (**a**) are computed at $t = 50$ ps (*dashed line*)

Fig. 4.9 CRZ system, schematic illustration

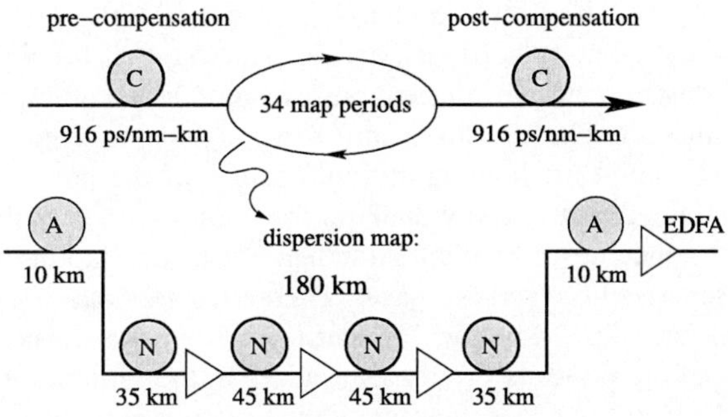

-2.5 ps/nm-km, indicated by the circles labeled N, preceded and followed by two 10 km spans of anomalous dispersion fiber with 16.5 ps/nm-km, labeled A. Third-order dispersion is of minor importance and we neglect it in this simulation, as we did

in the DMS system. The fibers have an attenuation of 0.2 dB/km and an effective area of $A_{\text{eff}} = 50\,\mu\text{m}$. The loss is compensated every 45 km by an erbium-doped fiber amplifier (EDFA) with a spontaneous emission factor of $n_{\text{sp}} = 2.0$. We use pre- and post-compensating fiber spans, labeled C, each of which has a total dispersion of 916 ps/nm. The signal pulses are co-polarized and have a FWHM duration of 45 ps with a bit-synchronously chirped raised-cosine shape of the form

$$u(t) = \left(\frac{P_{\text{peak}}}{2}\left\{1 + \cos\left[\pi \sin\left(\frac{\Omega t}{2}\right)\right]\right\}\right)^{1/2} \exp(iA\pi \cos \Omega t), \tag{4.56}$$

where $u(t)$ is the envelope of the optical field at time t, $\Omega = 2\pi/T_{\text{bit}}$ with the bit spacing $T_{\text{bit}} = 100$ ps, and the chirp parameter is A = -0.6 [145]. The initial optical peak power is $P_{\text{peak}} = 1$ mW before entering the pre-compensating fiber. We transmit 32 bits, corresponding to a pseudorandom bit sequence of $2^5 - 1 = 31$ bits, plus an additional zero bit, thereby exhausting all possible bit patterns of length five, with a Fourier vector length N_{FFT} equals to 4,096. At the receiver, we model an ideal square-law detector followed by a fifth order Bessel filter with a one-sided 3-dB bandwidth of 4.3 GHz. Figure 4.13 shows the narrow-band filtered noise-free receiver current after 6,100 km of transmission.

Results and Discussion

Figure 4.10 shows the phase-space portraits of the CRZ system analogous to Fig. 4.4, shown in Sect. 4.2.4.1 on p. 31. All graphs are produced by a simulation of a single CRZ pulse centered at t = 0 in an 800 ps time window. Figure 4.10a shows the six lines of the pulse portraits at the beginning of each fiber span in the first dispersion map period. It also shows the phase portrait of the pre-compensation. The latter is strongly chirped and has an S-shape, indicated by the arrow. Figure 4.10b is the same as (a), except that we have added four copies of the lines in (a), each offset in time by a multiple of 100 ps. The copies correspond to pulses centered at 100, 200 ps, and so on. Note that the phase-space portrait of a signal consisting of five adjacent pulses would look completely different from Fig. 4.10b, since the phase of the sum of two signals does not equal the sum of the phases. The point of showing multiple copies of the portraits in one graph is to demonstrate that the individual portraits of different pulses at a given transmission distance never overlap. Although pulses overlap in the time domain, they are still separated in phase space. Therefore, the phase space picture contains more information. A vertical line at t = 200 ps cuts through the portraits of five pulses, and the portraits of adjacent pulses in the same fiber span are separated by about 10 GHz. As the pulses evolve, their portraits rotate counter-clockwise and contract, as shown in Fig. 4.10c for the maps 16–19. In the middle of the total transmission distance near map 17, the pulses are maximally compressed and separated in time, but strongly chirped. After map 17, they expand again and the final state at map 34 in Fig. 4.10d has a mirror symmetry with Fig. 4.10b.

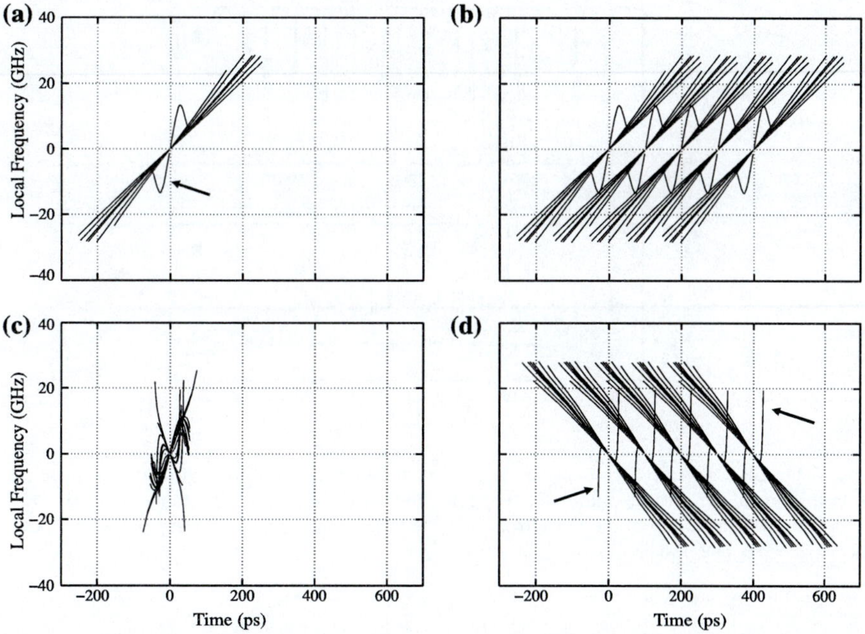

Fig. 4.10 Phase space portraits for the CRZ system at the beginning of each fiber span. **a** *Arrow* pre-compensation portrait, *oblique lines* portrait during the first map period; **b** Same as (**a**), but five portraits are superimposed with time offsets of 100 ps; **c** Portraits of the maps 16–19; **d** Same as (**b**), but for the last map. *Arrows* post-compensation portrait

The almost vertical curves, indicated by the arrows, correspond to the pulse after post-compensation which is slightly narrower than the launched pulse.

Figure 4.11a shows the optical power $|u_0|^2$ of the noise-free 32-bit signal at the end of the transmission in the time domain. The tic marks indicate the boundaries of the bit slots. Figure 4.11b shows the optical power spectrum at the end of the transmission as the circles. The 10 GHz tones and their harmonics are clearly visible. The dots show the average power spectrum of the noise as obtained from $|a_k|^2 + |a_{k+N}|^2 = \mathcal{K}_{k,k} + \mathcal{K}_{k+N,k+N}$; the number of modes in $\mathcal{K}$ is $N = 140$. The ratio of the noise power at frequency $f = 0$ to the power at $f = 22$ GHz is 2.1 and is due to parametric gain. The OSNR, defined as the ratio of total signal power over total noise power in the shown bandwidth, is 14.83, or 11.7 dB.

Figure 4.12a shows three slices through the RR part of the covariance matrix $\mathcal{K}$ similar to Fig. 4.7a, where the inset gives a pictorial representation of a matrix whose elements are located at the grid points. The oblique lines show the location of the three slices. The solid line runs along the principal diagonal, the circles run on the secondary diagonal, and the asterisks show elements that lie on a parallel to the principal diagonal where the cross-correlations are particularly large. As in the DMS system, most cross-correlations are negative.

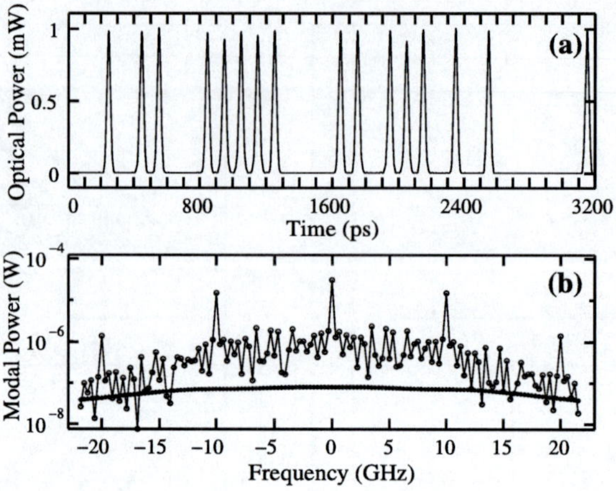

Fig. 4.11 CRZ system. **a** Optical power $|u_0|^2$ of the noise-free 32-bit signal at the end of the transmission in the time domain. **b** *Dots* optical power spectrum on a logarithmic scale. *Asterisks* average power spectrum of the noise from $\mathcal{K}$. The ratio of the noise power at $f = 0$ to the power at $f = 22$ GHz is 2.1. The OSNR is 11.7 dB

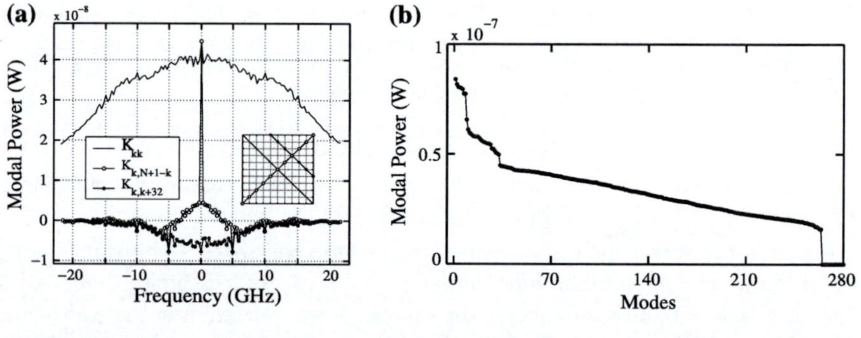

Fig. 4.12 CRZ system. **a** Three slices through the RR part of the reduced covariance matrix $\mathcal{K}$, indicated by the *inset* cartoon; **b** Eigenvalue spectrum of $\mathcal{K}$. The eigenvalues are sorted by magnitude

Figure 4.12b shows the eigenvalue spectrum of $\mathcal{K}$, analogous to Fig. 4.7b. Due to numerical imprecisions a few eigenvalues are vanishingly small and must be increased before the pdfs are computed, as described in Sect. 4.2.5.2.

Figure 4.13 shows the narrow-band filtered noise free receiver current after transmission over 6,120 km. The 32 partial pdfs are computed at the points in time indicated by the dots. The variation in the peak power is due to nonlinear pulse-to-pulse interactions during the transmission, highlighting the importance of bit patterns.

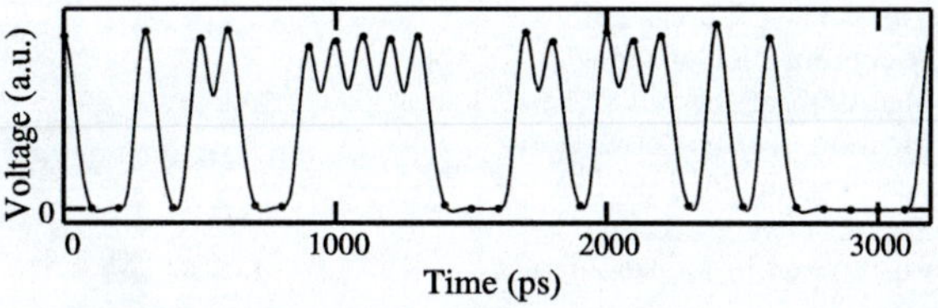

Fig. 4.13 CRZ system. Narrow-band filtered noise-free current $y_{nf}(t)$ of the signal after transmitting 6,120 km. The 32 partial pdfs are computed at the points in time indicated by the *dots*

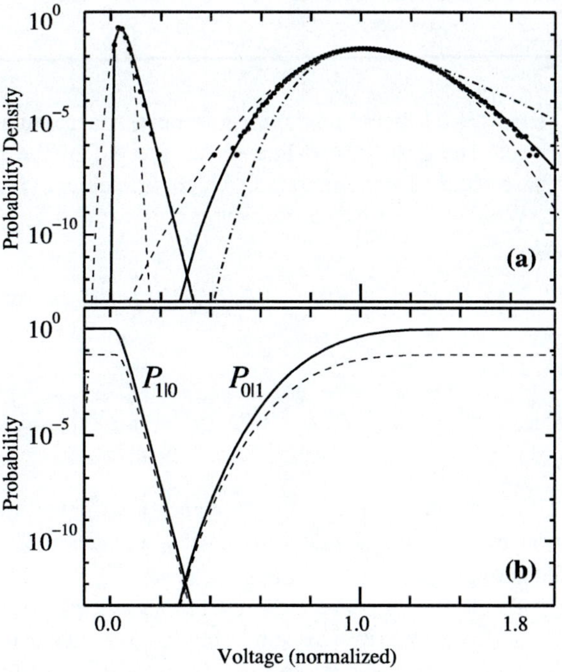

Fig. 4.14 CRZ system. **a** *Solid lines* average pdfs from the covariance matrix method; *dots* histogram from a Monte Carlo simulation; *dashed lines* Gaussian fit to the dots using their mean and variance; *dash–dotted line* average pdf from the covariance matrix method without phase jitter separation. **b** *Solid lines* error probabilities $P_{1|0}$ and $P_{0|1}$ corresponding to the solid lines in (**a**); *dashed lines* $P_{1|0}$ and $P_{0|1}$ for the worst noise-free mark and space only

Figure 4.14a shows the average pdfs as defined in (4.54) of the receiver current that result from the covariance matrix method as solid lines in comparison with a histogram from a traditional Monte Carlo simulation, consisting of 86,000 noise realizations represented by the dots. The current is normalized to the mean of the pdf of the marks. The dashed lines show a Gaussian fit to the Monte Carlo data, using the mean and variance. The large deviation between the solid and dashed curves is obvious, especially in the spaces. On the other hand, the agreement between the covariance matrix method and the Monte Carlo results is excellent. By integrating the pdfs, one obtains optimal BERs of 1.7×10^{-12} from the covariance matrix method and 9.9×10^{-12} from the Gaussian fit of the Monte Carlo data. The latter corresponds to a Q-factor of 6.71. Note that the relatively small difference between these two BERs occurs because the Gaussian fit overestimates the pdf of the marks and underestimates it in the spaces, and

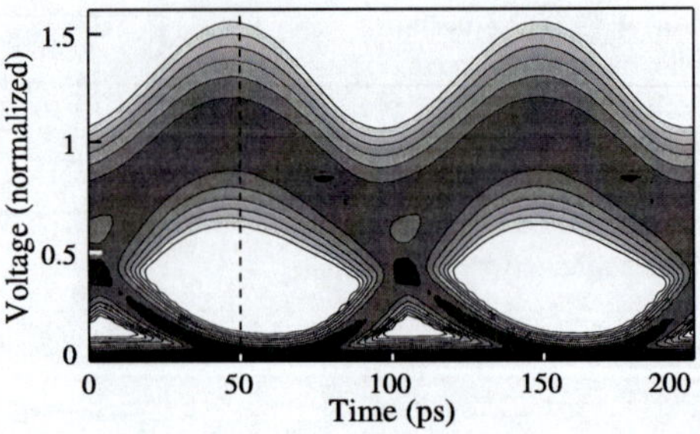

Fig. 4.15 CRZ system. A contour plot of the logarithm of the pdf as a function of time showing the eye diagram. The pdfs in Fig. 4.14 were taken at $t = 50$ ps (*dashed line*)

hence the Q-factor method relies on the accidental partial cancellation of two errors. The dash–dotted line shows the pdf of the marks that one obtains if the phase jitter is not separated, which is clearly wrong.

We define the two probabilities

$$P_{1|0}(y) = \int_{y}^{\infty} f_y(y', t_1)\mathrm{d}y', \tag{4.57a}$$

$$P_{0|1}(y) = \int_{0}^{y} f_y(y', t_0)\mathrm{d}y'. \tag{4.57b}$$

The quantity $P_{1|0}(y)$ is the probability of detecting a mark when a space was transmitted, using the decision level y, and analogously $P_{0|1}(y)$ is the probability of detecting a space when a mark was sent.

The left solid curve in Fig. 4.14b shows the error probability $P_{1|0}$ of detecting a "1" when a "0" was sent, using a given decision level as defined in (4.57a). This curve corresponds to the left pdf in Fig. 4.14a. The right solid curve is the probability $P_{0|1}$ of detecting a "0" when a "1" was sent. The dashed lines show the same, except that only the mark with the lowest current *in the noise-free signal* and the space with the highest current are taken into account. The bit errors near the optimum decision level are dominated by the worst mark and space. This result indicates that in the CRZ system it is sufficient to apply the linearization method only to the patterns that exhibit the worst behavior in the absence of noise to obtain a good approximation of the average pdfs and the BER.

Figure 4.15 shows the corresponding eye diagram of all 32 bits. The probability density is displayed as a contour plot. The pdf is only plotted over a range of about four orders of magnitude to make it look like an eye diagram, although we are able to accurately compute the probability density at any point in the diagram. The pdfs in Fig. 4.14 are taken at $t = 50$ ps, indicated by the dashed line.

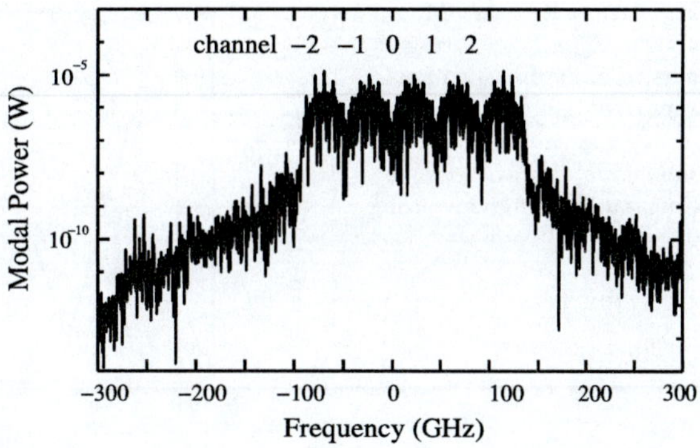

Fig. 4.16 CRZ DWDM system with five channels. Noise-free optical power spectrum $|A_k|^2$ at the end of the transmission line. The channel spacing is 50 GHz, and the channels are indicated by numbers

4.2.4.3 WDM System

We now discuss the application of the covariance matrix method to a WDM CRZ system. This system is an extension of the single-channel CRZ system that we described in the previous section. We launch identical bit sequences in five channels, spaced 50 GHz apart. This channel spacing is narrow, compared to the channel bandwidth of about ±23 GHz, and hence corresponds to a dense wavelength-division multiplexed (DWDM) system. While real-transmission systems have many more channels [142], work by Yu et al. [148] shows that it is possible to simulate a dense WDM system with a limited number of channels.

Figure 4.16 shows the optical power spectrum at the end of the transmission line. The channel spacing is 50 GHz and the channels are indicated by indices running from −2 to +2.

In the following, we focus on the central channel (channel 0). The system parameters $N_{\mathrm{FFT}} = 4{,}096$, $N = 140$, and $T = 3200$ ps, as well as all pulse parameters are identical to the single-channel system. The propagation algorithm of the covariance matrix, as shown in Sect. 4.2.5.2, is identical to the single-channel case, except that after step 1, we apply an artificial optical bandpass filter with a square shape and a bandwidth of ±25 GHz that only passes channel 0. The artificial filtering is necessary because the neighboring channels would otherwise disturb the phase jitter separation. However, we simulate the transmission of u_0 with all five channels. Thus, we model the signal–noise interactions of the signal in any channel with the noise in the central channel, but we ignore the *inter-channel* noise–noise correlations. In other words, we neglect off-diagonal matrix blocks in an extended covariance matrix that spans multiple channels. This simplification is physically reasonable, since the noise–noise correlations decay with frequency separation, and it is validated by Monte Carlo simulations.

As in the single-channel case, we ran a Monte Carlo simulation to validate our results. Since we expected that the neighboring channels might distort the pdf slightly and the linearization method might yield inaccurate results, we ran 100,000 noise realizations in a simulation that took 51 days to complete.

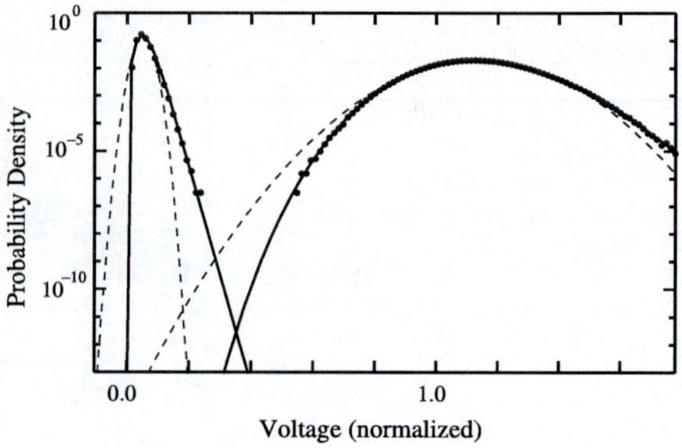

Fig. 4.17 CRZ DWDM system. *Solid lines* average pdfs from the linearization approach; *dots* histogram from a Monte Carlo simulation; *dashed lines* Gaussian fit to the dots using the mean and variance

Figure 4.17 shows the average pdfs of the receiver current that results from the covariance matrix method as solid lines in comparison with a histogram from a traditional Monte Carlo simulation represented by the dots. The voltage is normalized to the mean of the pdf of the marks. The dashed lines show a Gaussian fit to the Monte Carlo data, using the mean and variance. The agreement between the linearization approach and the Monte Carlo results is excellent. By integrating the pdfs we obtain an optimal BER of 4.7×10^{-12} from the linearization approach compared to 4.1×10^{-11} from the Gaussian fit of the Monte Carlo data; the latter corresponds to a Q-factor of 6.50.

As an additional test, we record the *inter-channel* regions of the extended covariance matrix, i.e., in addition to averaging over the noise realizations to obtain $\mathscr{K}^{(r)} = \mathscr{K}^{(r)}_{[0,0]}$, we also compute the matrices $\mathscr{K}^{(r)}_{[1,0]} = \mathscr{K}^{(r)}_{[0,1]}$ and $\mathscr{K}^{(r)}_{[1,1]}$, where $\mathscr{K}^{(r)}_{[i,j]}$ refers to the inter-channel covariance matrix defined by a generalization of (4.32)

$$\mathscr{K}^{(r)}_{[i,j]} = \left\langle r_{[i]} r_{[j]}^{T} \right\rangle, \tag{4.58}$$

where $r_{[k]} = (r_{-N/2+n,R}, \ldots, r_{N/2-1+n,R}, r_{-N/2+n,I}, \ldots, r_{N/2-1+n,I})^{T}$. Here, the offset $n = k|\Delta f_{\mathrm{ch}}|/\Delta f_{\mathrm{FFT}}$ is the number of frequency modes between the center of channel k and channel 0 with the channel spacing $|\Delta f_{\mathrm{ch}}| = 50\,\mathrm{GHz}$, and $\Delta f_{\mathrm{FFT}} = 1/T$ is the frequency resolution of the Fourier transform. We find that $\mathscr{K}^{(r)}_{[0,0]} = \mathscr{K}^{(r)}_{[1,1]}$ to within the computational accuracy set by the split-step algorithm and $\left\|\mathscr{K}^{(r)}_{[1,0]}\right\|_{\max} / \left\|\mathscr{K}^{(r)}_{[0,0]}\right\|_{\max} = 0.076$, where $\|\cdot\|_{\max}$ is the matrix maximum norm. Inter-channel noise–noise correlations are of little relevance since only the noise in the central channel contributes to the eye diagram.

In summary, the generalization of the covariance matrix to WDM does not seem to pose major difficulties. The simulation time increases relative to a single-channel system, and the size of this increase depends exclusively on the step size in the split-step algorithm when simulating multiple channels.

4.2.5 Appendix

4.2.5.1 Mathematical Tools

Simultaneous Diagonalization

The theorem of simultaneous diagonalization (see [45, p. 106]) holds for $\mathscr{K}^{-1}$ (real symmetric and positive definite) and $\mathscr{W}$ (symmetric) and states that in this case there is a real $2N \times 2N$ matrix $\mathscr{C}$ with $\det \mathscr{C} \neq 0$ and

$$(\mathscr{C}^{-1})^T \mathscr{K}^{-1} \mathscr{C}^{-1} = \mathscr{I} \qquad \text{and} \tag{4.59a}$$

$$(\mathscr{C}^{-1})^T \mathscr{W} \mathscr{C}^{-1} = \Lambda = \text{diag}(\lambda_1, \ldots, \lambda_{2N}), \tag{4.59b}$$

$$\mathscr{C}^{-1} = \mathscr{U}\mathscr{D}\mathscr{V} \qquad \text{with} \tag{4.59c}$$

$$\mathscr{U}^T \mathscr{K}^{-1} \mathscr{U} = \text{diag}(\varepsilon_1, \ldots, \varepsilon_{2N}), \qquad \mathscr{U}^T \mathscr{U} = \text{I}, \tag{4.59d}$$

$$\mathscr{D} = \text{diag}(\varepsilon_1^{-1/2}, \ldots, \varepsilon_{2N}^{-1/2}), \tag{4.59e}$$

$$\mathscr{V}^T \mathscr{H} \mathscr{V} = \Lambda, \qquad \mathscr{V}^T \mathscr{V} = \mathscr{I}, \tag{4.59f}$$

$$\mathscr{H} = \mathscr{D}^T \mathscr{U}^T \mathscr{W} \mathscr{U} \mathscr{D}, \tag{4.59g}$$

where all the λ_i are real numbers. This diagonalization is equivalent to solving the generalized real eigenvalue problem $\mathscr{W}\mathscr{C}^{-1} = \mathscr{K}^{-1}\mathscr{C}^{-1}\Lambda$. Note that there is no need to ever compute the matrix $\mathscr{C}$ explicitly from $\mathscr{C}^{-1}$. We found that the implementation of (4.59a–g) is preferable to using a general-purpose generalized eigensystem solver because the matrix $\mathscr{U}\mathscr{D}$ only depends on $\mathscr{K}$ and has to be calculated only once, while the matrix $\mathscr{H}$ depends on $\mathscr{W}$ and hence on the time t. In particular when calculating eye diagrams that require many pdfs, this approach yields a computational gain.

4.2.5.2 Algorithms

The computation of the electrical pdfs can be broken down into the optical transmission part, yielding the covariance matrix $\mathscr{K}$, the computation of the electrical pdf from $\mathscr{K}$, and additional post-processing such as drawing eye diagrams and computing the optimum BER. Without going into the details of the programming, we just state that the transmission part is implemented as a stand-alone C++ program, and the computation of the BER and any post-processing is implemented as Matlab™ code. We will now describe three parts of the algorithm, namely:

1. The main loop of the transmission simulation routine.
2. The propagation of the covariance matrix.
3. The computation of the electrical pdf.

The nonlinear propagation equation (4.14a) in all simulations in this subsection is solved with a third-order split-step algorithm [129].

The Main Loop of the Transmission Simulation

The transmission program can be operated in four different modes:

1. Noise-free propagation, yielding $u_0(t)$.
2. Traditional Monte Carlo simulation, yielding noise signal realizations $u(t)$ with options to generate traditional eye diagrams and compute jitter averages.
3. Extended Monte Carlo: same as in mode 2., but with the additional computation of $\mathscr{K}$ by averaging over noise realizations according to (4.32).
4. Deterministic simulation: compute $u_0(t)$ and $\mathscr{K}$, using the method described in Sect. 4.2.3.4.

The goal in setting up the simulation this way was to facilitate the comparison of the different pdfs resulting from modes 2 to 4. The algorithm in the main loop of the simulation can be itemized as follows:

1. Setup: Read in a parameter file that contains all physical and technical parameters, such as the number of bits, the pulse durations, the peak power, the random generator seed. Check the consistency of the parameters. Allocate and initialize all arrays.
2. In Monte Carlo modes 2 and 3, start a loop over different noise realizations:
   a. Launch the optical signal $u_0(t, 0)$.
   b. In all modes, start a loop over the dispersion map periods, possibly preceded by a pre-compensating fiber span:
      i. Compute the signal evolutions element by element of the dispersion map such as fiber spans, amplifiers, optical filters. Add ASE noise in amplifiers in the Monte Carlo modes 2 and 3. In mode 4, propagate $\mathscr{K}$ through all dispersion map elements.
      ii. At regular intervals during the transmission, for instance every five dispersion maps, draw eye diagrams, compute various fluctuations such as timing jitter, and accumulate data to compute $\mathscr{K}(z)$ at the local point.
   c. Write all data to the disk every 30 min.
3. Write the final field $u_0(t)$, the matrix $\mathscr{K}$, and all eye diagrams and statistics to the disk.

Deterministic Propagation of the Covariance Matrix

The deterministic propagation of the covariance matrix $\mathscr{K}$ is described in Sect. 4.2.3.4 and is implemented in the C++ routine `propagator()`. At every amplifier, Gaussian white noise power is added to the principal diagonal of $\mathscr{K}$. The action of an optical filter on $\mathscr{K}$ is given by $\mathscr{K}_{\text{out},kl} = \mathscr{K}_{k,l} H_k H_l$, where H_i is the real-filter function at frequency mode i.

We now summarize the algorithm for propagating $\mathscr{K}$ over one fiber span from $z =$ to $z = L$, followed by an EDFA. First, propagate the field $u_0(t, 0)$ using a standard Fourier split-step algorithm, yielding $u_0(t, L)$. Save $u_0(t, L)$, return to $z = 0$ and repeat the following for each k:

1. Compute $u^{(k)}(t, L)$ by perturbing $u_0(t, 0)$ in the kth frequency mode and propagating it to $z = L$.
2. Separate the pulses in the signal by passing $u^{(k)}(t, L)$ through a linear and lossless fiber with total dispersion $-D(L)$.
3. Compute the perturbation vector $a^{(k)}$ and apply the phase jitter separation (4.37a and b) individually to each pulse, yielding the vector $\widetilde{a}^{(k)}$ of the dispersion compensated signal.
4. Invert step 2 to compute $\widetilde{a}^{(k)}$ at point L. Evaluate the propagator matrix elements Ψ_{jk} for all j.

Finally, compute $\mathscr{K}(L)$ according to (4.34).

Computation of the Electrical pdf from the Covariance Matrix

The computation of the electrical pdf from the covariance matrix is implemented in Matlab™.

1. Read in $\overline{A}$, the noise-free optical field vector in frequency domain, and $\mathscr{K}$, the covariance matrix at the receiver.
2. To increase the numerical accuracy, scale $\overline{A}$ and $\mathscr{K}$ so that the maximum element of $\mathscr{K}$ becomes 1.
3. As Figs. 4.7b and 4.12b show, some of the eigenvalues of $\mathscr{K}$ can become very small or even negative during the propagation, due to numerical inaccuracies. These small eigenvalues would inhibit the inversion of $\mathscr{K}$. If there are any eigenvalues of $\mathscr{K}$ that are smaller than $\epsilon = 1/300$ times the maximum eigenvalue, diagonalize $\mathscr{M}^{\mathrm{T}} \mathscr{K} \mathscr{M} = \Lambda_{\mathscr{K}}$ where $\Lambda_{\mathscr{K}}$ is the diagonal matrix of the eigenvalues, increase the small eigenvalues to the threshold value, and compute a corrected covariance matrix $\mathscr{K}' = \mathscr{M} \Lambda'_{\mathscr{K}} \mathscr{M}^T$. We verified that the resulting pdf is independent of ϵ over a wide range.
4. Invert $\mathscr{K}'$.
5. Compute the matrix $\mathscr{U}\mathscr{D}$ by applying (4.59d and e).
6. Compute the Hermitian matrix $\mathscr{W}_{kl}$ as described in (4.40).
7. Start a loop over all points in time $t = t_k$ for which the pdf must be computed:

   a. If timing jitter was separated and must be reintroduced, start a loop over the n sample times t_i with $1 \leq i \leq n$:

i. Compute $\mathscr{W}(t_{k,i})$ from F.
ii. Compute the matrix $\mathscr{H}$ by evaluating (4.59g), then compute $\mathscr{V}$ and Λ.
iii. Compute the values $s_l(t_{k,i}) = \lambda_l Q_l^2 = \sum_m \bar{A}_l \mathscr{W}_{lm}(t_{k,i}) \bar{A}_m$ and $y_{\mathrm{nf}}(t_{k,i}) = \sum_l s_l$.
iv. Invert step 2.
v. Evaluate the characteristic function $\Phi_{\tau=0}(\zeta, t_{k,i})$ for zero timing jitter according to (4.49), using s_k.
vi. Prepare the discrete Fourier transform to compute $f_{y,\tau=0}(y, t_{k,i})$: If $i = 1$, optimize the grid increment $\Delta\zeta_k$ to yield the best resolution in $f_{y,\tau=0}(y, t_{k,i})$ This $\Delta\zeta_k$ is then used for all i.
vii. Compute $f_{y,\tau=0}(y, t_{k,i})$ by taking the discrete Fourier transform of $\Phi_{\tau=0}(\zeta, t_{k,i})$.
viii. Verify that $f_{y,\tau=0}(y, t_{k,i})$ is normalized and $f_{y,\tau=0}(0, t_{k,i}) = 0$, then save it.

b. Compute $f_y(y, t_k)$ from the $f_{y,\tau=0}(y, t_k)$.
c. Save $y_{\mathrm{nf}}(t_k)$, $\Delta\zeta_k$, and $f_y(y, t_k)$.

8. Write all $y_{\mathrm{nf}}(t_k)$, $\Delta\zeta_k$, and $f_y(y, t_k)$ to the disk.

It turns out that steps 2, 3, and 6 significantly improve the numerical accuracy of the resulting pdf. The computation can fail altogether if step 3 is omitted. Both steps 2 and 6 improve $f_y(y, t)$ by lowering the roundoff floor. According to our experience, $f_y(y, t)$ will never span more than about 16 orders of magnitude when using `double` floating point arithmetic (64 bits). We use a Fourier transform vector length of 1,024 in step 7(a)vi.

Split-Step Accuracy

All simulation in this subsection are based on the third-order split-step algorithm [129]. In order to compare traditional Monte Carlo simulations to the deterministic method, one has to verify that the numerical accuracy is sufficient. We propagate one single Monte Carlo noise realization in both the 8-bit DMS system over 24,000 km as described in Covariance Matrix Computation and in the 32-bit CRZ system over 6,100 km as described in Sect. 4.2.4.2, while varying the local relative error bound $\delta = \|u_c - u_f\|/\|u_f\|$, where u_c and u_f are the coarse and the fine solutions, respectively [129]. The noise is only added in the N lowest frequency modes and the random generator's seed is the same in all simulations.

In Fig. 4.18 we show the global relative error $\epsilon = \|u_{(\delta)} - u\|/\|u\|$ versus the local relative error δ in double-logarithmic scale, where $u_{(\delta)}$ is the optical field at the receiver that was obtained using the error-bound δ, and u is the solution for $\delta = 10^{-12}$. The circles in Fig. 4.18a show simulation values of $\epsilon(\delta)$ for the DMS system, and the line is a fit with the function $q\delta^m$, where $q = 1.737 \times 10^4$ and $m = 1.594$. Figure 4.18b displays the same for the CRZ system, where $q = 33.19$ and $m = 1.258$. For a global error goal of $\epsilon \leq 10^{-5}$ that we used in the Monte

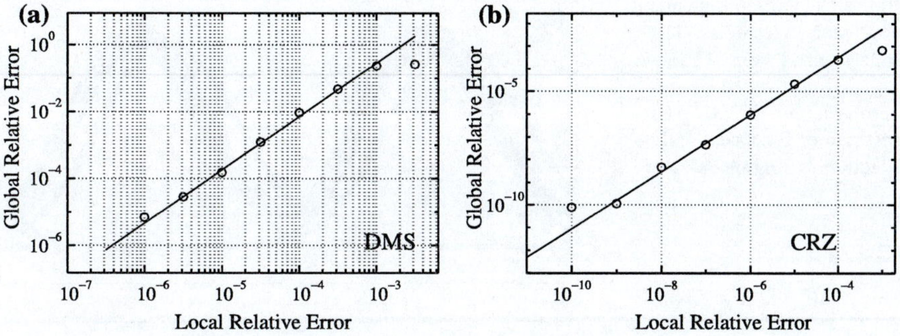

Fig. 4.18 Simulated values of global versus local relative error $\epsilon(\delta)$ shown by the circles for the **a** DMS system and **b** CRZ system. The lines are fits to a power law

Carlo simulations, the fit yields the local relative errors of $\delta = 1.6 \times 10^{-6}$ and $\delta = 6.6 \times 10^{-6}$ for the DMS and CRZ systems, respectively. For $\epsilon = 10^{-4}$, one finds $\delta = 6.8 \times 10^{-6}$ and $\delta = 4.0 \times 10^{-5}$.

4.3 Momentum Methods

4.3.1 Collision-Induced Timing and Amplitude Jitter

4.3.1.1 Introduction

Nonlinear effects in optical fibers cause distortion of optical signals in fiber communications lines thus imposing an upper limit on the signal power [8, 75, 76, 78]. From a system designer's perspective, the nonlinear effects lead to a power penalty, which can be partially overcome by reducing the input power to the system. Thus nearly all modern systems operate at powers at which the signal evolution is almost linear [105]. However, there always exist small nonlinear interactions, leading to a small signal distortion that accumulates during transmission over long distances and can introduce a significant system penalty [18, 53, 80, 139]. Calculating the BER in long-haul systems and finding the optimal power level requires an accurate model of the nonlinear interactions.

The main challenge in characterizing the nonlinear penalty is that it is a statistical quantity. The amount of distortion that an optical pulse suffers depends on the particular pattern of surrounding pulses, which is effectively random because these pulses represent the information bits, and the information sequence of bits is quasirandom. This effect is often referred to as the nonlinear pattern-dependence effect [78, 44]. In single-channel transmission, dispersion leads to the spread of optical signals, causing approximately three to seven adjacent pulses to interfere

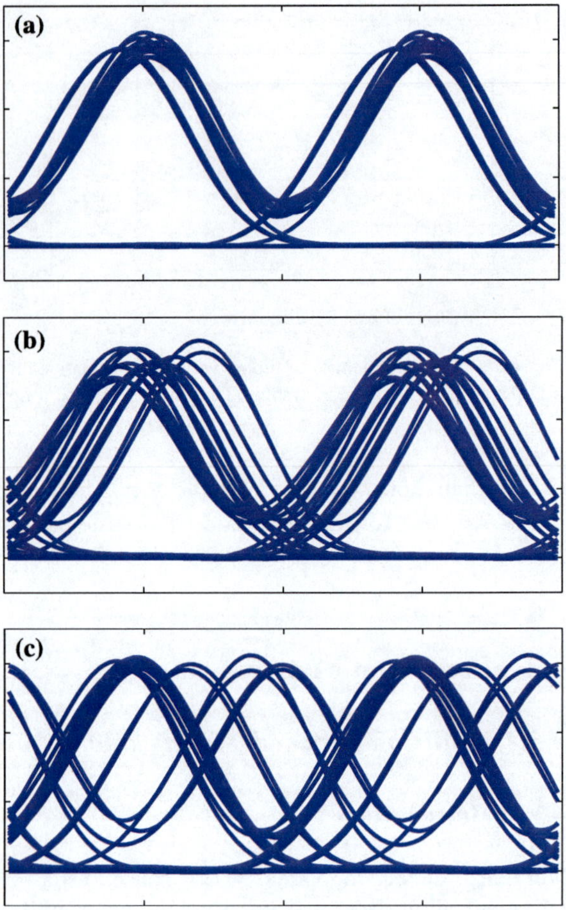

Fig. 4.19 Pattern dependent nonlinear effect in a WDM RZ system. (From [126], Fig. 1. Reproduced by permission of © 2007 The Institute of Electrical and Electronics Engineers)

[52, 105]. Therefore, a common approach to account for pattern-dependent nonlinear effects is to use a pseudo-random sequence of bits, which is typically 2^3–2^7 bits, to find the worst-case bit in the sequence. When we consider a multi-channel system, this approach is inappropriate since there are many more pulses interacting with each other due to the dispersive walkoff between the frequency channels. As an illustration, Fig. 4.19 shows three simulated optical eye diagrams of a noise-free signal in the center channel of a 10 Gb/s wavelength-division multiplexed (WDM) RZ system after propagating over 5000 km. We used nine co-polarized channels spaced by 50 GHz and the average power was approximately −0.7 dBm/channel. We used three different sets of bit patterns in different WDM channels, while the bit pattern in the center channel remained unchanged. As we move from Fig. 4.19a to 4.19c, it is apparent that the eye changes from being almost completely open to completely closed, which corresponds to the best (a), an intermediate (b), and the

worst (c) bit pattern, respectively. In the presence of WDM channels, the total number of different interaction patterns grows exponentially with the number of channels. Therefore, finding the worst-case performance becomes prohibitively time-consuming. Moreover, it is also not very useful since the likelihood of the worst-case pattern is negligibly small. Therefore, a probabilistic approach is necessary to treat this problem. One can use biased Monte Carlo simulations to estimate the pdf with the required accuracy. Monte Carlo simulations are, however, time consuming and even with modern-day computers it is computationally expensive to obtain an estimate of the pdf over many orders of magnitude even with biasing. Hence there is a need for deterministic, computationally efficient techniques.

Typically, the dominant nonlinear effect in modern high-speed systems operating at 10 GB/s, is cross-phase modulation [35, 53, 57, 80, 94, 109]. The manifestation of this effect depends on the light modulation format. In NRZ transmission, the signal distortion appears in the form of amplitude jitter [69, 137, 121, 94], while in the RZ systems, the dominant nonlinear effect is typically collision-induced timing-jitter [3, 57, 73, 125, 130]. This fact requires the development of completely different approaches to account for the nonlinear effects in these two types of systems. Because the NRZ modulation format has been the dominant modulation for many years, techniques have been developed to characterize the nonlinear effects in WDM NRZ transmission [13, 28, 29, 34, 36, 69, 74, 79, 94, 121, 122, 137, 138], and the BER calculations based on these techniques agree well with experimental results. The basic idea in these approaches [13, 28, 29, 69] is to utilize a pump-probe method, in which the cross-phase modulation-induced distortion is treated as an additive perturbation. An exception is [36] where the authors treat the distortion as multiplicative. In order to determine the influence of the nonlinearity on the system performance, one further assumes that the XPM-induced distortion may be treated as additive Gaussian noise and a correction to the Q-factor is calculated [13, 69, 74, 79, 93, 94, 138].

It has been discovered that the RZ-modulated signal undergoes less intersymbol interference in the receiver and is more robust to fiber nonlinearities and thus became widely used in long-haul data transmission [12, 26, 52, 87, 107, 130]. The major nonlinear effect in WDM RZ systems is collision-induced timing jitter. Therefore, the pump-probe approach just described cannot be directly applied to the RZ systems. Collision-induced timing jitter has, however, been extensively studied in both soliton and linear systems [1–4, 9, 57, 73, 89, 134, 135]. From this work, it is known how to calculate the time shift that results from a collision of a pair of pulses and to calculate the standard deviation of the time shift. Further, one can calculate the complete pdf of the collision-induced time shift and use it to accurately calculate BER [125, 127]. This approach takes into account the ASE noise and the inter-channel nonlinear bit-pattern effect due to the collision-induced timing jitter.

In the following sections, we describe a semianalytical approach for calculations of collision-induced timing jitter and its pdf. First, we need to determine the evolution of the pulse shape with distance. For solitons, this could be done analytically [3]. For quasilinear systems with arbitrary pulse shapes, there is no exact solution for the pulse evolution; however, for timing jitter calculations, one can use approximations [1–4, 9]. Although analytical approximations may sacrifice accuracy and simplicity, they may provide important insights into the system behavior and provide important scaling laws. We determine the pulse evolution numerically, which is simpler to implement, allows one to achieve desired accuracy, and is computationally highly efficient. We stress that we need to determine the evolution of only one pulse rather than a pulse sequence. Also, we need to process the bandwidth of only one channel rather than the whole WDM bandwidth. Second, we can characterize timing jitter by its mean and variance or by its complete pdf. While mean and variance can provide important information on the system behavior and could even be used as design criteria [57], the complete pdf enables us to calculate the BER. Therefore, we will focus on calculation of the pdf of collision-induced time shift.

4.3.1.2 Collision-Induced Timing Jitter

Calculation of Collision-Induced Time Shift

In order to obtain the pdf of the collision-induced time shift and the BER, we start by calculating the time shift function [128, 9]. We refer to the pulse v_T for which we are calculating the time shift function as the target pulse and the channel in which it is located as the probe channel. The other WDM channels are referred to as the pump channels. We define the time shift function $\tau(\Delta f_k, l) \equiv \tau_{kl}$ as the time shift of the target pulse v_T after a propagation distance L due to a collision with a pulse v_{kl} that is initially located in the lth bit slot of the kth pump channel, where Δf_k is the frequency offset of the pump channel from the probe channel.

The starting point is a version of the propagation equation [57], in which the nonlinear part includes only the effects of self-phase modulation and interchannel cross-phase modulation (XPM):

$$\frac{\partial v_T}{\partial z} + i\frac{\beta''}{2}\frac{\partial^2 v_T}{\partial t^2} - i\gamma\left(|v_T|^2 + 2\sum_{k,l}\alpha_{kl}|v_{kl}|^2\right)v_T - g v_T = 0, \tag{4.60}$$

where z is the physical distance, t is the retarded time with respect to the probe channel, β'' is the local dispersion, γ is the nonlinear coefficient, g is the fiber loss and gain coefficient, and $\alpha_{kl} = 1$ if the lth bit slot in the kth channel contains a pulse, corresponding to a digital 1, and is zero otherwise. While including higher-order dispersion poses no difficulty in principle, its effect on the system under study is negligible and we set it to zero for simplicity. Equation 4.60 is derived assuming a large channel separation and representing the field envelope in the kth channel as

$$v_k = \sum_l \alpha_{kl} v_{kl}, \tag{4.61}$$

where v_{kl} is the field envelope of the pulse located in the lth bit slot of the kth channel. We made an assumption that pulses in one channel are well-separated in time during the propagation when the nonlinearity is important so that for each channel k we can use the approximation,

$$|v_k|^2 = \left|\sum_l \alpha_{kl} v_{kl}\right|^2 \approx \sum_l \alpha_{kl} |v_{kl}|^2. \tag{4.62}$$

Originally, this assumption was used in the theory of collision-induced timing jitter for solitons and was well-justified because in soliton communications systems, pulses in one channel do not overlap. In modern quasilinear RZ systems, a pulse overlaps with many of its neighbors due to a large dispersive spread. However, in many systems, the approximation (4.62) is still reasonable because the peak power of a pulse reduces when it disperses, thus reducing the nonlinear interactions. Another argument to advocate the use of (4.62) is the following: After some cumbersome calculations, one can derive the standard deviation of the collision-induced time shift without using approximation (4.62). One can show that many cross products resulting from the term $\left|\sum_l \alpha_{kl} v_{kl}\right|^2$ tend to cancel out in the calculation of the standard deviation. As a consequence, the resulting exact standard deviation is equal to the standard deviation obtained using approximation (4.62). This can formally explain the fact that approximation (4.62) has proven to yield an accurate calculation of collision-induced timing jitter [57] and an accurate pdf of the time shift [125].

We define the central time and frequency of the target pulse v_T as

$$\tau = \frac{1}{E_T} \int_{-\infty}^{\infty} t |v_T|^2 \mathrm{d}t \tag{4.63}$$

$$\Omega = \frac{1}{E_T} \int_{-\infty}^{\infty} \mathrm{Im}\left(\frac{\partial v_T^*}{\partial t} v_T\right) \mathrm{d}t, \tag{4.64}$$

where

$$E_T = \int_{-\infty}^{\infty} |v_T|^2 \mathrm{d}t. \tag{4.65}$$

Using definitions (4.63)–(4.65) and (4.60), we obtain the dynamic equations for collision-induced time and frequency shifts [3, 57, 73, 125, 130]

$$\frac{\mathrm{d}\tau}{\mathrm{d}z} = \beta''(z)\Omega(z) \tag{4.66}$$

$$\frac{\mathrm{d}\Omega}{\mathrm{d}z} = -\frac{2\gamma}{E_T(z)} \int_{-\infty}^{\infty} |v_T(z,t)|^2 \frac{\partial P(z,t)}{\partial t} \mathrm{d}t, \tag{4.67}$$

where

$$P(z) = 2\sum_{k,l} \alpha_{kl}|v_{kl}(z,t)|^2. \quad (4.68)$$

If we assume that the pulse shape in all the channels is identical, then the time position of the pulse v_{kl} is determined by the dispersive walkoff as

$$v_{kl}(z,t) = v_T\left(z, t - \int_0^z 2\pi\Delta f_k \beta''(x)\mathrm{d}x + T_{\mathrm{bit}} l\right), \quad (4.69)$$

where T_{bit} is the bit period and $l = 0$ corresponds to the bit slot of the target pulse.
The total time shift of the target pulse is then given by

$$T_{\mathrm{total}} = \sum_{k,l} \alpha_{kl}\tau_{kl}, \quad (4.70)$$

where the time shift τ_{kl} is given by

$$\tau_{kl} = \int_0^L \Delta\Omega_{kl}(z)\beta''(z)\mathrm{d}z, \quad (4.71)$$

and $\Delta\Omega_{kl}(z)$ is the collision-induced frequency shift, whose evolution is given by

$$\frac{\mathrm{d}\Delta\Omega_{\mathrm{kl}}}{\mathrm{d}z} = -\frac{2\gamma}{E_T(z)} \int_{-\infty}^{\infty} |v_T(z,t)|^2 \frac{\partial |v_{kl}(z,t)|^2}{\partial t}\mathrm{d}t. \quad (4.72)$$

Probability Density Function of the Time Shift

We assume that the α_{kl} are independent, identically distributed random variables, each having probability 1/2 of being 1 or 0. Thus the total shift of the target pulse T_{total} is a random variable, which is a linear combination of independent binary random variables. We compute the pdf of T_{total} using the characteristic function [125] $w(\xi)$, which is given by

$$w(\xi) = \langle \exp(i\xi T_{\mathrm{total}})\rangle = \left\langle \exp\left(i\xi \sum_{k,l} \alpha_{kl}\tau_{kl}\right)\right\rangle, \quad (4.73)$$

where $\langle\cdot\rangle$ denotes the statistical average. Since the random variables α_{kl} are independent and the probability that α_{kl} equals either 1 or 0 is 1/2, we may write

$$w(\xi) = \prod_{k,l}\left\{\frac{1}{2}[1 + \exp(i\xi\tau_{kl})]\right\}. \quad (4.74)$$

Then the pdf of the time shift is simply the Fourier transform of the characteristic function,

$$p_T(t) = \frac{1}{2\pi} \int_{-\infty}^{\infty} w(\xi) \exp(-i\xi t) \mathrm{d}\xi. \tag{4.75}$$

One can in principle accommodate more complex α_{kl}, e.g., pseudorandom, by using the appropriate characteristic function. At this point it is also straightforward to calculate the standard deviation of collision-induced time shift σ_T^2 as

$$\sigma_T^2 = \langle T_{\text{total}}^2 \rangle - \langle T_{\text{total}} \rangle^2. \tag{4.76}$$

Using (4.70) and the fact that data α_{kl} are independent with a probability of being either 1 or 0 equal to 1/2, we obtain

$$\sigma_T^2 = \sum_{k,l,m,n} < \alpha_{kl}\alpha_{mn} > \tau_{kl}\tau_{mn} - \left(\sum_{k,l} < \alpha_{kl} > t_{kl} \right)^2 = \frac{1}{4} \sum_{k,l} \tau_{kl}^2. \tag{4.77}$$

As we mentioned earlier, it is possible to derive σ_T^2 without assumption (4.62) using similar formalism as (4.77). However, due to the absolute square of the sum of the fields $\alpha_{kl}v_{kl}$ in (4.62), the expression for σ_T^2 will contain the fourth-order products of the form $\alpha_{kl}\alpha_{mn}\alpha_{pq}\alpha_{rs}$. After evaluating those terms and also assuming that the pulse shape v_T (and v_{kl}) is an even function of time, one can show that final expression for σ_T^2 will be the same as (4.77). This important result also justifies the use of approximation of pairwise collisions (4.62) even when the dispersive pulse overlap is nonnegligible. The calculation of the probability density function of the time shift without assumption (4.62) would be prohibitively complex.

Another interesting question is whether the pdf of the collision-induced time shift is Gaussian or not. The number of pulse collisions is large, so that based on the central limit theorem one can assume that the distribution p_T of the time shift is Gaussian. However, as we showed in [127], the Gaussian curve deviates significantly from the true pdf in the tails since the number of pulse collisions during the propagation in the system is finite and thus there exists a worst-case time shift.

We also note that even if the number of pulse collisions becomes large in the presence of hundreds of channels, the Gaussian approximation of the time shift pdf is still inaccurate in the tails because one of the conditions of the central limit theorem does not hold. In particular, in order for a normalized sum

$$Z_n = \frac{1}{s_n} \sum_{k=1}^{n} X_k \tag{4.78}$$

of independent random variables X_k with zero mean and variance σ_k^2 where

$$s_n^2 = \sum_{k=1}^{n} \sigma_k^2 \tag{4.79}$$

to converge to the normal pdf, it is required that for a given $\varepsilon > 0$ there exists an n such that [113]

$$\sigma_k < \varepsilon s_n, k = 1, \ldots, n. \tag{4.80}$$

In our case, the time shifts τ_{kl} given by (4.71) are inversely proportional to the square of the wavelength separation [128], or, equivalent, to the channel index k if the channel count starts from the target channel. Hence the variance for each individual random variable $\alpha_{kl}\tau_{kl}$ is proportional to k^4. Since the number of bits in a given channel that collide with the target pulse is proportional to k due to the dispersive walkoff, the total variance for n channels is proportional to $\sum_{k=1}^{n} 1/k^3$, which is limited by a constant for any number of channels n. Thus condition (4.80) of the central limit theorem does not hold and the limiting distribution of the time shift is not necessarily Gaussian. Physically, what happens in this case is that as k increases, the contribution of the individual bits to the timing jitter of the pulses in the target channel falls off so rapidly that it is as if on average the target pulses were only interacting with a finite number of pulses in the other channels, even though the number really tends toward infinity and even though the worse case timing jitter also tends logarithmically with the number of channels toward infinity. We will illustrate the difference between the Gaussian pdf and the true pdf with a numerical example in Validation.

4.3.1.3 Calculation of the Pulse Amplitude Distortion and the Bit Error Rate

In this section, we first calculate the pdf of the received current that is due to the nonlinear interactions in the absence of the noise from the amplifiers. We review the model based on the timing jitter alone and we then introduce a reduced model that accounts for the nonlinearly induced amplitude jitter in WDM systems that is not due to the timing jitter. We then validate this reduced model with IS simulations. Finally, we show how to apply our probabilistic model of nonlinear interactions to calculate the BER in the presence of amplifier noise.

Model of Signal Degradation Due to Collision-Induced Timing Jitter

In order to calculate the distortion of the received current signal that is due to the timing jitter, we use a simplified method, in which we calculate the pulse shape $i(t)$ at the receiver using a full propagation model based on (4.60) with $\alpha_{kl} = 0$ for all k and l, and a receiver model that includes an optical filter, an ideal square-law photodetector, and an electrical filter. We then use this pulse shape to determine the value of the sampled current I given a time shift ΔT by using the expression

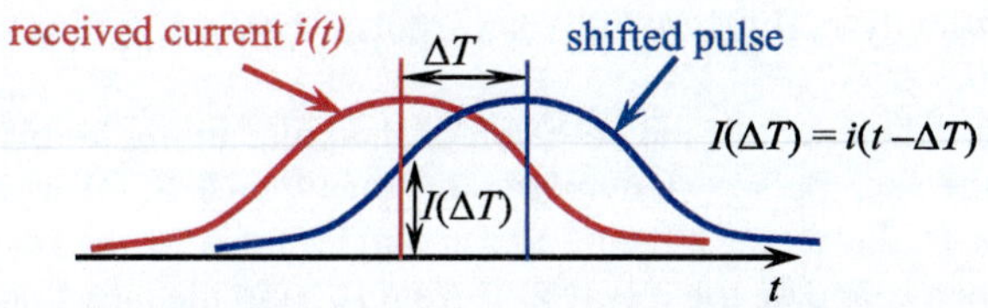

Fig. 4.20 Conversion of time shift to the current distortion. (From [126], Fig. 2. Reproduced by permission of © 2007 The Institute of Electrical and Electronics Engineers)

$$I(\Delta T) = i(T_0 - \Delta T), \tag{4.81}$$

where T_0 is the central time of the pulse, as illustrated in Fig. 4.20. We then obtain the pdf of the current using the cumulative distribution function of the time shift $F_{\Delta T}(t) = \int_{-\infty}^{t} p(\tau)d\tau$, where $p(\tau)$ is the pdf of the time shift,

$$\begin{aligned} F_I(x) &= \Pr[i(t) < x] = \Pr(\Delta T < T_1 \cup \Delta T > T_2) \\ &= F_{\Delta T}(T_1) + 1 - F_{\Delta T}(T_2), \end{aligned} \tag{4.82}$$

where $T_1 < T_2$ are the solutions of the equation

$$i(t - T_0) = x. \tag{4.83}$$

The pdf of the received current $p_I, \mathrm{TJ}(x)$ is then given by

$$p_{I,\mathrm{TJ}}(x) = \frac{dF_I(x)}{dx}. \tag{4.84}$$

Multipulse Interactions and Amplitude Jitter

Up to this point, we have treated pulses of the same frequency channel as if they do not overlap. In reality, in the system under study, each pulse overlaps with a maximum of four of its neighbors in the prototypical system that we consider due to dispersive spreading. This intra-channel pulse interaction combined with inter-channel nonlinear crosstalk leads to an increase in the amplitude jitter that must be accounted for in the target channel, in addition to the timing jitter, in order to obtain accurate results. Hence we treat the electric field in the target channel as a sum of the electric fields of individual pulses rather than simply adding powers as we did previously (4.62), so that

$$|v_0|^2 = \left|\sum_l \alpha_{0l} v_{0l}\right|^2. \tag{4.85}$$

In the other channels, we continue to neglect pulse overlap. Since in our system only five pulses in the same channel overlap due to dispersion, we consider a pseudorandom bit sequence (PRBS) of length $2^5 = 32$ bits in the target channel that contains all patterns of five bits. We then treat this PRBS as a superpulse and consider a two-body collision of this superpulse with a single pulse in a neighboring channel.

To determine the effect of the collision of the pulse in the lth bit slot of the kth channel with the superpulse, we numerically solve the NLS, for which the input is the superpulse in the target channel and a single pulse in the lth bit of the kth channel. We then calculate the received current $I_{kl}(t)$ in the target channel after the electrical filter. We repeat this numerical procedure for all k and l.

In the next step, we remove the time shift τ_{kl} since we previously accounted for it, and we do not want to count it twice. Since pulses at the receiver are well-separated and the time shifts τ_{kl} are small, we can remove the time shifts by translating individual pulses by a corresponding time determined by the pre-calculated time shift function τ_{kl} (4.71).

Finally, we determine the received current distortion due to the two-body collisions relative to the unperturbed solution. As an unperturbed solution, we calculate the value of the current $I_T(t)$ of a single target channel with the same PRBS of length 32, in the absence of the neighboring channels. We then obtain the current distortion δI_{kl} using the expression

$$\delta I_{kl}(t) = I_{kl}(t) - I_T(t). \tag{4.86}$$

We assume that for an arbitrary bit pattern, the total distortion $\delta I(t)$ can be represented as a sum of the individual contributions δI_{kl} from pairwise interactions, so that

$$\delta I(t) = \sum \alpha_{kl}\delta I_{kl}(t), \tag{4.87}$$

where $\alpha_{kl} = 1$ or 0.

In order to obtain the pdf of the amplitude deviation $\delta I(t)$, we apply the characteristic function method, as we did for the timing jitter

$$v(\xi,t) = \langle \exp[i\xi\delta I(t)] \rangle = \left\langle \exp\left(i\xi \sum_{k,l} \alpha_{kl}\delta I_{kl}(t) \right) \right\rangle, \tag{4.88}$$

where $\langle\cdot\rangle$ denotes the statistical average taken over the ensemble of all bit patterns. Using the independence of random variables α_{kl} and the assumption that the probability of α_{kl} is equal to either 1 or 0 is 1/2, we obtain

$$v(\xi,t) = \prod_{k,l} \frac{1}{2}\{1 + \exp[i\xi\delta I_{kl}(t)]\}. \tag{4.89}$$

Then the pdf of current is simply the inverse Fourier transform of the characteristic function,

$$p_{\delta I}(I,t) = \frac{1}{2\pi} \int_{-\infty}^{\infty} v(\xi,t)\exp(-i\xi I)d\xi. \tag{4.90}$$

In order to obtain the pdf p_INL of the current due to nonlinear distortion, we first convolve the pdf (4.84) of the current at the center of the pulse and (4.90) due to the amplitude distortion assuming that the two processes are independent and then average over all bits so that

$$p_{I,\mathrm{NL}}(I) = \frac{1}{N}\sum_{k=1}^{N}\int p_{I,\mathrm{TJ}}(x, t_k)p_{\delta I}(I - x, t_k)\mathrm{d}x, \tag{4.91}$$

where N is the number of bits, and t_k is the central time of the kth bit.

Validation

To validate our reduced deterministic method, we compared the results of the calculations based on (4.84) and (4.91) with the corresponding results from the full statistical model based on IS simulations. We applied our method to a prototypical undersea system with a bit rate of 10 Gb/s and a propagation distance of approximately 5,000 km [125] that is similar to an experimental system reported in [25]. The transmission part includes 100 periods of a dispersion map consisting of 34 km of D_+ fiber and 17.44 km of D_- fiber followed by an amplifier. The values of dispersion, effective core area, nonlinear index, and loss are 20.17 ps/nm-km, 106.7 $\mu\mathrm{m}^2$, 1.7×10^{-20} m^2/W, and 0.19 dB/km for the D_+ fiber and −40.8 ps/nm-km, 31.1 $\mu\mathrm{m}^2$, $2.2 \times 10^{-20}\mathrm{m}^2/\mathrm{W}$, and 0.25 dB/km for the D_- fiber, respectively. The average map dispersion is −0.5 ps/nm-km, and the amount of pre- and post-compensation is 1,028 and 1,815 ps/nm, respectively. We used 35-ps raised-cosine pulses with a peak power of 5 mW, and we launched nine co-polarized channels separated by 50 GHz, each carrying a 32-bit sequence. We verified with both a full simulation model and the reduced models described here that a further increase in the number of channels and number of bits per channel had a negligible effect on the system performance. The receiver included a 30 GHz super-Gaussian optical demultiplexer and a photodetector. We did not consider amplifier noise in this part of the study, in which we validated our reduced model of nonlinear interactions.

To validate the proposed reduced model we used the standard IS method [23, 32, 131, 143], which we describe in detail in Appendix I. We also used the multicanonical Monte Carlo method [16, 146, 66, 86, 92, 110, 65, 144, 114] to verify that the results from the IS simulations are correct.

First, we compare the pdf of collision-induced time shift p_T obtained from (4.75) to the results of importance-sampled Monte Carlo simulations in Fig. 4.21. The agreement between the semianalytical momentum method and the numerical simulations is excellent. We also plot the Gaussian pdf with σ_T defined by (4.77). Note that the tails of the Gaussian pdf deviate significantly from the exact pdf as we discussed in Probability Density Function of the Time Shift. Thus Gaussian pdf cannot be used to accurately predict the BER.

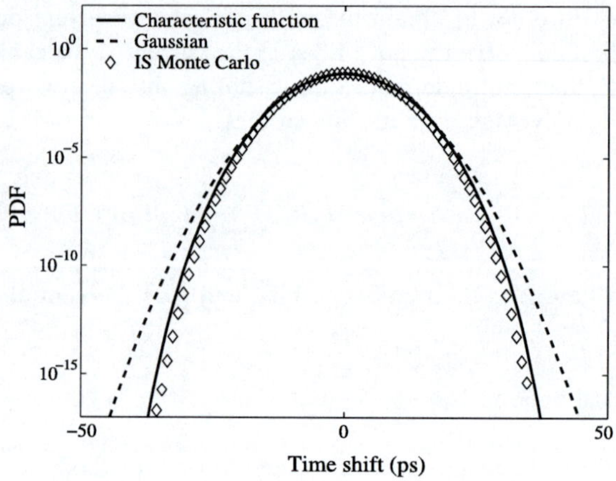

Fig. 4.21 Probability density function of the collision-induced time shift. (From [127], Fig. 1. Reproduced by permission of © 2006 The Institute of Electrical and Electronics Engineers)

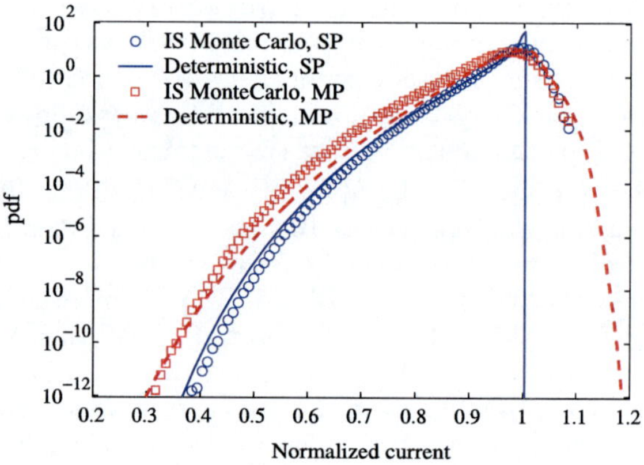

Fig. 4.22 Probability density function of the current at the detection point in the receiver due to the nonlinear distortion with multiple pulses (MP) in the target channel compared to a single pulse in the target channel (SP). (From [126], Fig. 3. Reproduced by permission of © 2007 The Institute of Electrical and Electronics Engineers)

Next, in Fig. 4.22 we show the pdf $p_I, \mathrm{NL}(I)$ of the received current at the center of the pulse and the pdf obtained using IS simulations with multiple pulse in the target channel. For comparison, we also show the pdf $p_I, \mathrm{TJ}(x)$ and the pdf obtained from the corresponding IS simulations with a single pulse in the target channel. We immediately notice the difference that multipulse interactions make in the low-current tail of the pdf. A deterministic model of the pdf of the current at the decision point of the receiver that only includes timing jitter alone agrees well with IS simulations when there is only a single pulse in the target channel. However, the agreement is no longer good when there are multiple pulses in the target channel. It is necessary to include nonlinearly induced amplitude jitter in the deterministic model, and we then see a good agreement between our deterministic model and IS simulation. The discrepancy in the pdf is less than an order of magnitude over the entire range of interest.

Calculation of the BER

Now we show how to use the deterministic theory that we developed for the evaluation of the nonlinear penalty to calculate the BER in the presence of the ASE noise from optical amplifiers. We assume that optical noise is additive white Gaussian noise (AWGN) at the entry to the receiver, and we take into account the actual pulse shape, as well as the frequency-dependent optical and electrical filtering, using an approach described by Forestieri [43]. For this calculation, we used an 8-GHz electrical fifth-order Bessel filter and we set the optical signal-to noise ratio to 15 dB, calculated over a 25 GHz bandwidth. For the purpose of this work, the choice of the noise model is not restrictive since it does not affect the methodology that we propose and we only need the distribution of the noise-induced variation of the received current. In principle, one could also use a noise model that takes into account the nonlinear signal–noise interaction, for example, models described in [22], [27], [62], [67], [68], [114], [119], [120].

We will use the assumption that the nonlinear penalty is statistically independent of the ASE noise since it greatly simplifies the analysis. In principle, these two processes are not independent, as one may envision that the noise affects the signals, and that in turn affects the way the signals interact with each other. However, as the noise is a zero-mean process, the effect of noise will result in either reduction or increase of the optical power of a pulse and, consequently, of the nonlinear interference that this pulse generates. Averaging over a large statistical ensemble tends to extinguish the correlation between the noise and the nonlinear penalty. In addition, we treat the contributions of timing jitter and amplitude jitter to nonlinearly induced signal distortion as independent.

In order to compute the pdf $p(I, t)$ of the received current I at the sampling time t due to both nonlinear signal distortion and noise, we use the pdf of the collision-induced time shift $p_T(t)$ given by (4.75), the pdf of the nonlinearly induced amplitude distortion $p_{\delta I}(I,t)$ given by (4.90), and the pdf of the noise-induced distortion $p_{\text{noise}}(I,t)$. We calculate $p_{\text{noise}}(I,t)$ by propagating a single-channel signal carrying a PRBS of length 32 through the system and using an additive white Gaussian noise model [43]. The resulting pdf $p(I, t)$ is given by

$$p(I,t) = \int_{\zeta=-\infty}^{\infty} \left[\int_{\tau=-\infty}^{\infty} p_{\text{noise}}(\zeta, t-\tau) p_T(\tau) d\tau \right] \cdot p_{\delta I}(I-\zeta, t) d\zeta. \tag{4.92}$$

We show the results of our calculations in Fig. 4.23. The dot–dashed curves show the pdf of the current in the marks and spaces only due to noise. The collision-induced timing jitter (TJ) increases the BER by over three orders of magnitude. The corresponding pdfs are shown with the dashed curves. Finally, the nonlinearly induced amplitude jitter (AJ) degrades the BER by more than one order of magnitude and the total pdfs are displayed with the solid curves. Since the nonlinear interactions can degrade the BER by many orders of magnitude, it is important to model them accurately. Furthermore, even though the collision-

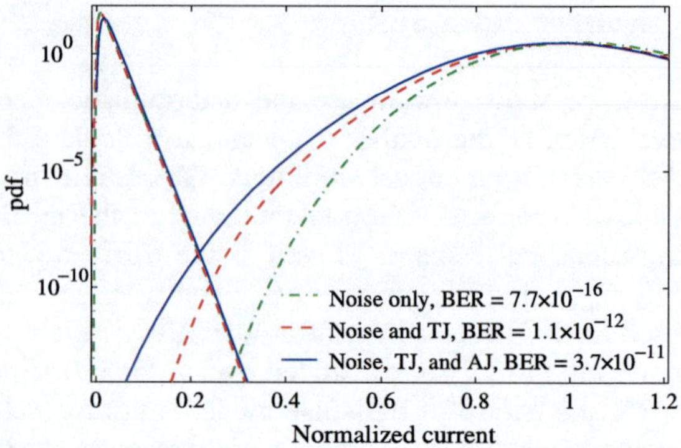

Fig. 4.23 Pdf of the current in the marks and spaces at the decision time. (From [126], Fig. 4. Reproduced by permission of © 2007 The Institute of Electrical and Electronics Engineers)

induced timing jitter is the dominant nonlinear effect in WDM RZ systems, it is necessary to consider the intra- and inter-channel nonlinearly induced amplitude distortion as it leads to further performance degradation.

4.3.1.4 Summary

We have presented a reduced deterministic approach to calculate the BER in WDM RZ systems due to nonlinear distortion and ASE noise. The method is based on finding the complete pdf of the nonlinearly induced penalty, and it accounts for both nonlinearly induced amplitude and timing jitter.

First, we evaluated the pdf of the received current due to nonlinear effects, neglecting noise. We used the pdf of the time shift to obtain the pdf of the received current at the detection point using a pulse shape at the receiver [125]. With a single pulse in the target channel, the agreement between the reduced deterministic and full statistical approaches is excellent [127]. This agreement demonstrates that the collision-induced timing jitter is the dominant nonlinear effect in this type of systems.

Then we showed that when considering multiple pulses in the target channel, we must also account for the amplitude jitter that is induced by the inter- and intra-channel nonlinear interactions, which does not arise due to timing jitter. To calculate the pdf of the current due to this additional amplitude jitter, we applied an approach based on pairwise interactions, similar to what we did for the timing jitter. In this case, we considered the interaction of a pseudo-random pulse sequence (a superpulse) in the target channel with single pulses in the neighboring channel. Assuming additivity of the pulse distortions at the receiver, we have calculated the pdf of the current using the characteristic function approach. Then we combined the pdfs of the current that are due to collision-induced timing jitter and nonlinearly induced amplitude jitter, assuming the statistical independence of the two. Despite the approximations, we have achieved agreement to within an order of magnitude between this reduced deterministic approach and a full statistical approach over the entire range of the pdfs that we calculated.

Finally, we showed how to calculate the BER in the presence of both nonlinear signal distortion and ASE noise. As we showed, the nonlinear effects can significantly degrade the system performance, and in order to take these effects into account, one must use a probabilistic approach.

4.3.1.5 Appendix

The importance sampling (IS) method is well-known in statistics and has been used to model a wide variety of systems, including optical communications systems [21, 23, 32, 88, 102, 131, 143]. Therefore, we will only describe the basic approach of IS and the specific features of the IS method that we used in our work—in particular the choice of biasing distributions.

Suppose one has a random vector $\mathbf{z} = (z_1, z_2, ..., z_n)$, defined on a sample space Ω. In our case, the vector $\mathbf{z}$ is discrete, and it represents the set of bits in all the channels: $z_j = 0$ or 1, where $j = 1, ..., n$. Note that one can establish a one-to-one mapping between z_j and α_{kl}. Our goal is to compute the average

$$P = E[f(\mathbf{z})] = \sum_{\Omega} f(\mathbf{z}_k) p(\mathbf{z}_k), \tag{4.93}$$

where $E(\cdot)$ denotes the statistical average with respect to $p(\mathbf{z})$, $f(\mathbf{z})$ is an indicator function that is equal to 1 if some quantity $X(\mathbf{z})$ falls into a desired subspace Γ of Ω:

$$f(\mathbf{z}) = I_\Gamma[X(\mathbf{z})]. \tag{4.94}$$

In our studies, the quantity $X(\mathbf{z})$ is either the collision-induced time shift or the received current. As was mentioned earlier, calculating (4.93) exactly is not possible since $f(\mathbf{z})$ is a complicated nonlinear function of the signal and system parameters. We estimate (4.93) in a set of Monte Carlo simulations,

$$\hat{P} = \frac{1}{M} \sum_{m=1}^{M} f(\mathbf{z}_m), \tag{4.95}$$

where M is the number of Monte Carlo samples. In a standard Monte Carlo simulation, one draws samples $\mathbf{z}_m$ from a distribution $p(\mathbf{z})$. In this case, the estimator $\hat{P}$ is just the number of samples in the subspace of interest S divided by the total number of samples. It follows from (4.95) that $E(\hat{P}) = P$. Typically, we are interested in rare events, for example, large nonlinear distortions that lead to transmission errors, and a standard Monte Carlo simulation would require us to use a prohibitively large number of samples to observe the small number of events in the region of interest in the sample space that lead to errors. In standard importance sampling (IS), we use a modified probability distribution $p^*(\mathbf{z})$ to draw samples,

$$P = \sum_{\Omega} f(\mathbf{z}_k) \frac{p(\mathbf{z}_k)}{p^*(\mathbf{z}_k)} p^*(\mathbf{z}_k) = E^* \left[f(\mathbf{z}_k) \frac{p(\mathbf{z}_k)}{p^*(\mathbf{z}_k)} \right]. \tag{4.96}$$

The function

$$L(\mathbf{z}) = \frac{p(\mathbf{z})}{p^*(\mathbf{z})} \tag{4.97}$$

is called the *likelihood ratio*, and the distribution $p^*(\mathbf{z})$ is called the *biasing distribution*. Then the estimator of P, denoted $\hat{P}^*$, becomes

$$\hat{P}^* = \frac{1}{M}\sum_{m=1}^{M} f(\mathbf{z}_m)L(\mathbf{z}_m), \tag{4.98}$$

where the samples are now drawn from the biasing distribution. Intuitively, we can sample the region of interest Γ in the sample space more efficiently if we choose the biasing distribution such that $p^*(\mathbf{z}) > p(\mathbf{z})$. Then the number of samples falling into Γ will be larger.

Next, we describe how we choose the biasing distribution for our simulations. We are interested in the pdf of the received current. Since the dominant nonlinear effect is collision-induced timing jitter, the large signal distortion is correlated with large time shifts. Therefore, we bias the distribution of the α_{kl} toward large values of the collision-induced time shift, using our knowledge of the time shift function τ_{kl}.

Suppose we have a set of random variables x_m with the distribution,

$$p(x_m = \sigma_m) = \frac{1+p_m}{2}, p(x_m = -\sigma_m) = \frac{1-p_m}{2}. \tag{4.99}$$

We want to bias the distributions of x_m in such a way that the sum $z = \sum_m x_m$ has a mean near a desired value z_0. For the sake of simplicity, let us assume that z is a random variable that is approximately Gaussian-distributed with the mean z_0 and standard deviation S:

$$\begin{aligned} p(z) &= \frac{1}{\sqrt{2\pi S}}\exp\left[-\frac{(z-z_0)^2}{2S}\right], \\ S &= \sum_m \sigma_m^2. \end{aligned} \tag{4.100}$$

We then use Bayes rule,

$$p(x_m \mid x) = \frac{p(x_m)p(x - x_m)}{p(x)}, \tag{4.101}$$

so that

$$p(x_m = \sigma_m \mid z) = \frac{g_+}{g_+ + g_-}, \tag{4.102}$$

where

$$g_\pm = \frac{1}{\sqrt{2\pi(S-\sigma_m^2)}}\exp\left[-\frac{(z \mp \sigma_m)^2}{2(S-\sigma_m^2)}\right]. \tag{4.103}$$

Then the expected value of x_m is given by

$$\begin{aligned} E(x_m) &= \sigma_m p(x_m = \sigma_m \mid z) \\ &\quad - \sigma_m p(x_m = -\sigma_m \mid z) \\ &= \sigma_m \tanh\left(\frac{\sigma_m z}{S - \sigma_m^2}\right). \end{aligned} \tag{4.104}$$

On the other hand, from Eq. 4.99 we have

$$E(x_m) = \sigma_m \frac{1 + p_m}{2} - \sigma_m \frac{1 - p_m}{2} = \sigma_m p_m. \tag{4.105}$$

From Eqs. 4.104 and 4.105 it follows that

$$p_m = \tanh\left(\frac{\sigma_m z}{S - \sigma_m^2}\right). \tag{4.106}$$

Now we transform the set of variables x_m to the set $\alpha_{kl}\tau_{kl}$ using the mapping

$$\begin{aligned} \beta_{kl} &= 2\alpha_{kl} - 1, \\ x_m &= \frac{1}{2}\beta_{kl}\tau_{kl}. \end{aligned} \tag{4.107}$$

The biasing probability mass function for α_{kl} becomes

$$\begin{aligned} p^*(\alpha_{kl} = 1) &= \frac{1 + p_{kl}}{2}, \\ p^*(\alpha_{kl} = 0) &= \frac{1 - p_{kl}}{2}, \end{aligned} \tag{4.108}$$

where

$$p_{kl} = \tanh\left(\frac{\frac{1}{2}\tau_{kl} T_{\text{goal}}}{\frac{1}{4}\sum_{k,l}\tau_{kl}^2 - \tau_{kl}^2}\right) \tag{4.109}$$

and T_{goal} is the target value of the mean of the total time shift.

Despite the use of approximation (4.100) that the total time shift is Gaussian-distributed, our simulations show that the choice of biases (4.109) is efficient for both estimating the pdfs of both the collision-induced time shift and the received current.

In order to sample the entire pdf efficiently we used multiple importance sampling with the balance heuristics [21, 143]. We used the values of T_{goal} in (4.109) from −50 to 50 ps with a 5 ps interval. We calculated the error of the pdf estimate from IS [21, 143], and it did not exceed 10% for all values of the time shift.

4.3.2 Noise-Induced Amplitude and Timing Jitter

4.3.2.1 Background

ASE noise and interchannel interference cause fluctuations of the time and energy of signal pulses. The temporal fluctuations are referred to as timing jitter, while the pulse energy fluctuations are referred to as amplitude jitter. These fluctuations degrade the phase and amplitude margins, respectively, leading to errors. Fundamental parameters determining the timing and amplitude jitter are the variance of the pulses central time and the variance and average of the pulse energy. The time variance was calculated for the first time by Gordon and Haus [54, 61] for a hyperbolic secant soliton pulse propagating in a fiber with uniform dispersion. One of the key ideas used in [54] and [61] was linearization of the problem about the analytically known soliton solution. The linearization was feasible because the noise power is much weaker than the signal power. In dispersion- managed fibers one cannot directly apply the Gordon–Haus theory in most cases because the pulse shapes may differ significantly from the analytically assumed hyperbolic-secant shape. For DMSs, the pulse shapes range from hyperbolic-secant to Gaussian to flattop, depending on the strength of the dispersion management [31, 132], and oscillate periodically. For RZ and NRZ signals, the pulse shapes differ even more from the hyperbolic secant shape and evolve continually. For DMSs, it was predicted in [132] that a corrected Gordon–Haus formula reduced by the enhancement factor (the ratio of the DMS energy to the energy of a standard soliton in fibers with equal path-average dispersion) would successfully describe the timing jitter. This prediction is roughly correct [31], but it is intuitively clear that the timing jitter must be sensitive to pulse bandwidth: the greater the bandwidth of the pulse, the greater the amount of noise the pulse can incorporate. Different pulse shapes have different bandwidths. Hence, the Gordon–Haus formula reduced by the enhancement factor will not yield exactly accurate results. Recently, a new approach was suggested [49], [81], [140] that takes into account the more general DMS pulse shape. However, this work still presumes an analytically fixed pulse shape with quadratic chirp that can change only its duration and amplitude. This approach yields only a rough approximation of the jitter for solitons and cannot describe the jitter at all for RZ and NRZ signals. Another approach that has been used for NRZ signals is to linearize the noise around an assumed continuous wave signal to calculate the noise power [70], [71], [96], [98]. This approach is fairly successful when the pulse evolution is not too large, but it cannot determine the effect of timing jitter. Direct Monte Carlo simulations can be used to fill this gap, but processing the large number of different realizations of random noise that is required to obtain an accurate solution can be very numerically time-consuming. Thus, developing a general approach that allows us, first, to calculate the timing and amplitude jitter for arbitrary pulse shapes and, second, to avoid time-consuming Monte Carlo simulations, is vitally important. We stress that the linearization approach, used in the Gordon Haus theory [54], [61], has a much broader range of applicability in optical fiber communications than just for hyperbolic-secant solitons in uniform dispersion fibers, continuous wave signals [70], [71], [96], [98], or other analytically known pulse shapes. The basic idea of the present approach is to use the linearization approximation around an arbitrary, numerically determined solution of the NLS. In effect, we break the problem into two steps. In the first, we determine the signal evolution in the absence of noise. While this step requires computational methods for arbitrary pulse shapes, it is fully deterministic and thus does not require the use of many realizations. In the second step, we linearize around the numerically determined solution, using statistical properties of the ASE noise to derive dynamic equations for the mean and variance of a pulses central time and energy. We show that in the limits of the linearization approximation the pulses central time is Gaussian-distributed with the variance and average calculated by our method. The situation with the amplitude jitter is a bit more complex. In order for the linearization approximation to be valid, the spontaneous–spontaneous beat noise of a single noise mode must be small compared to the signal spontaneous beat noise. However, the number of noise modes is approximately equal to $B_{\mathrm{opt}}T_w$, where B_{opt} is the optical bandwidth and T_w is the bit window. When this number becomes large, it is not possible to neglect the spontaneous–spontaneous beat noise even when the linearization approximation is valid. When the number of noise modes is small, so that the sum of the spontaneous–spontaneous beat noise contributions from all the noise modes can be ignored, the signals central energy is Gaussian-distributed with the variance and average calculated by our method. When the number of noise modes is large, and the amount of energy that goes into each noise mode is the same, the signal energy is Rician-distributed [71], [96], [116]. More generally, the energy is chi-square-distributed [116]. In typical cases that we considered, the variance of the noise energy is about 1/100 of the signal energy, while $B_{\mathrm{opt}}T_w \lesssim 10$, so that it is reasonable to ignore spontaneous–spontaneous beat noise which is what we do in this approach. When either timing jitter or amplitude jitter becomes the dominant source of errors, it is then possible to calculate the BER which we show how to do. We compare our theoretical approach to simulations for a wide range of pulse formats, including DMS, RZ, and NRZ pulses. The remainder of this part is organized as follows. In Sect. 4.3.2.2, we derive the basic equations for the variance of the central time and energy of the pulse in the presence of ASE noise. In Sect. 4.3.2.3, we validate our approach by comparing the timing and amplitude jitter predicted by theory to the results of Monte Carlo simulations over a large number of realizations of the ASE noise. In Sect. 4.3.2.4, we show how we estimate the BER probability knowing the variances of the central time and the energy of the signal pulses.

4.3.2.2 Basic Equations

We start from the NLS written in Langevin form

$$i\frac{\partial u}{\partial z} + \frac{1}{2}[D(z) - ib(z)]\frac{\partial^2 u}{\partial t^2} + C|u|^2 u = ig(z)u + \hat{F}(z,t) \tag{4.110}$$

using a formulation due to [61] that is particularly useful for noise problems. In this formulation $|u|^2$ is the photon flow, t is unnormalized time, $z = |\beta_0''|Z$ is the product of a scaling dispersion β_0'' and the unnormalized distance Z, while $D(z)$ and

$b(z)$ are the local dispersion and filtering normalized with respect to $|\beta_0''|$. The quantity $C = n_2\hbar\omega^2/cA_{\text{eff}}|\beta_0''|$ is the nonlinear coefficient, where n_2 is the Kerr coefficient, ω is the signal central frequency, A_{eff} is the effective fiber cross section, and c is the speed of light. The gain $g(z)$ may be written

$$g(z) = \begin{cases} g_0 & \text{if } z_n < z < z_n + L_{\text{amp}} \\ -\Gamma & \text{otherwise} \end{cases}$$

where g_0 and Γ are respectively the gain and loss coefficients in the optical fiber, z_n is the position of the nth amplifier, and L_{amp} is the amplifier length. The noise contribution $\hat{F}$ from the amplifiers has the autocorrelation function

$$\langle \hat{F}(z,t)\hat{F}^*(z',t')\rangle = 2g_0\theta(z)\delta(z-z')\delta(t-t') \tag{4.111}$$

where $\theta(z)$ is the spontaneous emission factor n_{sp} when $z_n < z < z_n + L_{\text{amp}}$, and we set $\theta(z) = 0$ elsewhere since there is no noise contribution outside the amplifiers. We also define the central pulse time t_p, central frequency Ω, and photon number U in the pulse as

$$\begin{aligned} t_p &= \frac{1}{U}\int_{-\infty}^{\infty} t|u|^2 \mathrm{d}t \\ \Omega &= \frac{1}{2iU}\int_{-\infty}^{\infty} (u_t u^* - u_t^* u)\mathrm{d}t \\ U &= \int_{-\infty}^{\infty} |u|^2 \mathrm{d}t \end{aligned} \tag{4.112}$$

where the subscript t designates the partial time derivative, so that $u_t \equiv \partial u/\partial t$.

Timing Jitter in Unfiltered Systems

Differentiating t_p, Ω, and U in (4.112) with respect to z and combining them with (4.110) where $b(z) = 0$, we derive the following dynamic equations for the central time t_p and central frequency Ω of a signal pulse,

$$\begin{aligned} \frac{\mathrm{d}t_p}{\mathrm{d}z} &= D\Omega + \frac{i}{U}\int_{-\infty}^{\infty} (t-t_p)\left(u\hat{F}^* - u^*\hat{F}\right)\mathrm{d}t \\ \frac{d\Omega}{\mathrm{d}z} &= i\frac{\Omega}{U}\int_{-\infty}^{\infty} (u^*\hat{F} - u\hat{F}^*)\mathrm{d}t + \frac{1}{U}\int_{-\infty}^{\infty} (u_t^*\hat{F} + u_t\hat{F}^*)\mathrm{d}t \end{aligned} \tag{4.113}$$

Implicit solutions of (4.113) are

$$t_p = \mathscr{F} + \mathscr{S} \tag{4.114}$$

where

$$\mathcal{F} = \int_0^z D\Omega \mathrm{d}z'$$

$$\mathcal{S} = i\int_0^z \left\{ \frac{1}{U} \int_{-\infty}^{\infty} (t - t_p)\left[q^* \hat{F}\exp(-i\Omega t) - q\hat{F}^* \exp(i\Omega t)\right]\mathrm{d}t \right\} \mathrm{d}z' \tag{4.115}$$

$$\Omega = \Omega_0 + \int_0^z \left\{ \frac{1}{U} \int_{-\infty}^{\infty} \left[q_t^* \hat{F}\exp(-i\Omega t) + q_t \hat{F}^* \exp(i\Omega t)\right]\mathrm{d}t \right\} \mathrm{d}z'.$$

Quantity $q = u\exp(-i\Omega t)$ is a new field shifted such that its central frequency is zero, and Ω_0 is the central frequency of the input pulse. As is seen from (4.114) the time position deviation is a superposition of the time shift $\mathcal{F}$ induced by the frequency shift and the time shift $\mathcal{S}$ due to the direct impact of the noise on the pulse. We represent the solution of (4.110) as $u = \bar{u} + \delta u$ where δu is a small noise contribution such that second- and higher order corrections to $\mathcal{F}$, $\mathcal{S}$, and Ω are negligible. Since $\delta u \propto \hat{F}$, it follows that one can neglect the contribution of δu to the right-hand sides of (4.115). Using (4.111), (4.114), and (4.115), we can calculate the variance of the central pulse time t_p

$$\sigma_t^2 = \langle t_p^2 \rangle - \langle t_p \rangle^2 = A + B + C \tag{4.116}$$

where, defining a scalar product

$$(\delta q_i, \delta q_j) = \int_{-\infty}^{\infty} (\delta q_i \delta q_j^* + \delta q_i^* \delta q_j)\mathrm{d}t \tag{4.117}$$

and the components

$$\begin{aligned} \delta q_1 &= 2\sqrt{g_0 \theta}\frac{q_t}{U} \\ \delta q_2 &= 2i\sqrt{g_0 \theta}(t - t_0)\frac{q_t}{U} \end{aligned} \tag{4.118}$$

we find that

$$\begin{aligned} A &= \langle \mathcal{F}^2 \rangle = \int_0^z D(z_1)\mathrm{d}z_1 \int_0^{z_1} D(z_2)\mathrm{d}z_2 \int_0^{z_2} (\delta q_1, \delta q_1)\mathrm{d}z_2' \\ B &= 2\langle \mathcal{F}\mathcal{S} \rangle = \int_0^z D(z_1)\mathrm{d}z_1 \int_0^{z_1} (\delta q_1, \delta q_2)\mathrm{d}z_1' \\ C &= \langle \mathcal{S}^2 \rangle = \frac{1}{2}\int_0^z (\delta q_2, \delta q_2)\mathrm{d}z'. \end{aligned} \tag{4.119}$$

Equations 4.116 and 4.119 have a simple physical interpretation. We note that one can always expand the noise in a series of orthogonal functions such that

$$\hat{F}(z,t)\exp(-i\Omega t) = \delta q_1(z,t) + \overline{\delta q_2}(z,t) + \delta q_r(z,t) \tag{4.120}$$

where $\overline{\delta q_2} = a\delta q_1 + \delta q_2$ is made orthogonal to δq_1 by setting

$$a = \frac{-(\delta q_1, \delta q_2)}{(\delta q_1, \delta q_1)}$$

and δq_r is a remainder that is orthogonal to both δq_1 and $\overline{\delta q_2}$. In this case, substituting (4.120) into the right-hand side of (4.115) one can see that δq_1 generalizes what Gordon and Haus [54] refer to as the noise field phasor component that shifts the central frequency Ω. Similarly, δq_2 generalizes the component of the noise field that directly shifts the time. We note that no reference to inverse scattering theory or soliton perturbation theory is required to obtain either noise component. All the functions on the right-hand sides of (4.119), q, U, t_p, and Ω are determined by the unperturbed solution $\bar{u}$ of (4.110) with no noise contribution. From the equation for Ω in (4.115) we also find that the variance of the frequency in one amplifier is

$$<\delta\Omega^2> = [(G-1)/(2g_0)](\delta q_1, \delta q_1) = 2\theta(G-1)\int_{-\infty}^{\infty} |q_t|^2 \mathrm{d}t/U^2 \tag{4.121}$$

where $G = \exp(2g_0 L_{\mathrm{amp}})$ is the total amplifier gain and L_{amp} is the amplifier length. Equations (4.119) and (4.121) are valid for any arbitrary pulse shape and represent a key result. Integrals on the right-hand sides of (4.119) can easily be calculated numerically once $\bar{u}$ is known computationally. Although (4.116)–(4.119) appear complex at first sight, they have a simple physical meaning. The terms A, B, and C are responsible for the timing jitter due to the frequency shift, timing jitter due to the pulse chirp, and the timing jitter due to direct time offset. Equations (4.116)–(4.119) and (4.121) differ from the Gordon–Haus theory [54, 61] in two respects. First, the variance of the frequency shift after an amplifier is proportional to the ratio of the square of the pulse bandwidth divided by the pulse energy. This result is physically intuitive because when the bandwidth of a pulse becomes larger, the pulse will incorporate a larger amount of the noise radiation. On the other hand, when the pulse energy becomes larger, then the influence of the incorporated noise becomes smaller. This frequency shift translates into a shift of the central time which is the contribution of the A term. Second, the B term represents a contribution to the timing jitter caused by the pulse chirp. This term contains a new physical effect that is not in the Gordon–Haus calculation because standard solitons are unchirped. To understand this effect, we recall that there exist two sources of the time shift. The first is the frequency shift, represented by the term $\mathscr{F}$ in (4.114), and the second is the time offset, represented by the term $\mathcal{S}$ in (4.114). Consequently, the term A represents the timing jitter induced by the frequency shift alone, the term C represents the timing jitter induced by the direct time offset alone. The term B represents the interference between the two effects. Physically, the term B has the following origin. If a pulse has a chirp, then its local frequency depends on time, so that a time offset leads to a frequency shift that in its turn translates into an additional shift in the pulses central time. The interference term B can either increase or decrease the total timing jitter depending on the sign of the chirp. For example, if we consider a chirped pulse $q = |q|\exp[i\alpha(t - t_p)^2]$, then the scalar cross product becomes

$$(\delta q_1, \delta q_2) = 16\alpha \frac{g_0 \theta}{U^2} \int_{-\infty}^{\infty} (t - t_p)^2 |q|^2 \mathrm{d}t \tag{4.122}$$

Hence, it follows from (4.119) that when the product $D\alpha$ is positive then is positive, increasing the jitter; however, if the product $D\alpha$ is negative then the opposite occurs.

Timing Jitter in Filtered Systems

As in the previous section, we differentiate t_p, Ω, and U in (4.112) with respect to z and combine them with 4.110 to derive the dynamic equations for the central time and central frequency of a signal pulse. In this case, $b(z) \neq 0$ which allows us to include filtering. We note that only quadratic filters are considered in (4.110), but our results can be generalized without much difficulty to include more general filters. We now obtain

$$\begin{aligned} \frac{\mathrm{d}t_p}{\mathrm{d}z} &= D\Omega + \frac{i}{U} \int_{-\infty}^{\infty} (t - t_p) \left[ib|q_t|^2 - b\Omega(qq_t^* - q^* q_t) + \hat{F}_1 \right] \mathrm{d}t \\ \frac{d\Omega}{\mathrm{d}z} &= \frac{1}{U} \int_{-\infty}^{\infty} \left\{ -b\left[2\Omega|q_t|^2 + \frac{i}{2}(q_t q_{tt}^* - q_t^* q_{tt}) \right] + \hat{F}_2 \right\} \mathrm{d}t \end{aligned} \tag{4.123}$$

where

$$\begin{aligned} \hat{F}_1 &= q\exp(i\Omega t)\hat{F}^* - q^*\exp(-i\Omega t)\hat{F} \\ \hat{F}_2 &= q_t^*\exp(-i\Omega t)\hat{F} + q_t\exp(i\Omega t)\hat{F}^*. \end{aligned} \tag{4.124}$$

Implicit solutions of (4.123) are again

$$t_p = \mathcal{F} + \mathcal{S} \tag{4.125}$$

where

$$\mathcal{F} = \int_0^z \hat{D}\Omega \mathrm{d}z'$$
$$\mathcal{S} = -\int_0^z \left\{ \frac{1}{U} \int_{-\infty}^{\infty} (t - t_p)\left[b|q_t|^2 + i\hat{F}_1\right]\mathrm{d}t \right\} \mathrm{d}z' \tag{4.126}$$

and

$$\hat{D} = \left[D - \frac{ib}{U} \int_{-\infty}^{\infty} (t - t_p)(qq_t^* - q^* q_t)\mathrm{d}t \right] B_f$$
$$\Omega = \Omega_0 + \int_0^z \left\{ \frac{1}{2U} \int_{-\infty}^{\infty} \left[ib(q_t^* q_{tt} - q_t q_{tt}^*) + 2\hat{F}_2 \right] \mathrm{d}t \right\} \mathrm{d}z' \tag{4.127}$$
$$B_f = \exp\left[-2 \int_0^z b(z')\Omega_B^2(z')\mathrm{d}z' \right]$$
$$\Omega_B^2 = \frac{1}{U} \int_{-\infty}^{\infty} |q_t|^2 \mathrm{d}t.$$

Typically, the filtering term is much smaller than the dispersion term. Substituting (4.127) into (4.126) and then (4.126) into (4.125), linearizing (4.125) around the unperturbed solution of (4.110), taking the square of (4.125), and using (4.111) and (4.117), we find the variance of the central time,

$$\sigma_t^2 = \langle t_p^2 \rangle - \langle t_p \rangle^2 = A + B + C \tag{4.128}$$

where now A, B, and C are

$$A = \langle \mathcal{F}^2 \rangle = \int_0^z \hat{D}(z_1)\mathrm{d}z_1 \int_0^{z_1} \hat{D}(z_2)\mathrm{d}z_2 \int_0^{z_2} (\delta\hat{q}_1, \delta\hat{q}_1)\mathrm{d}z_2'$$
$$B = 2\langle \mathcal{F}\mathcal{S} \rangle = \int_0^z \hat{D}(z_1)\mathrm{d}z_1 \int_0^{z_1} (\delta\hat{q}_1, \delta q_2)\mathrm{d}z_1' \tag{4.129}$$
$$C = \langle \mathcal{S}^2 \rangle = \frac{1}{2}\int_0^z (\delta q_2, \delta q_2)\mathrm{d}z'$$

and where

$$\delta\hat{q}_1 = \frac{2\sqrt{g_0\theta}}{UB_f} q_t \tag{4.130}$$

We note that all the functions on the right-hand sides of (4.128) and (4.129) that define σ_t^2 are deterministic.

Amplitude Jitter

Differentiating U in (4.112) with respect to z and combining it with (4.110) we obtain

$$\frac{dU}{dz} = (2g - b\Delta^2)U + i\int_{-\infty}^{\infty}(u\hat{F}^* - u^*\hat{F})dt \tag{4.131}$$

where we define the mean square central frequency of the pulse

$$\Delta^2 = \frac{1}{U}\int_{-\infty}^{\infty}|u_t|^2 dt.$$

An implicit solution of 4.131 is

$$U = \left[U_0 + i\int_0^z\int_{-\infty}^{\infty}(u\hat{F}^* - u^*\hat{F})dt\frac{dz'}{\mathcal{G}(Z')}\right]\mathcal{G}(z) \tag{4.132}$$

where U_0 is the input photon number of the pulse, and

$$\mathcal{G}(z) = \exp\left[\int_0^z(2g - b\Delta^2)dz'\right] \tag{4.133}$$

represents the mean gain. Linearizing (4.132) and averaging the square of (4.132) using (4.111), we obtain the variance of the photon number in the signal pulse

$$\sigma_U^2 = \langle U^2\rangle - \langle U\rangle^2 = 4U_0 g_0\mathcal{G}^2(z)\int_0^z\frac{\theta(z)}{\mathcal{G}(z')}dz'. \tag{4.134}$$

It will be useful to define a normalized energy variance

$$\rho^2 = \frac{\sigma_U^2}{\langle U\rangle^2} = \frac{\sigma_U^2}{U_0^2\mathcal{G}^2(z)}. \tag{4.135}$$

In a system with periodically distributed lumped amplifiers, in which $\mathcal{G} = 0$ after each amplifier, (4.135) reduces to

$$\rho^2 = 2n_{sp}\frac{G-1}{U_0}K \tag{4.136}$$

where K is the number of the amplifiers. Equation 4.136 implies that the increase of the normalized energy variance is proportional to the number of the noise photons per mode $n_{sp}(G-1)K$ added to the pulse by K amplifiers. We note that due to filter losses $b\Delta^2$ in

4.3.2.3 Comparison with Monte Carlo Simulations

To validate the use of linearization to calculate the timing and amplitude jitter, we simulated the propagation of signal pulses with different signal formats—RZ, NRZ, and DMS—in a dispersion-managed fiber with alternating spans of normal dispersion D_1 of length L_1 and anomalous dispersion D_2 of length L_2. We used the split-step method to solve (4.3) along with the Monte Carlo method to generate the ASE noise. In this study, we used the unbiased Monte Carlo method. We repeated the same calculations for a hundred different realizations of the ASE noise, except where stated, and collected their statistics to find the variance and the average of the pulses central time and energy at the midpoint of anomalous dispersion span for each set of the system parameters. The loss coefficient of the fiber was 0.21 dB/km.

Dispersion-Managed Solitons

For DMS calculations we propagated single pulses. Figure 4.24 compares the timing jitter calculated using Monte Carlo simulations to the results of our linearization approach for DMS pulses with a pulse duration $t_{\mathrm{FWHM}} = 20$ ps at the midpoint of the anomalous span. For a given average dispersion, the energy of the DMSs is larger than that of standard solitons, and this increase in energy is referred to as the enhancement factor. A simple theoretical approach to calculate the timing jitter is to use the Gordon–Haus formula for the timing jitter and to reduce it by the square root of the enhancement factor [132]. One can see in Fig. 4.24 that if the profile of the DMS is nearly Gaussian, the difference between the reduced Gordon–Haus formula [132] and the Monte Carlo simulations, though visible at 10 Mm, remains small. In the case of Fig. 4.24a the energy enhancement factor is as large as 2.17. However, for more strongly dispersion-managed fibers with an energy enhancement factor of 8.16, shown in Fig. 4.24b, the shape of the DMS becomes more like a flattop and its time-bandwidth product becomes larger, $\Delta\nu\Delta t = 0.661$ as compared to $\Delta\nu\Delta t = 0.424$ for a Gaussian. As the pulse bandwidth increases for a fixed pulse duration, the frequency shift increases in accordance with (4.121). The larger frequency shift translates into a larger contribution to the term which at long distances makes the largest contribution to the timing jitter. The Gordon–Haus theory, even with the standard reduction factor, does not take into account the dependence of the jitter on the pulse profile, as it is based on a hyperbolic-secant pulse shape. That is why the deviation from the reduced Gordon–Haus formula becomes considerable in the case of Fig. 4.24b. On the other hand, there is excellent agreement with our theory. For strong dispersion management, DMSs can exist with zero and normal path average dispersion [15, 33, 55, 82, 83, 111, 141]. Our calculations show that despite the complete

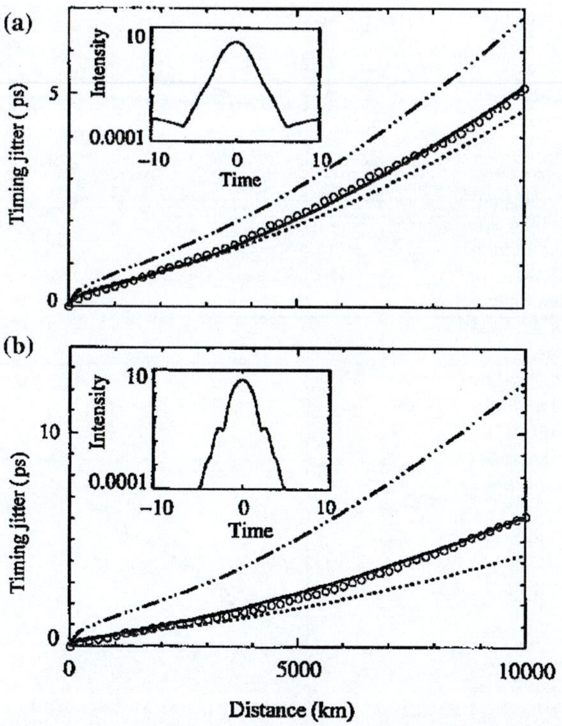

Fig. 4.24 The root-mean-square central time of the pulse versus distance for a dispersion-managed soliton system. Inserted graphs are intensity profiles of the dispersion-managed solitons plotted on a logarithmic scale. *Circles* are results of Monte Carlo simulations, *solid lines* are results of our theory, and the *broken lines* and *dots* represent the Gordon–Haus theory and reduced Gordon–Haus theory, respectively; $t_{FWHM} = 20$ ps, losses are 0.21 dB/km; **a** dispersion coefficients are $\beta_1'' = -3.0\text{ps}^2/\text{km}$ and $\beta_2'' = 2.8\text{ps}^2/\text{km}$, the amplifier distance is 50 km, amplifiers are placed at the edges and at the midpoints of normal and anomalous dispersion spans, span lengths are $L_1 = L_2 = 100$ km, peak power at the midpoint of the anomalous dispersion span is 4.12 mW and **b** dispersion coefficients are $\beta_1'' = -3.75\text{ps}^2/\text{km}$ and $\beta_2'' = 3.55\text{ps}^2/\text{km}$, amplifiers are placed at the edges of the spans, amplifier distance is 100 km, span lengths are $L_1 = L_2 = 100$ km, peak power at the midpoint of the anomalous span is 1.39 mW. Here, β_j'' is the dispersion coefficient and L_j is the length of span j. (From [56], Fig. 1. Reproduced by permission of © 1999 The Institute of Electrical and Electronics Engineers)

dispersion compensation there is a residual timing jitter for zero average dispersion. In this case, however, the frequency jitter term A no longer makes the largest contribution to the timing jitter.

Figure 4.25 shows dependence of the timing jitter on distance and the different contributions of the A, B, and C terms for zero average dispersion. We note that, as follows from (4.119), A and C are always positive. However, for this configuration the term B is larger then either A or C and negative, which means that the chirp-induced central time shift significantly suppresses the timing jitter caused by the

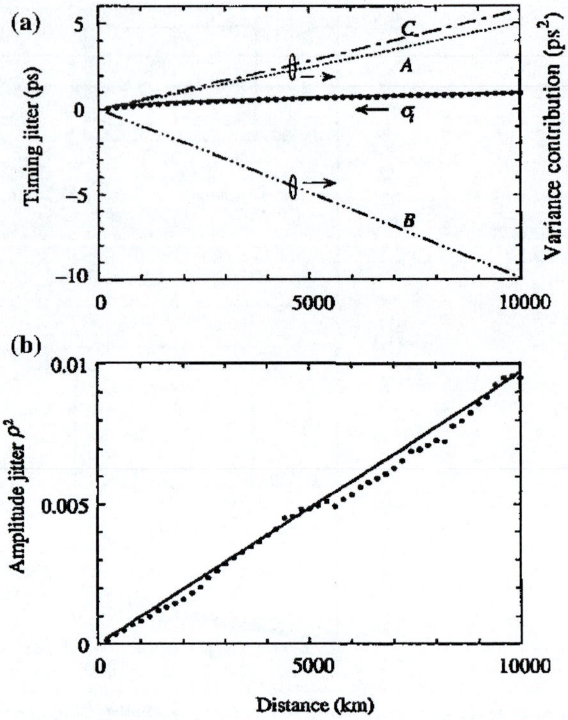

Fig. 4.25 **a** Timing jitter σ_t and **b** amplitude jitter ρ^2 versus distance for a dispersion-managed soliton at zero average dispersion. In (**a**), we also show the individual terms *A*, *B*, and *C*. *Solid lines* are results of the theory, *dots* are the results of Monte Carlo simulations; the peak power at the midpoint of the anomalous dispersion span is 4.33 mW and $t_{\mathrm{FWHM}} = 20$ ps, $\beta_1'' = -\beta_2'' = -11.0\mathrm{ps}^2/\mathrm{km}$, $L_1 = L_2 = 100$. Amplifiers are placed at the edges and the midpoints of the spans every 50 km. (From [56], Fig. 2. Reproduced by permission of © 1999 The Institute of Electrical and Electronics Engineers)

frequency shift and time offset. We note that the inner integral for *B* in (4.119) is proportional to the chirp α due to (4.122). Hence, amplifiers located at the midpoints make a small contribution to *B* because the chirp at the midpoint is small. However, amplifiers located at the ends of the spans always yield a negative contribution to *B*. Thus, locating the amplifiers at the ends of the spans reduces the timing jitter. The resultant timing jitter agrees perfectly with the Monte Carlo simulations.

Figure 4.25b shows the amplitude jitter, i.e., the normalized variance of the photon number in the pulse, versus distance. It grows linearly and remains well below 0.028 up to 10 Mm. From [96], we infer that $\rho^2 = 0.028$ corresponds to a BER of 10^{-9}.

Figure 4.26 illustrates the results for two different arrangements of the amplifiers. In the first, the amplifiers are placed at the edges of the spans, and, in the second, the amplifiers are placed at the midpoints. The peak power and the pulse duration at the midpoint of the anomalous span are 1.2 mW and 20 ps, respectively. Since the chirp is small at the midpoints, the term *B* is small in Fig. 4.26b in contrast to Fig. 4.26a. The resultant timing jitter is larger in Fig. 4.26b than in Fig. 4.26a because the term *B* does not partially cancel out the *A* and *C* terms.

Finally, we compared our linearization approach to Monte Carlo simulations for the configuration reported in the experiment of Carter and Jacob [30] on

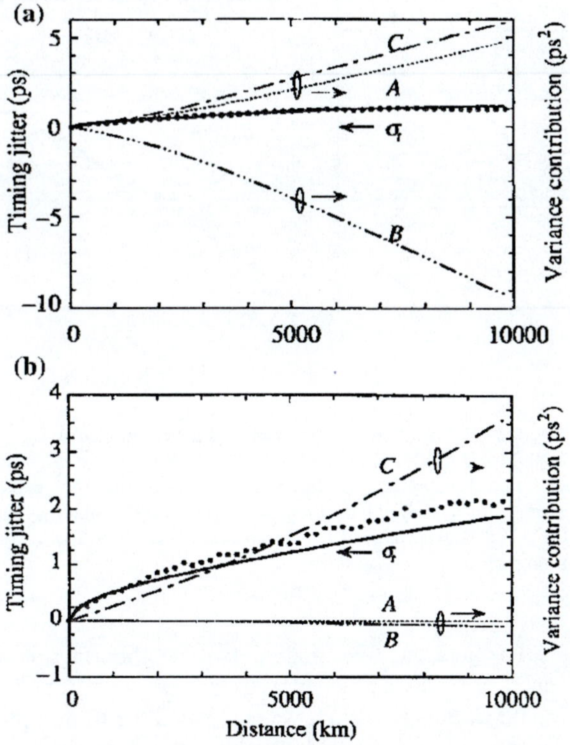

Fig. 4.26 **a** Dependence of the timing jitter σ_t and the contributions of the terms A, B, and C to the central time variance on distance for the dispersion-managed soliton at zero average dispersion in a system with $L_1 = L_2 = 50$, peak power in the midpoint of the anomalous dispersion span of 1.21 mW, amplifier spacing of 50 km and other parameters being the same as in Fig. 4.24 for the amplifier arrangements at **a** the edges and at **b** the midpoints of the spans. (From [56], Fig. 3. Reproduced by permission of © 1999 The Institute of Electrical and Electronics Engineers)

ultra-long-distance propagation of DMSs in a filtered recirculating fiber loop consisting of 100 km of fiber in the normal dispersion regime and about 7 km of fiber in the anomalous dispersion regime. Figure 4.26 shows the timing jitter in the loop as a function of propagation distance. The pulse durations in the midpoints of the anomalous and normal dispersion spans were 9.5 and 12.5 ps, respectively. The solid curve in Fig.4.27 was calculated based on (4.128) and (4.129), the black circles show the result of Monte Carlo simulations with 200 realizations of the ASE noise, and the open circles show the experimental results. We note that the timing jitter grows more slowly than in the unfiltered case of Fig. 4.24. We find a remarkable agreement of our approach with the Monte Carlo simulations and with the experimental data over the entire propagation distance of 22 Mm.

RZ Pulses

We simulated propagation of RZ signals by launching raised-cosine prechirped pulses of the form $u(0,t) = A[1 - \cos(t/t_0)]\exp(-it^2/t_c^2)$ with $t_0 = 32$ ps, $t_c^2 = 1{,}000\text{ps}^2$, and a peak power of 1.08 mW at the beginning of the anomalous

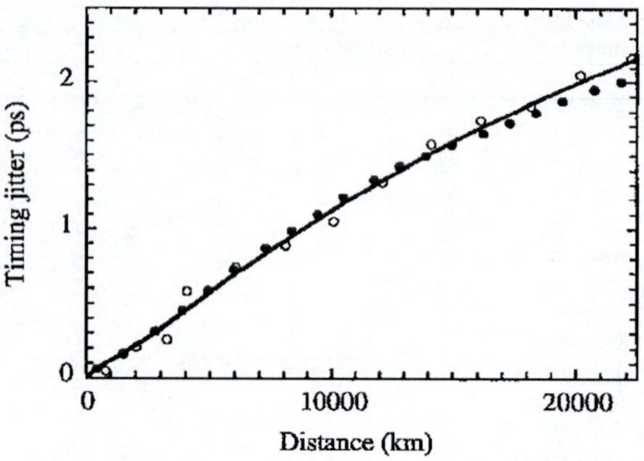

Fig. 4.27 Comparison of the timing jitter calculated using linearization approach (*solid line*) with Monte Carlo simulations (*black circles*) to the experiment of Carter and Jacob [30] on ultralong-distance propagation of the dispersion-managed solitons in a recirculating loop with an average dispersion of 0.04 ps/nm-km. The experimental results are plotted as *open circles*. (From [56], Fig. 4. Reproduced by permission of © 1999 The Institute of Electrical and Electronics Engineers)

dispersion span. We studied two neighboring pulses, with periodic boundary conditions, which allowed us to include the effect of interpulse interference. In our Monte Carlo simulation, we calculated the timing jitter and amplitude jitter using the later pulse. Figure 4.28 shows the dependence of the timing and amplitude jitter on distance. We show the RZ pulse evolution in the presence of noise in Fig. 4.29. The dynamics of the timing jitter is close to that of Fig. 4.25 for the DMS, although the final jitter is a little larger. As in the previous cases there is a good agreement between the linearization approach and Monte Carlo simulations. In this case, timing jitter remains small and the major source of errors is the amplitude jitter. There is some drift of the amplitude jitter at large distances which may indicate that the linearization approach is breaking down.

NRZ Pulses

There are significant pattern dependences in the nonlinear pulse evolution of the marks ("1") in the NRZ format due to intersymbol interference. Thus, we must explore different combinations of the marks and spaces ("0"). Anderson and Lyle [10] have shown that in many cases, the behavior is dominated by the eight combinations: 0-0-0, 0-0-1, 0-1-0, 0-1-1, 1-0-0, 1-0-1, 1-1-0, and 1-1-1. Thus, in our study of amplitude and timing jitter, it is appropriate for us to focus on the strings 0-1-0, 0-1-1-0, and 0-1-1-1-0. We simulated propagation of these signals by launching ideal rectangular pulses with a power of 0.35 mW using the same dispersion map and amplifier arrangement as in Fig. 4.25. The time duration is

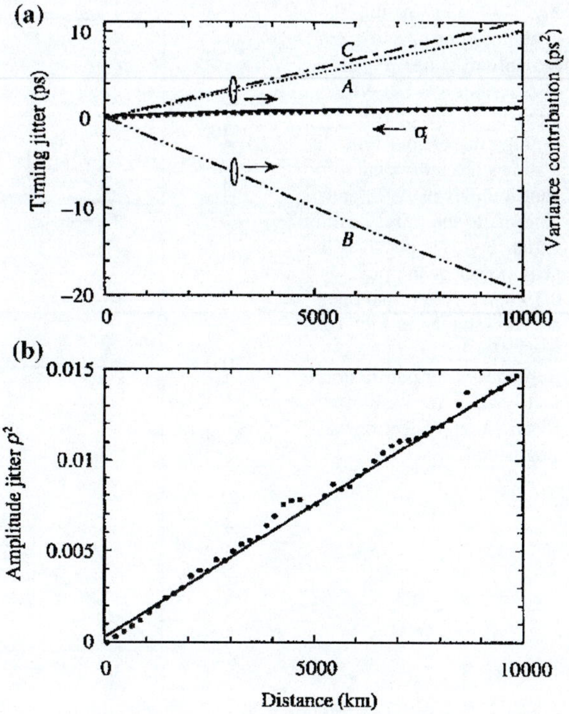

Fig. 4.28 Plots of the **a** timing jitter σ_t and **b** amplitude jitter ρ^2 versus distance for an RZ pulse at zero average dispersion. Plot **(a)** shows the individual contributions of *A*, *B*, and *C* terms to the pulse position variance versus distance. *Solid lines* are results of the theory; *dots* are the results of the Monte Carlo simulations; pulse duration and the peak power at the midpoint of the anomalous dispersion span are $t_{\mathrm{FWHM}} = 20$ ps and 1.08 mW, $\beta_1'' = -\beta_2'' = -11\mathrm{ps}^2/\mathrm{km}$, $L_1 = L_2 = 100$. Amplifiers are placed at the edges and the midpoints of the spans every 50 km. (From [56], Fig. 5. Reproduced by permission of © 1999 The Institute of Electrical and Electronics Engineers)

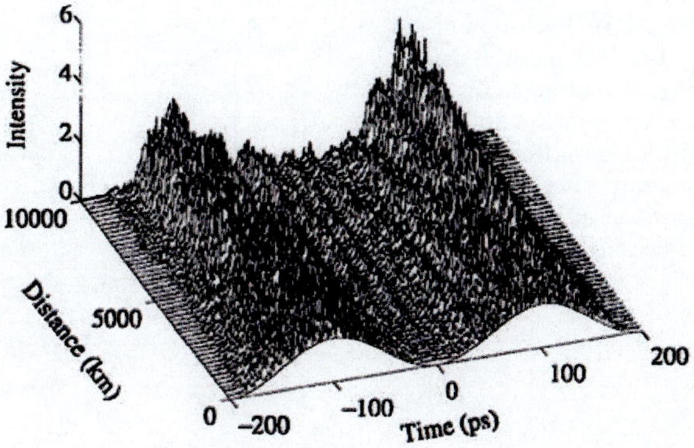

Fig. 4.29 Evolution of the RZ pulse in the presence of noise for the system in Fig. 4.28. (From [56], Fig. 6. Reproduced by permission of © 1999 The Institute of Electrical and Electronics Engineers)

100 ps per bit, corresponding to a 10 Gb/s bit rate. While real NRZ signals have rounded edges, the use of rectangular pulses provides a more stringent test of our approach. The results are shown in Figs. 4.29, 4.30, 4.31, 4.32, 4.33 and 4.34. The frequency jitter given by the *A* term is significantly larger for the pattern 0-1-0 than for the patterns 0-1-1-0 and 0-1-1-1-0 because the bandwidth is largest and the energy is smallest in this case. Similarly, the *B* and *C* terms decrease in magnitude as we move from the pattern 0-1-0 to the pattern 0-1-1-0 to the pattern 0-1-1-1-0.

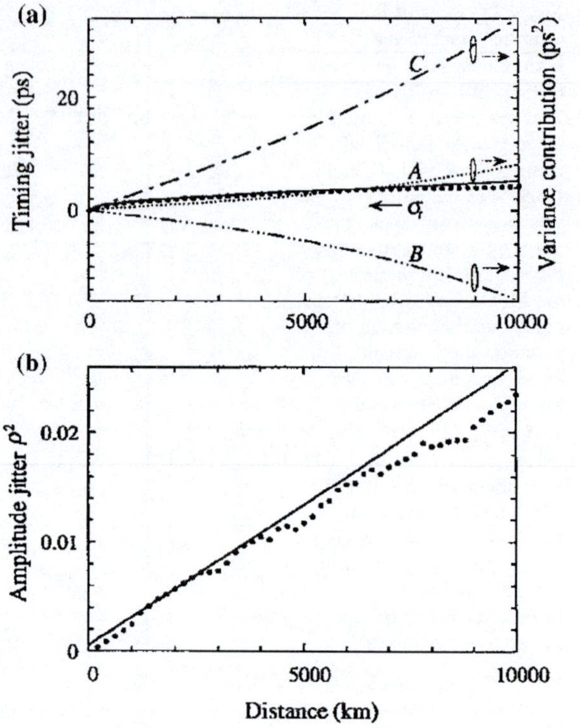

Fig. 4.30 Plots of the **a** timing jitter σ_t and **b** amplitude jitter ρ^2 versus distance for a 0-1-0 NRZ pulse of 100 ps at zero average dispersion. Plot **a** shows the individual contributions of *A*, *B*, and *C* terms to the pulse position variance versus distance. The peak power of the pulse is 0.35 mW. System parameters are the same as in Fig. 4.26 (From [56], Fig. 7. Reproduced by permission of © 1999 The Institute of Electrical and Electronics Engineers)

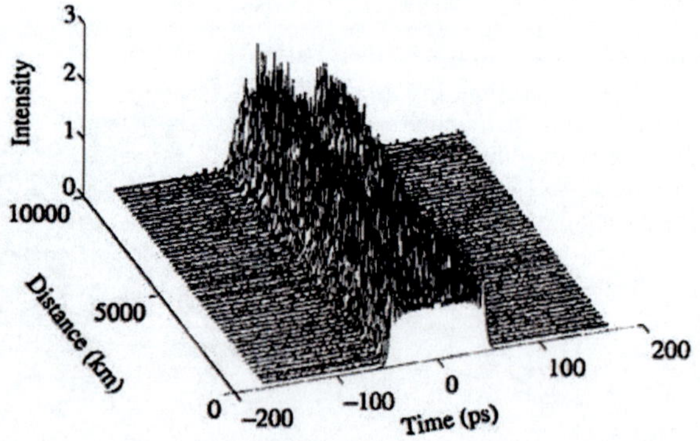

Fig. 4.31 Evolution of a 0-1-0 NRZ pulse in the presence of noise for the system in Fig. 4.30. (From [56], Fig. 8. Reproduced by permission of © 1999 The Institute of Electrical and Electronics Engineers)

However, the resultant timing jitter is a few picoseconds in all three cases. The amplitude jitter is the largest for the pattern 0-1-0 and decreases for 0-1-1-0 and 0-1-1-1-0. However, this result has been normalized to the total pulse energy. If we examine instead the amplitude jitter per bit, we find that this value is constant as we change the pattern. We find good agreement between the linearization approach and Monte Carlo simulations, although it appears that the linearization approach may be beginning to break down beyond 5,000 km.

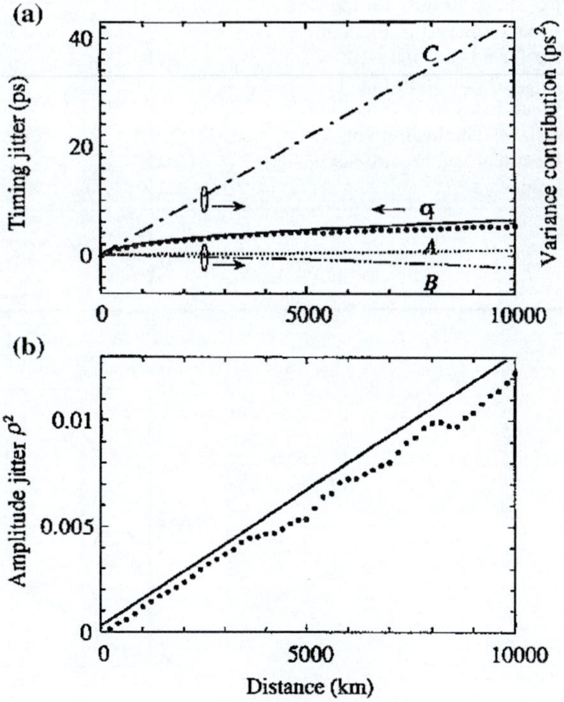

Fig. 4.32 Results for the same system as in Fig. 4.30 but for a 0-1-1-0 NRZ pulse. (From [56], Fig. 9. Reproduced by permission of © 1999 The Institute of Electrical and Electronics Engineers)

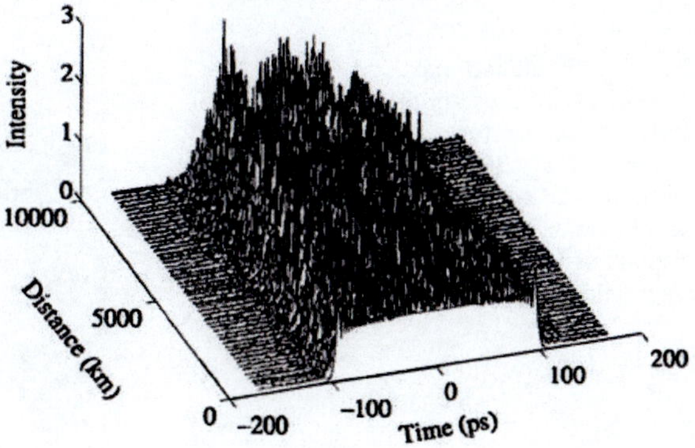

Fig. 4.33 Evolution of a 0-1-1-0 NRZ pulse in the presence of noise for the system in Fig. 4.32. (From [56], Fig. 10. Reproduced by permission of © 1999 The Institute of Electrical and Electronics Engineers)

4.3.2.4 Estimates of the Bit Error Rate

Calculation of the timing and amplitude jitter leads to important insights into the physical sources of errors in communication systems. Moreover, these quantities can be measured experimentally, allowing us to compare theory and experiment [30]. However, the BER is more important than either of these quantities for designing systems, and one would ideally like to use these calculations to infer the

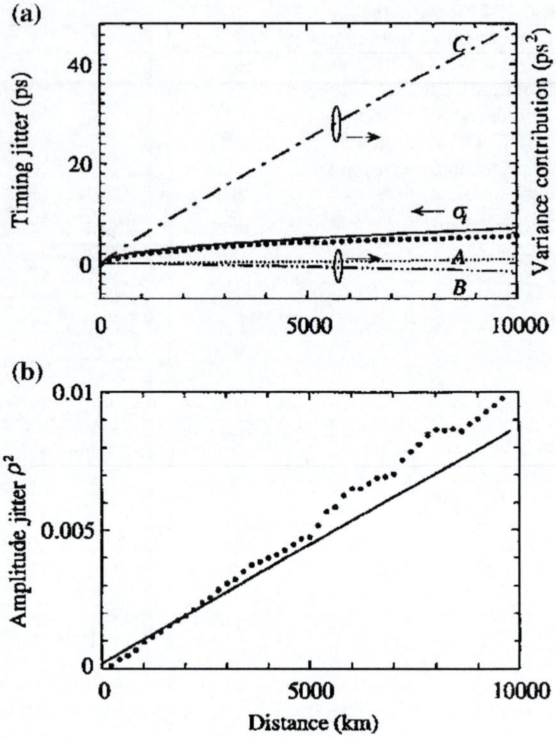

Fig. 4.34 Results for the same system as in Fig.4.30 but for a 0-1-1-1-0 NRZ pulse. (From [56], Fig. 11. Reproduced by permission of © 1999 The Institute of Electrical and Electronics Engineers)

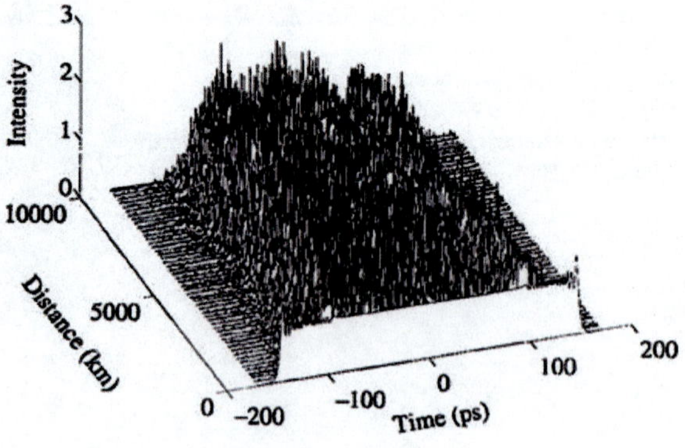

Fig. 4.35 Evolution of a 0-1-1-1-0 NRZ pulse in the presence of noise for the system in Fig. 4.30. (From [56], Fig. 12. Reproduced by permission of © 1999 The Institute of Electrical and Electronics Engineers)

bit error rate. In principle, this is difficult to do accurately. First, one is typically interested in bit error rates that are 10^{-9} or even lower, implying that one must accurately know the tails of the distribution functions for the time shifts and the amplitudes. Even when the jitter, which corresponds to the variance of the distribution function is determined accurately using the linearization assumption, the tails may not be [46, 48, 97]. Second, actual bit errors depend critically on the receiver design. Appropriate design of both the optical and electrical filters can play a crucial role in reducing the errors [110].

In this section, we will show how to estimate the bit error rate, using a simple integrate-and-dump receiver. We will assume that spontaneous–spontaneous beat noise is negligible throughout most of this section, although we will briefly consider its effect at the end. Even in this case we cannot completely determine the bit error rate from just the timing and amplitude jitter because the amplitude jitter is typically accompanied by pulse distortion. For example, in the case of standard solitons, the pulse duration is proportional to $U^{1/2}$ [60]. However, it is often the case that either the timing or the amplitude jitter becomes the dominant source of errors. When the timing jitter dominates, then it is possible to calculate the BER due to it while neglecting amplitude jitter and the accompanying pulse distortion. Conversely, when amplitude jitter dominates over timing jitter and the time window is large enough to include the entire pulse, then pulse distortion does not effect the calculation of the bit error rate due to amplitude jitter.

We begin by making a key observation: The Langevin noise source $\hat{F}(Z,t)$ can be treated as a Gaussian-distributed white noise source [90] in each point z. The Gaussian noise sources in different points z are statistically independent. Since $\mathrm{d}t_p/\mathrm{d}z$ and $dU/\mathrm{d}z$ and are a linear superposition of contributions from $\hat{F}$ and $\hat{F}^*$, as shown in (4.113) and (4.131), it follows that t_p and U must be Gaussian-distributed [114]. Thus, we may write their probability distribution functions explicitly as

$$P(t_p) = \frac{1}{\sqrt{2\pi}\sigma_t}\exp\left[-\frac{(t_p - \langle t_p\rangle)^2}{2\sigma_t^2}\right] \tag{4.137}$$

$$W(U) = \frac{1}{\sqrt{2\pi}\sigma_U}\exp\left[-\frac{(U - \langle U\rangle)^2}{2\sigma_U^2}\right] \tag{4.138}$$

where we recall that $\langle t_p\rangle$ and $\langle U\rangle$ are respectively the central time and photon number in the absence of noise. Errors occur when the energy of a mark inside the time window falls below a threshold U_{th} which is determined by the background optical noise in the spaces and electrical noise in the receiver. We will designate the timing window $-T_w/2 < t - \langle t_p\rangle < T_w/2$. Since receivers use clock recovery to locate the timing window, the window will be centered about the overall average $t = \langle t_p\rangle$. The number of photons in the timing window may be written

$$U_w\left(t_p - \langle t_p\rangle\right) = \int\limits_{\langle t_p\rangle - T_w/2}^{\langle t_p\rangle + T_w/2} |\bar{u}(t - t_p)|^2 \mathrm{d}t = \int\limits_{-T_w/2}^{T_w/2} |\bar{u}(t' - t_p - \langle t_p\rangle)|^2 \mathrm{d}t'. \tag{4.139}$$

There is some value of $|t_p - \langle t_p\rangle|$ beyond $U_w\left(t_p - \langle t_p\rangle\right) < U_{\mathrm{th}}$. We will designate this value as $t_p, \mathrm{th} - < t_p >$. Then the BER is given by

$$\begin{aligned}\mathrm{Prob}\left[U_w\left(t_p - \langle t_p\rangle\right) < U_{\mathrm{th}}\right] &= 2\int_{t_p} , \mathrm{th} - \langle t_p\rangle^{\infty} \frac{1}{\sqrt{2\pi}\sigma_t}\exp\left(-\frac{t^2}{2\sigma_t^2}\right)\mathrm{d}t \\ &= \mathrm{erfc}\left(\frac{t_p}{,}\mathrm{th} - \langle t_p\rangle\sqrt{2}\sigma_t\right).\end{aligned} \tag{4.140}$$

The factor of two preceding the integral in (4.140) accounts for the two-sided nature of the distribution, i.e., errors occur both when $t_p - \langle t_p\rangle > t_p, \mathrm{th} - \langle t_p\rangle$ and when $t_p - \langle t_p\rangle < -\left(t_p, \mathrm{th} - \langle t_p\rangle\right)$. While the specific shape of the final pulse plays an important role in determining the dependence of $t_{p,\mathrm{th}}$ on U_{th}, the BER is given by (4.140) regardless of the shape given that σ_t is fixed. This is a consequence of (4.137).

We may now calculate the BER due to amplitude jitter, and we find

$$\begin{aligned}\mathrm{Prob}[U < U_{\mathrm{th}}] &= \int_0^{U_{\mathrm{th}}} \frac{1}{\sqrt{2\pi}\sigma_U}\exp\left[-\frac{(U - \langle U\rangle)^2}{2\sigma_U^2}\right]\mathrm{d}U \\ &= \frac{1}{2}\mathrm{erfc}\left(\frac{\langle U\rangle - U_{\mathrm{th}}}{\sqrt{2}\sigma_U}\right),\end{aligned} \tag{4.141}$$

which again has a very simple form. In obtaining (4.141), we set $\langle U\rangle/\sigma_U \to \infty$. Since the energy cannot be negative, it is physically meaningless that the probability distribution function given by (4.138) allows negative energies. This result is a consequence of neglecting the spontaneous–spontaneous beat noise and will not affect (4.140) as long as it is genuinely negligible at the threshold value U_{th}. However, when there are many noise modes, then the spontaneous–spontaneous beat noise may make a significant contribution to the total energy even though the contribution of a single mode is negligibly small. Marcuse [95] included this noise, assuming that the noise energy was equally divided among all the noise modes. The number M of the complex modes is given by $M = B_{\mathrm{opt}}T_w$, where B_{opt} is the optical bandwidth. Thus, if for example, we consider a system in which a single optical wavelength channel occupies a bandwidth of 100 GHz and the receiver window is 100 ps, then there are 10 complex modes. In this case, the number of photons per mode U_{mode} is given by

$$U_{\mathrm{mode}} = n_{\mathrm{sp}}(G - 1)K \tag{4.142}$$

where G is the amplifier gain and K is the number of inline amplifiers just like in (4.134). One then finds that the probability distribution function for the energy in a space is given by [95]

$$P_0(U) = \frac{1}{(M-1)!}\frac{U^{M-1}}{U_{\mathrm{mode}}^M}\exp\left(-\frac{U}{U_{\mathrm{mode}}}\right) \tag{4.143}$$

while the probability distribution for the energy in a mark is given by [95]

$$P_1(U) = \frac{1}{U_{\text{mode}}}\left(\frac{U}{\langle U\rangle}\right)^{(M-1)/2} \exp\left(-\frac{U+\langle U\rangle}{U_{\text{mode}}}\right) I_{M-1}\left(2\frac{\sqrt{U\langle U\rangle}}{U_{\text{mode}}}\right) \tag{4.144}$$

where I_{M-1} is the modified Bessel function of order M–1. Marcuse [95] has discussed in some detail the consequence of replacing (4.144) with (4.138). For the RZ and NRZ systems that we studied in Sect. 4.3.2.3, it is possible to estimate that the error made in determining the threshold value at which the error rate reaches 10^{-9} is about 10–20%, assuming $M = 10$.

A more refined estimate would require a method such as the covariance matrix method discussed in Sect. 4.2.3 because a set of modes that is originally orthogonal is mixed together by the nonlinearity so that they do not remain orthogonal. Physically, the signal can parametrically pump some noise modes due to the four-wave mixing (FWM). Thus, the assumption that the noise energy is equally divided among the modes in the case when spontaneous–spontaneous beat noise becomes significant is suspect at best, particularly in real systems with more complicated receiver filters than we have considered. This assumption is also false in systems with inline filters, as is the case in many soliton systems [30].

4.3.2.5 Summary

We have developed an approach, based on linearizing the modified NLS (4.110) around a numerically determined noise-free solution. This approach allows us to determine the timing jitter and the amplitude jitter for arbitrary pulse shapes. Its principal advantage is that it allows us to avoid numerically time-consuming Monte Carlo simulations. We have applied this approach to DMS, RZ, and NRZ pulse formats, and we have validated it by comparison to Monte Carlo simulations. We have shown that within the limits of validity of the linearization approximation, the timing jitter is Gaussian-distributed. When, in addition, spontaneous–spontaneous beat noise may be neglected, the energy is also Gaussian-distributed. We have used this result to separately estimate the BER due to timing jitter and amplitude jitter when one or other is the dominant source of errors.

Acknowledgments The authors gratefully acknowledge valuable contributions to this work by Prof. Curtis R. Menyuk, Prof. Gary M. Carter, and Prof. John Zweck with the University of Maryland Baltimore County (UMBC), Baltimore, MD, as well as Prof. William L. Kath, Northwestern University, Chicago, IL.

References

1. Ablowitz MJ, Ahrens C, Biondini G, Chakravarty S, Docherty A (2004) Reduction of collision-induced timing shifts in dispersion-managed quasi-linear systems with periodic-group-delay dispersion compensation. Opt Lett 29(20):2354–2356
2. Ablowitz MJ, Biondini G, Biswas A, Docherty A, Hirooka T, Chakravarty S (2002) Collision-induced timing shifts in dispersion-managed soliton systems. Opt Lett 27(5):318–320

3. Ablowitz MJ, Biondini G, Chakravarty S, Horne RL (1998) On timing jitter in wavelength-division multiplexed soliton systems. Opt Commun 150(1–6):305–318
4. Ablowitz MJ, Docherty A, Hirooka T (2003) Incomplete collisions in strongly dispersion-managed return-to-zero communication systems. Opt Lett 28(4):1191–1193
5. Ablowitz MJ, Hirooka T (2000) Resonant nonlinear intrachannel interactions in strongly dispersion-managed systems. Opt Lett 25(24):1750–1752
6. Abramowitz M, Stegun IA (1965) Handbook of mathematical functions. Dover, New York
7. Agrawal GP (1995) Nonlinear fiber optics, 2nd edn. Academic Press, London
8. Agrawal GP (1997) Fiber-optics communication systems, 2nd edn. Wiley, New York
9. Ahrens C, Ablowitz MJ, Docherty A, Sinkin OV, Zweck J, Grigoryan VS, Menyuk CR (2006) Asymptotic analysis of collision-induced timing shifts in return-to-zero quasi-linear systems with pre- and post-dispersion compensation. Opt Lett 31(1):5–7
10. Anderson CJ, Lyle JA (1994) Technique for evaluating system performance using Q in numerical simulations exhibiting intersymbol interference. Electron Lett 30(1):71–72
11. Arnold L (1992) Stochastic differential equations: theory and applications. Krieger Publishing Company, Malabar. (Reprinted from Wiley, New York, 1974)
12. Bakhshi B, Vaa M, Golovchenko EA, Patterson WW, Maybach RL, Bergano NS (2001) Comparison of CRZ, RZ, and NRZ modulation formats in a 64 × 12.3Gb/s WDM transmission experiment over 9,000 km. In: Proceedings of IEEE/OSA optical fiber communication conference (OFC), paper WF4, Anaheim, CA
13. Bellotti G, Varani M, Francia C, Bononi A (1998) Intensity distortion induced by cross-phase modulation and chromatic dispersion in optical-fiber transmissions with dispersion compensation. IEEE Photon Technol Lett 10(12):1745–1747
14. Bennetin G, Galgani L, Strelcyn JM (1976) Kolgomorov entropy and numerical experiments. Phys Rev A 14(6):2338–2345
15. Berg BA (1998) Algorithmic aspects of multicanonical simulations. Nucl Phys Proc Suppl 63(1–3):982–984
16. Berg BA, Neuhaus F (1992) Multicanonical ensemble: a new approach to simulate first-order phase transitions. Phys Rev Lett 68(1):9–11
17. Bergano NS, Kerfoot FW, Davidson CR (1993) Margin measurements in optical amplifier systems. IEEE Photon Technol Lett 5(3):304–306
18. Bigo S, Bellotti G, Chbat M (1999) Investigation of cross-phase modulation limitation over various types of fiber infrastructures. IEEE Photon Technol Lett 11(5):605–607
19. Bigo S, Idler W, Antona J-C, Charlet G, Simonneau C, Gorleir M, Molina M, Borne S, de Barros C, Sillard P, Tran P, Dischler R, Poehlmann W, Nouchi P, Frignac Y (2001) Transmission of 125 WDM channels at 42.7 Gbit/s (5 Tbit/s capacity) over 12 × 100km of TeraLight Ultra fibre. In: Proceedings of European conference on optical communications (ECOC) 01, postdeadline paper PDM11, Amsterdam, The Netherlands
20. Biondini G, Kath WL, Menyuk CR (2002) Importance sampling for polarization-mode dispersion. IEEE Photon Technol Lett 14(3):310–312
21. Biondini G, Kath WL, Menyuk CR (2004) Importance sampling for polarization-mode dispersion: techniques and applications. IEEE/OSA J Lightwave Technol 22(4):1201–1215
22. Bosco G, Carena A, Curri V, Gaudino R, Poggiolini P, Benedetto S (2000) A novel analytical method for the BER evaluation in optical systems affected by parametric gain. IEEE Photon Technol Lett 12(2):152–154
23. Bratley BL, Fox BL, Schrage LE (1987) A guide to simulation. Springer-Verlag, New York
24. Cai J-X, Foursa DG, Liu L, Davidson CR, Cai Y, Patterson WW, Lucero AJ, Bakhshi B, Mohs G, Corbett PC, Gupta V, Anderson W, Vaa M, Domagala G, Mazurczyk M, Li H, Jiang S, Nissov M, Pilipetskii AN, Bergano NS (2005) RZ-DPSK field trial over 13,100 km of installed non-slope-matched submarine fibers. IEEE/OSA J Lightwave Technol 23(1):95–103
25. Cai J-X, Nissov M, Pilipetskii AN, Davidson CR, Mu R-M, Mills MA, Xu L, Foursa D, Menges R, Corbett PC, Sutton D, Bergano NS (2001) 1.28Tb/s(32 × 40Gb/s) transmission

over 4,500 km. In: Proceedings of European conference on optical communications (ECOC) 01, postdeadline paper PDM12, Amsterdam, The Netherlands
26. Cai J-X, Nissov M, Davidson CR, Cai Y, Pilipetskii AN, Li H, Mills MA, Mu R-M, Feiste U, Xu L, Lucero AJ, Foursa DG, Bergano NS (2002) Transmission of thirty-eight 40 Gb/s channels (> 1.5 Tb/s) over transoceanic distance. In: Proceedings of IEEE/OSA optical fiber communication conference (OFC), postdeadline paper FC4, Anaheim, CA
27. Carena A, Curri V, Gaudino R, Poggiolini P, Benedetto S (1997) New analytical results on fiber parametric gain and its effects on ASE noise. IEEE Photon Technol Lett 9(4):535–537
28. Cartaxo A (1998) Impact of modulation frequency on cross-phase modulation effect in intensity modulation-direct detection WDM systems. IEEE Photon Technol Lett 10(9):1268–1270
29. Cartaxo A (1999) Cross-phase modulation in intensity modulation direct detection WDM systems with multiple optical amplifiers and dispersion compensators. IEEE/OSA J Lightwave Technol 17(2):178–190
30. Carter GM, Jacob JM (1998) Dynamics of solitons in filtered dispersion-managed systems. IEEE Photon Technol Lett 10(4):546–548
31. Carter GM, Jacob JM, Menyuk CR, Golovchenko EA, Pillipetskii AN (1997) Timing-jitter reduction for a dispersion-managed soliton system: experimental evidence. Opt Lett 22(8):513–515
32. Chen JC, Lu D, Sadowsky JS, Yao K (1993) On importance sampling in digital communications—part I: fundamentals. IEEE J Sel Areas Commun 11(3):289–299
33. Chen Y, Haus HA (1998) Dispersion-managed solitons with net positive dispersion. Opt Lett 23(13):1013–1015
34. Chiang TK, Kagi N, Marhic ME, Kazovsky LG (1996) Cross-phase modulation in fiber links with multiple optical amplifiers and dispersion compensators. IEEE/OSA J Lightwave Technol 14(3):249–260
35. Ciaramella E (2002) Nonlinear impairments in extremely dense WDM systems. IEEE Photon Technol Lett 14(6):804–806
36. Ciaramella E, Forestieri E (2005) Analytical approximation of nonlinear distortions. IEEE Photon Technol Lett 17(1):91–93
37. Coelho LD, Molle L, Gross D, Hanik N, Freund R, Caspar C, Schmidt ED, Spinnler B (2009) Modeling nonlinear phase noise in differentially phase-modulated optical communication systems. Opt Express 17(5):3226–3241
38. Cohen SD, Hindmarsh AC (1994) CVODE User Guide. Lawrence Livermore National Laboratory
39. Demir A (2007) Nonlinear phase noise in optical-fiber-communication systems. IEEE/OSA J Lightwave Technol 25(8):2002–2032
40. Dieci L, Eirola T (1994) Positive definitenness in the numerical solution of Riccati differential equations. Num Math 67(3):303–313
41. Doob JL (1954) The Brownian movement and stochastic equations. In: Wax N (eds) Selected papers on noise and stochastic processes. Dover, New York
42. Essiambre RJ, Foschini GJ, Kramer G, Winzer PJ (2010) Capacity limits of information transmission in optically-routed fiber networks. Bell Labs Tech J 14(4):149–162
43. Forestieri E (2000) Evaluating the error probability in lightwave systems with chromatic dispersion, arbitrary pulse shape and pre- and postdetection filtering. IEEE/OSA J Lightwave Technol 18(11):1493–1503
44. Forghieri F, Tkach RW, Chraplyvy AR (1997) Fiber nonlinearities and their impact on transmission systems. In: IP Kaminow, Koch TL (eds) Optical fiber telecommunications III A. Academic Press, London
45. Franklin J (1968) Matrix theory. Prentice-Hall, New Jersey
46. Georges T (1995) Bit-error rate degradation of interacting solitons owing to non-Gaussian statistics. Electron Lett 31(14):1174–1175
47. Georges T (1995) Perturbation theory for the assessment of soliton transmission control. Opt Fiber Technol 1(2):97–116

48. Georges T (1996) Study of the non-Gaussian timing jitter statistics induced by soliton interaction and filtering. Opt Commum 123(4–6):617–623
49. Georges T, Favre F, Guen DL (1998) Theoretical and experimental study of soliton transmission in dispersion managed links. Inst Electron Inf Commun. Eng Trans Electron E81-C(2):226–231
50. Gnauck AH, Winzer PJ (2005) Optical phase-shift-keyed transmission. IEEE/OSA J Lightwave Technol 23(1):115–130
51. Golovchenko EA, Jacob JM, Pillipetskii AN, Menyuk CR, Carter GM (1997) Dispersion-managed solitons in a fiber loop with in-line filtering. Opt Lett 22(5):289–291
52. Golovchenko EA, Pilipetskii AN, Bergano NS (2000) Transmission properties of chirped return-to-zero pulses and nonlinear intersymbol interference in 10 GBit/s WDM transmission. In: Proceedings of IEEE/OSA optical fiber communication conference (OFC), paper FC3, Baltimore, MD
53. Golovchenko EA, Pilipetskii AN, Bergano NS, Davidsen CR, Khatri FI, Kimball RM, Mazurczyk VJ (2000) Modeling of transoceanic fiber-optic WDM communications systems. IEEE J Sel Topics Quant Electron 6(2):337–347
54. Gordon JP, Haus HA (1986) Random walk of coherently amplified solitons in optical fiber transmission. Opt Lett 11(10):665–667
55. Grigoryan VS, Menyuk CR (1998) Dispersion-managed solitons at normal average dispersion. Opt Lett 23(8):609–611
56. Grigoryan VS, Menyuk CR, Mu RM (1999) Calculation of timing and amplitude jitter in dispersion-managed optical fiber communications using linearization. IEEE/OSA J Lightwave Technol 17(8):1347–1356
57. Grigoryan VS, Richter A (2000) Efficient approach for modeling collision-induced timing jitter in WDM return-to-zero dispersion-managed systems. IEEE/OSA J Lightwave Technol 18(8):1148–1154
58. Grigoryan VS, Veselka J, Sardesai HP (2008) Highly efficient method for BER modeling in quasi-linear fibers and its validation in a 40 Gb/s DWDM test bed. In: Proceedings of IEEE/OSA optical fiber communication conference (OFC), paper JThA53, San Diego, CA
59. Hahn PM, Jeruchim MC (1987) Developments in the theory and application of importance sampling. IEEE Trans Commun 35(7):706–714
60. Hasegawa A (1995) Solitons in optical communications. Clarendon, Oxford
61. Haus HA (1991) Quantum noise in solitonlike repeater systems. J Opt Soc Am B 8(5):1122–1126
62. Holzlöhner R (2003) A covariance matrix method for the computation of bit errors in optical transmission systems. Ph.D. thesis, University of Maryland Baltimore County. Baltimore, Maryland, USA
63. Holzlöhner R, Ereifej HN, Carter GM, Menyuk CR (2002) Experimental and theoretical characterization of a 40 Gb/s long-haul single-channel transmission system. IEEE/OSA J Lightwave Technol 20(7):1124–1131
64. Holzlöhner R, Grigoryan VS, Menyuk CR, Kath WL (2002) Accurate calculation of eye diagrams and bit error rates in long-haul transmission systems using linearization. IEEE/OSA J Lightwave Technol 20(3):389–400
65. Holzlöhner R, Mahadevan A, Menyuk C, Morris J, Zweck J (2005) Evaluation of the very low BER of FEC codes using dual adaptive importance sampling. IEEE Commun Lett 9(2):163–165
66. Holzlöhner R, Menyuk CR (2003) The use of multicanonical Monte Carlo simulations to obtain accurate bit error rates in optical communications systems. Opt Lett 28(20):1894–1896
67. Holzlöhner R, Menyuk CR, Kath WL VSG (2003) A covariance matrix method to compute bit error rates in a highly nonlinear dispersion-managed soliton system. IEEE Photon Technol Lett 15(5):688–690
68. Hui R, Chowdhury D, Newhouse M, O'Sullivan M, Poettcker M (1997) Nonlinear amplification of noise in fibers with dispersion and its impact in optically amplified systems. IEEE Photon Technol Lett 9(3):392–394

69. Hui R, Demarest KR, Allen CT (1999) Cross-phase modulation in multispan WDM optical fiber systems. IEEE/OSA J Lightwave Technol 17(6):1018–1026
70. Hui R, O'Sullivan M, Robinson A, Taylor M (1997) Modulation instability and its impact in multispan optical amplified IMDD systems: theory and experiments. IEEE/OSA J Lightwave Technol 15(7):1071–1081
71. Humblet PA, Azizo glu M (1991) On the bit error rate of lightwave systems with optical amplifiers. IEEE/OSA J Lightwave Technol 9(11):1576–1582
72. Iannone E, Matera F, Mecozzi A, Settembre M (1998) Nonlinear optical communication networks. Microwave and optical engineering. Wiley, New York
73. Jenkins RB, Sauer JR, Chakravarty S, Ablowitz MJ (1995) Data-dependent timing jitter in wavelength-division-multiplexing soliton systems. Opt Lett 20(19):1964–1966
74. Jiang Z, Chongcheng F (2003) A comprehensive study on XPM- and SRS-induced noise in cascaded IM-DD optical fiber transmission systems. IEEE/OSA J Lightwave Technol 21(4):953–960
75. Kaminov, IP (eds) (1997) TLK optical fiber telecommunications IIIA. Academic Press, San Diego
76. Kaminov, IP (eds) (2002) Optical fiber telecommunications IVB. Academic Press, San Diego
77. Kaup DJ (1990) Perturbation theory for solitons in optical fibers. Phys Rev A 42(9):5689–5694
78. Kazovsky L, Benedetto S, Willner A (1996) Optical fiber comminication systems. Artech House, Boston
79. Killey RI, Thiele HJ, Mikhailov V, Bayvel P (2000) Prediction of transmission penalties due to cross-phase modulation in WDM systems using a simplified technique. IEEE Photon Technol Lett 12(7):804–806
80. Kovsh DI, Liu L, Bakhshi B, Pilipetskii AN, Golovchenko EA, Bergano NS (2002) Reducing interchannel crosstalk in long-haul DWDM systems. IEEE J Sel Topics Quant Electron 8(3):597–602
81. Kumar S, Hasegawa A (1997) Quasi-soliton propagation in dispersion-managed optical fibers. Opt Lett 22(6):372–375
82. Kutz JN, Evangelides SG (1998) Dispersion-managed breathers with average normal dispersion. Opt Lett 23(9):685–687
83. Lakoba TI, Yung J, Kaup DJ, Malomed BA (1998) Conditions for stationary pulse propagation in the strong dispersion management regime. Opt Commun 149(4–6):366–375
84. Le Guen F., Del Burgo S, Moulinard ML, Grot D, Henry, M, Favre F, Georges T (1999) Narrow band 1.02 Tbit/s (51 × 20 Gbit/s) soliton DWDM transmission over 1,000 km of standard fiber with 100 km amplifier spans. In: Proceedings of IEEE/OSA optical fiber communication conference (OFC), postdeadline paper PD4, Washington, DC
85. Lee JS, Shim CS (1994) Bit-error-rate analysis of optically preamplified receivers using an eigenfunction expansion method in optical frequency domain. IEEE/OSA J Lightwave Technol 12(7):1224–1229
86. Lima AO, Lima IT, CR M (2005) Error estimation in multicanonical Monte Carlo simulations with applications to polarization-mode-dispersion emulators. IEEE/OSA J Lightwave Technol 23(11):3781–3789
87. Lima IT, Lima AO, Sun Y, Jiao H, Zweck J, CR M (2005) A receiver model for optical fiber communication systems with arbitrarily polarized noise. IEEE/OSA J Lightwave Technol 23(3):1478–1490
88. Lima IT, Lima AO, Zweck J, Menyuk CR (2003) Efficient computation of outage probabilities due to polarization effects in a WDM system using a reduced Stokes model and importance sampling. IEEE Photon Technol Lett 15(1):45–47
89. Liu X, Wei X, Mollenauer LF, McKinstrie CJ, Xie C (2003) Collision-induced time shift of a dispersion-managed soliton and its minimization in wavelength-division-multiplexed transmission. Opt Lett 28(16):1148–1150

90. DeLong KW, Trebino R, Kane DJ (1994) Comparison of ultrashort pulse frequency-resolved optical gating traces for three common beam geometries. J Opt Soc Am B 11(9):1595–1608
91. Loudon R (1983) The quantum theory of light. Clarendon, Oxford
92. Lu T, Yevick D (2005) Efficient multicanonical algorithms. IEEE Photon Technol Lett 17(4):861–863
93. Luis RS, Cartaxo AVT (2005) Analytical characterization of SPM impact on XPM-induced degradation in dispersion-compensated WDM systems. IEEE/OSA J Lightwave Technol 23(3):1503–1513
94. Lyubomirsky I, Qui T, Roman J, Nayfeh M, Frankel M, Taylor MG (2003) Interplay of fiber nonlinearity and optical filtering in ultradense WDM. IEEE Photon Technol Lett 15(1):147–149
95. Mamyshev PV, Mamysheva NA (1999) Pulse-overlapped dispersion-managed data transmission and intrachannel four-wave mixing. Opt Lett 24(21):1454–1456
96. Marcuse D (1990) Derivation of analytical expressions for the bit-error probability in lightwave systems with optical amplifiers. IEEE/OSA J Lightwave Technol 8(12):1816–1823
97. Marks B (2000) Dispersion management in fiber optic communications. Dissertation, Northwestern University, Chicago
98. Mecozzi A (1994) Long-distance transmission at zero dispersion: Combined effect of the Kerr nonlinearity and the noise of the in-line amplifiers. J Opt Soc Am B 11(3):462–469
99. Menyuk CR (1995) Non-Gaussian corrections to the Gordon–Haus distribution resulting from soliton interactions. Opt Lett 20(3):285–287
100. Menyuk CR (1999) Application of multiple-length-scale methods to the study of optical fiber fiber transmission. J Eng Math 36(1–2):113–136
101. Metropolis N, Ulam S (1949) The Monte Carlo method. J Am Stat Assoc 44(247):335–341
102. Moore RO, Biondini G, Kath WL (2003) Importance sampling for noise-induced amplitude and timing jitter in soliton transmission systems. Opt Lett 28(2):105–107
103. Mu RM, Grigoryan VS, Menyuk CR, Carter GM, Jacob JM (2000) Comparison of theory and experiment for dispersion-managed solitons in a recirculating fiber loop. IEEE J Sel Topics Quant Electron 6(2):248–257
104. Mu RM, Menyuk CR (2001) Symmetric slope compensation in a long-haul WDM system using the CRZ format. IEEE Photon Technol Lett 13(8):797–799
105. Mu RM, Menyuk CR (2002) Convergence of the chirped return-to-zero and dispersion managed soliton modulation formats in WDM systems. IEEE/OSA J Lightwave Technol 20(4):608–617
106. Mu RM, Yu T, Grigoryan VS, Menyuk CR (2000) Convergence of the CRZ and DMS formats in WDM systems using dispersion management. In: Proceedings of IEEE/OSA optical fiber communication conference (OFC), paper FC1, Baltimore, MD
107. Mu RM, Yu T, Grigoryan VS, Menyuk CR (2002) Dynamics of the chirped return-to-zero modulation format. IEEE/OSA J Lightwave Technol 20(1):47–57
108. Narimanov EE, Mitra P (2002) The channel capacity of a fiber optics communication system: perturbation theory. IEEE/OSA J Lightwave Technol 20(3):530–537
109. Nelson LE, Jopson RM, Gnauk AH, Chraplyvy AR (1999) Resonances in cross-phase modulation impairment in wavelength-division-multiplexed lightwave transmission. IEEE Photon Technol Lett 11(7):907–909
110. Neokosmidis I, Kamalakis T, Chipouras A, Sphicopoulos T (2005) Estimation of the four-wave mixing noise probability-density function by the multicanonical Monte Carlo method. Opt Lett 30(1):11–13
111. Nijhof JHB, Doran NJ, Forysiak W, Knox FM (1997) Stable soliton-like propagation in dispersion managed systems with net anomalous, zero, and normal dispersion. Electron Lett 33(20):1726–1727
112. Ogawa K, Tzeng LD, Park YK, Sano E (1997) Advances in high bit rate transmission systems. In: Kaminow IE, Koch TL (eds.) Optical fiber telecommunications, vol IIIA, chap 11, Academic, San Diego, pp 336–372

113. Papoulis A (1991) Probability, random variables, and stochastic processes, 3rd edn. McGraw-Hill, New York
114. Pellegrini W, Zweck J, Menyuk CR, Holzlöhner R (2005) Computation of bit error ratios for a dense WDM system using the noise covariance matrix and multicanonical Monte Carlo methods. IEEE Photon Technol Lett 17(8):1644–1646
115. Pilipetskii AN, Mazurczyk VJ, Chen CJ (1999) The effect of dispersion compensation on system performance when nonlinearities are important. IEEE Photon Technol Lett 11(2):284–286
116. Proakis JG (1995) Digital communications. McGraw-Hill, New York
117. Risken H (1989) The Fokker–Planck equation, 2nd edn. Springer, Berlin
118. Schwartz L (1966) Théorie des distributions I. Hermann, Paris
119. Secondini M (2006) Optical communication theory and techniques for high bit-rate systems. Ph.D. thesis, Scuola Superiore Sant'Anna, Pisa, Italy
120. Secondini M, Forestieri E, Menyuk CR (2006) A novel perturbation method for signal–noise interaction in nonlinear dispersive fibers. In: Proceedings of IEEE/OSA optical fiber communication conference (OFC), paper OthD3, Anaheim, CA
121. Shapiro EG, Fedoruk MP, Turitsyn SK (2001) Numerical estimate of BER in optical systems with strong patterning effects. Electron Lett 37(19):1179–1181
122. Shtaif M, Eiselt M (1998) Analysis of intensity interference caused by cross-phase modulation in dispersive optical fibers. IEEE Photon Technol Lett 10(7):979–981
123. Sidje RB (1998) Expokit: A software package for computing matrix exponentials. ACM Trans Math Softw 24(1):130–156
124. Sinkin OV (2006) Calculation of bit error rates in optical fiber communications systems in the presence of nonlinear distortion and noise. Ph.D. thesis, University of Maryland Baltimore County, Baltimore, Maryland, USA
125. Sinkin OV, Grigoryan VS, Holzlöhner R, Kalra A, Zweck J, Menyuk CR (2004) Calculation of error probability in WDM RZ systems in presence of bit-pattern-dependent nonlinearity and of noise. In: Proceedings of IEEE/OSA optical fiber communication conference (OFC), paper TuN4, Los Angeles, CA
126. Sinkin OV, Grigoryan VS, Menyuk CR (2007) Accurate probabilistic treatment of bit-pattern-dependent nonlinear distortions in BER calculations for WDM RZ systems. IEEE/OSA J Lightwave Technol 25(10):2959–2968
127. Sinkin OV, Grigoryan VS, Zweck J, Menyuk CR (2006) Calculation of the bit-error ratio in wavelength-division-multiplexed return-to-zero systems when the nonlinear penalty is dominated by collision-induced timing jitter. In: Proceedings of IEEE/OSA optical fiber communication conference (OFC), paper JThB3, Anaheim, CA
128. Sinkin OV, Grigoryan VS, Zweck J, Menyuk CR, Docherty A, Ablowitz M (2005) Calculation, characterization, and application of the time shift function in wavelength-division-multiplexed return-to-zero systems. Opt Lett 30(16):2056–2058
129. Sinkin OV, Holzlöhner R, Zweck J, Curtis CR (2003) Optimization of the split-step Fourier method in modeling optical fiber communications systems. IEEE/OSA J Lightwave Technol 21(1):61–68
130. Sinkin OV, Zweck J, Menyuk CR (2003) Effects of the nonlinearly-induced timing and amplitude jitter on the performance of different modulation formats in WDM optical fiber communications systems. In: Proceedings of IEEE/OSA optical fiber communication conference (OFC), paper TuF5, Atlanta, GA
131. Smith NJ, Doran NJ, Forysiak W, Knox FM (1997) Soliton transmission using periodic dispersion compensation. IEEE/OSA J Lightwave Technol 15(10):1808–1822
132. Smith NJ, Forysiak NJ, Doran W (1996) Reduced Gordon–Haus jitter due to enhanced power solitons in strongly dispersion managed systems. Electron Lett 32(22):2085–2086
133. Strang G (1980) Linear algebra and its applications. Academic Publishing, New York
134. Sugahara H, Kato H, Inoue T, Maruta A, Kodama Y (1999) Optimal dispersion management for a wavelength division multiplexed optical soliton transmission system. IEEE/OSA J Lightwave Technol 17(9):1547–1559

135. Sugahara H, Maruta A, Kodama Y (1999) Optimal allocation of amplifiers in a dispersion-managed line for a wavelength-division-multiplexed soliton transmission system. Opt Lett 24(3):145–147
136. Sun H, Wu KT, Roberts K (2008) Real-time measurements of a 40 Gb/s coherent system. Opt Express 16(2):873–879
137. Thiele HJ, Killey RI, Bayvel P (1999) Influence of transmission distance on XPM-induced intensity distortion in dispersion-managed, amplified fibre links. Electron Lett 35(5):408–409
138. Thiele HJ, Killey RI, Bayvel P (2000) Simple technique to determine cross-phase modulation induced penalties in WDM transmission. In: Proceedings of IEEE/OSA optical fiber communication conference (OFC), paper ThM2, Baltimore, MD
139. Tsuritani T, Agata A, Morita I, Edagawa N, Shigeyuki A (2004) Ultra-long-haul 40-Gbit/s-based DWDM transmission using optically prefiltered CS-RZ signals. IEEE J Sel Topics Quant Electron 10(2):403–411
140. Turitsyn SK (1997) Theory of average pulse propagation in high bit rate optical transmission systems with strong dispersion. Lett J Exp Theor Phys 65(11):845–850
141. Turitsyn SK, Shapiro EG (1998) Dispersion-managed solitons in optical amplifier transmission systems with zero average dispersion. Opt Lett 23(9):682–684
142. Vaa M, Bakhshi B, Golovchenko EA, Chai Y, Heismann F, Li H, Arend M, Patterson WW, Simons AL, Harvey GT, Maybach RL, Bergano N (2001) Demonstration of a 640 Gbit/s × 7,000 km submarine transmission system technology ready for field deployment. In: Proceedings of IEEE/OSA optical fiber communication conference (OFC), paper WF5, Anaheim, CA
143. Veach E (1997) Robust Monte Carlo methods for light transport simulations. Ph.D. thesis, Stanford University, Stanford, USA
144. Yadin Y, Shtaif M, Orenstein M (2005) Bit-error rate of optical DPSK in fiber systems by multicanonical Monte Carlo simulations. IEEE Photon Technol Lett 17(6):1355–1357
145. Yang TS, Kath WL, Evangelides Jr SG (1999) Optimal prechirping for dispersion-managed transmission of return-to-zero pulses. In: Proceedings of IEEE/OSA optical fiber communication conference (OFC), paper ThQ4, Washington, DC
146. Yevick D (2002) Multicanonical communication system modeling—application to PMD statistics. IEEE Photon Technol Lett 14(11):1512–1514
147. Yu T, Mu RM, Grigoryan VS, Menyuk CR (1999) Energy enhancement of dispersion-managed solitons in optical fiber transmission systems with lumped amplifiers. IEEE Photon Technol Lett 11(12):75–77
148. Yu T, Reimer WM, Grigoryan VS, Menyuk CR (2000) A mean field approach for simulating wavelength-division multiplexed systems. IEEE Photon Technol Lett 12(4): 443–445

Chapter 5
Commercial Optical Communication Software Simulation Tools

Dwight H. Richards

Abstract System simulation tools can enhance and accelerate user modeling capabilities and provide real field design scenarios using extensive industry specifications. Users of these tools include optical component and equipment manufacturers, system integrators, service providers, as well as government labs and academic institutions. Modeling tools can be used to maximize performance, minimize costs, reduce time-to-market, fast-prototyping, and analyze multiple "what-if" scenarios for optical communication networks. In this chapter we briefly review three system level modeling tools that are commercially available, namely, OptSim™ from RSoft Design Group (http://www.rsoftdesign.com), VPItransmissionMaker™ Optical Systems from VPIphotonics, a division of VPIsystems (http://www.vpiphotonics.com), and OptiSystem™ from Optiwave Systems Corporation (http://www.optiwave.com). This is not an attempt to compare the tools and no recommendation should be implied from the order or content of the material presented, but rather a presentation of what resources are available to the scientific community.

5.1 RSoft Design Group: OptSim™ and ModeSYS™

5.1.1 Introduction

OptSim™ allows users to design and simulate current and next generation optical communication systems at the signal propagation level. OptSim™ is primarily used to model systems based on single-mode fiber, but users who are interested in

D. H. Richards (✉)
Department of Engineering Science and Physics
College of Staten Island/The City University of New York
Staten Island, NY 10314, USA
e-mail: Dwight.Richards@csi.cuny.edu

N. (Neo) Antoniades et al. (eds.), *WDM Systems and Networks*, Optical Networks,
DOI: 10.1007/978-1-4614-1093-5_5,

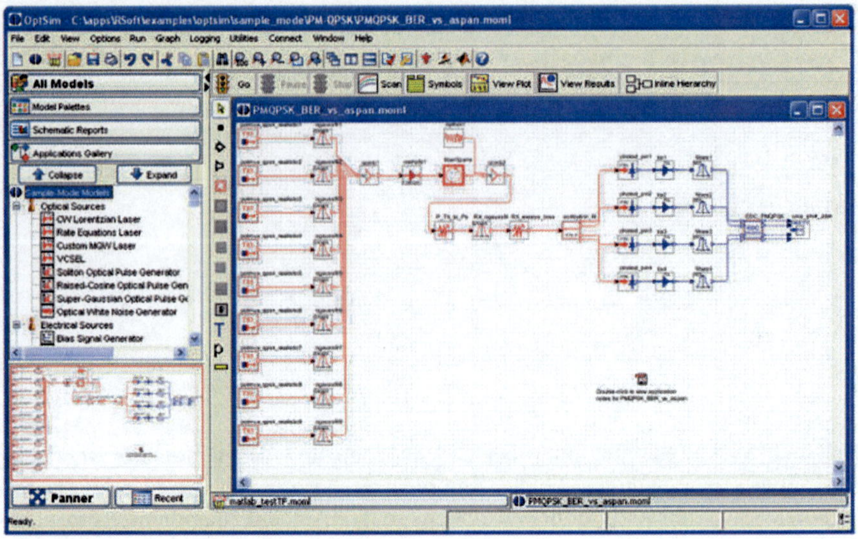

Fig. 5.1 OptSim™ GUI (Copyright 2010 RSoft Inc. Used by permission of RSoft Inc.)

multimode fiber system modeling can purchase ModeSYS™, the multimode system simulation tool, as an add-on to OptSim™ or as a standalone module. The graphical user interface (GUI), which is shared by OptSim™ and ModeSYS™, is shown in Fig. 5.1. The tool supports encryption of private data files and schematic designs. Network and standalone licensing configurations are also available.

OptSim™ is ideally suited for computer-aided design of optical communication systems including, but not limited to:

- DWDM/CWDM Amplified, e.g. EDFA, Raman, SOA, OPA
- Advanced Modulation Formats, e.g. D(Q)PSK, Duobinary, OFDM
- OCDMA/OTDM
- CATV Digital/Analog
- Optical Interconnects
- FTTx/PON
- EDC (Electronic Dispersion Compensation)
- FSO (Free-space Optics)
- Soliton

A brief summary of the key features in OptSim™ are:

- Multiple engines implementing both the Time Domain Split Step and the Frequency Domain Split Step for the most accurate and efficient simulation of an optical link architecture.
- MATLAB® interface makes it easy to develop custom user models using the m-file language and/or the Simulink® modeling environment.

- Interfaces with laboratory test equipment such as Agilent and Luna to merge simulation with experiment.
- Interfaces with device-level design tools such as BeamPROP™, GratingMOD, and LaserMOD™ from RSoft provide a powerful mixed-level design flow for optoelectronic circuits and systems.
- Interfaces with EDA tools such as Berkeley SPICE™, Cadence Virtuoso Spectre, and Synopsys HSPICE™ for a mixed-domain electrical and optical simulation.
- Application programming interface (API) for programming languages such as C/C++ for the development of custom user models.
- Best Fit Laser Toolkit™ makes customizing powerful rate-equation laser model parameters to fit desired performance characteristics easy.
- Extensive library of predefined manufacturer components makes it easy to model commercially available devices.
- Intuitive and flexible measurement post-processing graphical interface acts like a virtual laboratory instrument.

5.1.2 Twin Simulation Engines

The twin simulation engines in OptSim™ support two complementary simulation approaches. The block mode simulation engine performs simulations in which the signal data passed between components represents the entire simulated time in one block of data. The advantage of this approach is that component models and algorithms can easily work with the entire signal, transforming it back and forth between the time and frequency domains to operate on the data in the domain most convenient for the simulation algorithm. Nonlinear fiber is simulated using the Split Step Fourier technique in this mode [1]. The sample mode simulation engine performs simulations in which the signal data passed between components represents a single sample or time step at a time. Unlike the block mode simulation approach where one component model usually passes data to another component model once during the course of the simulation, covering the entire simulated time in one block of data, in the sample mode simulation approach a component model will pass new sample data to another component model at each time step in the simulation. One advantage of this approach is that simulations can be performed over an unlimited amount of simulated time covering an unlimited transmitted sequence length. Nonlinear fiber is simulated using the Time Domain Split Step technique in this mode [2, 3]. OptSim™ also includes capabilities for simulation of power transients due to adding and dropping of channels in optical amplifier chains and all-optical gain control techniques [4]. These capabilities make use of the Wavelength Domain Simulation approach [5] and through simulation iterations enable simulations of feedback paths and fast simulation of large time windows (on the order of seconds) which is not computationally efficient with time- and frequency-domain simulation technologies.

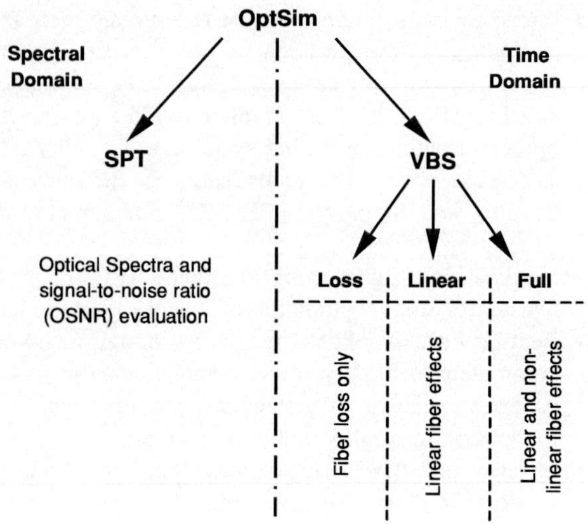

Fig. 5.2 Frequency slicing in the VBS simulation (Copyright 2010 RSoft Inc. Used by permission of RSoft Inc.)

5.1.3 The Design Process

OptSim^{TM} offers several different strategies for simulating a given optical system based on two different simulation techniques as illustrated in Fig. 5.2. In order of increasing accuracy and computational effort, these strategies are:

- Spectral propagation technique (SPT): signals are propagated in the network as power spectra, taking into account component loss and noise. Spectra, power levels, and optical signal to noise ratio (OSNR) can be evaluated at any point in the network.
- Variable bandwidth simulation (VBS): signals are propagated in the network as time-domain samples over a user-selectable bandwidth. VBS simulates the performance of fiber and other optical components. VBS can be used with different fiber models:
  - *Simplified Fiber Model*: VBS takes into account attenuator only (*Loss Only Fiber*), or attenuation and dispersion (*Linear Fiber*). In both cases the simulation is strictly linear.
  - *Full model*: VBS takes into account all fiber effects, linear and nonlinear. This is the most complete and powerful simulation strategy.

This suite of simulation strategies allows a tradeoff between accuracy and simulation speed. This is particularly useful when using OptSim^{TM} to choose between different designs of an optical link or network. Often the space of free parameters to be investigated is so large that a complete simulation taking into account all effects would be too time-consuming. A reasonable approach is as follows:

SPT Simulation	The user may first explore the space of all free parameters using the SPT simulation, which is computationally very efficient. As mentioned above, from the SPT simulation, one may evaluate OSNR and power levels at any point in the network. In doing so, you may reduce the free parameter space by eliminating all configurations having, for example, an OSNR below a given limit, or a fiber input power greater than a certain level.
VBS Simulation	The user may then run a simplified VBS simulation, by choosing which effects are to be taken into account for the fiber, and again eliminate all configurations that do not match your predefined criteria. Finally, on this further reduced set of parameters, you may run a full VBS simulation and settle on the configuration that gives the best results.

Besides being useful for this incremental approach to the design of an optical network, different simulation strategies are useful also in other contexts. For example, when designing optical networks of limited extension (LAN or MAN), fiber non-linearity may usually be neglected, while noise levels and signal crosstalk are clearly the fundamental issues. In this case, an SPT simulation to estimate OSNR levels, followed by a simplified VBS simulation may be sufficient.

To understand the utility of the variable bandwidth approach, let us consider the simulation of a dense WDM system with 64 channels. Further, suppose that a dispersion map has already been chosen so that the non-linear effects among spectrally distant channels are negligible. Under these conditions, we may draw the complete WDM network with all the 64 channels, but then run the simulation only on a smaller bandwidth which includes only 8–10 channels. In this way, the simulation runs much faster.

For example, as shown in Fig. 5.3, the SPT bandwidth should be set to include all WDM channels, while the VBS bandwidth can be set to simulate only channels 11–20 in the time domain.

5.1.4 ModeSYS™ Multimode Simulation Platform

OptSim™ can be fully integrated with ModeSYS™, RSoft's multimode simulation platform, for accurate simulation of multimode optical communication systems. With a primary focus on data communication applications, ModeSYS™ allows users to evaluate both temporal and spatial attributes of optical signal propagation [6, 7]. This capability enables the system-level analysis of standardized communication technologies such as 1 and 10 Gb Ethernet and Fiber Channel, as well as the study of proprietary data communication platforms. ModeSYS™ was the first commercial tool to address this challenge. Spatially dependent models included with ModeSYS™ enable the accurate analysis of a transverse electromagnetic spatial field starting with its generation by an optical

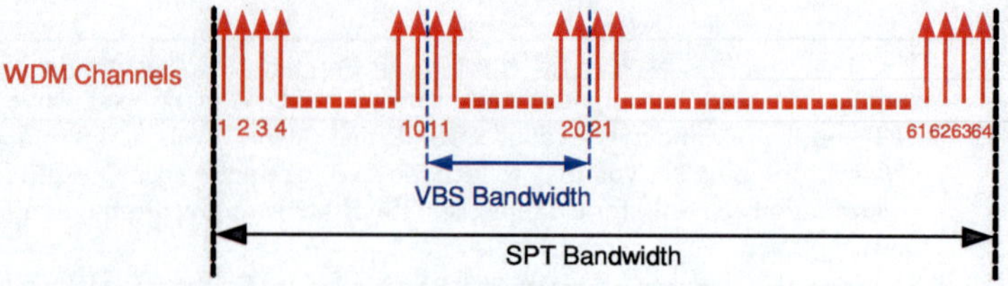

Fig. 5.3 Frequency slicing in the VBS simulation (Copyright 2010 RSoft Inc. Used by permission of RSoft Inc.)

source, continuing with its propagation through spatially-dependent components in a multimode optical link such as multimode fiber and connectors, and through to its detection at a photodiode. The wide range of multimode analyses that can be performed with ModeSYS™ include differential mode delay (DMD), encircled flux (EF), effective modal bandwidth (EMB), launch conditions, modal coupling, transverse optical field profiles, and transient analyses. In addition, eye diagrams, signal waveforms, frequency spectra, bit error rate (BER) curves, etc. are also provided. ModeSYS™ performs device-level modeling within a system simulation infrastructure, so extremely detailed spatial field calculations are implemented in a manner that allows a very straightforward system-level analysis. The platform shields the user from much of the complex optical physics that are necessary to analyze multimode systems, making it not only powerful, but also easy to use.

5.1.5 OptSim™ Models

OptSim™ has a large library of components, which are included in the component palettes, and model libraries at the left of the design area. User-defined models can also be added to the libraries. Various predefined *compound components* (*CCs*) such as the standard transmitter and receiver blocks for different modulation schemes, and other network elements like the multiplexers and demultiplexers are also included in OptSim™.

5.1.5.1 Transmitters

Optical sources: Various laser models are provided, including CW Lorentzian laser, rate equation laser, custom MQW and VCSEL models, pulse sources, and noise sources.

Electrical sources: Electrical waveform generators to generate various commonly used functions such as sine wave, square wave, saw tooth, and others and electrical noise generators are provided.

Logical sources: Logical sources of OptSim™ include data source (pseudo random bit sequence (PRBS) generator), which simulates a pseudo-random or a deterministic logical signal generator of arbitrary level (number of bits per symbol) and logical clock, which simulates a logical clock generator giving an alternated sequence of 0 and 1.

Optical modulators: These include Mach-Zehnder dual arm type that is used to simulate a dual arm Mach–Zehnder amplitude modulator, Two Mach-Zehnder single arm models, with sin^2 and linear electrically shaped input–output power voltage characteristic, electro-absorption (EA) modulator, phase modulator, and a polarization modulator that simulates a modulator where the state of polarization (SOP) of the optical signal is changed as a function of the electrical driving voltages.

Modulator drivers: Electrical drivers are available corresponding to all possible combinations of NRZ and RZ signal types with modulating shapes of rectangular, raised cosine, and super-gaussian. Transmitter modules for CRZ, CS-RZ, DPSK, IM-DPSK, CSRZ-DQPSK, Duobinary, Manchester encoding, Polarization Shift Keying (PolSK), OFDM, PolMux DQPSK, (D)8PSK, and 16QAM are also included.

5.1.5.2 Fiber Models

Nonlinear fiber model: The *optical fiber link* simulates the propagation of the optical signal through a fiber span. The implemented fiber-model fully takes into account the linear and nonlinear phenomena influencing propagation, as well as polarization-related effects. In particular, PMD and birefringence are simulated considering their interaction with the nonlinearities. PMD and birefringence are simulated by taking into account the statistical variation of the principal axes of the fiber. The nonlinear Kerr effect is simulated considering its *instantaneous* (SPM, XPM, FWM, PG) and *delayed* (Raman cross-talk and amplification) effects on the optical signal. The frequency variation of the dispersion is modeled using a fifth order polynomial expansion, and the statistical variation of the dispersion may also be specified. Fiber loss may be defined as a function of frequency (or wavelength) using a second-order polynomial. In addition to specifying polynomial coefficients, both fiber dispersion and loss may be defined through an ASCII file (measured values versus frequency or wavelength). The Raman profile may also be specified through an ASCII file. Raman amplification is simulated together with the generated in-line amplified spontaneous emission (ASE) noise. Single- and double-scattered ASE noise components are simulated. The Raman pumps can be specified as either co- or counter-propagating with respect to the signal propagation. The level of simulation accuracy obtained can be adjusted by turning on and off individually many of the effects mentioned above.

Bi-directional nonlinear fiber: This model provides a detailed implementation of a bidirectional fiber with all dispersive, nonlinear, PMD and Raman effects. A number of fiber propagation applications require the ability to solve for fields traveling in both directions. The most obvious and important is the simulation of Raman amplification of signals by CW pumps. Raman amplification is a bidirectional problem, first because the pumps can be launched from either ends, but

Fig. 5.4 Bidirectional fiber effects in Raman amplification. Incoming signal (*upper arrow*) is amplified by co-propagating and counter-propagating pumps (*lower arrows*). Deleterious effects include spontaneous emission (*double ended arrows*), Rayleigh backscatter of the spontaneous emission and double-Rayleigh backscatter of the signal (*inner arrows*)

also because the noise component to the field travels in both directions—a single pump produces spontaneous emission in both directions, and Rayleigh back-scattering couples field in both directions. These effects are summarized in Fig. 5.4. The forward traveling (red) signal is amplified by pumps (blue) traveling in both directions. The backward traveling signal (not shown in the figure) is also similarly amplified. Spontaneous emission is generated by the pumps (green crosses) and back-scattered. The signal also contributes directly to the noise through double Rayleigh back-scattering indicated by the red loop.

5.1.5.3 Optical Amplifier Models

Doped Fiber Amplifiers

OptSim™ includes the following doped fiber amplifier models:

- *Black box optical amplifier*: A simple physical model of an erbium-doped fiber amplifier (EDFA) supporting fixed gain, saturable gain, and fixed output power configurations.
- *Physical EDFA*: A full physical model of an EDFA that supports four levels of model implementation
  - Giles level
    - Deals with average level-population densities.
    - Uses measured gain and loss spectra that are weighted by the local doping and mode profiles.
    - Generally valid for strongly confined erbium-doping profiles.
  - Constant overlap and calculated overlap levels
    - Supports more parameterized description of the EDFA including absorption/emission cross-sections, doping profiles and meta-stable lifetime.
    - Factors for spatial overlap between erbium population densities and optical modes are either user-specified (constant overlap) or calculated based on doping and mode profiles (calculated overlap).

    - Generally valid for strongly confined erbium-doping profiles.

  - Cladding pumped level

    - Useful for cladding-pumped devices.
    - Overlap factor is calculated from the core area and the inner cladding area.
    - Overlap approach is used once the overlap factor is calculated.

  - Spatial level

    - Provides full spatial description of the interplay between the erbium population densities and optical mode profiles.
    - Options to select "lp," rectangular, Gaussian, and user-specified mode profiles.
    - Other mode shapes can be specified via the input file option.
    - Computationally intensive.
    - Useful when a detailed modeling of EDFA is required.

The following additional features are also supported by the EDFA model:

- Option to include/exclude thermal effects (temperature dependence).
- Higher order effects like homogeneous up-conversion and pair-induced quenching are included.
- Background loss can be excluded or included (constant(s) for 980- and 1,550-bands, or user-defined file).
- Support for encrypted data files.
- Options to optimize numerical solution (integration step, spectral step, number of radial points, speed and accuracy of convergence, etc.).
- Multitude of user-selectable plots of quantities of interest are automatically generated.

- *Physical erbium–ytterbium co-doped fiber amplifier* (*EYCDFA*): a full physical model of an EYCDFA

  - The erbium and ytterbium atomic manifolds are linked via the cross-relaxation rates.
  - Similar treatment and levels of model implementation like the ones described for EDFA except the new factors, parameters, and spectra due to the ytterbium co-dopant.

Semiconductor Optical Amplifiers

- *Semiconductor optical amplifier* (*SOA/RSOA*): a physical model of a SOA based on travelling wave approximation. Also includes reflective semiconductor optical amplifier (RSOA).
- *Controlled SOA*: a physical model of a controlled SOA with injection current as sum of pump and external control currents.

5.1.5.4 Networking Components

Some network simulation components that are available in OptSim™ include:

- Optical Switch (Ideal and Realistic)
- Optical Cross Connectors (OXC)
- Optical Circulator
- Optical Add and Drop Multiplexer (OADM)
- Optical Demultiplexer
- Optical Multiplexer
- WC-XGM (Wavelength Converter using cross-gain modulation)
- WC-XPM (Wavelength Converter using cross-phase modulation)

5.1.5.5 Receiver Components

PIN and APD photodiodes, sensitivity receiver, and receiver components for different modulation formats including PolSK, DPSK, PMQPSK and others, are included in OptSim™. OptSim™ also includes various N-channel WDM receiver models for different modulation formats. Optical direct-detection receivers with PIN/APD including receiver electronics and arbitrary filtering, optical balanced receiver to support differential, phase, and all types of multi-level modulation formats are available. Various digital signal processing (DSP) modules and electronic equalization modules based on decision feedback/feed-forward equalizers (DFE/FFE), maximum likelihood sequence estimation (MSLE), multi-symbol phase estimation (MSPE), and constant modulus algorithm (CMA) techniques are included in OptSim™. Several analyzer and measurement components such as power meter, optical spectrum analyzer, optical probe, electrical analyzer, electrical spectrum analyzer, scattering diagram viewer, BER estimator, BER counter, and logical probe among others are also available in OptSim™.

5.1.5.6 ModeSYS™ Modules

As mentioned earlier, the multimode add-on, ModeSYS™, includes a number of spatial models as listed below:

- *Spatial adder*: Attaches a spatial field to an optical signal.
- *Spatial direct modulated laser*: The spatially enabled version of the direct-modulated laser.
- *Spatial mode-locked laser*: The spatially enabled version of the mode-locked laser.
- *Spatial CW laser*: The spatially enabled version of the continuous-wave laser.
- *Spatial VCSEL*: The spatially enabled version of the vertical-cavity surface emitting laser.

- *Spatial LED*: The spatially enabled version of the light-emitting diode.
- *Thin lens*: Focus a spatial field.
- *Vortex lens*: Focus a spatial field with vortex phase transformation.
- *Spatial coupler*: Perform rotational and translational offsets as well as free-space propagation.
- *Spatial* BeamPROP™ *interface*: Provide interface to RSoft's BPM simulation tool BeamPROP™.
- *Multimode fiber*: Model for the multimode fiber.
- *Spatial aperture*: Pass spatial fields through an aperture, cropping when necessary.
- *Spatial Photodetector*: The spatially enabled version of the photodetector.
- *Spatial receiver*: The spatially enabled version of the photoreceiver.
- *Spatial analyzer*: Perform various calculations on spatial fields and produce spatial field plots.
- *EF analysis tool*: Calculate the EF of a laser/fiber pair.
- *DMD analysis tool*: Measure the DMD of a given fiber.
- *Signal band converter*: Collapse a linked list of spatially orthogonal fields into a single signal.
- *Gridded field converter*: Convert an optical signal's spatial mode profiles into two-dimensional gridded representations.

5.1.6 Signal and Noise Representations

Underlying all the physical modeling in OptSim™ are the representations of the different signals that represent binary information, electrical signals, and optical signals. The purpose of every model in OptSim™ is to create or manipulate signals, convert from one signal type into another, or display signals as outputs.

There are three basic signal types:

5.1.6.1 Logical Signals

The *logical signal type* is the simplest in OptSim™ and contains the binary information plus a few other auxiliary pieces of information that describe the nature of the binary data. The binary signal is naturally expressed as an arbitrary sequence of 0's and 1's. Various predefined formats are provided in the tool.

5.1.6.2 Electrical Signals

Electrical signals are intended to represent voltages and currents in the RF domain, for example the voltage used to drive a modulator or laser, or the current generated at a photodetector. Since voltages and currents are real quantities, an electrical signal in OptSim™ is a real-valued time-sampled function. In addition to the sampled vector, the electrical signal is characterized by the number of points, a sampling rate, a start time, and a flag indicating whether the signal is to be considered as a voltage or current.

5.1.6.3 Optical Signals

Since the majority of the optical components in OptSim™ are waveguide devices of some form or other, it is natural to expand the fields in terms of waveguide modes. Most components utilize only one or perhaps a few modes, so the problem is reduced to the evolution of a few mode amplitudes along the propagation path. As OptSim™ is a time-domain tool, the appropriate representation is to describe the amplitude of the electric field for a small window in time. The amplitude in this window evolves as the signal progresses along the spatial path. If we assume the field is linearly polarized along the x-axis we can write the electric field as

$$\mathbf{E}(t) = \hat{x} f(x, y) E(t) \tag{5.1}$$

in terms of the amplitude function $E(t)$and mode profile $f(x, y)$.

The main information specifying the optical signal is a vector of complex numbers accompanied by a center frequency f_0 or free-space wavelength $\lambda_0 = c/f_0$ and the sampling timeΔt. As was the case for electrical signals, it is useful to add a start time t_0 to allow for signal delays.

Single-Band and Multiple-Band Representations

In WDM optical simulations, we often encounter systems consisting of many optical channels whose spacing far exceeds their individual bandwidths. For a four-channel case, the amplitude function can be written as

$$A(t) = a_0(t) + e^{-i2\pi\,\delta f_1\, t} a_1(t) + e^{-i2\pi\,\delta f_2\, t} a_2(t) + e^{-i2\pi\,\delta f_3\, t} a_3(t) \tag{5.2}$$

where the amplitude functions $a_i(t)$ describe the individual channels and the detunings δ_i indicate their spacing from the first channel a_0. However, since the majority of the signal's spectral range is empty, simulating the propagation of this signal down a fiber results in a significant amount of the numerical work devoted to the frequency bands between the signals, effectively wasting significant effort. This representation of a signal containing several signals is referred to as the *single-band* mode.

An alternative treatment of this problem is to truly replace the single amplitude function by four separate functions. Thus rather than using Eq. 5.2, we replace the original expansion of the electric field by the expression

$$E(t) = \sum_{k=0}^{3} \gamma_k \left[A_k(t) e^{-2\pi j (f_0 + \delta_{k1})t} + A_k(t)^* e^{2\pi j (f_0 + \delta_{k1})t} \right] f_k(x, y) \tag{5.3}$$

and treat the four amplitudes $A_k(t)$ as separate entities. This increases the computational efficiency since the intervening portions of the spectrum are not modeled. This approach is referred to as the *Multiple-Band* representation.

There are cases where the single-band representation is absolutely necessary; for example, when four-wave-mixing (FWM) occurs we need to simulate the entire band to properly capture the newly generated spectral components.

5.1.6.4 Polarization

Polarization of the optical signal plays a critical role in system performance through such effects as polarization mode dispersion (PMD) and polarization-dependent loss (PDL), and it is essential to incorporate the vector nature of the field into the optical signal representation. For the vast majority of optical components in communications systems, it is acceptable to work with only the transverse components of the field $[E_x(t), E_y(t)]$—the longitudinal component $E_z(t)$is typically negligible and may be recovered from the transverse components anyway. Thus the (single-band) envelope representation for the field is extended to the vector form:

$$\mathbf{E(t)} = \gamma f(x, y) \{ \hat{x} [A_x(t) e^{-2\pi j f_0 t} + A_x(t)^* e^{2\pi j f_0 t}] + \hat{y} [A_y(t) e^{-2\pi j f_0 t} + A_y(t)^* e^{2\pi j f_0 t}] \} \tag{5.4}$$

and it is convenient to define the vector envelope quantity

$$\mathrm{A}(t) = \begin{bmatrix} A_x(t) \\ A_y(t) \end{bmatrix} \tag{5.5}$$

The two-component column in Eq. 5.5 is referred to as the "Jones vector" representation. Internally, the two components are represented by parallel complex arrays.

Since many models have no polarization dependence, by default OptSim™ creates optical signals with a nominal $\hat{x}$ polarization and reserves no memory for the $\hat{y}$ part of the signal. This saves the models from unnecessarily repeating identical calculations on both polarizations. The additional polarization component is created when an $\hat{x}$-polarized signal encounters either a Polarization Transformer model or an optical fiber with PMD effects activated. Certain optical sources can also create arbitrary polarization states.

5.1.6.5 Electrical Noise

Many components can introduce noise into an electrical signal. In the optical communications context the most obvious sources are shot noise in photodetectors and thermal noise in electrical amplifiers. Since in many systems, electrical noise is the dominant degradation, powerful representations of noise are critical to accurate system modeling. Electrical noise in OptSim™ is incorporated with the signal using one of two complementary representations:

Monte-Carlo Representation

This is the more direct way of representation, where the stochastic component is simply added to the sampled signal. So a "quiet" electrical signal could be converted into a noisy one as follows:

$$V_k^{\mathrm{nse}} = V_k^{\mathrm{qte}} + \sigma\hat{\xi}_k \tag{5.6}$$

where $\hat{\xi}_k$ is a real Gaussian random variable of zero mean and unit variance and σ is a suitable standard deviation. The initial and final signals might appear as in Fig. 5.5. This unified representation in which signal and noise are represented as a single sampled vector is very general since it allows any form of noise to be added. This is referred to as the Monte-Carlo (MC) description of noise in OptSim™. The principal disadvantage of the MC approach is that the signal and noise are intertwined permanently—once noise has been added to the signal there is no way of perfectly separating the two. Only by filtering the combined signal can the noise be manipulated further.

Quasi-Analytic Representation

To escape this permanent intertwining, it is sometimes useful to represent the noise as a separate component from the "pure" signal. A reasonable approach would be to introduce a normalized Gaussian noise source and assign a single standard deviation to represent the strength of the noise at all times t_k. A vector of discrete time noise standard deviations to complement the pure signal vector is created for this purpose. Thus the whole signal representation is written as

$$E_k = \begin{bmatrix} v_k \\ \sigma_k \end{bmatrix} \quad k = 0, \ldots, N-1 \tag{5.7}$$

in terms of the signal samples $v_k$and the noise samples σ_k, where N is the number of samples. Since the instantaneous noise is expressed as a standard deviation in Eq. 5.7, this can be considered an ensemble representation of the noise. Figure 5.6 shows the electrical signal produced by a receiver where the photodetector shot noise is the dominant impairment. The left panel shows the separate signal and noise vectors with an obvious correlation between the strength of the noise and signal. The right-hand panel shows the combined signal obtained from Eq. 5.8, confirming that the noise is stronger in the "ones" as expected.

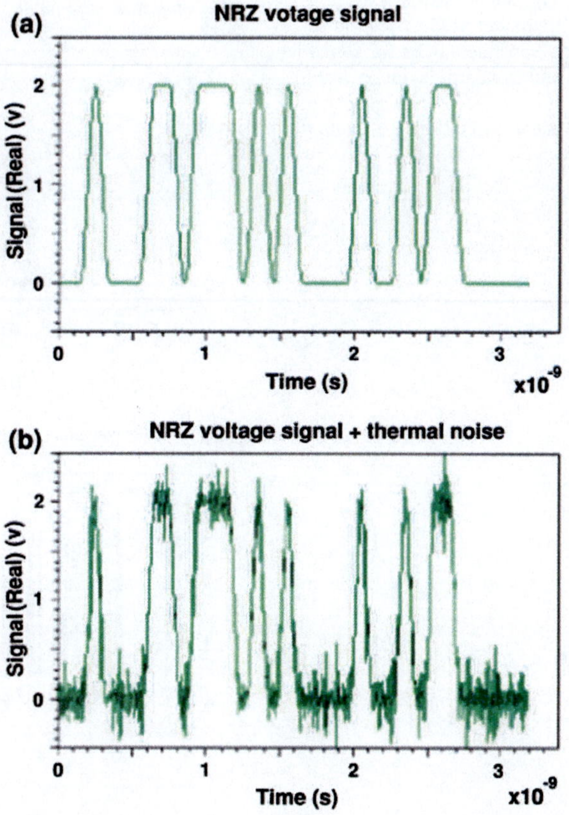

Fig. 5.5 **a** NRZ electrical signal emitted from signal generator (*top*), **b** the same signal after addition of thermal noise in Monte Carlo representation (*bottom*)

$$V_k = v_k + \sigma_k \xi_k \tag{5.8}$$

Separating the noise and signal in this fashion is very useful for certain BER calculations using the so-called "quasi-analytical (*QA*)" technique. With the Monte-Carlo representation, the extraction of bit error rates is considerably more involved and requires the simulation of a longer bit stream for equivalent accuracy, particularly if the system has strong inter-symbol interference (ISI) effects.

The separate representation of signal and noise has two important limitations. First, since the noise is essentially described as a power (or square root of power), no information about its correlation function is retained. It is therefore impossible to filter the noise in any sensible fashion. Second, it is only valid if the noise can truly be considered locally Gaussian, which is not always the case.

5.1.6.6 Optical Noise Representation

The principal source of optical noise in fiber links is ASE. This is a form of noise appearing in optical amplifiers. The spontaneous emission noise spectrum has a

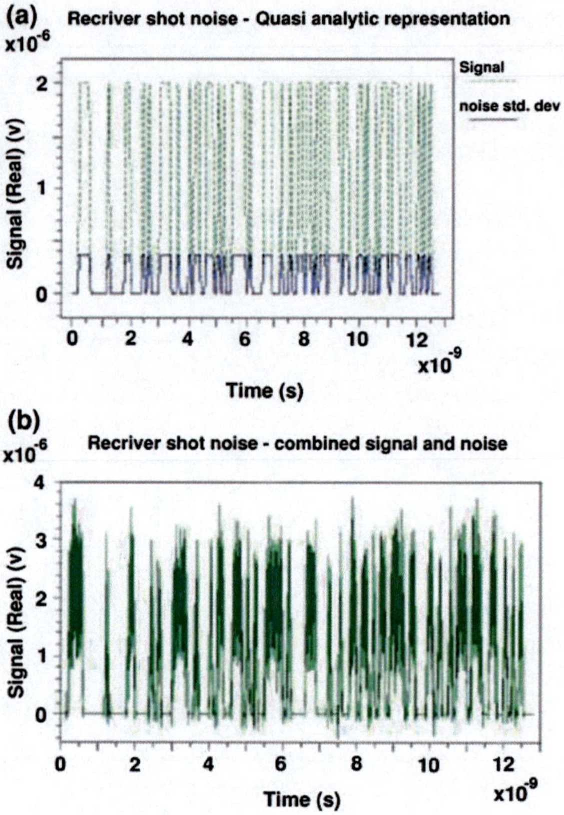

Fig. 5.6 Quasi-analytic treatment of electrical noise. **a** Separate values for signal and standard deviation (*top*), **b** combined signal generated according to Eq. 5.8 (*bottom*)

shape and strength dependent on both the amplifying medium and properties of the pump laser and has an extent of typically hundreds of nanometers.

Noise Bins Representation

The ASE is most naturally represented as a power spectral density function and can be sampled on a much coarser frequency grid than the signal channels, permitting huge savings in computation time and memory. This power representation on a coarse frequency grid is commonly referred to as a series of "noise bins".

Stochastic (Monte-Carlo) Noise Representation

It is sometimes necessary to treat the optical noise stochastically. To transfer ASE noise from the spectral density to a stochastic picture the following operation is performed:

$$\begin{bmatrix} A_x(f_k) \\ A_y(f_k) \end{bmatrix} \rightarrow \begin{bmatrix} A_x(f_k) + \sqrt{\frac{1}{2}P_{\mathrm{ASE}}(f_k)\Delta f}\hat{\eta}_k^x \\ A_y(f_k) + \sqrt{\frac{1}{2}P_{\mathrm{ASE}}(f_k)\Delta f}\hat{\eta}_k^y \end{bmatrix} \tag{5.9}$$

where $P_{\mathrm{ASE}}(f_k)\Delta f$ is the ASE noise power contained in a range Δfaround the signal sample frequency f_j, and $\Delta f$is the sampling rate of the *signal*. In Eq. 5.9, the ASE power $P_{\mathrm{ASE}}(f_k)\Delta f$ is shared equally between the two signal polarizations and the power in the noise bins is assumed to be un-polarized.

5.1.7 Performance Estimation Techniques

The BER of a fiber link is the most important measure of the faithfulness of the link in transporting the binary data from transmitter to receiver. In a laboratory or field experiment, BER can be easily and accurately measured using commercial BER testers, which are based on direct error counting by comparing the input and output digital streams. On the other hand, BER evaluation is usually a very difficult task in a software simulation environment for the following reasons:

- the reference BER for an optical communication system is very low (10^{-9} and lower)
- the signal at the output of the link is usually affected by ISI
- the number of simulated bits is limited due to the need to limit computation time

The most commonly used techniques in communication system simulations are direct error counting, Monte Carlo (*MC*) method and QA techniques. A detailed analysis of all these techniques is presented in [8]. The direct error counting technique is rarely used in the simulation of optical systems because it can be very time-consuming. In fact, a BER close to 10^{-9} requires the simulation of at least 10^{10} bits to achieve a reasonable accuracy [8].

OptSim™ provides both the QA and MC techniques for calculating the BER. Despite the shared name, the Monte-Carlo technique described below has nothing in common with the "Monte-Carlo" error counting method.

5.1.7.1 Quasi-Analytic BER Estimation

Several versions of QA or semi-analytical techniques are available. The common requirement to utilize all these techniques is knowledge of the exact noise distribution at the receiver. Whenever this is true, the semi-analytical techniques are very powerful and provide extremely accurate results even when a few bits are simulated [8]. Moreover, since the noise statistics do not have to be recovered from the signal, the semi-analytical techniques easily handle systems with strong ISI. OptSim™ includes among others, the Karhunen–Loève technique (KLT) [8–12].

The basis of the quasi-analytic technique is illustrated in Fig. 5.7. In this example, the eye diagram has a very simple structure with unambiguous levels for

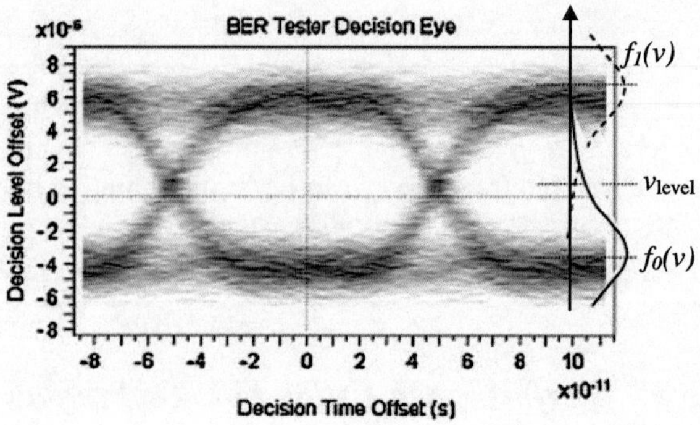

Fig. 5.7 Eye diagram and distribution functions for spaces (*zeros*)

the marks and spaces. Due to thermal noise, there is a significant spread in values around each of the levels. We are concerned with the distribution of values at the *decision point* indicated by the vertical line at the center of the eye.

The "Q-factor" is defined as

$$Q = \frac{v_1 - v_0}{\sigma_1 + \sigma_0} \tag{5.10}$$

where v_1, σ_1 (v_0, σ_0) are the mean and standard deviation of the received signal at the sampling instance when a logic "1" ("0") is transmitted. The BER takes the simple form

$$\mathrm{BER} = \frac{1}{2}\mathrm{erfc}\left(\frac{Q}{\sqrt{2}}\right) \tag{5.11}$$

Pattern Dependence and the QA Approach

In most realistic communication systems, eye diagrams show signs of "pattern dependence" or "ISI". Due to laser properties, dispersion, nonlinearity, or certain noise effects, the shape of a received bit is influenced by the bit values in its neighborhood. The problem with an eye exhibiting pattern dependence is that it is impossible to fit a simple Gaussian distribution to either the mark or space level.

The QA representation provides a solution to the pattern dependence problem. It provides an explicit value for the standard deviation of the noise at the decision time for each bit in the simulated sequence. Moreover, since the PRBS generator produces maximal length sequences, every possible sequence of marks and spaces up to a given length appears in the simulated bit train. Therefore, it is possible to independently evaluate the probability of an error for each of the N bits by the following equation:

$$\mathrm{BER}^{\mathrm{QA}} = \frac{1}{N}\sum_{k=1}^{N}\begin{cases} P_{0\to 1}(v_k, \sigma_k) & \text{bit k is a zero} \\ P_{1\to 0}(v_k, \sigma_k) & \text{bit k is a one} \end{cases} \tag{5.12}$$

where v_k and σ_k are the signal and standard deviation values at the decision point for bit k, and N is the number of bits in the sequence. The maximal length properties of the bit sequence guarantee that the various patterns receive an equal weighting in the sum in Eq. 5.12, and the BER for eyes with strong ISI can be successfully predicted.

5.1.7.2 Monte-Carlo BER Estimation

The QA approach solves the problem of pattern dependence, but is limited in the types of noise and operations it can handle. In particular, it cannot describe colored noise or noise filtering, or the nonlinear amplification of noise as it occurs for ASE in a nonlinear fiber. For systems in which these kinds of noise are present, the Monte-Carlo noise representation, where the noise is directly included as a stochastic contribution to the signal vector, must be used.

In the Monte Carlo approach all the stochastic processes are simulated by extracting a random variable from a random number generator. All noise sources are simulated by extracting a noise sample, adding it to each signal sample and propagating them together. The system parameters are simulated by extracting a set of random numbers at the beginning of the simulation. The digital signal sources are simulated using PRBS generators. The system performance is then estimated by calculating a few moments (typically mean and variance) of the received signal.

In the Monte Carlo approach the noise is propagated with the signal, reflecting the physical reality. If nonlinearities are important, the method is accurate since signal and noise can mix together nonlinearly. Furthermore, for very complex transmission systems, Monte Carlo is often the only available general-purpose approach because the noise statistics at the receiver can only be handled with a Gaussian approximation.

5.1.7.3 Simulated Number of Bits and Confidence Interval

The estimate of the system performance over a finite and typically small time frame using the Monte Carlo method is affected by statistical fluctuations of the results. The relation shown in Eq. 5.11 is exact only when:

- Noise at the receiver is Gaussian and
- Inter-symbolic interference is negligible

There is a widespread perception among the users of simulation tools of the scope and consequences of these two assumptions. On the contrary, it is much less understood that there is an intrinsic uncertainty in the evaluation of the Q-factor whenever it is measured over a finite number of bits. In fact for a given number of bits, the range in terms of BER increases while decreasing the reference BER. This is due to the fact that the BER versus Q function is steeper for smaller BER.

Therefore, a larger number of bits should be simulated to achieve the same confidence interval when the reference BER is lowered.

5.1.8 Interface with Third-Party Tools

OptSim™ also provides powerful capabilities for creating user models in both the block mode and sample mode simulation engines. User models may be written in a programming language such as C++ or FORTRAN and linked into the simulation tool utilizing the API provided with OptSim™. In addition, OptSim™ supports MATLAB® interfaces that enable users to write their user models in MATLAB® for full co-simulation with OptSim™ as well as use MATLAB® for post-processing analysis of OptSim™ simulation results.

OptSim™ also interfaces with a Spice simulation engine. The *custom component* for using with SPICE™ (CCS) is an OptSim™ component whose behavior during the simulation is described by a SPICE™ circuit file.

In addition, OptSim™ provides interfaces to third-party tools such as Thin Film Center's Essential MacLeod thin film design software. Also, OptSim™ provides interfaces to leading laboratory test equipment including Agilent's 86146B-DPC optical spectrum analyzer and time resolved chirp measurement solution and 81910A photonic all-parameter analyzer, and Luna Technologies' optical vector analyzer (OVAe).

5.1.9 Interface with Other RSoft Tools

5.1.9.1 Device-Level Tools

OptSim™ also provides interfaces to various device-level simulation tools including RSoft's BeamPROP™, GratingMOD™, and LaserMOD™.

5.1.9.2 Best Fit Laser Toolkit and SPICE™ Model Generator™

OptSim™ also provides the Best Fit Laser Toolkit™ for quickly and easily producing accurate rate equation laser models that are well-fitted to real-world laser performance characteristics as determined through measurements or data sheets provided by the manufacturer. Accompanying this capability is the SPICE™ interface that bridges the gap between the electronic circuit and optical system simulation domains. This includes the SPICE™ Model Generator™ that automatically generates a SPICE™ model deck based on OptSim's rate equation laser models for use in SPICE™ simulations. Using this capability, the laser's true driving conditions can be accurately modeled in SPICE™ simulations of the laser

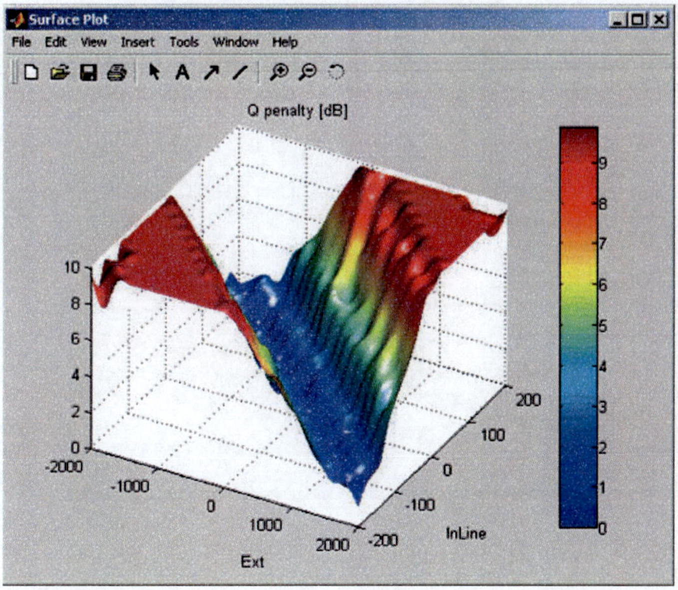

Fig. 5.8 Example of surface map (Copyright 2010 RSoft Inc. Used by permission of RSoft Inc.)

driver. Then the laser driver's accurate output waveform can be used to drive the laser model in OptSim™ to produce an accurate simulation of the system performance.

5.1.10 Visualization

5.1.10.1 Q Surfaces Toolkit

This toolkit plots the contour lines map and the surface map of $Q(var_1, var_2)$ expressed in *dB*. The Q values are extracted from the OptSim™ project results while var_1 and var_2 are two project variables on which the user has performed a parametric run. A series of pop-up menu windows guides the user through choosing the *Q* estimator block and the two variables, var_1 and var_2, to plot the Q evolution. The implementation of this toolkit utilizes the OptSim™ MATLAB® interface.

As an example, let us consider launching a 4-channel WDM optical signal into 1,000 km of fiber, where the link consists of 10 spans composed of 100 km of fiber (D = 4 ps/nm/km) with in-line dispersion compensated followed by an optical amplifier. A post-compensation unit is placed after the entire length of fiber. Both the *in-line* and *post-compensation* units are assumed to be ideal fiber grating blocks. Varying the *in-line* (var_1) and *post dispersion* (var_2) values it is possible to plot the Q penalty maps by means of the Q surfaces routine as shown in Fig. 5.8.

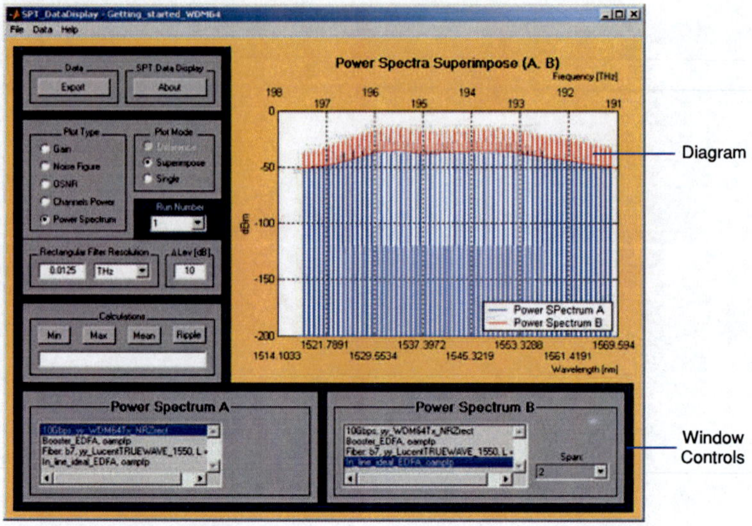

Fig. 5.9 The SPT data display dialog window (Copyright 2010 RSoft Inc. Used by permission of RSoft Inc.)

5.1.10.2 SPT Data Display Toolkit

The *SPT data display* toolkit is an interface to MATLAB® that performs some graphical analysis on the optical results of OptSim™ simulations based on the SPT. The toolkit retrieves the SPT simulation results from any pair of optical points of your current project schematic, it shows the optical power spectra and for each channel that is detected, it calculates the gain, the noise figure, and the OSNR between the points.

The *SPT data display* dialog window is shown in Fig. 5.9. It consists mainly of a group of controls (buttons, options, and lists) and a diagram that graphically displays the results.

5.1.11 Scripting and Batch Simulations in OptSim™

The scripting feature in OptSim™ provides the following advantages:

- Projects can be run in the background or can be scheduled for overnight execution.
- Simulation can be performed directly from the Linux console or the Windows OptSim™ console via the GUI.
- Simulation execution can be done without the GUI.

5.1.12 Higher Order Composites

The term higher order composite (HOC) describes a composite, hierarchical block that has the ability to replicate itself a number of times during execution of a schematic model. The HOC functionality permits a user to quickly create very large models with many repeating elements in a compact schematic layout.

5.1.13 Design-Friendly Features

- *Parametric scan*: This technique permits a sequence of simulations, which differ only in the numerical values of certain parameters/variables in the simulation.
- *Test component*: Component parameters may be tested prior to running a system simulation. This feature allows component parameters to be more easily specified given measured performance characteristics. For example, the L-I curve and small signal frequency response curves of a semiconductor laser are output based on the specified component parameters.
- *Repeat loops*: Repetition loops allow portions of a link topology to be specified as being repeated *n* times such that the simulation will behave as if there were *n* sets of the groups of component icons within the repetition loop icons attached end-to-end. Repetition loops may be nested, and may have multiple input and output signals. These are especially useful for defining spans of fibers and amplifiers in long-haul systems.
- *Compound components*: A CC can be thought of as a subsystem that is built from other models and is represented in a project as a single entity. OptSim™ supports a number of methods to easily create CCs (also can be referred to as hierarchy, composite actor, or superblock), which can either be reused in other schematics or, are used simply to reduce clutter in the original schematic.
- *Flexible export/import of data for post processing*: Several post-processing and simulation results analysis tools are provided including the WinPlot™ and RPlot™ simulation result plotting tools as well as the VirLab™ post-processing and simulation results analysis tool. These tools make it easy to interactively work with the simulation results to present them in the manner desired and analyze them further.
- *Decorative elements*: A number of decorative blocks are available in OptSim™, which allow the users to add specific graphical objects and images to their schematics.
- *Project report generation*: OptSim™ comes with a user-friendly way of web-ready report generation. The report is an HTML file that can be edited in external software like Microsoft Word, can be archived using external software, like WinZip, or can be converted into PDF using external software, like Acrobat. The report can contain screenshots of the project layout, user-entered description, results in the form of graphs and tables, parameters used in the project and their numerical values in tabular form.

5.2 VPIphotonics: VPItransmissionMaker™ Optical Systems

5.2.1 Introduction

VPIphotonics (http://www.vpiphotonics.com) through the VPItransmission-Maker™/VPIcomponentMaker product family offers an integrated environment for Photonic Design Automation (PDA). VPItransmissionMaker™ Optical Systems accelerates the design of new photonic systems from short-range via access and metro to long-haul optical transmission systems and allows technology upgrade and component substitution strategies to be developed for existing network plants. It can fully verify optical transmission link designs to identify cost savings, investigate novel technologies, or fulfill specialist requirements.

The combination of a feature-rich GUI (shown in Fig. 5.10), a sophisticated and robust simulation scheduler together with flexible optical signal representations and simulation domains enables an efficient modeling of any transmission system including bidirectional links and complex networks.

VPItransmissionMaker™ Optical Systems is widely used as an R&D tool to evaluate novel component and subsystems designs in a systems context, and investigate, compare and optimize diverse system technologies (such as coding, modulation, monitoring, compensation, regeneration). As part of VPIphotonics' suite for PDA solutions, it supports behavioral and detailed physical modeling of subsystems and components. Optical fiber-based amplifiers and high-power lasers created using VPI-componentMaker Optical Amplifiers, and active/passive photonic integrated circuit applications created using VPIcomponentMaker Photonic Circuits can be included and simulated together with VPItransmissionMaker™ Optical Systems. The design, analysis and optimization of almost every system concept is supported, for instance:

- PON techniques, FTTx distribution
- Aggregation and metro networks
- Core and transoceanic WDM systems
- High capacity/high-speed systems
- ROADMs and optical networking
- Short-range connections, free-space links
- Analog and digital video distribution services
- RF-over-Fiber, Microwave photonics
- Modulation and detection techniques
- Optical and electrical mitigation techniques
- Power transients and network dynamics

5.2.2 Summary of Numerical Models

VPItransmissionMaker™ Optical Systems provides libraries of models for all parts of an optical transmission system. Among them are:

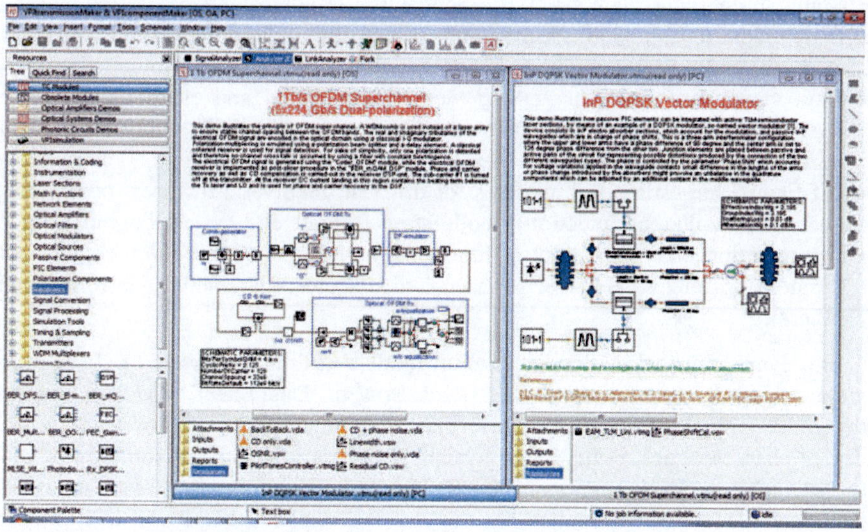

Fig. 5.10 VPItransmissionMaker^{TM}/VPIcomponentMaker GUI (Copyright 2010 VPIsystems. Used by permission of VPIsystems Inc.)

- Arbitrary code word generators (random, alternate, via file or direct code word input among others).
- Laser sources and optical modulators (Mach–Zehnder, EA, generic amplitude, phase, frequency, chirp, polarization modulator).
- Optical transmitter and transmitter arrays supporting various modulation formats: NRZ, (C)RZ, CS-RZ, (IM-)DPSK, CSRZ-DQPSK, Duobinary, Manchester encoding, OFDM, DP-QPSK, (D)mPSK and mQAM with $m = 2^n$ and others).
- Passive components (attenuator, combiner/splitter, mux/demux among others).
- Optical and electrical filter banks (physical: FBG, AWG, AOF, etc.; general: Bessel, Gauss, Trapezoid, etc.; measured: input from file).
- Fibers (single-mode, multi-mode, various physical models for considering dispersion, nonlinear effects, and polarization dependency).
- Static and dynamic optical amplifiers (SOA, EDFA, Raman, hybrid).
- Direct-detection receivers with PIN/APD including receiver electronics and arbitrary filtering.
- Balanced and coherent detection techniques (incl. impairments) to support differential, phase, and multi-level modulation formats.
- Receiver electronics and post-detection circuitry (electrical filters and amplifiers, clock recovery, thresholder, decider).
- Electronic (digital) signal processing as dispersion/nonlinearity pre-emphasis at the transmitter and using equalization techniques (DFE/FFE, MLSE, MSPE, CMA, ellipse fitting, and back propagation) at the receiver.

- Signal processing elements (arithmetical tools, control elements, matrix and logic operations, signal sources and sinks, digital communications, spectral analysis, and others).
- Data visualizers and analyzers (power meter, optical and electrical spectrum analyzers, oscilloscopes, eye diagrams, Poincaré sphere, link analysis, 1/2/3-D numerical data analyzer, and others).
- Bit-Error-Rate estimation modules for different receiver structures (direct, balanced, and coherent detection), modulation formats and system degradations.
- Optimization and signal control functions (signal repeat and chop, conditional iterator, optimizer, signal converter and selector, parameter controller and others).

For every physical device, simulation models at different levels of abstraction are provided. It is convenient to use 'Black-Box' or 'Data Sheet' models following the philosophy that measurable performance parameters are used to describe the device characteristics, rather than physical parameters, which are sometimes hard to obtain by optical systems designers. However, if information about physical parameters is available, detailed physical models can be used as well to study the impact of individual components along the transmission link.

A range of over 450 application demos is provided. These offer insight into simulation techniques and act as templates for standard problems in well-known industry applications allowing extensive investigation of advanced design problems and offering a significant tutorial value. An automated formal design process for WDM systems is also provided, which aids at every stage by accurate and efficient numerical techniques, presented as design assistants, simulation scripts, parameter, and module sweeps.

5.2.3 Signal Representation

VPIphotonics was the first to develop and utilize the concept of multiple-signal representations to support different abstraction requirements of different types of signals, pumps, noise, and degradations. For the first implementation concept [13], Optical Society of America (OSA) provided VPIphotonics with an award in 1998. VPIphotonics has constantly improved its flexible simulation domains and signal representations over the years. Figure 5.11 shows a graphical overview of the currently available representations [14].

The so-called *sample mode* representation is often used in cases where bidirectional interactions between closely coupled components need to be considered. It passes data bidirectionally on an iteration-by-iteration basis. The iterations represent sub-picosecond slices of time. Sample mode is used for modeling optical devices and integrated circuits in great detail. Sample mode is the most general signal representation, but is also the least efficient. Polarization effects are included however some modules use isotropic approximation modeling.

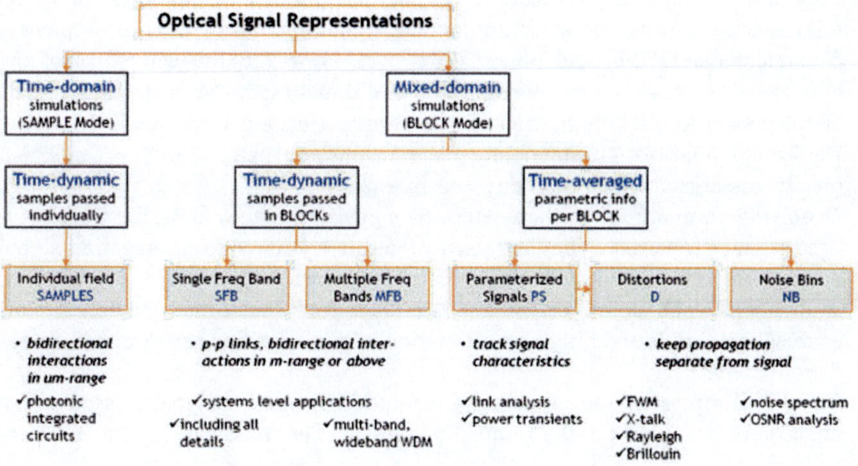

Fig. 5.11 Overview of VPIphotonics' award-winning concept of multiple optical signal representations (Copyright 2010 VPIsystems. Used by permission of VPIsystems Inc.)

In the so-called *block mode* (time blocks of data are passed between modules) five sub-representations with differing levels of efficiency are available. All include polarization information. They can be used simultaneously, to communicate between components. The representations either define signals as samples of the optical field (time-dynamic representation), or by statistical characteristics of the signals (time-averaged representation). The two optical field representations are closely related:

- A *single frequency band* (SFB) signal contains all the modulated optical carriers within a single, continuous frequency band. The band represents the total sampled optical field signal, in two polarizations.
- A *multiple frequency band* (MFB) signal is created when several signals with non-overlapping spectra, each described by an SFB, are combined into a single signal (at a WDM multiplexer for instance). In this case, each band is represented as a separate sampled optical field, tagged with its center frequency. The full detail of interactions between the bands can be calculated by approximate methods (e.g. frequency decomposition techniques). MFB representation of a WDM system is more efficient than using an SFB.

A single optical transmitter will produce a SFB. When two or more sources are combined, the individual bands will join into a new SFB if their simulation bandwidths overlap, or form an MFB signal, if their simulation bandwidths do not overlap. Additionally, an MFB signal can be transformed into an SFB using a variety of resample modules, or settings within the fiber modules.

The following statistical representations are available:

- *Parameterized Signals* are time-averaged descriptions of channels based on fundamental information about the optical signal, including: central frequency, equivalent bandwidth, and polarization state. However, the waveform of the optical pulse stream is not available for this signal type. Parameterized signals are also used to track more than 15 signal properties (such as CD, SPM, noise, DGD, and polarization state) along a transmission path, allowing for signal property analysis versus topology and frequency [15].
- *Distortions* are similar to parameterized signals, e.g. they describe time-averaged signal properties. They are used to describe backward propagating signal powers due to Rayleigh or Brillouin scattering, and to keep track of degradations from FWM. The definition of an individual representation for such signals enables the simple tracking and estimation of their system performance degradation effects.
- *Noise Bins* represent noise separately from the information signals, based on the notation of noise bins in the frequency domain. The noise bin representation is similar to that of the parameterized signals, including polarization. The main difference is that noise bins define the noise power density in a certain frequency range instead of the channel power, which is defined by parameterized signals.

All signal formats can be tracked, visualized, and analyzed separately from each other. Between the different signal formats, a variety of conversion and grouping routines are available.

5.2.4 Design Process

The end-to-end support of the design process is illustrated in Fig. 5.12. The GUI of VPItransmissionMaker™/VPIcomponentMaker supports flexible design requirements such as single user, multiple user, remote users, data and results sharing, and encryption. In addition, VPIphotonics tools include the concept of *macros* and *design assistants*. Macros synthesize system designs for given specifications and ease the creation and adaptation of schematics. Design Assistants can be created by adding interactive comments and mathematical calculations, or calling a third-party program to an existing design.

VPItransmissionMaker™ optical systems comes with an interactive simulation control that supports the design process by allowing multi-dimensional tuning and sweeping of any parameter type, and supporting automatic optimization as well as yield estimation including confidence estimation of final results. Control ranges are defined by intervals with increments, or arbitrary lists. Furthermore, they can be driven from random generators with uniform, Gaussian, and truncated Gaussian distributions (enabling Monte-Carlo experiments). Control definitions can be combined and manipulated using mathematical expressions before being applied to module or global simulation parameters.

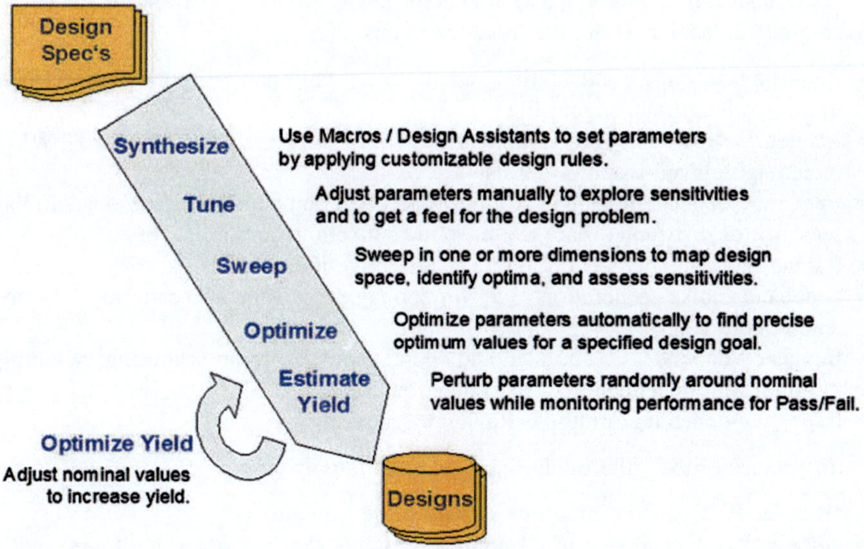

Fig. 5.12 End-to-end support of the design process (Copyright 2010 VPIsystems. Used by permission of VPIsystems Inc.)

Additionally, VPIphotonics offers with the simulation scripting capability an advanced automatic control of the simulation engine. Simulations can be performed automatically, parameters can be manipulated based on previous results, and files can be created and modified without supervision, an essential feature when simulating over large parameter spaces.

5.2.5 Transmission Fiber Modeling

VPIphotonics provides a Universal Fiber model suitable for simulating all aspects of transmission fiber propagation [16]. Using the single model, all physical phenomena can be combined, or considered separately (e.g. individual propagation effects can be turned on/off). It can take data from measured fibers and assess workable upgrade strategies in the presence of multiple degradation mechanisms.

The universal fiber model supports the modeling of wideband bidirectional signal and pump transmission in optical fibers. Signals and pumps can be represented as sampled waveforms or by time-averaged signal representations. All fiber parameters (attenuation, dispersion, effective mode area, nonlinear coefficient, Brillouin and Raman scattering responses, Rayleigh backscattering, birefringence profile and OTDR data) can be entered with their wavelength dependences as input files. Heterogeneous fibers comprising multiple spans of different fiber types spliced together can be described. The model uses stable and fast numerical algorithms with easy-to-use numerical controls.

Two modes of operation exist: full vector model (Jones) and scalar model (SOP averaging). In scalar mode, the model includes:

- frequency-dependent attenuation and any-order chromatic dispersion, described by a wavelength-dependent file
- frequency-dependent Kerr nonlinearities (SPM, XPM, FWM, I-XPM, I-FWM) including delayed-response terms
- frequency-dependent Raman scattering between pumps, signals and noise in the presence of distributed backscatter, point reflections and point losses
- Raman-induced changes in nonlinear refractive index
- wideband noise generation due to spontaneous Raman scattering at any temperature
- frequency-dependent stimulated and spontaneous Brillouin scattering including the linewidth-dependence of the reflection coefficient
- frequency-dependent multiple Rayleigh scattering

In vectorial mode, the model includes *additionally*:

- all-order PMD, using importance-sampling techniques
- polarization dependence of all nonlinear effects (Kerr, Raman, Brillouin), with co- and orthogonally polarized Raman responses accounted independently

5.2.6 Doped-Fiber and Raman Amplifier Modeling

For many transmission system applications, the provided EDFA *black-box* models in VPItransmissionMaker™ Optical Systems are well suited as they require parameters describing the behavior of EDFAs on a systems level, and do not need to be fed with information about the internal EDFA structure and details of the doped-fiber characteristics. Furthermore, those *black-box* models are computationally fast and support possibilities for incorporating internal automatic gain or power controls [17].

Characterization templates allow deriving behavioral EDFA characteristics (such as gain and noise figure as functions of wavelengths and pump powers) from detailed amplifier modeling in VPIcomponentMaker Optical Amplifiers using the Doped Fiber model—a comprehensive module for simulating erbium-, ytterbium- or co-doped fibers. The Doped Fiber model supports hosts such as Silica, Fluoride and Tellurite, and bi-directional pumping into the fiber core or cladding. If co-doped with Er/Yb, the energy transfer due to cross-relaxation can be considered. The gain shape description can be imported from file via Giles gain and attenuation or via cross-sections versus wavelength/frequency. With the information of Giles parameters and material information (such as overlap integral, dopant concentration, carrier lifetime), the cross-sections can be derived. The modeling of the supplementary effects (temperature dependence, ESA, SHB, up-conversion, pair-induced quenching, Rayleigh scattering, background loss and others) can be individually turned on/off. In conjunction with VPItransmissionMaker™ Optical

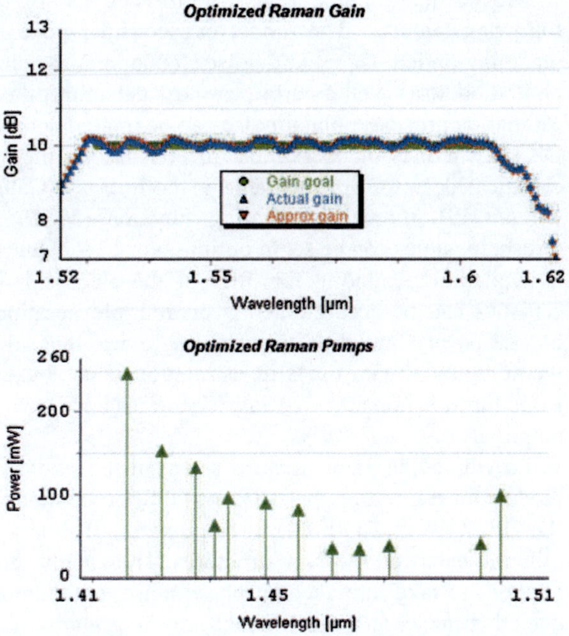

Fig. 5.13 Multi-pump Raman amplifier with optimized gain (*top*) over 80 nm and corresponding powers (*bottom*) of the counter-propagated 12 Raman pumps (Copyright 2010 VPIsystems. Used by permission of VPIsystems Inc.)

Systems, the propagation of time-dynamic optical signals (sampled bands) along the doped-fiber can be simulated. In this case, effects from the Kerr nonlinearities (e.g. FWM, SPM) can be modeled as well.

A unique Raman pump optimizer supports the rapid design of gain-flattened and gain-profiled Raman amplifiers with wide bandwidths, while incorporating degradations calculated by the universal fiber model in the optimization. To accelerate calculations and reduce processing time, the module pre-calculates a trial solution using faster semi-analytical solutions, before fine-tuning the solution in the presence of degradations [18]. In conjunction with the universal fiber model, the Raman pump optimizer supports arbitrary pumping and signal transmission schemes. It optimizes the wavelengths and powers of pumps into both ends of the fiber simultaneously. The target gains can be set arbitrarily, and channel powers specified individually. Various optimization constraints (such as max limit of single pump powers, or of overall pump power) are supported as well. Figure 5.13 shows exemplary optimization results for a 12-pump Raman amplifier.

5.2.7 *Bit-Error-Rate Estimation*

VPItransmissionMaker^{TM} Optical Systems provides different types of BER estimation methods, which support the various modulation formats and corresponding optical receiver structures [19–21].

BER estimation in systems with direct detection receivers is available with the following features: The model can consider polarization information about the incoming optical signal and noise. It supports the characteristics of PIN and APD photodiodes, as well as arbitrary post detection filter functions. During the BER estimation process, optical noise can be treated deterministically, or stochastically. Different modes of operations for considering the amplitude distributions at the electrical detector are available in both cases (Gauss or Chi2, possibly with or without ISI, moment-generating function (MGF) analysis). Sample time and threshold values can be set to optimum or fixed values. An adjustable sample range is available accounting for jitter of the electrical detection circuitry. The BER statistics can be accumulated over multiple iterations allowing a more effective utilization of simulation time. A flag exists that allows ignoring a subset of the incoming bit stream for BER estimation. This is useful for data streams suffering from turn-on transients of lasers, for instance, or dynamic gain fluctuations of amplifiers.

Providing the same feature set as listed above, VPIphotonics offers unique models for estimating the BER for DPSK-type modulation formats (such as NRZ/RZ-DPSK) and DQPSK-type modulation formats (such as NRZ/RZ-DQPSK) utilizing balanced receiver structures. They apply semi-stochastic or deterministic methods to take into account phase noise, optical amplifier noise, electrical post-detection noise, and ISI. Internal noise correlation that is, for instance, caused by tight optical filtering or nonlinear phase noise can be modeled as well (see for instance Fig. 5.14).

Furthermore, VPItransmissionMaker™ Optical Systems provides

- Bit error counting modules (to support Monte-Carlo experiments).
- BER estimation using error-counting or PDF-fitting for coherent transmission systems using multi-level modulation (optical mQAM/(D)mPSK) and DSP.
- BER estimation for generic differential-phase amplitude-shift-keying (DP-ASK) modulation formats with arbitrary receiver structures.
- Methods for BER estimation of subcarrier-modulated systems (such as OFDM using mQAM/(D)mPSK encoding of electrical subcarriers).
- BER estimation solutions for burst-mode operation.
- BER approximation techniques that work for Parameterized Signals, Noise Bins, and FWM distortions supporting different modulation formats.

5.2.8 Third Party Interfaces, Distribution Simulation Tasks, and Virtual Prototypes

Co-simulation allows parts of a simulation schematic to be modeled using third-party or in-house code. VPIphotonics' design tools provide live interfaces to MATLAB®, Python, and compiled code in a DLL (or shared library), so simulations seamlessly interact with models in these formats.

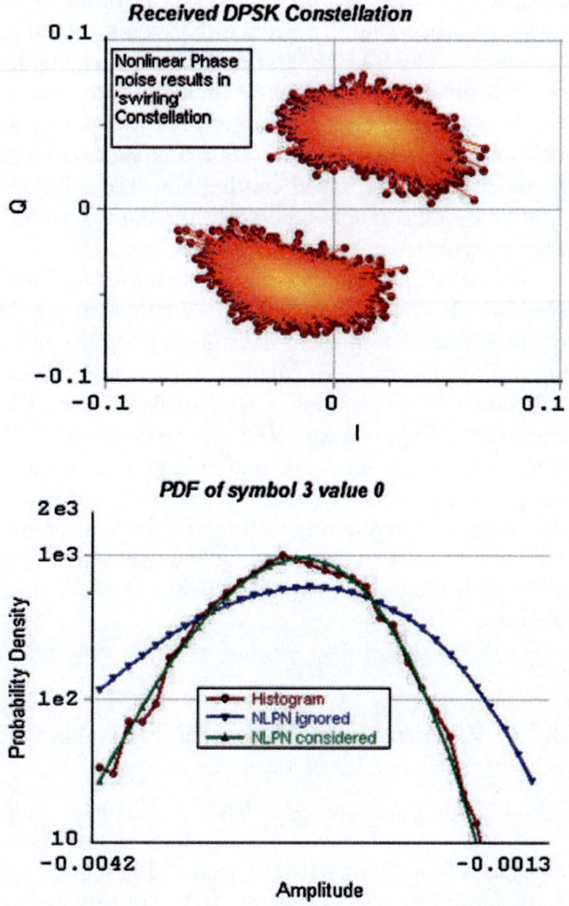

Fig. 5.14 Constellation diagram for a DPSK system experiencing nonlinear phase noise (NLPN), (see *top*) and corresponding PDF calculations (with and without consideration of optical noise correlation due to NLPN) (see *bottom*) (Copyright 2010 VPIsystems. Used by permission of VPIsystems Inc.)

Electronic circuit models from Agilent's advanced design system (ADS) can be called directly from VPItransmissionMaker^{TM}/VPIcomponentMaker allowing assessing the performance of high-speed electronics in photonic link designs. VPIphotonics' simulations can control directly parameters of an ADS project. Therefore, component characteristics modeled in detail by ADS projects can be automatically analyzed and optimized.

Furthermore, VPIphotonics simulations can import data from component characterizations and systems measurements. Various interfaces to all kinds of test and measurement equipments exist; new formats can be defined using customizable data type converters and macros.

Operating from one GUI, VPIphotonics' flexible licensing scheme supports the execution of multiple simulation processes on multi-core machines, and remote execution on external simulation servers. Results can be stored on the remote computer, or provided directly in proper order to the local GUI for data analysis and post-processing. To reduce the overall simulation time, calculations within certain modules (e.g. fibers) and for distinct modeling tasks (e.g. FFT) can be parallelized by making use of modern computer environments with multi-core CPUs and massively-parallel graphics processing units (GPUs) where available. For instance, using a GPU with 512 processors to simulate propagation over nonlinear fiber, the speed can be increased 50–100 times when compared to the simulation on a single-core CPU for high-capacity systems covering significant fiber distances.

VPIplayer is a patented communications tool that offers a format to exchange photonic designs while protecting intellectual property information. It empowers R&D engineers to present their ideas to colleagues and customers who have less specialized technical knowledge, and no need for detailed design and simulation software. VPIplayer runs a readymade dynamicDataSheet (dds)—exported and encrypted design from VPItransmissionMaker^{TM}/VPIcomponentMaker—on a fully featured photonic simulation engine. The dynamicDataSheet captures product parameters as a schematic for a live simulation that generates results on data analyzers, and supporting information such as company logo, contact details, and others. Interactive settings allow end users to adjust parameters using predefined sliders. VPIplayer is distributed freely and downloadable from the VPIphotonics web site.

5.2.9 Visualization, Data and Link Analysis and Design Rules

VPIphotonicsAnalyzer establishes a universal framework for data display and analysis. It allows maximum freedom in the display, arrangement, export, and analysis of simulation results from VPIphotonics' design tools and Third party software. It provides visualizers and analyzers that accurately represent laboratory test and measurement equipment for detailed results display, component characterization, and system performance analysis functions.

Incoming signals can be shown and analyzed in six modes: OSA, scope, eye, RFSA, Poincare, and numerical. The first five modes are designed to work with optical and/or electrical signals as provided by VPItransmissionMaker^{TM}/VPIcomponentMaker. The numerical mode is designed to work with ordinary numerical data in integer and float form. It offers one-, two- and three-dimensional display options supporting the creation of plots such as BER versus signal power, constellation diagrams, contour plots of 2-dimensional parameter sweeps and others. VPIphotonicsAnalyzer works with signals in both block and sample modes. Furthermore, dedicated characterization tools for optical amplifiers and lasers, power meters, a 2-port electrical signal analyzer, PMD test set, and histograms are provided.

The LinkAnalyzer offers means for displaying and analyzing signal properties that have been tracked along the fiber link (GVD, noise/distortions/signal power, Kerr, DGD, and derived values such as OSNR-, Q-, BER-limit and others) versus

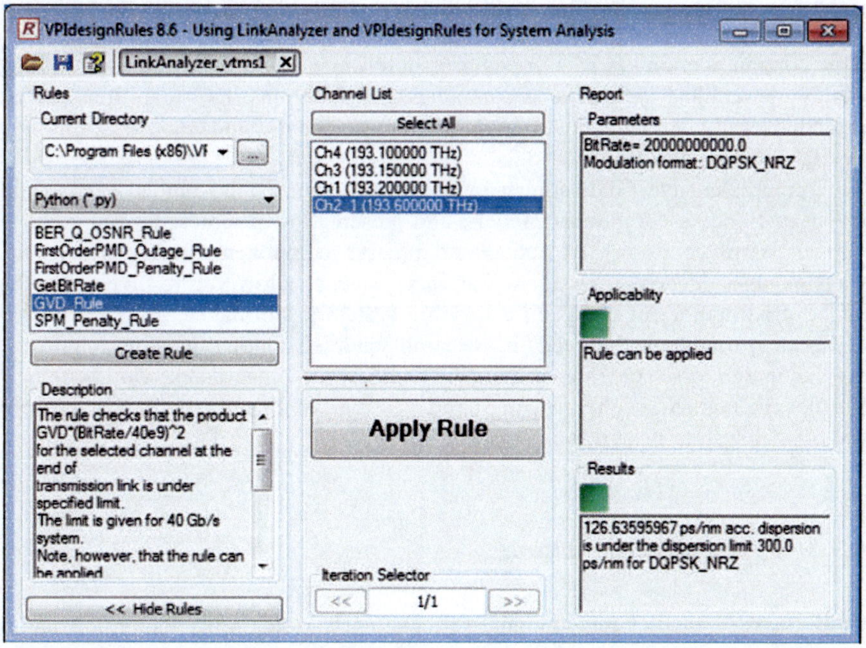

Fig. 5.15 User interface of VPIdesignRules, a framework for testing the WDM system performance based on built-in or user-supplied engineering rules (Copyright 2010 VPIsystems. Used by permission of VPIsystems Inc.)

position, distance, frequency, simulation iteration, or time instance [22]. Furthermore, tracked signal properties can be used to drive the VPIdesignRules—a framework that enables the user to perform quick and easy estimates of signal quality measures in WDM systems using built-in or user-supplied design (engineering) rules (Fig. 5.15). The design rule analysis consists of three steps: (a) a design rule has to be specified; (b) the channel(s) that should be analyzed are selected; (c) the specified design rule is applied to the selected channels delivering information about channel parameters, the applicability of the rule, and the result of the analysis.

5.3 Optiwave: OptiSystem™ Software

5.3.1 Introduction

OptiSystem™ (http://www.optiwave.com) is an innovative optical communication system simulation package that designs, tests, and optimizes virtually any type of optical link in the physical layer of a broad spectrum of optical networks, from analog video broadcasting systems to intercontinental backbones.

OptiSystem™ is a standalone product that does not rely on other simulation frameworks. It is a system-level simulator based on the realistic modeling of fiber-optic communication systems. It possesses a powerful simulation environment and a truly hierarchical definition of components and systems. Its capabilities can be extended easily with the addition of user components, and can be seamlessly interfaced to a wide range of tools.

A comprehensive GUI, shown in Fig. 5.16, controls the optical component layout and netlist, component models, and presentation graphics.

The extensive library of active and passive components includes realistic, wavelength-dependent parameters. Parameter sweeps allow you to investigate the effect of particular device specifications on system performance.

Created to address the needs of research scientists, optical telecom engineers, system integrators, students, and a wide variety of other users, OptiSystem™ satisfies the demand of the growing photonics market for a powerful and easy-to-use optical system design tool.

5.3.2 Signal Representation

OptiSystem™ handles mixed signal formats that add flexibility and efficiency to the simulation tool. It provides models at different abstraction levels, including the system, subsystem, and component levels. OptiSystem™ implements a hierarchical definition of components and systems, allowing the user to employ specific software tools for integrated and fiber optics at the component level and allowing the simulation to go as deep as the desired accuracy requires. Different abstraction levels imply different signal representations. There are four types of signals in the signal library: optical, electrical, binary, and M-ary. OptiSystem™ handles mixed signal formats for optical and electrical signals in the component library. It calculates the signals using the appropriate algorithms related to the required simulation accuracy and efficiency.

5.3.2.1 Optical Signal

Optical signals are generated by components such as lasers in the transmitters library. Optical signals accommodate different signal representations:

- Sampled signals: optical signals can accommodate any arbitrary number of signal bands. In the simplest case, there is one SFB when a single, continuous frequency band represents the waveforms of all the modulated optical carriers. A single optical source (for example a CW laser) produces a SFB. The band represents the complex sampled optical field of the signal in two polarizations.
- Parameterized signals: The signal description based on the sampled signals covers the majority of physical phenomena affecting the system design.

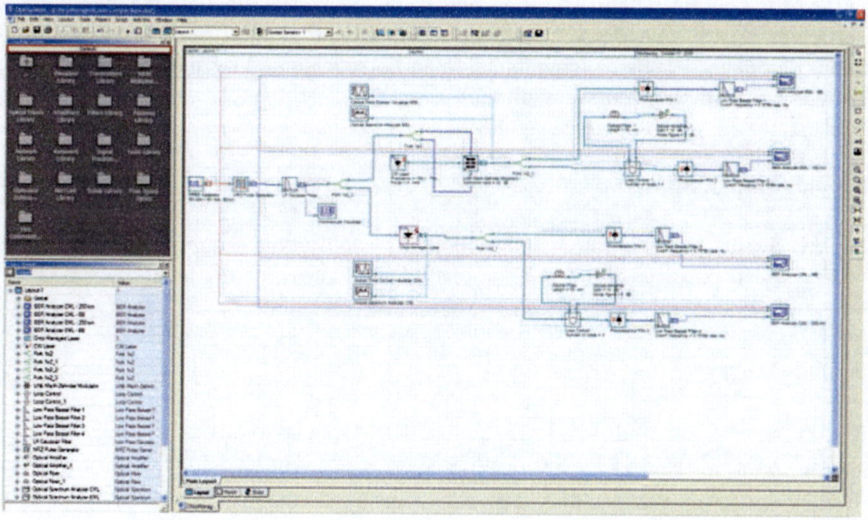

Fig. 5.16 The OptiSystem^{TM} GUI (Copyright 2010 Optiwave Systems Inc. Used by permission of Optiwave Systems Inc.)

When designing a system where the power budget analysis and fast signal-to-noise ratio estimations are the main performance evaluation results, signal channels can be approximated by their average power, assuming that the detailed waveform of their data streams are not important. Parameterized signals are time-averaged descriptions of the sampled signals based on the information about the optical signal (for example average power, central frequency, and polarization state).

- Noise bins: noise bins represent the noise by the average spectral density in two polarizations using a coarse spectral resolution. The resolution can be adapted to maintain the accuracy of the simulation. The main advantage of using noise bins is to cover the wide spectrum of the optical signals or to represent the noise outside the sample signal bandwidths. The noise bin representation is similar to the parameterized signals, including the polarization. The main difference is that noise bins are defined by the noise power spectral density and the bandwidth of each noise bin instead of by the average power.

In addition, optical signals can also be represented by individual samples, allowing for time-driven simulations.

5.3.2.2 Electrical Signal

Electrical signals are generated by components such as pulse generators and photodetectors. The electrical signal is represented by a sampled signal waveform in the time domain. The main properties of the electrical signal are the signal noise variances in the time domain and the noise power spectral densities in the frequency domain. Electrical signals can also be represented by individual samples, allowing for time-driven simulations.

5.3.2.3 Binary Signal

Binary signals are generated by components such as bit sequence generators. A binary signal consists of a sequence of ones and zeros, or marks and spaces.

5.3.2.4 M-Ary

M-Ary signals are multilevel signals used for special types of coding, such as PAM, QAM, PSK, and DPSK. M-Ary signals are similar to the binary signals. However, M-Ary signals can have any level instead of only high (1) and low (0) levels, or marks and spaces.

5.3.3 Simulation Methodologies

Creating an OptiSystem^{TM} project is very straightforward. OptiSystem^{TM} provides a hierarchically organized library of electrical, optical, and logical components. Components are added to the project using the familiar drag-and-drop technique and then connected. There are three types of connections: optical, electrical, and logical. Each component has a set of parameters that can be set to a constant value, swept through a range of values, determined by a script, read from a file (look-up table), or optimized. Some key features that enhance the design process in the OptiSystem^{TM} framework are:

- the user can create many designs using the same project file, which allows for quick and efficient creation and modification of designs. Each OptiSystem^{TM} project file can contain many design versions. Design versions are calculated and modified independently, but calculation results can be combined across different versions, allowing for comparison of the designs.
- the *calculation scheduler* controls the simulation by determining the order of execution of component modules according to the selected data flow model. The main data flow model that addresses the simulation at the transmission layer is the component iteration data flow (CIDF). The CIDF domain uses run-time scheduling, supporting conditional execution, data-dependent iteration, and true recursion.
- to make a simulation tool flexible and efficient, it is essential to provide models at different abstraction levels, including the system, subsystem, and component levels. OptiSystem^{TM} features a truly hierarchical definition of components and

systems, enabling you to employ specific software tools for integrated and fiber optics at the component level, and allowing the simulation to be as detailed as the desired accuracy dictates.

5.3.4 Category of Models

The OptiSystem™ component library includes hundreds of components that enable you to enter parameters that can be measured from real devices. It integrates with test and measurement equipment from different vendors. Users can incorporate new components based on subsystems and user-defined libraries, or utilize co-simulation with a third party tool such as MATLAB® or SPICE™.

5.3.4.1 Transmitters

An extensive library of single and multimode optical laser and transmitter components supporting direct and external modulation and different modulation formats is available as follows:

- Optical sources: single and multimode optical sources models are available for simulation of DFB lasers, Fabry–Perot lasers, VCSEL lasers, LEDs, white light sources, and others.
- Optical modulators: OptiSystem™ has a library of modulator models for amplitude, phase and frequency modulation such as Mach-Zehnder and electroabsorption modulators.
- Modulation formats (including multilevel and OFDM)—OptiSystem™ allows the simulation of a broad range of modulation formats such as:
  - Intensity modulation formats—NRZ, RZ, CRZ, CSRZ, VSB-CSRZ, Duobinary, and others.
  - Differential phase modulation—NRZ-DPSK, RZ-DPSK, DQPSK, DP-QPSK, OD8PSK, OFDM, and others.
  - Phase and quadrature amplitude modulation with coherent detection—QPSK, polarization-multiplexed QPSK, 16-QAM, and others.
  - Multi-carrier modulation—direct-detection optical OFDM, coherent optical OFDM, and others.

5.3.4.2 Optical Fiber

Several optical fiber models are available as listed below:

- Single-mode fiber model: Sophisticated single-mode fibers with unidirectional and bidirectional models considering wavelength-dependent attenuation, chromatic dispersion, Kerr nonlinearities (SPM, XPM, FWM), SRS, SBS (bidirectional model), multiple Rayleigh scattering (bidirectional model), any order PMD, polarization-dependence of all nonlinear effects, splice loss/reflection. The above allow the simulation of DWDM, CWDM systems in unidirectional and bidirectional configurations.

- Time-domain fiber model: A time-domain fiber model that employs a time-domain split-step numerical method is also available in OptiSystem™.
- Multimode Fiber model (MMF): A sophisticated MMF that accounts for laser-fiber coupling, modal group delays, using spatial representation of modes and built-in mode solvers are included with OptiSystem™.

5.3.4.3 Optical Amplifiers

The OptiSystem™ optical amplifiers library contains detailed bidirectional models of fiber (Raman fiber amplifiers, rare-earth-doped fiber amplifiers) and semiconductor optical amplifiers. The available optical fiber amplifier models are:

- EDFA: the EDFA models simulate bidirectional erbium-doped fibers considering excited state absorption (ESA), Raleigh scattering, ion–ion interactions, and temperature dependence effects. There are models for single-mode and multimode amplification analysis that can consider the use of double-clad fibers EDFAs. The Dynamic EDFA models can also simulate power transients due to adding and dropping of channels in optical amplifier chains and optical gain control techniques.
- YDFA: the YDFA models simulate bidirectional ytterbium-doped fibers, single-clad or double-clad. The component solves numerically the rate and propagation equations for the steady-state and can take into account nonlinear phase changes caused by SPM and XPM effects by propagating the signal using the nonlinear Schrödinger equation. There are models for single-mode and multimode amplification analysis and also models that can be used for the simulation of high power amplifiers.
- EYDFA: the EYDFA model simulates a bidirectional erbium–ytterbium co-doped fiber. The model solves numerically the rate and propagation equations for the steady-state case and can take into account nonlinear phase changes caused by SPM and XPM effects by propagating the signal using the nonlinear Schrödinger equation.
- TmDFA: this model simulates a bidirectional Thulium-doped fiber, single-clad or double-clad. The component solves numerically the rate and propagation equations for the steady-state case.
- SOA (including RSOA): the SOA models are based on the numerical solution of a set of coupled differential equations that describe the interaction between the carrier density and photon rates. The RSOA model includes the dynamic dependence between electrical and optical input signals.

- Raman: the Raman fiber amplifier models can simulate lumped and distributed RFAs taking into consideration propagation effects during amplification.
- Black box: this model can represent the EDFAs pumped by 980 nm or 1,480 nm. It requires just the experimental characterization of the practical device, such as the gain spectrum and noise figure under non-saturated and saturated conditions. Details about erbium-doped fiber specifications and elements in the layout are not required to perform the simulations.

5.3.4.4 Receivers

The OptiSystem™ receivers library contains a set of components for photodetection of optical signals (PIN, APD, and spatial photodetectors), regeneration, and demodulation among others. In particular, the following models are included:

- Coherent receivers: Optisystem™ has models (photodetectors, optical couplers, and others) that can be used to design coherent (homodyne and heterodyne designs) and differentially coherent receivers.
- Incoherent receivers: simple photodiodes like PIN and APD, or a more sophisticated combination of components can be used to simulate different incoherent receivers.

5.3.4.5 Performance Estimation

OptiSystem™ has flexible means for displaying and analyzing signal quality measures, such as optical spectrum analyzer for display of polarization-resolved amplitude, dispersion, phase, delay and normalized Stokes parameters; visualizers for display of binary, electrical and optical signals in time-domain, including eye-diagram, BER patterns, histograms; Poincare sphere, polarization analyzer, power meters, and single and dual port WDM analyzers.

There are also flexible means for displaying and analyzing multimode signals, such as DMD analyzers, spatial visualizers, and EF analyzers. Examples of OptiSystem™ analyzers are:

- *Eye diagram analyzer*: This visualizer allows the user to calculate and display the electrical signal eye diagram automatically. It can calculate different metrics from the eye diagram, such as Q factor, eye opening, eye closure, extinction ratio, eye height, mask violation ratio and others. It can also display histograms and standard eye masks.
- *Constellation diagram*: Displays the In-Phase and Quadrature-Phase electrical signals in a constellation diagram. It can also display the polar diagram and estimate the probability of symbol error for M-ary signals.
- *Optical spectrum analyzer*: This visualizer allows the user to calculate and display optical signals in the frequency domain. It can display the signal intensity, power spectral density, phase, group delay, and dispersion for both polarizations.

- *Optical time domain visualizer*: This visualizer allows the user to calculate and display optical signals in the time domain. It can display the signal intensity, frequency, phase, and alpha parameter for both polarizations. It also provides enhanced features to characterize ultra short pulses such as autocorrelation and Frequency Resolved Optical Gating (FROG).

Estimation Techniques

- *Monte Carlo*: OptiSystem™ allows Monte Carlo simulation to evaluate system sensitivity to stochastic phenomena (e.g. PMD, pattern effects, dispersion statistical variation, ASE noise).
- *Semianalytical*: OptiSystem™ calculates parameters such as BER and Q-factor using numerical analysis or semi-analytical techniques for systems limited by inter-symbol interference and noise.
- *BER counter*: OptiSystem™ can perform BER calculation by direct-error-counting using visualizers and/or a virtual BER test set.

5.3.4.6 Interfaces to Third Party Tools

- OptiSystem™ is compatible with design and modeling tools for amplifiers, waveguides, FBGs, LPGs, AWGs, modulators and fibers. Interfaces with Optiwave photonic device design and opto-electronic simulation tools, like OptiSPICE, OptiBPM, OptiGrating and OptiFiber are available.
- MATLAB® and Scilab co-simulation is available. The flexible user defined component allows the user to specify the number of inputs, outputs, and parameters. Co-simulation in time, frequency and spatial domain are possible.
- Agilent EEsof simulation link and SPICE™ interface based on common simulation data format standard.
- OptiSPICE is a circuit design software for analysis of integrated circuits including interactions of optical and electronic components. It allows for the design and simulation of opto-electronic circuits at the transistor level, from laser drivers to trans-impedance amplifiers, optical interconnects, and electronic equalizers. With the imminent coexistence of electrical and optical components at the chip and board levels, it is important to provide designers with a reliable simulation framework that can accurately and efficiently predict signal behavior in opto-electronic integrated circuits and boards. OptiSPICE produces self-consistent solutions of opto-electronic circuits that contain feedback spanning both optical and electrical parts. It is a fully integrated solution for parameter extraction, schematic capture, circuit simulation, and waveform analysis.

5.3.4.7 Other Key Features

- Measurement tools interface—interfaces with test and measurement equipment to import files with measurements, including wavelength dependence of the Jones matrix and time resolved chirp.
- Parameter sweeps and optimizations
  - Simulations can be repeated with iterated variation of parameters. Any parameter can be optimized to reach a minimum, maximum, or target value. Multiple parameter sweeps (deterministic and stochastic) and multiple optimizations can be combined. Optimizations are also available through user-script.
  - Monte Carlo simulation to evaluate system sensitivity to stochastic phenomena (e.g. PMD, pattern effects, dispersion statistical variation, ASE noise).
- Script editor
  - By using standard Microsoft VBScript language users can create designs, change parameters, run simulations, access results, import and export data from/to other Windows applications, such as MS Excel, MATLAB®, and MS Word.
  - VBScript language for automated synthesis and analysis of optical systems and performing repetitive design tasks.
- Bill of materials: provides a cost analysis of the system being designed in a table, arranged by system, layout, or component. These data can be exported to another application or to a spreadsheet.
- Data monitors: the user can select component ports to save the data and attach monitors after the simulation ends. This allows data processing after the simulation ended without recalculating. The user can attach an arbitrary number of visualizers to the monitor at a given port.
- Report page: a fully customizable report page allows you to display any set of parameters and results available in the design. The generated reports are organized into resizable and moveable spreadsheets, text, 2D, and 3D graphs. It also includes HTML export and templates with pre-formatted report layouts.

Acknowledgments The author acknowledges the contributions and assistance of a number of engineers and scientists: RSoft: Sriharsha Kota Pavan, Jigesh Patel, Pablo Mena, and Enrico Ghillino; VPIsystems: Andre Richter; and Optiwave: Bryan Tipper.

References

1. Agrawal GP (1995) Nonlinear fiber optics, 2nd edn. Academic Press, London
2. Carena A, Curri V, Gaudino R et al (1997) A time-domain optical transmission system simulation package accounting for nonlinear and polarization-related effects. IEEE J Select Areas Commun 15(4):751–765

3. Jeruchim MC, Balaban P, Shanmugan KS (1992) Simulation of communication systems. Plenum Publishing Corporation, New York
4. Richards DH, Jackel JL, Ali M (1997) A theoretical investigation of dynamic all-optical automatic gain control in multichannel EDFAs and EDFA cascades. IEEE J Selected Topics Quant Electr 3(4):1027–1036
5. Roudas I, Antoniades N, Richards DH, Wagner RE, Jackel JL, Habiby SF, Stern TE, Elrefaie AF (2000) Wavelength-domain simulation of multiwavelength optical networks. IEEE J Selected Topics Quant Electr 6(2):348–362
6. Morikuni J, Mena P, Whitlock BK, Scarmozzino R (2002) Simulation of modal delays for 10 Gb/s multimode fiber applications. In: Proceedings of IEEE/LEOS annual meeting, Glasgow, Scotland, pp 903–904
7. Morikuni J, Mena P, Whitlock BK, Scarmozzino R (2003) Multimode system simulation as an alternative to spreadsheet analysis for the study of Gb/s optical communication systems. In: Proceedings of IEEE/LEOS annual meeting, paper M04:27, Tucson, Arizona
8. Papoulis A (1991) Probability random variables and stochastic processes. Mc-Graw Hill, New York
9. Gnauck AH, Winzer PJ (2005) Optical phase-shift-keyed transmission. IEEE/OSA Lightwave Tech 23(1):115–130
10. Kazovsky L, Benedetto S, Willner A (1996) Optical fiber communication systems. Artech House, New York
11. Mathai AM, Prevost SB (1992) Quadratic forms in random variables, chap 3. Marcel Dekker, New York
12. Press WH, Teukolsky SA, Vetterling WT (1992) Numerical recipes in C. Cambridge University Press, Cambridge, UK
13. Lowery AJ, Lenzmann O, Koltchanov I, Moosburger R, Freund R, Richter A, Georgi S, Breuer D, Hamster H (2000) Multiple signal representation simulation of photonic devices, systems, and networks. IEEE J Select Topics Quant Electronics 6(2):282–296
14. VPIsystems (2011) VPItransmissionMaker^{TM} optical systems user's manual, chaps 4 and 5, p 83, VPItransmissionMaker^{TM} optical systems, version 8.6
15. Richter A, Louchet H, Poloyko I, Karelin N, Farina J, Koltchanov I (2009) A parametric approach to optical systems design, optimization and validation. In: Proceedings of European conference on network and optical communications (NOC), Valladolid, Spain, pp 361–367
16. VPIsystems (2011) Photonic modules reference manual: universal fiber, vol 1, pp 8-185, VPItransmissionMaker^{TM} optical systems, version 8.6
17. Richter A, Devatine R, Koltchanov I, Lowery A, Khomchenko D, Yevseyenko D, Moar P (2002) Virtual product prototyping of erbium doped fiber amplifiers for applications in dense WDM systems. In: Proceedings of IEEE/OSA national fiber optics engineers conference (NFOEC), paper 153, Anaheim, CA
18. Richter A, Koltchanov I, Myslivets E, Khilo A, Shkred G, Freund R (2005) Optimization of multi-pump Raman amplifiers. In: Proceedings of IEEE/OSA optical fiber communication conference (OFC/NFOEC), paper NTuB4, Anaheim, CA
19. Richter A, Koltchanov I, Kuzmin K, Myslivets E, Freund R (2005) Issues on bit-error rate estimation for fiber-optic communication systems. In: Proceedings of IEEE/OSA optical fiber communication conference (OFC/NFOEC), paper NTuH3, Anaheim, CA
20. Louchet H, Kuzmin K, Koltchanov I, Richter A (2009) BER estimation for multilevel modulation formats. In: Proceedings of IEEE Asia communications and photonics conference (ACP), vol 7632, paper 7632–7114, Shanghai, China
21. Louchet H, Kuzmin K, Richter A (2008) Efficient BER estimation for electrical QAM signal in radio-over-fiber transmission. IEEE Photon Tech Lett 20(2):144–146
22. Richter A, Karelin N, Louchet H, Koltchanov I, Farina J (2010) Dynamic events in optical networks—emulation and performance impact analysis. In: Proceedings of IEEE military communications conference (MILCOM), San Jose, CA, pp 1083–1087

Part II
Implementations

Chapter 6
Optical Interconnects

Nicholas Madamopoulos

Abstract We present an overview of optical interconnection systems by first defining the need and requirements for such systems and then presenting and reviewing the state of the art of the necessary components and processes, as well as the different technologies and architectures involved in their design. The design simplifications achieved by using optical interconnects compared to electrical solutions are also examined. Nevertheless, the view presented here is that the two worlds of optical and electrical interconnects can converge and successfully lead to solutions that are even beyond our imagination at this point in time.

6.1 Introduction

In communication and computing environments, there is a need for interconnecting different devices, instruments, and/or equipment with each other. This means that the communication is not between two points anymore but between multi-point to multi-point entities. This interconnection of the devices has started becoming more and more of a critical issue as performance limitations from electrical interconnects are creating a bottleneck. In particular, as the performance of microprocessors continues to improve, the data flow to and from the processors becomes an increasingly predominant bottleneck for the overall system performance. The limitations of electrical interconnects were identified back in the 1970s and potential solutions based on optical approaches were proposed for the optical

N. Madamopoulos (✉)
Department of Electrical Engineering
City College of New York/The City University of New York
New York, NY 10031, USA
e-mail: nmadamopoulos@ccny.cuny.edu

N. (Neo) Antoniades et al. (eds.), *WDM Systems and Networks*, Optical Networks,
DOI: 10.1007/978-1-4614-1093-5_6, © Springer Science+Business Media, LLC 2012

distribution of clock signals across different sections of microprocessor chips in the early 1980s [1]. Optical interconnects is the method of communication between two points through optical paths (e.g., glass fiber, free space, etc.). The history of optical interconnects is strongly coupled with the field of optics and photonics [2, 3]. Optical interconnects have revolutionized telecommunications over the last few decades. The early efforts on providing optical solutions for digital logic have been instrumental in the development of optical/photonic devices that were and are currently used in a variety of applications other than optical interconnects.

In the microelectronics domain the devices are interconnected by electrical wires, i.e. wire bonds. In the last few decades, the advances in integrated circuit design techniques and microprocessor architectures have led to an unprecedented increase in clock speeds and increased densities. Moore published in an article in 1965 that the number of transistors that can be placed inexpensively on an integrated circuit has doubled approximately every year and predicted that this would continue in the next few years [4]. Later he revised his prediction and stated that the rate would be doubled every two years [5]. It is true that this trend has continued pretty much with these rates for more than half a century and is not known when it is expected to stop. This trend though has created chips containing millions of transistors occupying smaller chip sizes. As the number of devices per chip continues to increase, the electrical interconnects start reaching their limitations as it pertains to higher speeds, packaging, fan-out, and power dissipation. As design rules drop below 90 nm, a variety of challenges emerge such as *RC* delay, electromigration resistance, and heat dissipation exacerbated by increased chip power [6]. Several technological improvements have been applied to reduce these limitations [6, 7]. Future directions include the three-dimensional (3D) integration of heterogeneous technologies such as memory and logic together with optical I/O, where the chip footprint is being drastically reduced compared to 2D integration. In this case, long horizontal wires are replaced with short vertical wires and the interconnect length is being reduced resulting in *R*, *L*, *C* being reduced, thereby contributing to power and delay reductions. Currently, electrical interconnects are the dominant approach for inter- and intra-chip interconnections simply because the optical approach has not reached the cost effectiveness and maturity required for these applications.

Nevertheless, optical interconnections offer a variety of advantages compared to their electrical counterparts, such as immunity to electromagnetic interference (EMI), independency from impendence mismatch, lower power consumption, and high speed of operation. Looking ahead, optical interconnects can be one of the dominant solutions to overcome the issues of speed, space, and power that are critical in such applications.

6.2 The Interconnect Challenge

The International Technology Roadmap for Semiconductors (ITRS) defines the interconnect as an "electrical wiring system which distributes clock and other signals, and provides power/ground to and among the various circuits/systems functions on a chip" [8]. During the first four decades of the semiconductor industry, system performance was limited almost exclusively by transistor delay and power. Component size reduction has decreased transistor delay and power by over three orders of magnitude, while interconnect performance has generally been negatively impacted by dimensional scaling [8]. The replacement of Al with Cu and SiO_2 with low-κ material property can in principle lead to a reduction in delay. Nevertheless, the resistivity of metals in small dimensions increases much more rapidly due to sidewall and grain boundary scattering, which leads to an increase in *RC* delay even for local interconnects with aggressive material scaling. This effect had not had a significant impact on system performance a decade or more ago. However, interconnect delay is now directly impacting system performance. In the older 1.0 μm Al/SiO_2 technology generation the transistor delay was 20 psec and the *RC* delay for a 1 mm line was 1.0 psec, while in a projected 35 nm Cu/low-κ technology generation the transistor delay will be 1.0 psec and the *RC* delay for a 1 mm line will be 250 psec [9]. In the past decade, the industry has actively reduced interconnect delay by breaking up long global lines into shorter interconnect segments using transistor-driven repeaters. The use of repeaters effectively decreases the interconnect line length so that the delay per segment is roughly equal for the transistor and interconnect components. However, this has a limit and the electrical interconnect challenge comes from the increased number of wires that are required to transmit and receive messages among the system components. As the number of messages gets larger, the associated number of wires gets very large, while memory and input/output (I/O) bandwidth (pin count × bit rate/pin) is limited by the number of available pins. The bit-rates are limited by the device speed, and the pin counts are not increasing mainly due to space limitations. It is expected that the total off-chip I/O bandwidth (BW) (pin count times bit rate per pin) will increase by roughly 2.7 times, while the internal chip performance improves by four times [10, 11]. Within the next years, this difference in improvement rates would result in an increasing need for technologies that provide substantially improved chip-to-chip interconnect capabilities. More complex 3D integration techniques of heterogeneous nature, as well as technologies based on Carbon Nanotubes (CNT) and nanowires (CNW), Graphene Nanoribbons, Superconductors and Wireless interconnects are now becoming competing technologies [12, 13].

In addition, signal attenuation (or insertion loss—IL) over long wires is a limiting factor in electrical interconnects. Such IL is unavoidable in wires because of the resistance of the conductors and other factors such as dielectric loss. To make matters worse, IL becomes worse at higher frequencies. As a result, signals of different frequencies will have different propagation losses as well as different phase shifts. Hence this creates distortion of the signals. Compensation techniques (e.g., signal amplification and/or equalization) can be applied to overcome some of these limitations. Nevertheless, this leads to more complicated designs and hardware intensive modules. In today's and future VLSI information processing systems real estate is limited and will be increasingly limited for the near future. In particular, in high-speed multi process devices a significant amount of this real estate is occupied by the interconnect wires. Additional space is required by the wires delivering power to the device and wires required to remove heat from the device. For this reason, information processing devices are limited by wiring density and power. To overcome the limitation of space the designer has to move to smaller (thinner) wires to save space, nevertheless, thinner wires have higher resistance and thus higher attenuation and reduced information capacity [3].

Optical interconnects have emerged as a very promising approach. Optics have many fundamental advantages in terms of bandwidth, suitability for multiplexing, low crosstalk, immunity from EMI, distance scalability, and less power consumption [14]. In analogy to what has happened in the optical communication arena, where copper wires were replaced with optical fibers and wavelength division multiplexing (WDM) technology in the long haul, metro optical networks and recently in the access networks, it is expected that optical interconnects will, and in some cases, are already finding applications in the rack-to-rack, card-to-card, on-card, intra-chip, and inter-chip interconnectivity. Optical solutions can handle the increased aggregate bandwidth and communication distance, while decreasing the power per bit by overcoming the limitations imposed by losses in present package interconnects (metal and dielectric), and by avoiding or minimizing the need for high power equalization and pre-emphasis.

Comparisons of different technologies show that optical interconnections have specific advantages that make the technology suitable for on-chip as well off-chip applications. For on-chip applications, the latency, energy per bit, power density and bandwidth density were compared for Cu, carbon nano-tubes (CNT), and optical interconnects approaches [15, 16]. Optical interconnects and CNTs have the lowest latency for global and semi-global levels, respectively. Because of a fundamental difference in the nature of power dissipation (static for optics versus dynamic for CNT/Cu) and "where" this power is dissipated (end devices for optics versus the medium itself for Cu/CNT), optical interconnects are more power efficient for higher switching circuits and longer lengths as shown in Fig. 6.1a [16]. For off-chip applications, comparing power and bandwidth, beyond a critical length, power optimized optical interconnects dissipate lower power compared to high-speed electrical signaling schemes. Beyond the 32 nm technology node, with its commensurate bandwidth requirement, optical interconnects become favorable for distances as little as 10 cm as shown in Fig. 6.1b [16]. These distances correspond to inter-chip communication on a board.

However, the use of optical interconnect technologies in commercial products has been limited for the most part to direct replacement of electrical wires or cables by optical fibers to support longer link lengths. As link bit rates have steadily increased, optical cables have replaced electrical cables for shorter cable

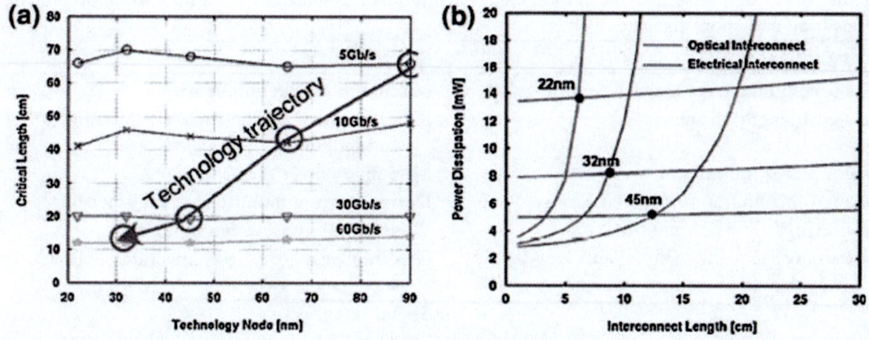

Fig. 6.1 **a** Critical length versus technology node for modulator/detector capacitance of 25 fF. The critical length sharply reduces with bit rates making optics overwhelmingly advantageous at high bit rates needed for future technology nodes. This is primarily due to the debilitating high-frequency effects in the case of electrical interconnects, which compromise the signal integrity and require additional power to satisfy a target bit error rate criteria. **b** Power comparison between electrical and optical interconnects for modulator/detector capacitance of 10 fF. The critical length is less than 10 cm for 32 and 22 nm technology nodes, which puts optics strongly in contention for inter-chip applications (From [16], Figs. 10 and 11. Reproduced by permission of © 2008 The Institute of Electrical and Electronics Engineers)

lengths too [17]. In the recent years, optical interconnect technology has become less costly and better integrated with additional functionality. Hence, there is increasing interest in applying optical interconnects for shorter links as those in rack-to-rack, board-to-board, intra-chip, and inter-chip applications. Such an approach would have substantial impacts on system packaging, interconnect topology, communication traffic patterns, and other aspects of server architectures [18]. Optical technology combined with cheaper component manufacturing processes associated with traditional complementary metal oxide semiconductor (CMOS) techniques has the potential to deliver the I/O bandwidth and system speed to overcome the electrical as well as optical interconnect limitations [2, 19]. However, optical interconnect technologies will have to overcome significant challenges to reach the required power, cost, and density. These challenges will require innovations in packaging, test, and reliability. Target energies of ~1 pJ per bit for backplane connections, and ~100 fJ per bit for shorter connections to chips and for global on-chip wiring are required. Such targets, if met, would give sufficient headroom to justify seriously considering changing to optical interconnects at these shorter distances [2, 3, 8].

6.3 Optical Interconnect Categorization

Optical interconnects cover a wide range of applications. In principle, any connection between two points through an optical path can be considered an optical interconnect. One can claim that optical interconnects have been used for

Table 6.1 Comparison of the reasons and requirements for the use of optics to communicate over long and short distances [3]

Long distance	Short distance
Maximum bit/s over longest possible span	Maximum connection density
Longest length of connection	Smallest cross-section for a given bandwidth and length
Total power not critical	Minimize power dissipation
Design for minimum received energy	Design for minimum total energy per bit
Wavelength division multiplexing	Wavelength division multiplexing
Reuse one fiber for many channels. Allow wavelength switching and routing	Give higher density of connections. Allow wavelength switching and routing
	Signal integrity
	Improved timing precision
	Reduced reflections, cross-talk, voltage isolation.

thousands of years, since the first fire-based signals messages used in ancient times (e.g., frictories in ancient Greece [20, 21]). In the modern times, optical interconnections or links between cities (long-haul optical networks) and within cities (metro network) have been used for several years. Nowadays, we start seeing the application of optical interconnects in access networks coming closer to our homes through the FTTx technology.

It is the next logical step to see the optical interconnects making (and in many cases it is already happening [22]) their entrance in applications within data centers (rack-to-rack) and further into the computing systems (card-to-card, inter-chip, and intra-chip). Optical interconnection applications mainly differ in the length of the optical interconnect (e.g., from hundreds of km in the ultra-long haul, to sub-mm distances in the intra-chip connections). At the same time, the number of interconnect lines changes as we go from a few fibers in the ultra-long haul case to hundreds of thousands in the case of intra-chip connections. However, additional differences and performance characteristics come into play depending on the particular application. Features that are not important in long-haul interconnection links (e.g., thermal management) are very critical in intra- or inter-chip applications. Table 6.1 highlights some of the requirements and rationale for using optics in long and short distances. It can be seen that the requirements in some cases are very different, nevertheless, optics do offer significant advantages.

We divide optical interconnects into the following types: on-chip or intra-chip (<few cm), chip-to-chip or inter-chip (<1 m), shelf-to-shelf (<2 m) and rack-to-rack (mostly <100 m), and data centers. For these interconnects, shown in Fig. 6.2, important factors for research and development are the possible architectures and technologies (both optical and electrical) that enable aggregate speeds of 100 Gb/s and beyond, and line speeds between 10 and 20 Gb/s. In particular, for chip-to-chip applications signal processing complexity, technology maturity and power consumption are critical parameters.

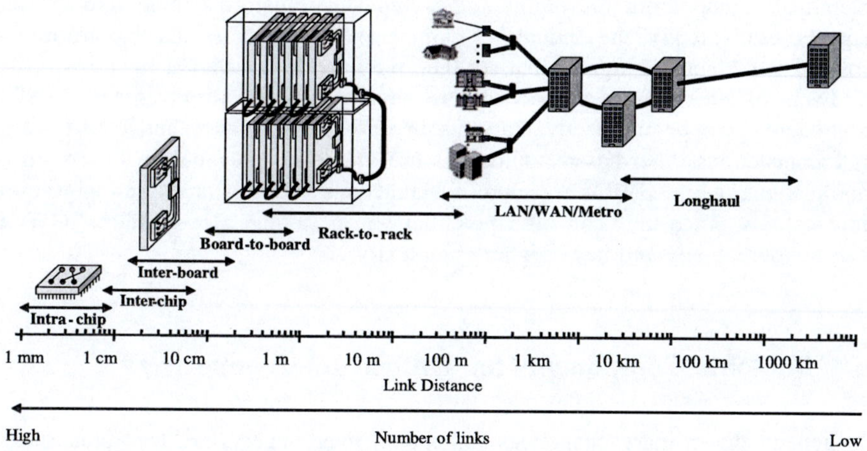

Fig. 6.2 The typical optical interconnection distances and the associated required number of links for different applications

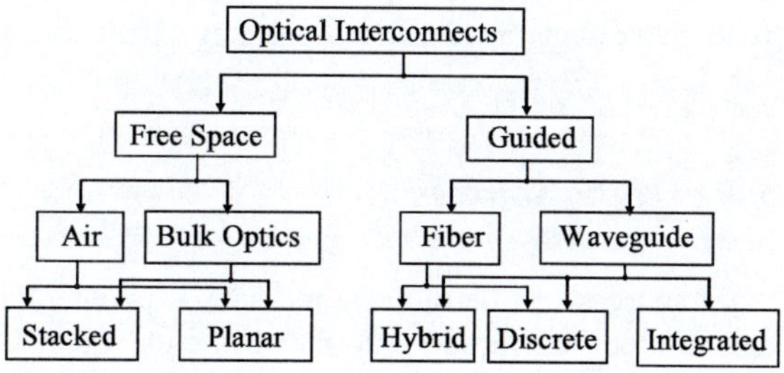

Fig. 6.3 Optical interconnect categorization based on mode of propagation (free space/guided)

Apart from the division of optical interconnects based on the application space (or equivalently the interconnection distance) they can also be divided based on the means of propagation. There are two main categories, which can be further divided into subcategories, as shown in Fig. 6.3. The first one is the Free Space Optical Interconnects (FSOI) and the second one is the Guided Wave Optical Interconnects (GWOI). In FSOI systems, the light is routed from transmitter to receiver by way of discrete optical components (e.g., lenses and mirrors) through free space. Note that even though the term "Free space" is used, many times, the signal propagation is performed through bulk optical material (e.g., glass, plastic). In GWOI systems the light is guided from transmitter to receiver by way of waveguides or optical fibers. The GWOI can be further divided into integrated and discrete architectures. In integrated architectures the optical components are integrated into a photonic integrated circuit, and part or all of the communications between the chip and the external world are done through optical signals. In the discrete architecture an optical I/O chip is used. This optical chip receives electrical signals from the digital logic chip and transforms them into optical signals. Equivalently, the optical I/O chip receives optical signals that are transformed into electrical signals that are then provided to the digital logic chip [8].

Each of the above approaches has advantages and disadvantages. FSOI approaches can be more advantageous in optical backplanes and card-to-card interconnections, fiber-based approaches are more advantageous in rack-to-rack applications, while GWOI are more advantageous in intra- and inter-chip communications, since they can offer potential power savings. Nevertheless, GWOI can be more costly and have higher complexity.

6.4 Photonic Components for Optical Interconnects

In general the primary challenges for optical interconnects are to produce cost effective, low-power-consumption components. FSOI have the challenge of optical alignment and real estate, while GWOI and in particular waveguide-based (GaAs or Silicon) have the challenge of optical loss and crosstalk. Before we move into the description of the different optical interconnect architectures and describe their critical parameters, we will briefly describe some of the most often used components. Note that this description is not exhaustive due to space limitations.

6.4.1 Light Sources

Light sources can be directly modulated or operated at CW. In the first case, the light sources are turned on/off with the electrical signal of interest. In the second case, the light source is operated at CW and an external modulator that can be controlled by electrical signals is used to encode the electrical information onto the optical signal. From a location perspective, lasers can be off-chip (package or board) or on-chip. Key parameters of the lasers are the output power, efficiency, cost, thermal stability, cooling, and modulation speed (in the case of directly modulated sources). Wavelength stabilization can be done once away from the relatively hostile temperature environment of the surface of a chip, and multi-wavelength sources could be used [23]. Typical sources include edge-emitting semiconductor diode lasers, vertical cavity surface emitting lasers (VCSELs), and quantum dot lasers. There are advantages as well as disadvantages for each of the technologies, as they pertain to the implementation of optical interconnects. For example, conventional edge-emitting lasers cannot meet such low energy targets [8]. VCSELs could likely be adequate for 1 pJ/bit systems [8]. New laser development, such as nanocavity lasers would be required for lower energy systems. Increased interest on VCSELs exists due to the capability of high integration

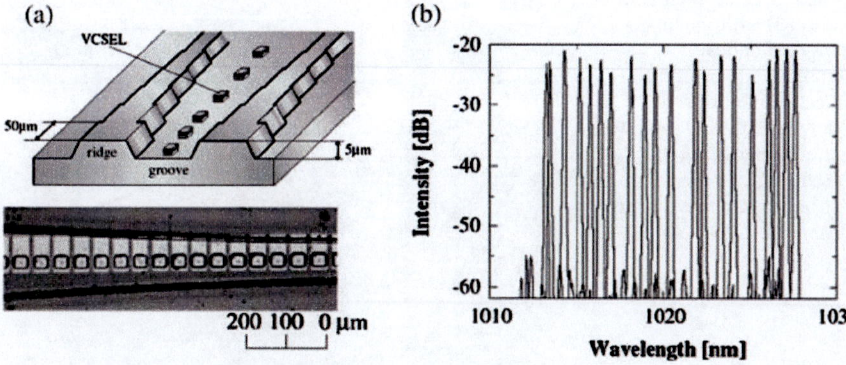

Fig. 6.4 **a** Densely integrated multiwavelength GaInAs/GaAs VCSEL array **b** Lasing spectra of 20 channel multiwavelength VCSEL array on a patterned substrate (From [34], Figs. 1,2 and 4. Reproduced by permission of © 2003 APEX/JJAP)

density in arrays as shown in Fig. 6.4 [24]. Iga et al. first proposed it in 1979 [25] and since then we have seen great research and development efforts from many teams in the world [26].

Implementing laser on silicon, for an all-optical integrated solution, has its limitations. Nevertheless, hybrid approaches have been reported recently that may eventually lead to an all-optical silicon platform for on-chip optical interconnections. In the recent years, silicon lasers have long been the goal for semiconductor scientists and a number of important breakthroughs have focused attention on silicon as a photonic platform [27]. The most recent progress in this field includes low-threshold silicon Raman lasers with racetrack ring resonator cavities [28, 29], germanium-on-silicon lasers operating at room temperature [30, 31], and hybrid silicon microring and microdisk lasers [32, 33]. These laser approaches are potentially the ones along with VCSEL architectures that will be integrated with optical interconnects for all-optical processing applications.

6.4.2 Photodetectors

Photodetectors can be integrated either on-die or off-die. Metal–semiconductor-metal photodetectors have received significant attention since they have the potential for being CMOS compatible [35–37], and can be fabricated using Ge, Si, and SiGe [38–45]. Key technical parameters include the device responsivity, operation voltage, input capacitance, ratio of photocurrent to dark current, ratio of photocurrent to input capacitance, light coupling efficiency, and dimensions. A key design parameter is the coupling of light into the photodetector. Ge/Si avalanche photodiodes (APD) are of interest since they have high gain-bandwidth product

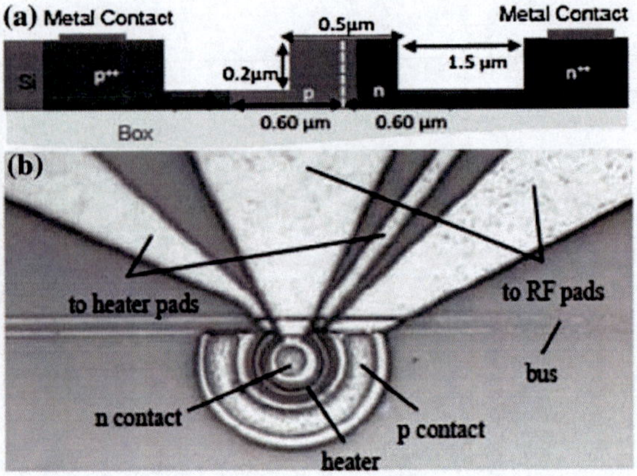

Fig. 6.5 **a** Cross section of a microring-based modulator, **b** top-view optical microscope image of a fabricated microring modulator with ring radius of 5 μm (From [50], Fig. 1. Reproduced by permission of © 2010 The Optical Society of America (OSA))

and better sensitivity than pin-type photodiodes. There are two approaches based on mesa-type and waveguide-type structures. The second type has several advantages over the mesa-type structure [46]. The most commonly recognized benefit is to decouple the optical absorption length and the carrier transit time, which is crucial for high data rate applications beyond 10 Gb/s. In addition, it allows for monolithic integration of the APD with other optical components [47]. Recently, detectors that take advantage of plasmons to enhance the coupling of light into the photodetector have been proposed [48].

6.4.3 Modulators

Modulators are used in combination with a light source operated in CW mode. A large variety of CMOS compatible modulators have been proposed in the literature, including microring resonators [48–50], electro-absorption modulators [51, 52], and Mach–Zehnder modulators [53]. The cross section and top view of a micro-ring resonator and a Mach–Zehnder modulator are shown in Figs. 6.5 and 6.6, respectively. The key performance parameters include coupling efficiency, operation voltage, switching time, waveguide loss, overall power, modulation depth/extinction ratio, and area. Modulators and lasers can be integrated in the same platform and this provides space and performance advantages [54].

The microring resonators are very compact and have excellent modulation depth. However, due to the nature of the ring structure, the optical bandwidth of the ring modulator is generally small (e.g., <1 nm), hence, multiple or cascaded rings or more complicated structures are required for optical interconnection systems that use multiple wavelengths (e.g., WDM). In addition, the device is sensitive to temperature fluctuations, bias condition, and fabrication tolerance [55]. Any small manufacturing error can result in the resonant shift; hence, extra components such as thermal heaters are usually applied for tuning and/or stabilization of the device.

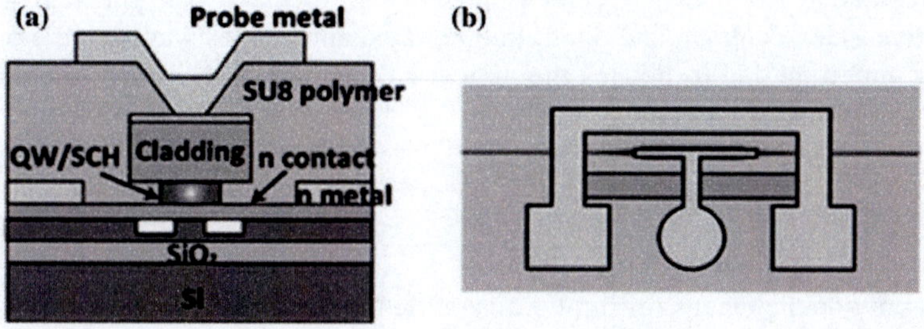

Fig. 6.6 **a** Cross section of both EAM and MZM at the hybrid section. **b** Schematic top view of an EAM (From [47], Fig. 13. Reproduced by permission of © 2010 The Institute of Electrical and Electronics Engineers)

The electro-absorption modulator (EAM) based on III–V materials has been a very common device used in optical communications. By applying bias on the material, the bandgap can be manipulated, thus resulting in a red shift of the absorption spectrum. Nevertheless, Silicon has an indirect bandgap and the absorption efficiency is weak. To overcome this limitation Germanium (Ge) grown on silicon [56, 57] and III–V bonded to silicon [58] has been used. Small size and low energy consumption per bit can be accomplished [56], though the additional absorption caused by the indirect bandgap of Ge introduces a higher propagation loss at zero bias. The second approach consists of III–V material bonded with patterned silicon on insulator (SOI) wafers such that the active devices can be implemented using III–V material property while low-loss waveguides can be made on SOI.

The third option is the Mach–Zehnder interferometer (MZI). Its operational principle is based on the interference of two beams propagating through the two arms of a waveguide-based MZI. By changing the phase on one or both arms of the MZI, either constructive or destructive interference is achieved, and consequently intensity modulation can be accomplished [59]. In order to introduce a phase shift, electro-optic effects (e.g., carrier injection, carrier depletion) can be used. In silicon these devices are required to be in the millimeter scale because of the weak electro-optic effect of the silicon [60].

6.4.4 Waveguides

Waveguides provide the means for light propagation on the chip with minimum loses. They also need to enable implementation of “bends” and “turns”, as well as an efficient coupling of the light into the detectors. A large refractive index contrast between the waveguide and the surrounding materials enables tight turn radii and small pitches, but at the expense of lower speed, which decreases with the inverse of the effective refractive index. Reported on-chip waveguides using materials already common in the industry include Si, Si_3N_4, or $Si_3O_xN_y$ cores on

SiO_2 cladding [61]. Key technical parameters include loss per unit length, refractive index contrast, and pitch. In integrated approaches waveguide crossover is also important and it dictates the optical crosstalk performance of the devices.

6.4.5 Fibers–Fiber Arrays

Fibers provide the means for light propagation in longer distances, e.g., from chip-to-chip, to rack-to-rack and beyond. Fiber can be used to form fiber arrays (e.g., 1D or 2D) and allow for parallel optical signal transport and processing. "Bends" and "turns" can be performed within the limits allowed by the optical fibers. Single-mode fiber (SMF), multi-mode fiber (MMF), or plastic-optical fiber (POF) can be used depending on the application. High-density fiber optic connectors can be used to interconnect boards, shelves, or racks. A high-density connector is the Tyco Electronics' high-density PARA-OPTIX™ Cable Assembly that uses MT-ferrule technology to provide the maximum fiber density for parallel optical applications [62]. By terminating these ferrules in MPO-style connectors, up to 72 fibers can be bundled and connectorized in a single MPO-style connector. Note that the performance of high-density connectors lags behind that of single-fiber connectors. This is largely due to the technology maturity; however, statistics also come into play. Unlike single-fiber ferrules that are inspected one by one to define the limits of the manufacturing distribution at any desired eccentricity values to limit insertion loss, with multifiber ribbons and MT ferrules, all fiber holes need to be located within a few microns of their ideal position. The more holes there are, the higher the tolerance for the worst hole. If one ferrule does not meet the spec the entire connector may be thrown away.

Novel designs of connectors have been proposed and demonstrated that can allow 90° turns of the optical signal (i.e., perpendicular from an optical board). Such a connector is the US Connect PRIZM™ LightTurn™ connector [63], which provides passive alignment and novel retention features allowing multiple matings perpendicular to the printed circuit board. The 7.4 mm × 5.7 mm PRIZM™ LightTurn™ connector consists of a multi-fiber ferrule with a photonic turn total internal reflection (TIR) lens array accepting cleaved fibers and a single outer housing. The termination is simple and low cost requiring no polishing of fibers and easy testing, yet providing excellent coupling to high bandwidth mini-parallel optic modules.

6.4.6 Free Space/Bulk Optics Optical Elements

Free space, as well as bulk optics (e.g., glass), provide the means of optical propagation for chip-to-chip interconnections. These paths have very low loss. "Bends" or turns" are accomplished through optical elements (e.g., prisms,

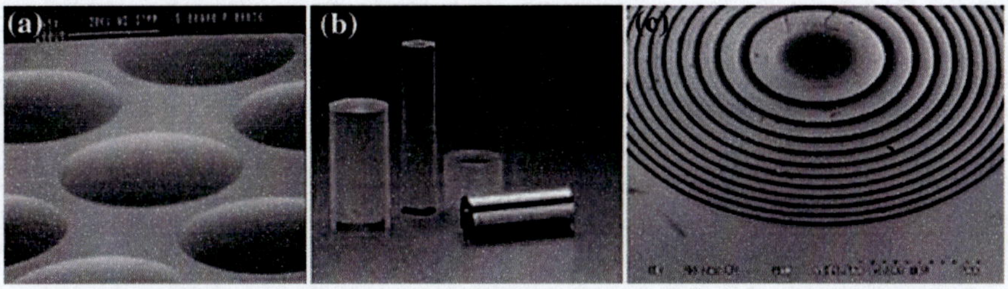

Fig. 6.7 Photographs of typical **a** spherical microlens (From [79], Fig. 6.1**a**. Reproduced by permission of © 1997 The Optical Society of America (OSA).) **b** GRIN lens (From [80]. Reproduced by permission of © 2011 Go!Foton Corporation) and **c** diffractive microlens (From [81], Fig. 3. Reproduced by permission of © 2002 The Optical Society of America (OSA))

mirrors). Free space/bulk optics provide the capability of multiple optical beam processing. Note that many of the first fiber-optic components used in optical networking applications for interconnecting two points were based on bulk optics approaches; examples include the optical isolators and circulators. In an isolator the light is delivered from an optical fiber to a microlens that couples the light from the guided mode in the fiber to the free space, a polarizer (typically birefringent wedges, made e.g., of rutile TiO_2) is used to separate spatially the two orthogonal beams, which after passing through a Faraday rotator element are combined again through a second polarizer [64, 65] and any back reflection to the input and due to the non reciprocal function of the Faraday rotator will be spatially displaced by the polarizer and hence no signal will be coupled back into the input fiber. Other bulk optic fiber-optic components were the first optomechanical switches, delay lines, and many other components used in optical network elements and equipment.

6.4.7 Microlenses

Microlenses are a subset of the above category of components. They are very important in optical interconnection applications; hence we chose to describe them in more detail. Microlenses are lenses in an array with an aperture of less than a few millimeters [66]. The microlenses can have a variety of aperture shapes (e.g., circular, hexagonal, etc.) and sizes. The surface of the lens can be flat, convex, or concave. Microlens arrays can be arranged in 1D or 2D architectures. The microlenses fall under three typical categories, which are shown in Fig. 6.7. These are (a) spherical, (b) GRIN (gradient index), (c) diffractive microlenses. A spherical lens is a small lens formed directly on materials such as plastics or glass based on a variety of techniques [67–72]. GRIN lenses are fabricated by ion exchange methods and have been used extensively in many optics and photonics components and in particular laser diode to fiber coupling, fiber optic collimators, and fiber-to-fiber coupling. The GRIN lens is in the form of a cylinder in which the index of

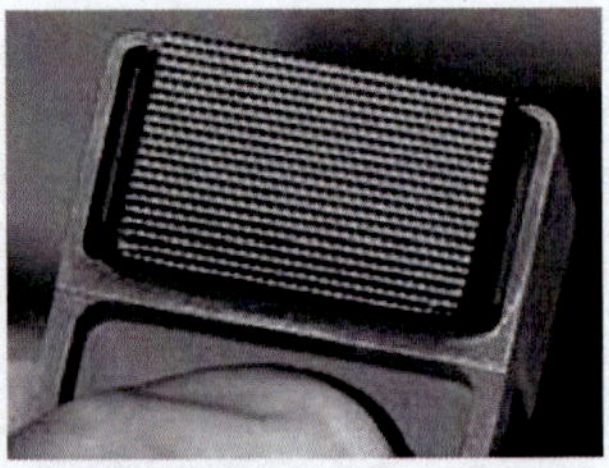

Fig. 6.8 Photograph of the microlens array. (From [83], Fig. 8. Reproduced by permission of © 2007 APEX/JJAP)

refraction follows a gradient, which bends the propagating light. Selfoc® trademarked GRIN lenses produced by Nippon Sheet Glass Company Ltd (NSG) have had success in telecommunication systems due to their unique shape, i.e., cylindrical, which allows for good settlement with holders or V-grooves, and their surfaces are flat and provide compatibility with optical fibers [73–75]. Several early experimental system demonstrations were based on the use of GRIN lenses mounted on optomechanical mounts for increased alignment flexibility [76], and in addition most of the early fiber-optic components (e.g., optical isolators, circulators, switches etc.) used GRIN lenses for coupling light from/to fibers. Finally, the diffractive microlenses or diffractive optical elements (DOE) allow for greater real estate savings and device miniaturization. DOEs have attracted attention because of their relatively low aberration and increased latitude in design, thus allowing integration in free space, planar optics, and integrated platforms. DOEs are composed of uneven surfaces and have a refractive index and amplitude distribution that depends on the material used, allowing effective control of wave planes [71, 77, 78]. DOEs are of particular interest in planar optical interconnects.

The development of large micromirror arrays for the implementation of 2D as well as 3D optical switches has led to the development of high-density microlens arrays based on a variety of approaches. Figure 6.8 shows the microlens array for the Calient 3D optical MEMS switch [82].

6.4.8 Micromirrors

Micromirrors are mainly based on micro-opto-electro-mechanical systems (MOEMS or Optical MEMS) a technology that has advanced significantly in the recent years. Many optical devices and systems based on movable structures have been demonstrated and are now commercial products. Compared with macro-scale opto-mechanical devices, the micromechanical devices are smaller, lighter, faster (higher resonant frequencies), and more rugged. Very efficient light modulators, switches, broadly tunable semiconductor lasers, detectors, and filters can be realized by the optical MEMS technology [84]. These optical systems show great potential for use in several industry segments, including displays, biomedicine, and telecommunications [85, 86]. Optical MEMS technologies are based on

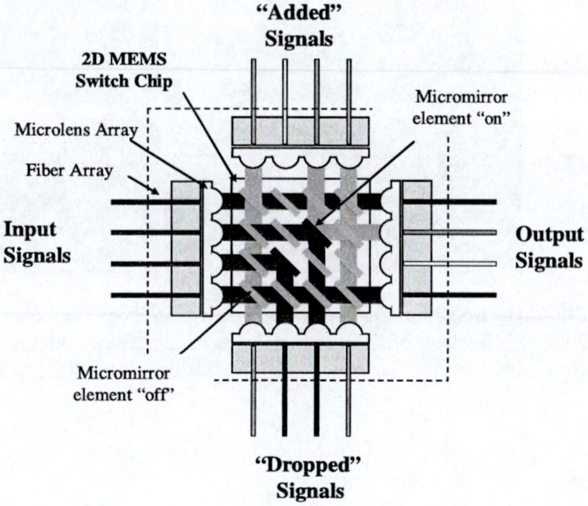

Fig. 6.9 Schematic of 2D MEMS optical switches

micromachining, a process that is also used in IC (integrated circuit) fabrication. Micromachining involves the manufacture of mechanical structures in the micron to millimeter range. The micromirrors can be tilted or moved based on different techniques (e.g., electromagnetic, electrostatic) and can reflect the incoming beam to different directions [87]. MEMS mirror switching systems are divided into two categories, the two dimensional (2D) and the 3D architectures.

In 2D MEMS, every mirror has two possible states corresponding to two different positions: active (perpendicular) or non-active (parallel to the substrate plane to which the mirrors are hinged) as shown in Fig. 6.9. Collimated light beams propagate parallel to the substrate plane. When a mirror is activated, it is lifted up into the optical path of the beam and reflects the light to one of the outputs or to another mirror. The mirrors are typically positioned at an angle of 45° relative to the beam propagation direction in a typical crossbar configuration as that shown in Fig. 6.9. Alternative designs have also been proposed that reduce the required number of moving mirrors to accomplish the same switch port count and functionality [88]. Figure 6.10 shows a real 2D MEMS switch fabric implementation based on a crossbar configuration of the micromirrors [89, 90].

On the other hand, in a 3D MEMS architecture, there is a dedicated movable mirror for each input and each output fiber port. The incoming optical beam is steered by selecting the proper mirror position. There are two types of micromirror operation modes. The first one is a bistable mode where the mirror can be positioned in two predefined angles [91]. Figure 6.11a shows an example of this approach. Incoming light is incident on the micromirror at normal incidence and hence it is retro-reflected back. For a different position of the mirror the incident light is reflected towards a different direction. This type of approach has been used for add-drop switching in optical network elements [92]. Figure 6.11b shows a SEM photograph that shows the perspective view of a micromirror array.

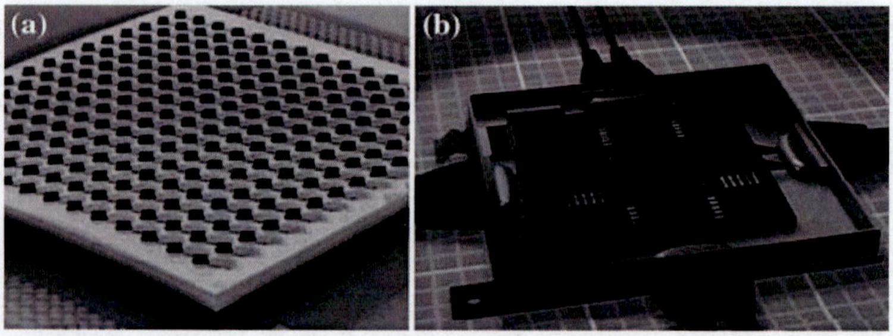

Fig. 6.10 a SEM of OMM's 16 × 16 switch (From [89], Fig. 3. Reproduced by permission of © 2002 The Institute of Electrical and Electronics Engineers.) b Photograph of the packaged switch (From [90], Fig. 4. Reproduced by permission of © 2003 SPIE and with author permission)

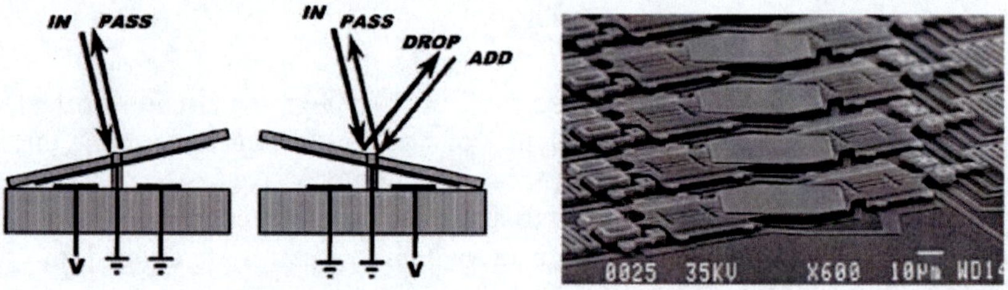

(a) (b)

Fig. 6.11 a Principle of operation of a micromirror element for reflecting light into two different directions, b SEM photograph that shows the perspective view of a micromirror array (From [92], Figs. 2 and 4. Reproduced by permission of © 1999 The Institute of Electrical and Electronics Engineers)

Another mode of operation is based on micromirrors that can move continuously, similar to scanners, and hence can direct or steer the incoming beam in a wider range of angles. Figure 6.12 shows that a connection path is established by independently tilting a set of two mirrors (one from an input array and another from an output array) to steer the light beam from an input port to a selected output port. Hence, a connection between any input fiber to any output fiber can be established. Placing two arrays of micromirrors in an optical system, with the help of either bulk lenses or microlens arrays, has resulted in optical 3D MEMS switch implementations [82, 93, 94]. A 3D perspective drawing of the optical 3D MEMS switch is also shown in Fig. 6.12b [95]. Different implementations have been proposed and demonstrated. The most common one is shown in Fig. 6.12 and it is a transmissive design that uses two sets of fiber ports and micromirror arrays facing each other. Figure 6.13 shows a reflective design, where there is one set of micromirrors and fiber ports leading to savings in real estate and hardware.

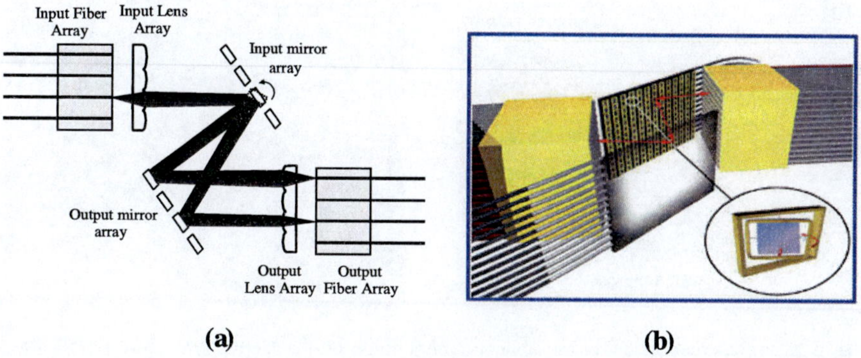

Fig. 6.12 **a** Operational principle of optical 3D MEMS switch. Any mirror in the input mirror array can steer the incoming optical beam to any of the mirrors on the output mirror array. **b** 3D conceptual drawing of the optical 3D MEMS switch showing the fiber array, the holding mount, the microlens array, the two micromirror arrays and the associated output fiber array and micromirror array (From [95], Fig. 5. Reproduced by permission of © 2009 The Institute of Electrical and Electronics Engineers)

The applicability of the 2D or 3D approaches depends on the application and the advantages of each of the approaches. Figure 6.14 shows that 2D designs do not scale as the 3D ones, mainly because of the increase in the optical insertion loss of the system. On the other hand, 3D architectures have been shown to scale to hundreds [82] of ports for commercial products and thousands of ports [93] for laboratory demonstrations. However, the 3D beam-steering approach requires analog control of the mirrors to establish each connection and often feedback loop controls as well as optical monitoring to keep the optical connection stable, hence it requires a more complex system driver. As a result, the 2D MEMS architectures are of interest when a lower port count (e.g., $<100 \times 100$) is required [96] and the 3D MEMS are of interest when a high port count is required.

6.4.9 Optical Switches

In the previous section, we presented the micromirror elements and gave the basic architecture for building micromirror optical switching fabrics. A variety of other optical switch technologies have been widely explored too. These include large-port-count optical switches through free-space-type optical switch architectures [98, 99], Mach–Zehnder interferometer (MZI)-based switches [100], electro-optics- [101], thermo-optics- [102], liquid–crystal technology- [103], bubble-jet technology- [104], acousto-optics-based [105], and others.

Early implementations were based on free space/bulk optic approaches. For example, a "rearrangeable" 128×128-channel optical switch based on a Benes

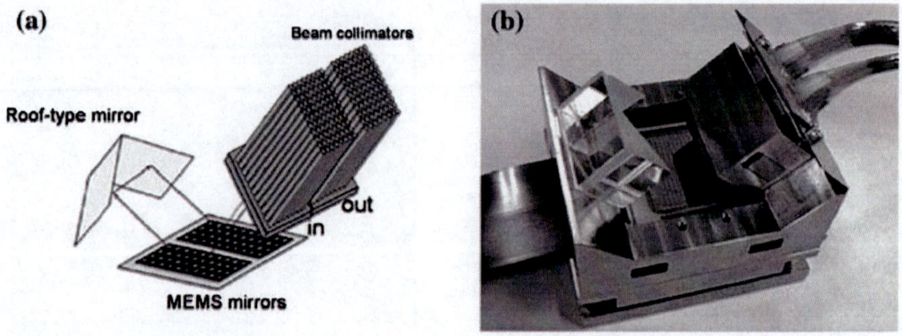

Fig. 6.13 **a** Schematic view of the optical configuration of a reflective type optical 3D MEMS switch. The input and output fiber collimators are placed from the same side of the structure and signals are reflected through a folding prism assembly achieving size reduction and use of a single mirror array, **b** photograph of the reflective type optical 3D MEMS switch (From [97], Figs. 4 and 12. Reproduced by permission of © 2005 The Institute of Electrical and Electronics Engineers)

Fig. 6.14 Comparison of the required number of switch elements in 2D and 3D switches (From [95], Fig. 4. Reproduced by permission of © 2009 The Institute of Electrical and Electronics Engineers)

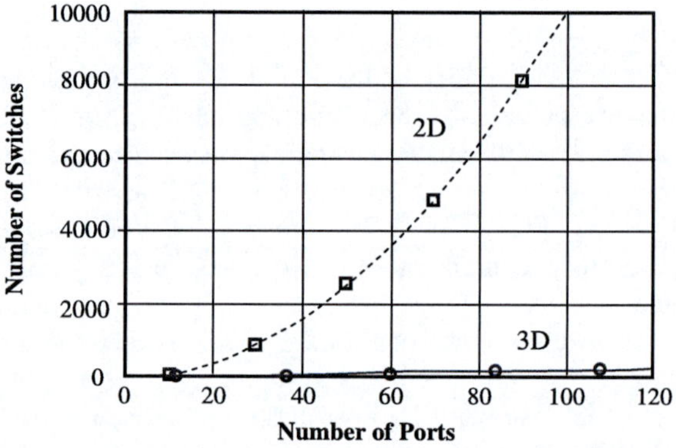

multistage network configuration used liquid crystal light modulator (LCLM) arrays with less-than 0.5 dB average insertion loss and less-than −30 dB average crosstalk. The system's average loss and crosstalk were demonstrated at 7.9 and −21.2 dB, respectively, as shown in Fig. 6.15 [98]. Figure 6.16 depicts these implementations which were based on polarization-based switching (i.e., controlling the state of polarization of the incoming beam) and polarization-dependent components (e.g., polarization beam-splitter-PBS) stacked at different configurations.

Switches based on Mach–Zehnder interferometers can be built on hybrid silicon platforms [106], where ridge waveguides on the silicon-on-insulator (SOI) wafer enable low-loss light propagation and III–V epitaxial layers provide the necessary phase shift. Carrier depletion in the doped quantum wells is utilized for the hybrid

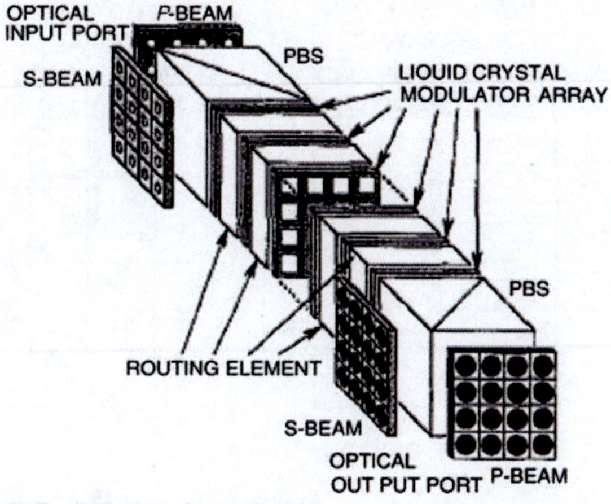

Fig. 6.15 Schematic of the free-space optical switch structure based on liquid crystal polarization switches and PBSs (From [98], Fig. 2. Reproduced by permission of © 1991 The Institute of Electrical and Electronics Engineers)

silicon switch to introduce efficient index change as well as to keep power consumption low by reverse biasing the device. Two adiabatic tapers are placed between passive and hybrid sections to minimize reflection and increase coupling efficiency. Trenches etched around the MMI coupler offer better extinction ratio and lower crosstalk [100]. Further improvement in the device speed is accomplished by incorporating a traveling wave electrode (TWE) [107, 108].

Optical switches based on ring resonator (RR) approaches have been implemented [109–111]. Figure 6.17 shows two possible RR-based optical switch fabrics. On the top of Fig. 6.17, the 1×2 optical switch fabric with crossing is shown. In the off-resonance case the signal is not coupled into the RR, when it passes by the RR-to-waveguide coupling region. On the other hand, in the on-resonance the signal is coupled into the RR at the first RR-to-waveguide coupling region, and then it is coupled from the RR to the second waveguide at the other RR-to-waveguide coupling region. In this approach and due to the waveguide crossing, there is channel crosstalk, or optical leakage from one waveguide to the other. Hence, this can have an impact on the optical performance of an optical interconnect. Figures 6.17c and d show the 1×2 optical switch fabrics without waveguide crossing. In this approach, there is no crossing between the two waveguides; hence, no additional crosstalk is generated. Note that in both cases the RR-to-waveguide coupling region is also a point of possible crosstalk.

6.4.10 Couplers/Splitters

Couplers are used to bring light from an external source into the package and die. The key merit metrics are coupling efficiency, cost, and alignment requirements.

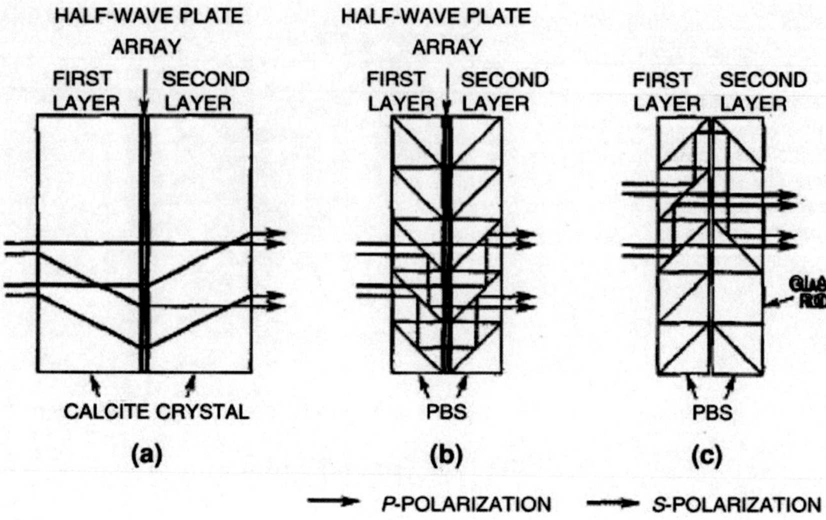

Fig. 6.16 Possible structures for the routing elements, **a** calcite-type, **b** PBS-HWP type, and **c** PBS type (From [98], Fig. 4. Reproduced by permission of © 1991 The Institute of Electrical and Electronics Engineers)

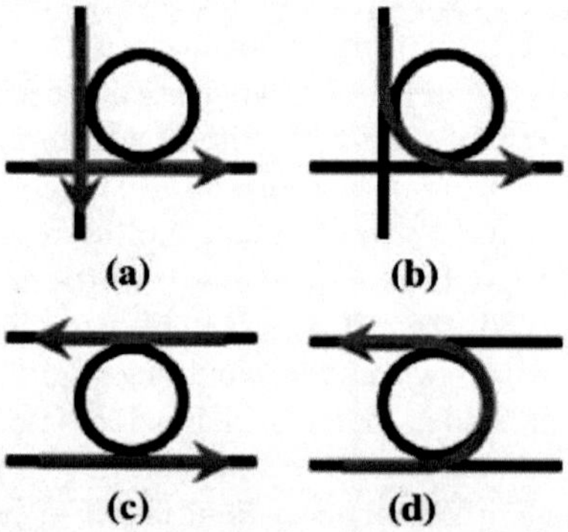

Fig. 6.17 1 × 2 optical switch fabrics based on the ring-resonator approach: **a** Off-resonance propagation with crossing. **b** On-resonance propagation with crossing. **c** Off-resonance propagation without crossing. **d** On-resonance propagation without crossing (From [112], Fig. 5. Reproduced by permission of © 2010 The Institute of Electrical and Electronics Engineers)

Power losses in couplers can potentially dominate the optical budget in optical systems. Splitters are used to divide a light source (laser or single waveguide) into two or more waveguides. Power losses and size are the two most important quality metrics of splitters.

6.5 Requirements for Optical Interconnects

In order for the optical interconnects to have a competitive advantage over the electrical interconnects, they must have similar or better bandwidth, cost, power, and latency as the copper solution, while being capable of very high volume manufacturing.

The consumed power in an electronic link scales with the bandwidth and length. On the other hand, in an optical link, this dependence of the power consumption is less due to high-bandwidth and low-loss optical fibers. However, due to necessary electronic-to-optical and optical-to-electronic conversions the power consumption can significantly increase. Miller gives a detailed analysis on the device requirements for optical interconnects [113]. In his study, he concludes that the following requirements must be met for the optical interconnects to be a viable competitor to their electronic counterparts:

a. To be competitive with the current state of the art in electrical off-chip interconnects, the system energy per bit should be <1 pJ, and to offer sufficient energy advantage for optics, it should be $\sim$100 fJ/bit or lower.
b. To meet the demands of off-chip interconnects out to the ITRS projections of 2022, system energies per bit of 100 fJ/bit may be sufficient, but to sustain the number of bytes/FLOP in the later years will require 50 fJ/bit or lower system energy.
c. To be competitive with near-future electrical global on-chip interconnects, the system energy per bit should be $\ll$50–200 fJ/bit.
d. To meet global on-chip interconnect demands out to 2022 will require system energies per bit of $\sim$30 fJ/bit on ITRS projections (assuming the global on-chip bandwidth is 5 times the off-chip bandwidth), and to sustain the number of bytes/FLOP in the later years will require $\sim$10 fJ/bit system energies.

Another important aspect of optical interconnects is the required interconnect density and the available (cross-sectional) area for it. The required number of channels is on the order of thousands to tens of thousands, if we assume that next-generation system architectures operating at data rates of 1,020 Gbps would require around 780 Tb/s per chip [113, 114]. Such a huge number of channels may drain the available space to implement the optical connections (or physical paths) on the chip level. More advanced and integrated approaches need to be developed and are currently under investigation [115].

Optical-interconnect density as well as power dissipation depends on the technology of the optical components to be used. In particular, the material or medium that will be the basis of the integration and the optical transmission. For long-length optical interconnections the choice is low-loss optical fiber or free space. Nevertheless, as the lengths of the optical interconnects become smaller and smaller (e.g., inter- or intra-chip interconnections), material platforms based on silica or InP are becoming potential candidates.

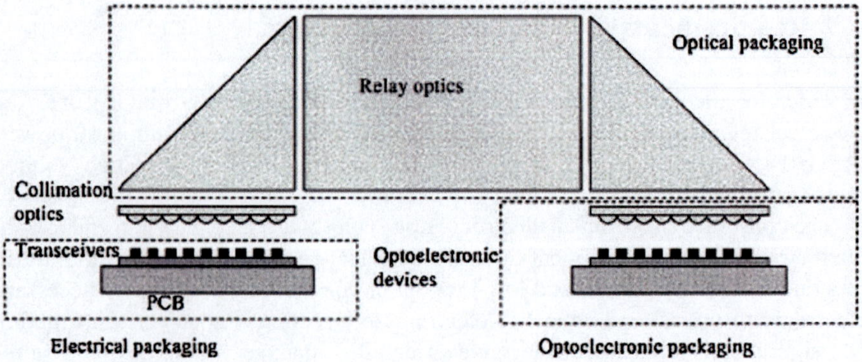

Fig. 6.18 A generic FSOI architecture. Two PCBs are interconnected via transceiver (transmitter/receiver pairs) and optical relay system. (From [124], Fig. 5. Reproduced by permission of © 2003 The Institute of Electrical and Electronics Engineers)

Regardless of the technology or approach used, other important physical performance characteristics include the optical insertion loss (IL) of the individual elements incorporated in the optical interconnect and the noise leakage performance of some of these components (e.g., filter, switches). The IL includes not only device passive losses, but also coupling losses between interfaces. Optical IL can limit the network scalability and link throughput performance, because it reduces the optical power of the signals as they propagate through the system. Hence, by the time the signal arrives at the destination, it may have been severely attenuated and may be below the sensitivity of the photodetector. On the other hand, noise leakage can occur at the individual elements (e.g., switches, filters, waveguide crossing points etc.) and eventually can be recombined with the signal at the end of the path and impinge on the photodetector. Note that this noise leakage can be either on the same wavelength (e.g., intra-channel crosstalk) or on a different wavelength (e.g., inter-channel crosstalk). This noise leakage interferes with the signal and deteriorates the signal transmission. Hence, the goal of the designer of the optical interconnect is to design and/or select processes, devices, and/or architectures that minimize optical IL and noise leakage of each element as well as the entire network.

6.6 Free Space Optical Interconnects (FSOI)

From a historical perspective, free space optical interconnects (FSOI) were the first to be proposed and studied extensively. A generic FSOI is shown in Fig. 6.18, where transceivers from two different PCB are interconnected with the use of microlens arrays and relay optics. Optical approaches for communications and digital computing were proposed soon after the development of the first lasers. The first studies showed that optical approaches were not a good alternative for electrical interconnects, mainly because of the increased energy they would require [116]. Nevertheless, interest in optical switching and optical gates continued [117, 118] and led to the belief that lower powers could be achieved and hence enable optical logic [119]. The developments in the area of semiconductor devices for optics in the 1980s eventually led to the idea of optical parallel processing through optical interconnects that use 2D arrays of semiconductor devices. At the same time other technologies (e.g., liquid crystal devices) were used to create 2D arrays of LC-pixels that can be used for parallel optical beam processing [120, 121]. In the 1980s increased research interest from different groups around the world proposed optical solutions for the distribution of clock signals across different sections of microprocessor chips. The idea for optical interconnection of very large-scale integration (VLSI) electronics was proposed and analyzed by Goodman et al. [122]. Following that, different FSOI designs have been proposed and experimental demonstrations have been reported for all system levels, from rack-to-rack applications to inter- and intra-chip applications [123].

In the first demonstrations and/or implementations, optical interconnects were the means of relaying light from plane to plane within the system. Transmitters located on one PCB were used to transmit optical signals, which after proper collimation from microlens arrays were relayed through free-space/bulk-optics paths, after multiple reflections from reflecting elements (e.g., total internal reflection prism, mirrors etc.). Receivers at the other PCB detected the incoming signal, which through the elements of the microlens array focuses on the associated receiver element. In some applications, a signal would be required to reach multiple locations to other PCB(s). In such a case, DOEs were used to replicate the optical signals and at the same time route them to different directions. Hence, DOEs, which could generate fan-out beams that replicated the incoming signal, were powerful techniques that along with arrays of bistable switches could be used to manipulate optical signals traveling in 2D arrays of optical beams [125–130]. Again, in these systems, the optical interconnect served as a simple point-to-point (or point-to-multipoint) link of large arrays of beams and also performed part of the processing operation [131–134]. The potential of achieving a high number of free space optical beams interconnects was realized in the 1990s. An example is shown in Fig. 6.19 [135]. This demonstration showed the feasibility of dense free-space optical array connections.

With the advent of optical switching technology, and the promising large-port-count free-space-type optical switch architectures the trend changed from just a relaying system to switched architectures, where a signal could actively be switched to one or another route [136–138]. Their major advantage, compared with waveguide-type switch architectures, was the higher density connectivity they provided for spatial parallelism. Combining 2D integration of (N) input and output fiber arrays, the dimensions of switch fabrics were reduced to the square root of the

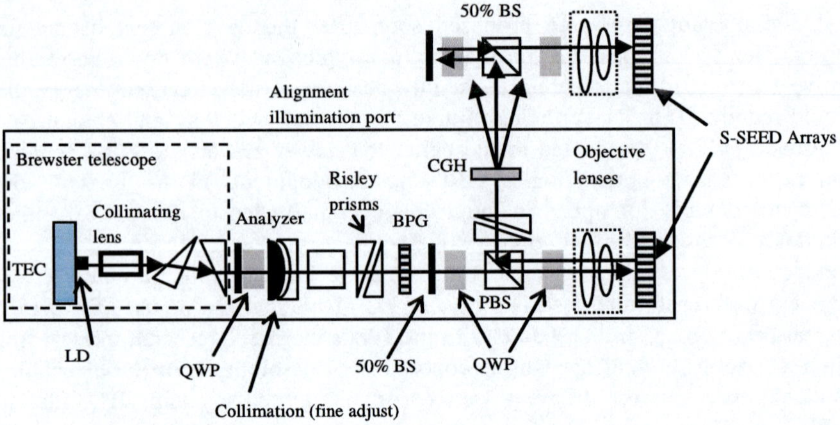

Fig. 6.19 The building block of the six-stage photonic switching system of Ref [135]. The system as shown has more than 60,000 light beams, with arrays of light beams generated from each laser using diffractive optical elements. Figure drawn based on data in [135] (QWP: Quarter wave plate, BS: Beamsplitter, PBS: Polarizing beanspitter cube, BPG: Binary Phase Grating, CGH: Computer Generated Hologram)

port count $\left(\sqrt{N}\right)$, and the total number of switch elements was equal to twice the port count (2*N*).

In general we can divide the FSOI into three subcategories:

a. stacked optics architectures,
b. planar optics architectures, and
c. micromachined 3D architectures.

6.6.1 Stacked Optics FSOI

The stacked optics approach was first introduced by Iga et al. in the early 1980s [139]. The approach was based on using 2D arrays of optical components separately fabricated on planar optical substrates to build integrated imaging systems. The different planar optical subassemblies are stacked and attached on top of each other with the help of optical adhesives as shown in Fig. 6.20. For a number of parallel optical systems the alignment task is reduced to the alignment of the 2D component arrays and not the individual components. The systems are arranged so that each planar substrate contains only one type of optimized elements (e.g., microlenses, microprisms, etc.). Alignment is based both on mechanical alignment and fixing fixtures, as well as optical alignment marks [140]. The advantage of this technique is that the different layers can be fabricated using different materials and processes. Thus, optimized performance from each layer can be achieved.

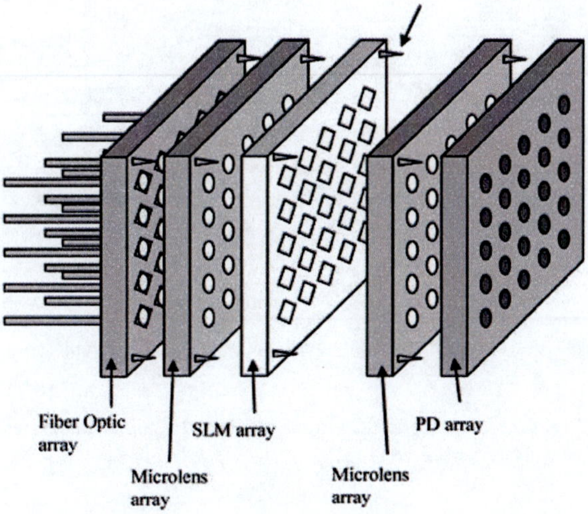

Fig. 6.20 Stacked optics FSOI is based on stacking individual planar 2D arrays of components. Each array can be built based on optimized processes appropriate for the individual functionality of each stack. (SLM: Spatial light modulator, PD: Photodiode)

However, for many applications, the different behavior of the substrates under changing environmental conditions (e.g., different thermal expansion coefficients) may cause problems. Furthermore, with increased number of stack layers and more complex optical systems the alignment difficulty increases and it becomes impractical. Hence, this approach has mostly been used for small number of stacked-layer systems with low complexity.

6.6.2 Planar Optics FSOI

The planar optics approach is one step further in the integration of microoptical systems. Unlike the stacked optics, where each layer is built independently, in the planar optics approach the system is integrated monolithically in one substrate as shown in Fig. 6.21. To accomplish this, the optical system (and associated optical paths) is folded in such a way that light follows a "zig-zag" path inside the substrate [141]. Light is coupled in and out of the planar substrate by transmissive components that deflect the light into the propagation direction. Note that even though the propagation is not performed in free space, the light is not confined in the direction perpendicular to the direction of propagation (as is done in waveguides). As the optical signal propagates along the optical path it impinges on a variety of microoptical elements, which are integrated in the substrate as reflective elements and allow the beam to be directed to the proper path. These microoptical reflective elements can act as plain mirrors, reflective diffraction gratings that fan out the incoming beam to multiple replicas, etc. Optoelectronic component arrays (e.g., VCSELs, photodiodes) can also be integrated in the system for extra functionality [142]. Due to the 2D layout of the approach, lithographic processes can be

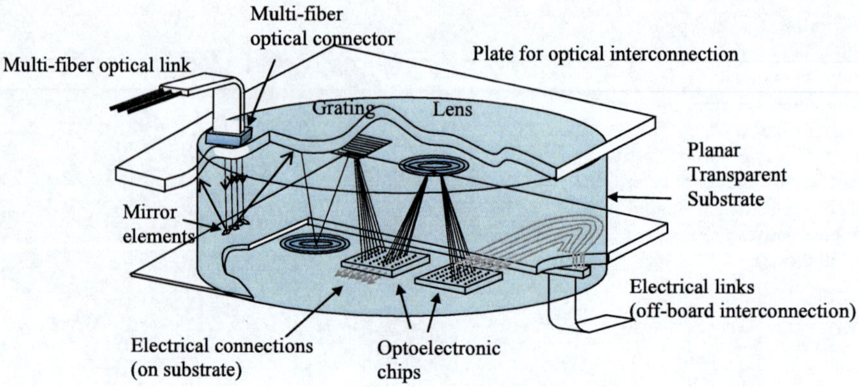

Fig. 6.21 Schematic drawing that shows the planar-integrated photonic microsystems based on a concept proposed in [143]

used for the fabrication and alignment of the optical components. Submicron alignment precision is achieved in lithographic processes. Hence, no additional opto-mechanical alignment is required. This approach also allows for hybrid integration of components that are not compatible with the manufacturing process and/or the materials of the substrate. For example, laser chip arrays may be able to be positioned using a motherboard–daughter board approach to form a hybrid system.

As it is immediately obvious from the discussion above, the attraction of the FSOI approach is that a multi-channel interconnect can be implemented with only one optical component. However, there are equally obvious issues such as the degree of alignment precision required between transmitter and receiver modules, the performance requirements of the optical imaging element, and the scalability of this approach that must be addressed.

6.6.3 Micromachined FSOI

In the micromachined FSOI architectures, surface micromachining is used to fabricate individual microoptical components using appropriate etching techniques (i.e., RIE) on the substrate. With this approach, it is possible to fabricate 3D structures in polysilicon, such as free-standing cantilevers, moving mirrors etc., as shown in Fig. 6.22. Hence, microoptical benches that can include a variety of different opto-mechanical components can be fabricated [87].

A combination of the micromachined optical bench and the 2D arrays of beam-steering micromirrors presented earlier in Sect. 6.4.8 can provide large-port-count 3D FSOI. The micromachined 3D architectures can take advantage of the large-scale fabrication of MEMS mirrors. 3D MEMS beam-steering technologies

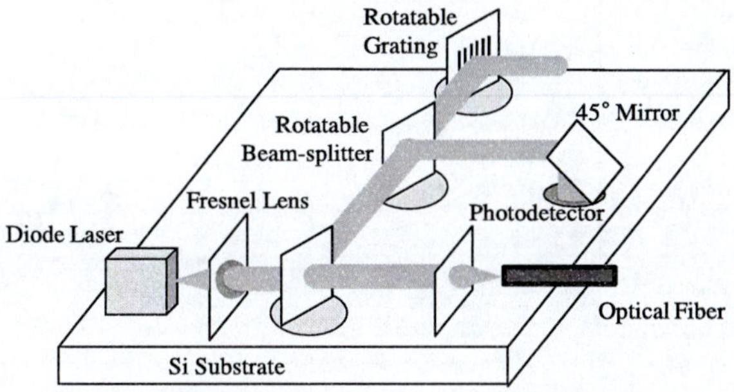

Fig. 6.22 Concept of the micromachined optical bench approach for FSOI. Microoptical elements, micropositioners, and microactuators are monolithically integrated on silicon substrate by surface micromachining (From [87], Fig. 6. Reproduced by permission of © 1997 The Institute of Electrical and Electronics Engineers)

have been proposed for use in free-space-type optical switch architectures [82, 93]. An important characteristic of the MEMS technologies is the feasibility of various forms of optical integration; for example, a 2D array of optical elements could be constructed or multiple functions can be integrated on the system incorporating other elements (e.g., array waveguide gratings (AWGs), multiplexers/demultiplexers, DOE for fan out, detectors for monitoring etc.) to form optical signal processing systems [144, 145]. Therefore, 3D-MEMS beam-steering technologies based on a 2D array of integrated micromirrors are much more suitable for application in optics and promise to enable high-density, compact packaging of optical switch fabrics.

6.6.4 FSOI Design Issues

The design of the FSOI architectures is based on light propagation in bulk medium (e.g., free-space, polymer, glass). There are many different methodologies to calculate the propagation of light. These depend on the degree of precision required, the propagation distance, the ratio of any material dimensions to the wavelength of light, and whether a scalar or a vectorial solution is required. In this chapter, we have limited space to describe a full treatment of light propagation. We will only attempt to give a quick overview highlighting the important issues and offering additional references. There are many classic books that describe the basic concepts of light propagation [146–148]. One of the most widely used approximations is the Fresnel–Kirchhoff diffraction integral which is reasonably accurate for regions which are many wavelengths from the source region. Such an analysis shows that any wavefront of finite extent, like the ones emanating from a laser or a microlens, will diffract and spread out as it travels through space. Therefore all FSOIs, except those over very trivial distances, require lenses, mirrors, or other optical components to prevent the propagating waves from

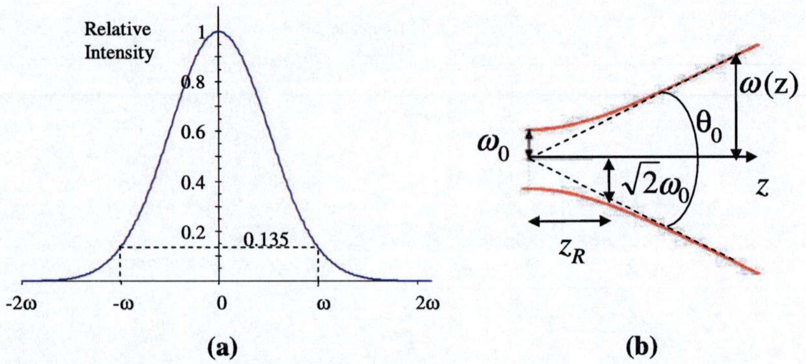

Fig. 6.23 **a** Gaussian beam intensity profile, **b** Gaussian beam propagation in free space causes the beam to expand

spreading. One other source amplitude distribution for which we can find an analytical model for propagation is the Gaussian distribution [147]. Note that optical fibers and many lasers used in FSOI emit beams that are more or less Gaussian. Hence, Gaussian Optics formalism has been used extensively in designing FSOI systems. A Gaussian beam is symmetrical around the optical axis (z-axis) and its intensity profile is described by

$$I(r) = I_o \exp\left(-\frac{2r^2}{\omega^2}\right) \tag{6.1}$$

where ω is the beam radius and represents the radius at which the beam amplitude falls to $1/e^2$ (or 13.5%) of its peak value as shown in Fig. 6.23a. As the beam travels through space (along the z-axis) the beam radius expands according to

$$\omega(z) = \omega_0 \left(1 + \frac{z^2}{z_R^2}\right)^{\frac{1}{2}} \tag{6.2}$$

where ω_0 is the initial beam waist and z_R is a parameter known as the Rayleigh range, given by

$$z_R = \frac{\pi \omega_0^2}{\lambda} \tag{6.3}$$

where λ is the wavelength of light. At position z_R the radius is $\sqrt{2}\omega_0$. The behavior of the Gaussian beam is entirely determined by the initial beam waist and the wavelength. The more tightly confined the initial beam is, the more rapidly it will diverge. When the axial distance is several times the Rayleigh range we approach an asymptotic regime where the beam appears to diverge linearly, with divergence angle $\theta_0 = \lambda/\pi\omega_0$ (Fig. 6.23b). Note that beams that have a small beam waist will diverge rapidly whereas those with a large beam waist will diverge slowly.

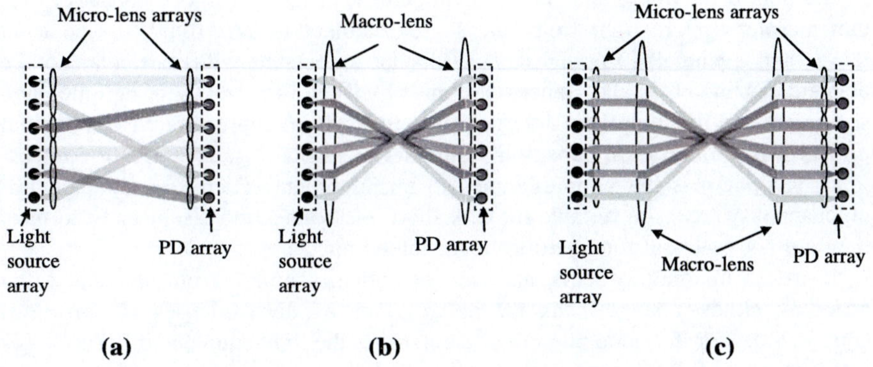

Fig. 6.24 Schematic diagram of the FSOI **a** micro-, **b** macro-, and **c** hybrid-imaging approaches

The Gaussian beam divergence can be controlled or modified with proper selection of imaging techniques. Note that many times the optical beams follow folded paths where the beams are reflected by mirror elements. In any case the Gaussian beam propagation formalism can be used in the FSOI design. If a lens of focal length f is positioned so that the initial beam waist ω_0 is a distance x_1 from the object focal point, then a new beam waist ω_1 is obtained at position x_2 from the image focal point, according to the equation

$$x_2 = \frac{x_1 f^2}{x_1^2 + z_R^2} \tag{6.4}$$

The new beam waist is given by

$$\omega_1^2 = \frac{f^2 \omega_0^2}{x_1^2 + z_R^2} \tag{6.5}$$

from which we can see that depending on the relative magnitude of the initial Rayleigh range z_R and the focal length we can either focus the beam to a smaller beam waist (and increase its divergence) or expand it (and reduce its divergence).

We can divide the imaging systems into three categories [149, 150] depending on the type of lenses used. The three categories are the micro-, macro-, and hybrid-imaging systems. In micro-imaging systems, shown in Fig. 6.24a, a microlens is assigned in each input light source and each photodetector (PD). The light from the source is collimated and directed to an arbitrary direction. The advantage of this approach is that each lens operates with a field of view of a single source, rather than the entire array. The array size can be increased simply by adding extra microchannels/elements, which does not require a modification of the lens design. Nevertheless, due to the small aperture size, the system is limited by diffraction effects and hence the interconnection distance is limited to small distances.

In addition, if the microlens pitch is small, and the beam divergence is larger than the microlens pitch, light from one source element leaks through the adjacent lenses and eventually impinges on a PD not associated with particular source element causing crosstalk. Hence, care must be taken to correctly design the pitch separation and the microlens divergence. In this system approach, due to potential optical leakage the beam density is limited. The second disadvantage of a micro-imaging system is the poor tolerance to misalignment. Therefore although microchannel systems are suitable for very short interconnection distances (<20 mm) they are not practical for board-to-board interconnects.

In macro-imaging systems, an array of optical sources is imaged onto a PD array by either a single lens (2*f* imaging) or a pair of lenses (4*f* imaging) (Fig. 6.24b). For the same number of elements in the light source and PD arrays as well as the same divergence angle for the much larger than the size of the sources, the 2*f* imaging system has a larger aperture size. Note that if the light source pitch is much larger than the size of the source, the higher level of resolution that can be accomplished with this technique is wasted. However, the macrolens approach has the simplest design; the entire system has a single aperture stop. The two planes are mutually imaged onto each other with unit magnification and inversion. Since the system is telecentric, small axial misalignments of the system components result in defocus without changing the magnification of the system. This is a significant advantage as it ensures that light transmitted from the sources will be coincident with the receivers (assuming no other misalignment). However, in more complex systems, where other optical components are involved, care should be taken to minimize other aberrations. Many experimental demonstrations have been reported based on these approaches [151–153]. Although the macro-imaging approach is simple to design and construct with off-the-shelf components, macro-optical systems do not scale as easily [149, 150, 154]. In most cases and when a small area needs (e.g., 1 cm^2) to be imaged over a long distance (e.g., 10 cm) lenses with multiple elements are required to correct aberrations and will not fit within the required space. Hence, there is a tradeoff between the size of the array that can be imaged and the optical throw (distance between input and output arrays). One alternative that has been used to partially overcome this limitation is to use GRIN lenses as relay elements [155], however, these also become excessively long when used for array sizes of more than 2 mm.

The hybrid-imaging system combines the two previous approaches (Fig. 6.24c). In particular, it combines the relatively long optical throw and misalignment tolerance of macro-imaging systems with the scalability and moderate field-of-view requirements of micro-imaging systems. First, the light is collimated by an element in the microlens array. The collimated beams are then imaged by a pair of macrolenses. Since the beams are collimated, the macrolens aperture is smaller than the one required in the macrolens approach. In the design process, the constraints are typically (a) the spacing of the two device planes (i.e., the optical throw) and (b) the maximum size of the device plane [124, 156, 157]. Within these constraints the designer balances various parameters such as microelement array density (which determines the number of parallel channels), lens focal length

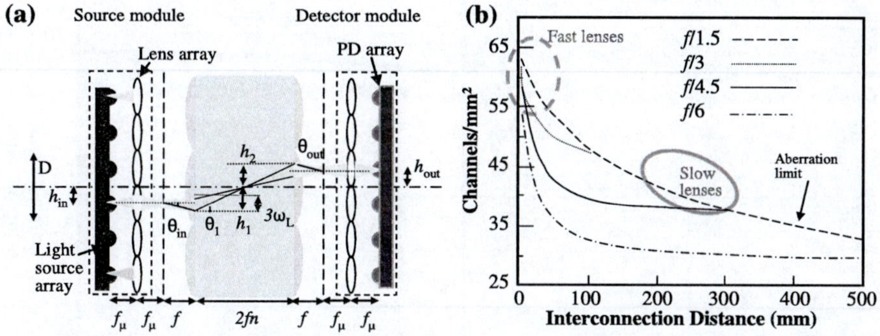

Fig. 6.25 **a** Schematic of the three modules used to model a hybrid-imaging based optical interconnect similar to the one in Ref [157]. **b** Maximum channel density for different *f*/# values at different interconnection distances using a source pitch of 125 μm based on data from Ref [157]

f (which determine the costs and practicality of the lenses), and alignment tolerance (which will determine the degree of difficulty in assembling the system and its potential manufacturability).

From the discussion above, we see that there is a range of parameters the FSOI designer will have to study and evaluate to achieve an optimum design. These important parameters include the diffraction, geometric aberrations, speed of the lens, and tolerance of element manufacturing processes and alignment. These effects determine the limits on the spacing of the two device planes (i.e., the optical throw) in the FSOI system, the number of interconnected channels and the associated channel density, the pitch of the channels, and the crosstalk performance Figs. 6.25 and 6.26.

Tolerance analysis (commonly referred to as tolerancing) is also very important. Tolerancing is critical both for the optimization of each element array, and involves the design and manufacturing process of the individual arrays, as well as the tolerance in the alignment of the individual arrays with each other within the system. Most FSOI are multi-element systems that require the mutual alignment of many different components in different planes. As described earlier, planar and micromachined FSOI approaches are based on lithographic approaches and as such they can be aligned with much better precision. Furthermore, the individual arrays used in the stacked-architecture can be fabricated with high precision. Nevertheless, positioning, alignment, assembly, and packaging is based on mechanical techniques and hence the tolerance to misalignment between different modules should be as large as possible. When input and output planes consist of single mode fiber optical arrays the tolerancing requirements are more stringent. The misalignment of the input/output arrays will have a significant effect on the optical insertion loss of the system. Gaussian beam propagation can be applied to evaluate the effects of misalignment in the optical insertion loss [158]. There are three misalignment types between two microlens elements and are shown in Fig. 6.27. These are the:

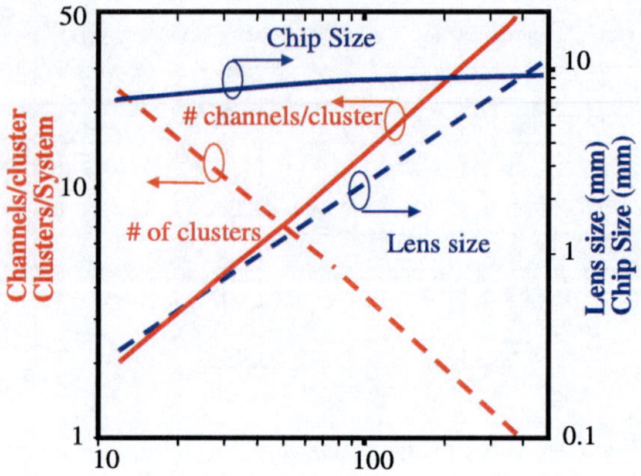

Fig. 6.26 Optimum link parameters for a system having 2500 channels on a 125 μm pitch relayed by *f*/6 lenses based on data from Ref [157]

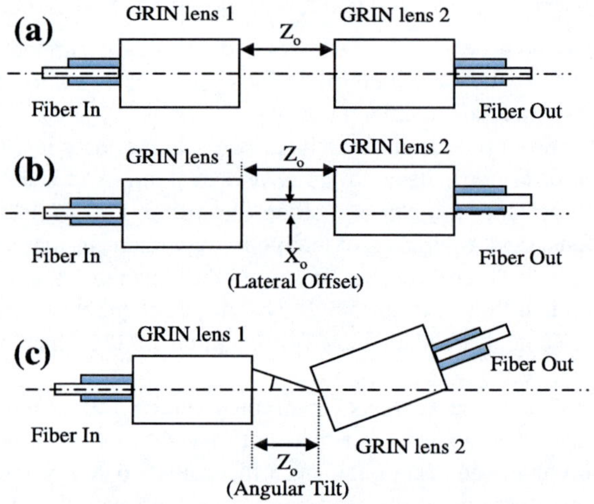

Fig. 6.27 **a** Separation misalignment, **b** lateral offset misalignment, and **c** angular (tilt) misalignment between two GRIN lenses

a. separation misalignment from the optimum operating distance,
b. lateral offset, and
c. angular offset misalignment.

A general formula for determining the coupling loss between two single-mode fiber collimators with the simultaneous existence of separation, lateral offset, angular tilt misalignments, and spot-size mismatch has been theoretically developed and experimentally verified. Utilizing that formula the individual contribution of the misalignments to the insertion loss can be evaluated. The analysis was based on GRIN lenses but it can be adapted for other microlens designs as well. Figures 6.28 and 6.29 show the results of this study. There is good agreement between the model and the experiment up to a particular value in each case, and

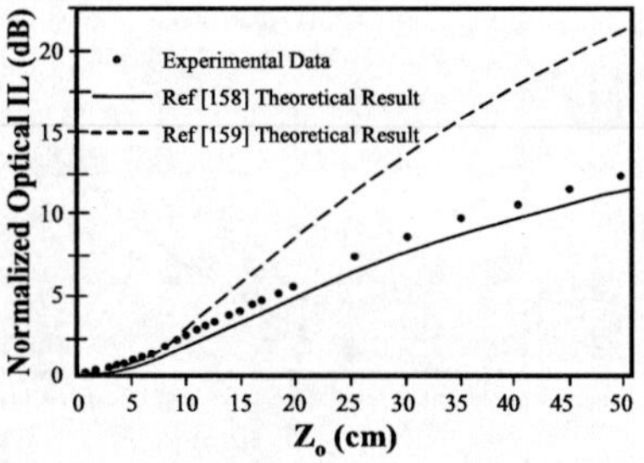

Fig. 6.28 Experimental and theoretical results, based on the data in Ref [158], for the normalized insertion loss that is due to the misalignment from the optimum separation between two fiber collimators. A comparison with another model in [159] is also shown

beyond that value there are deviations that are due to the fact that the model was based on Gaussian beam propagation, while in practice optical beams from real lasers and/or fiber optic collimators are only approximately Gaussian. For example, real laser beams diverge faster than would be predicted by a Gaussian beam analysis due to the departure of the wavefront from the ideal Gaussian.

One last issue that must be considered for an FSOI design is the aperture size(s) of the different optical elements the optical beam passes through or reflected from without being clipped. Since the Gaussian beam intensity profile only goes exponentially to zero, any aperture will cause some clipping. The degree of that clipping is important, because it directly affects the optical insertion loss of the system as well as the propagation of the optical beam. Typically, if the aperture is large enough to transmit/reflect at least 99% of the power in the beam the impact of the clipping effect is minimal. If the beam is clipped more severely than this then the departure from the Gaussian intensity profile will be too great and it will no longer be possible to use the Gaussian beam model to predict the propagation of the beam.

Kirk et al. presented a very detailed study dealing with the issues of FSOI misalignment, tolerance, and sensitivity [135]. They describe the basic design rules for FSOI. We have to note here that for more complex systems a more rigorous investigation and tolerance and sensitivity study is required. The incorporation of different and multiple optical elements (reflective, refractive, and diffractive) would pose different limits to the designer. A variety of tools (optical design software) and techniques have been developed to solve these challenges. Some of these tools integrate the optical design with mechanical (e.g., vibration) and thermal stability effects.

6.7 Guided Wave Optical Interconnects

In contrast to the FSOI approach, where the optical signal travels in free space or bulk optical material unrestricted in the lateral dimension, in the GWOI the signal is confined in the lateral dimension. GWOI can be implemented using either fiber optic- or waveguide-based paths. In the previous section, we described the design issues of FSOI. One of the most important disadvantages of FSOI is the optical alignment issue; specifically the fact that high-density and complex systems

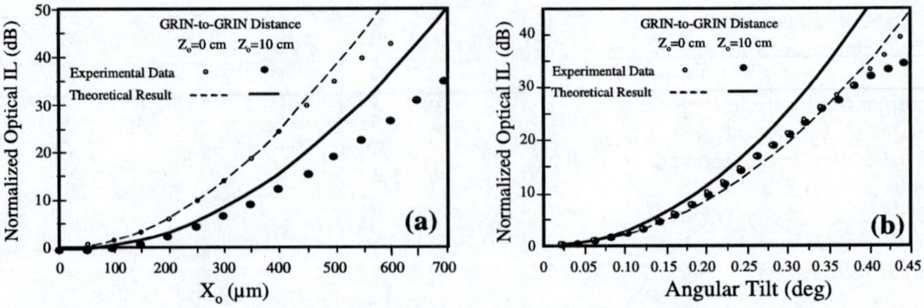

Fig. 6.29 Experimental and theoretical results, based on the results of Ref. [158], for the normalized insertion loss that is due to the **a** lateral misalignment at operating distance of 0 and 10 cm, **b** angular misalignment at operating distance of 0 and 10 cm

require hardware intensive passive, as well as active alignment techniques, which can make the manufacturability of FSOI very difficult and lead to prohibitively expensive approaches. We also described that FSOI is not the best option for intra-chip applications due to the dimensions and the mechanical limitations of the different components.

Optical interconnections using guided approaches have been investigated since the invention of the laser and fiber optics. The initial work focused on switching and routing systems and motivated the hybrid integration of VLSI with detectors, modulators, and vertical cavity lasers. The great potential of CMOS technology to provide large-scale manufacturing advantages to photonic systems has been the driver for the integration of the elemental building blocks of optical interconnects in silicon [160]. The key to unlocking this potential came in the late 1980s with the work of Soref et al., in their investigation of waveguides and silicon's electro-optic effect [161, 162]. Since then, active and passive photonic elements have been demonstrated [163–165]. The main driver for these approaches was to miniaturize the devices and also reduce the fabrication cost of the optoelectronic components via integration in a similar way that happened in the microelectronics industry. Hence, the birth of *silicon photonics* happened and it opened up a great application space and potential for the fabrication of different photonic structures and elements for optical communications, optical interconnects, as well as other applications using the standard fabrication technology and infrastructure [166–174]. Note that silicon photonics can capitalize on the advancements of DWDM with a large number of wavelengths traveling simultaneously through the same physical path (e.g., waveguide) to further improve area and energy efficiency. Monolithic integration has had its obstacles for the implementation of full optical interconnection systems, nevertheless, most optical elements required for the implementation of optical interconnects have been realized on silicon [175]. Alternative to the silicon platform is monolithic integration on III–V optical materials. Nevertheless, complicated epitaxial re-growth steps needed for III–V materials significantly reduce yield and increase cost.

Optical devices are limited in size by the wavelength of light routed through them. Due to the high refractive index of silicon significantly smaller devices as compared to those made in a low index medium (e.g., silica) can be accomplished. Nevertheless, there are challenges such as increased scattering losses due to sidewall roughness, increased sensitivity to Polarization Dependant Loss (PDL), and increased coupling losses that need to be addressed [176, 177].

Current proposals for chip-to-chip optical interconnects can be grouped into three categories:

a. monolithic integration of emitters and detectors into silicon,
b. off-chip hybrid integration, where light sources and integrated photodetectors are interconnected to the chip via fiber, and
c. on-chip hybrid integration, where a III–V emitter and detector are placed on top of the silicon die.

6.7.1 Monolithic Integration

Monolithic integration is the most challenging approach due to the limitations of silicon-based light emission. The most common definition of monolithic integration refers to a process by which all of the components (e.g., electronic circuits, light sources, photodetectors, modulators, waveguides, and multiplexers) are manufactured on the same piece of semiconductor substrate as shown in Fig. 6.30. In Fig. 6.30 a Si-based light source is shown. Recent developments in Si-based light source show the potential of this approach. Monolithic integration techniques that allow a mixed-material monolithic process such as the inclusion of III–V compounds or polymers (either grown or deposited) processed on a silicon platform are also possible. This approach allows for different materials to be combined during processing and hence provides greater flexibility in the choice of devices to be integrated. This would be an option for integrating a non-silicon-based light source on an otherwise all-silicon chip. To fully realize mixed-material integration, compromises in processing may need to be made which could severely affect both performance and yield. Despite these concerns, researchers continue to work on both single-material and mixed-material monolithically integrated planar optical chips [178–180].

6.7.2 Off-chip Hybrid Integration

Off-chip hybrid integration is accomplished by interconnecting the laser source(s), modulators and the photodetector to the chip via optical fibers. The advantage of this approach lies in the ability to independently choose the best option for the individual components and assemble them in a module or system. Nevertheless,

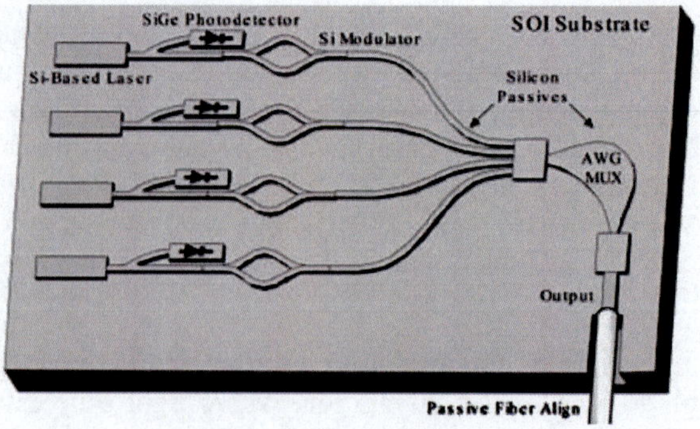

Fig. 6.30 Example of monolithic integration on silicon (From [181], Fig. 11. Reproduced by permission of © 2004 Springer Inc)

this approach can have increased optical insertion loss due to the fiber-to-chip interface loss as well as the need for the design of tapers that transform the optical fiber mode into one for the optical waveguides on the chip. The use of fiber and the requirements for the minimum bending radius to avoid increased IL, as well as the requirement on the stress on the fiber can set limits on the minimum size of the module.

6.7.3 On-chip Hybrid Integration

On-chip hybrid integration is accomplished by assembling disparate parts onto one common platform. This approach is shown in Fig. 6.31. For example, the base material can be SOI. Common hybrid optical components are those in which III–V compound light sources and detectors are attached onto SOI, silica, or polymer platforms [47]. The advantage of hybrid integration lies in the ability to combine the best performing devices onto one common chip, despite the different technologies used to manufacture each individual device [182]. An alternative hybrid approach is when the laser is located off the chip and coupling of the optical signal is accomplished either through an optical fiber or free space coupler (e.g., fiber Bragg grating).

6.7.4 GWOI Design Issues

One of the most important issues in integrated photonics approaches is determining the impact of integration on the component performance. Note that even though some individual devices can have very good optical performance when they are fabricated as single elements, in the case where they are integrated with other photonic elements their performance becomes worse. Hence, the challenge is

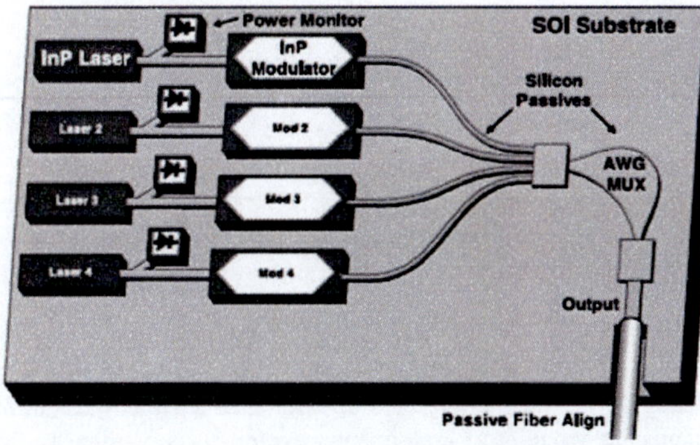

Fig. 6.31 Example of on-chip hybrid integration on silicon (From [181], Fig. 12. Reproduced by permission of © 2004 Springer Inc)

to integrate different components onto the same platform. Doing so there are many tradeoffs that must be performed in the integration process. Note that the tradeoffs and tolerance general issues are similar to the discussion of tolerance and alignment presented earlier for the FSOI.

In terms of alignment accuracy, monolithic integration on silicon has advantages as compared to hybrid integration. Misalignment between devices in the monolithic circuit is defined by the accuracy of the lithography tool. Integration of components that are based only on the silicon waveguide platform (e.g., AWGs) is defined in a single step and is not subject to misalignment at all. This minimizes or eliminates the losses between the individual devices. Losses between interfaces can also be reduced if the mode is kept in a constant refractive index material with constant dimensions. However, monolithic integration has non-optimum device performance due to constraints placed on material selection and compatibility. This limitation is critical for devices that actively manipulate light, such as emitters, modulators, and detectors [181]. Most of these devices operate near the band edge of the material. Since photodetection only becomes efficient when the wavelength is well above the bandgap, a silicon-based detector with a much smaller bandgap than the silicon-based emitter has to be integrated.

On the other hand, hybrid integration offers more degrees of freedom for the design of each optical component. Each individual component can be fabricated based on the most optimized technique and then integrated on the same platform with other components. In this case, the additional interfaces and the more difficult alignment are the disadvantages of the approach. Care should be taken to design interfaces with appropriate antireflection coatings in order to minimize reflection IL. In addition, proper mode coupler adaptors (e.g., tapers) must be implemented so that mode mismatch losses are minimized [183].

As we have described earlier, a complete optical interconnect network requires light sources that provide the optical carrier, modulators that encode the data onto the carrier, waveguides to carry signals, switches that route signals through the network, and photodiodes to detect the data. Hence, it is imperative that the optical components be built in a single CMOS-compatible process to achieve a significant reduction of the manufacturing cost. The last limitation on achieving this goal has

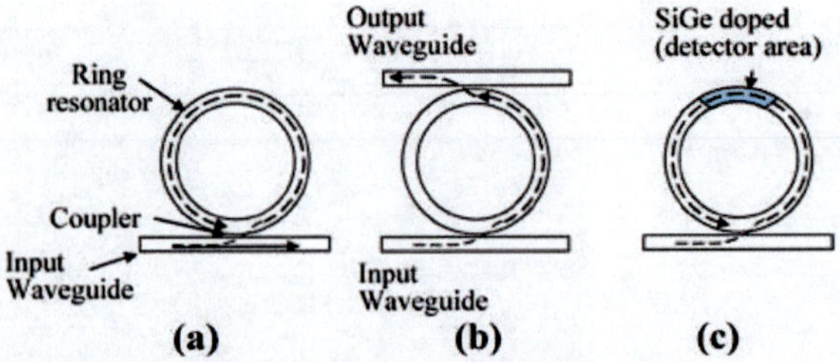

Fig. 6.32 Schematic of RR-based photonic building blocks for optical interconnects on silicon photonics platform. **a** Off-resonance Modulator: the ring resonator is coupled to a waveguide through evanescent coupling. Off-resonance wavelengths (*solid line*) are transmitted through; On-resonance Modulator: a resonant wavelength is coupled in the ring (*dashed line*). By switching between the on- and off-resonance state one can achieve modulation (or diversion) of a continuous-wave laser. **b** Injector: A resonant wavelength in the input (*lower*) waveguide is coupled into the ring and out through the output (*top*) waveguide. **c** Detector: A resonant wavelength is coupled in the ring and absorbed by a SiGe detector coupled to the ring

been the optical source. In terms of a laser source, VCSELs are an option, nevertheless, they are built using III–V compound semiconductors, and they cannot be easily integrated in a CMOS-compatible process. An alternative is on-stack mode-locked lasers, which generate a comb of phase-coherent wavelengths at equally spaced wavelengths [185], and the racetrack mode-locked silicon evanescent laser [186]. As mentioned in Sect. 6.4.9 the ring-resonator structure can be used to implement optical switches. In addition, the RR can be used to form filters, modulators, as well as photodetectors as shown in Fig. 6.32. Thus the RR can be the elemental building block for integrated photonic circuits, since they render possible the addition or extraction of signals from a waveguide based on wavelength in a WDM flow. The use of standard SOI technology leads to high compactness (structures with radii as small as 4 μm have been reported) and the possibility of low-cost photonic integration.

In terms of the interconnection architecture there is an important interworking of both the interconnection topology and the physical layer implementation that can satisfy the performance requirements. The real challenge in demonstrating optical interconnects has been the absence of a good implementation of photonic buffers and registers. There has been interest in slow light techniques for the implementation of optical buffers and this is an active area of research [187–189]. Nevertheless, at this point the obtained buffering time does not meet the requirements of a microprocessor network. Techniques that are based on multiple opto-electronic (OE) and electro-optic (EO) conversions have been proposed. Nevertheless, such approaches pose a power limitation on the chips, and because of the multiple OEO conversions the associated heat dissipation limits the viability of the solution. Alternatively, the architecture can be designed so that electronic and photonic functionality complement each other. An approach that takes advantage of this is the Data Vortex [190, 191]. In the data vortex architecture, when processing and routing decisions are required, high-speed digital electronic circuitry is employed in a way that complements the semiconductor optical amplifier (SOA) wide-band photonic switching elements. This allows the high-bandwidth optical payload, which is encoded on multiple wavelengths in order to maximize transmission capacity, to transparently traverse the network. A model for high-capacity low-latency interconnection networks based on the data vortex architecture is described in [191]. The architecture uses the bandwidth provided by optical signals encoded and transmitted over contemporary fiber-optic components and in particular SOA-based buffers. A comprehensive discussion of the data vortex topology, switching node design, packet routing, and control is also described in [191]. The study also looks at the optical performance requirements and characteristics of the approach.

Hence, to correctly design and implement a GWOI network, an analysis of the physical layer performance of the associated optical links (e.g., optical connection from the source to destination) has to be included. There are two main optical interconnection network categories, the ones using wavelength-selective routing [186, 192, 193] and the ones using space routing [194, 195]. The difference between the two networks is in the way the source-to-destination path in the photonic medium is performed. In the first category, switching is performed through wavelength selective switches or filters. The filters are tuned in such a way so as to allow each signal to be routed from source to destination through the selection of specific wavelengths. This form of routing ensures low latency but it is not able to leverage the full throughput that optics can provide. Space routing is based on multi-wavelength transmission to enable messages with high-aggregate bandwidth. In such an approach actively controlled broadband switches are used to route all wavelengths (or channels) from the source to destination [195].

The parameters that play an important role in the final optical performance of the GWOI network are insertion loss, optical crosstalk (or optical leakage from the devices), and electrical power requirements for controlling the optical interconnect elements. Chan et al. performed a detailed physical layer architectural study of chip-scale photonic interconnection network designs [195]. They examined the impact of physical parameters, namely, insertion loss (IL) and crosstalk on the network scaling and signal performance.

Figure 6.33 shows the different topologies studied in [195]. These are based on the (a) 4×4 photonic Torus, (b) 4×4 Non-Blocking Torus, (c) 4×4 TorusNX, and (d) Square Root topology. Typical insertion loss and crosstalk performance numbers, as well as projected target numbers are used to study the scalability, throughput, and error performance of the different networks. In addition, typical optical powers for the multiple wavelengths are used such that optical nonlinearity is avoided in the propagation medium, while adequate optical power impinges on the photodetector. Important findings of this study are the fact that the most significant optical loss contribution comes from the waveguide crossings. This is particularly true in the Torus and Non-Blocking Torus architectures, where a large number of crossings are realized due to the structure of the elemental switching elements as shown in Fig. 6.33a and b. By changing the architecture to Torus NX

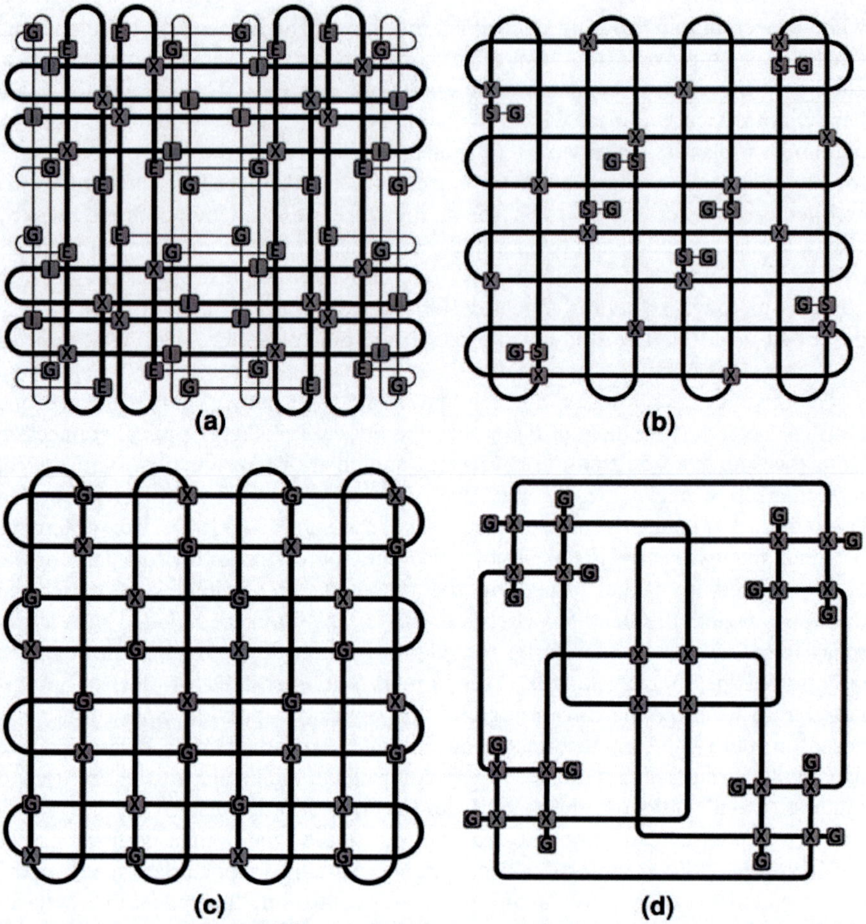

Fig. 6.33 **a** 4 × 4 photonic Torus with 16 access points. Switching points and access points are labeled X and G, respectively. I and E represent switches used to inject and eject messages into and out of the network respectively. **b** 4 × 4 non-blocking torus with eight access points. Labels indicate combined injection–ejection switching points. **c** 4 × 4 TorusNX network with 16 access points, **d** 4 × 4 Square Root topology (From [195], Figs. 1, 2, 7 and 9. Reproduced by permission of © 2010 The Institute of Electrical and Electronics Engineers)

and Square Root architectures, which are based on switch fabric designs that eliminate the crossings, the insertion loss impact on off-resonance message traversal is reduced while keeping similar switching functionality as shown in Fig. 6.33c and d. Figure 6.34 shows the insertion loss performance for the 4 different architectures, where the IL contribution of the crossing points is overwhelming in some of these cases, especially as the optical interconnect network scales up.

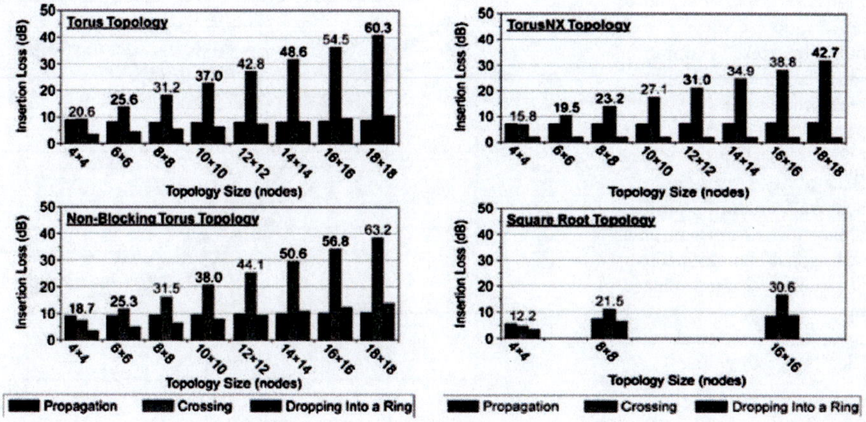

Fig. 6.34 Maximum possible network-level insertion loss by component for varying sizes the Torus, Non-Blocking Torus, TorusNX, and Square Root architectures using typical insertion loss parameters. Labeled values represent the peak cumulative insertion loss (in dB) for the network (From [195], Figs. 3 and 10. Reproduced by permission of © 2010 The Institute of Electrical and Electronics Engineers)

As mentioned earlier the noise leakage affects the signal integrity. The optical signal to noise ratio (OSNR) is a measure of the signal integrity. Noise can be any of the following: relative intensity noise (RIN) from the laser source, inter-channel crosstalk and intra-channel crosstalk. Figure 6.35 shows that for large messages, none of the networks is able to achieve signal integrity beyond the BER $= 10^{-12}$. Hence, either smaller messages must be used or higher network-layer error correction schemes should be implemented.

Several studies on the impact of the optical interconnect topology have been reported. Some are based only on the architectural differences with their electronic counterparts, while others focus on the details of the physical layer performance. For example, Vantrease et al. [110] report on the Corona architecture, which uses optically connected memories (OCMs) that have been architected for low power and high bandwidth. A set of 64 OCMs provide 10 Tbps of memory bandwidth through 128 fibers using dense wavelength division multiplexing. Corona uses a photonic crossbar with optical arbitration to fully interconnect its cores, providing near uniform latency and 20 Tbps of on-stack bandwidth. They report that systems using optically connected memories and an optical crossbar between cores could perform 2 to 6 times better on memory-intensive workloads than systems using only electrical interconnects, while dissipating much less interconnect power.

6.8 Comparison Between FSOI and GWOI

The key challenge for optical interconnection networks is to fully leverage the immense bandwidth of the fiber-optic components and in particular the great

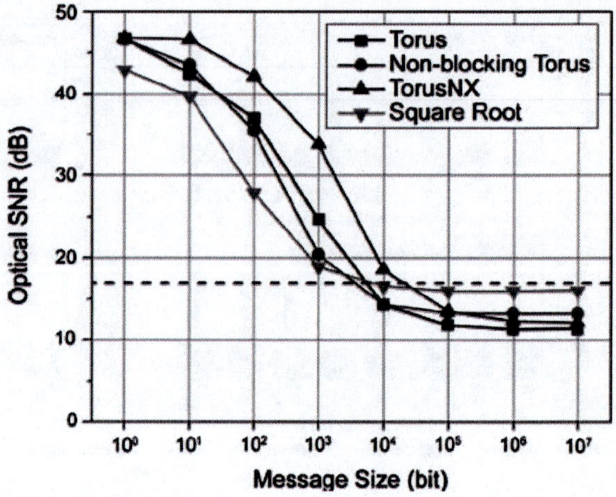

Fig. 6.35 Optical signal to noise ratio for varying message sizes, assuming saturated network load, measured at the photodetectors. The line at OSNR = 16.9 dB is where a BER of 10^{-12} can be achieved, assuming an ideal binary receiver circuit and orthogonal signaling (From [195], Fig. 12. Reproduced by permission of © 2010 The Institute of Electrical and Electronics Engineers)

potential of the parallel processing that WDM provides. Both FSOI and GWOI do offer these advantages. However, in contrast to guided wave optical interconnects where waveguide loss, cross-section, minimum bend radius, and crosstalk performance dominate the design process, FSOI has a different set of tradeoffs. These include the way in which the optical aperture is partitioned, the number of optical relay stages, and the degree of tolerance to misalignment.

The physical paths in GWOI can be constructed by integrating either fibers or waveguides into the planar substrate. The GWOI can provide small form factor and robust systems, especially in the intra-chip configurations. In particular, techniques that are based on the CMOS platform can provide fully integrated approaches, which include optical sources and detectors along with electronics components. The major drawbacks of fiber-based waveguides at the chip-to-chip and board-to-board levels are:

1. Lack of bending or looping flexibility,
2. Crosstalk between adjacent waveguides or at crossing points, and
3. Severe coupling losses at the different interfaces.

WDM approaches can be used to increase the interconnection density. Densely-connected networks that require many interconnection lines have a serious problem with bending and looping fibers, due to radiation losses induced by bending. Approaches that are based on photonic crystals [196, 197] as well as total internal reflection from waveguide to free space interfaces are being investigated and have shown great potential.

On the other hand, FSOI exploit free space (or bulk optic) paths for the signal propagation. Lenses, mirrors, prisms, DOE are used to route the signals. Free space interconnects have proven very suitable for chip-to-chip and board-level interconnects because they:

1. have the potential to fully utilize the large spatial bandwidth and parallelism of optics,
2. have no mutual interference (light beams can cross in space without affecting each other),
3. result in flexible network configurations because the interconnects are not confined to physical paths, and,
4. can use simple imaging optics, which results in light-power efficiency, very low signal skew, and simpler architectures, and permits chip-like integration of optoelectronic circuits.

The main disadvantages include increased alignment and tolerance accuracy, as well as more stringent optomechanical mounts.

In the near future, some of the applications can be satisfied with FSOI or hybrid approaches where some optoelectronic conversion is performed. Note that FSOI techniques can be integrated along with GWOI and use the advantages of each. However, there seems to be a consensus that in order to achieve low-cost integrated approaches the associated components are required to be CMOS compatible, a technology that is very mature and can provide high-volume manufacturing that can keep the price low.

6.9 Optical Interconnect Applications

In the following sections, we will describe the optical interconnects from the interconnection distance and application point of view. We will group them in the following classifications:

1. Data centers
2. Rack-to-rack
3. Shelf-to-shelf and board-to-board
4. Inter-chip and intra-chip.

6.9.1 Data Centers

In data centers and high-performance computing, energy efficiency, low cost, ability to integrate with electronics, price/performance ratio, and reliability are key parameters in the design, operation, and installation [198]. Energy efficiency has at least three aspects:

a. energy consumption in transmitter and multiprocessor,
b. heat generation and dissipation in transmitter, processor, transmission medium, and
c. air conditioning and air circulation in the environment.

Cost associated with energy consumption and cooling is one of the largest operating costs for modern data centers. It is expected that under the current trend of increase in the processed capacity in data centers, soon we will be unable to provide adequate electrical power and heat dissipation to operate the data centers. The lower power requirements of optical fiber networks give them a big advantage over copper when it comes to operating costs, provided that EO and OE conversions are avoided. A 10GBASE-SR optical transceiver uses less than 1 Watt compared to about 6 Watt used by a 10GBASE-T transceiver in a copper system. Note that a rule of thumb is that for each Watt of electrical power being consumed, it will take another Watt for removing the heat being generated and dissipated into the room. Thus, if we consider a data center with thousands of transceivers, the power savings become considerable. This has a great impact not only on the economics of the operation and maintenance of the data center but also on the environment because of the reduced energy consumption requirements.

In addition, optical cables are more lightweight and occupy less space than electrical cables (e.g., a duplex 2.0 mm fiber jumper occupies about 1/8 of the space of a single CAT 6A cable). This makes the installation of fiber cables in a data center an easier task. Note also that since optical cables take less space, cable congestion can be avoided allowing for more free space among racks, and shelves, and hence better airflow and cooling efficiency. Data center interconnections can be accomplished either through SM or MM glass fiber. Approaches that incorporate plastic optical fibers are also considered, and have been demonstrated with high bandwidth-distance products of above 1 GHz km and low loss in the 850 nm and 1300 nm regions. MM fiber offers lower total system cost by allowing usage of low cost VCSELs and high tolerance to alignment and connection offset.

The above approaches are simple transmitter–receiver links. The use of large 3D MEMS optical switches in optical interconnect systems can provide additional advantages to the setup, provisioning, and operation of the data center. For example, automated and remote test and measurement can be performed by personnel physically located at a remote site, with access to the test and measurement instrumentation for data acquisition and system configuration. At the same time, detailed and automated records for the health of the system, possible reconfigurations, or new connections can be logged in a database. Automated and remote testing eliminates fiber tip cleaning requirements for operational time savings and allows for faster setup and provisioning of the end customer [199]. In addition, advanced fiber management approaches that incorporate optical switching can provide monitoring ports and test equipment connected to the interconnection network, through the switch, which can be cycled through the switch connections to determine potential issues compared to the original performance data. Large-scale 3D MEMS optical switches are the ideal candidates for such applications due to the large number of fibers involved in a data center environment. Automation of reconfiguring patch panel connections or routing signals to test and measurement instrumentation without the need for fiber tip cleaning and manual connections can significantly reduce the service time. Flexible fiber management with test and monitoring capabilities also enables proactive monitoring in which all fibers can be continuously polled and compared to historical data in order to identify deterioration of system performance. Eliminating manual fiber handling means that the fiber path integrity is not jeopardized by the introduction of contamination during manual handling. Remotely located personnel can have an instantaneous view of all fiber paths and a real-time inventory of all fiber paths and path changes by means of a real-time software control for multiple data centers [199].

6.9.2 Rack-to-rack

Within a data center, interconnection of racks can be accomplished either using electrical or optical cables. Similar advantages as in the case of data center interconnection can be accomplished using optical fiber cables. Active optical cables, which are assemblies of tranceivers pigtails with fiber cables, can interconnect racks with each other. Two main types of active optical cables are used within data centers today, the first with a form factor known as QSFP [200, 201] which has four parallel lanes (at 10 Gbps each) in each direction, and the second with a form factor known as CXP [202] which has 12 parallel lanes (at 10 Gbps/ 12.5 Gbps each [203]) in each direction thus achieving an aggregate bandwidth of up to 125 Gbit/s in full-duplex applications.

An alternative rack-to-rack interconnection, with higher fiber count, can be implemented with transceivers on both cards and interconnection through optical fibers on OptiFlex circuit technology [204, 205]. The concept of the OptiFlex technology was invented and developed to form arbitrarily complex fiber-optic fabrics capable of transporting optical signals between termination points in optical network equipment with almost negligible loss [204]. Through manufacturing automation this technology can lead to cost-effective manufacturing and assembly efficiency similar to the ones in the semiconductor and PCB industry. The OptiFlex technology can be used to lay multiple fibers on a piece of plastic laminate sheet, providing low-loss distribution, compact, and flexible overlays. The process is very accurately controlled, via automated equipment, and is used to design, construct, and terminate a flexible optical fiber fabric that is used as a board, backplane, or other interconnection assembly [204–206]. The technology uses sequential layering of planar processes to allow the application of successive operations to build a circuit of optical fibers along the vertical axis [207]. Hence, it provides sets of fibers with accurate control on their length. Figure 6.36 shows a typical OptiFlex circuit previously used for optical backplanes. Note that the fibers in the Optiflex circuit can be terminated with appropriate transceivers (e.g., CXP, QSFP). Potential deleterious effects such as crossover that can cause microbending losses are minimized in the Optiflex circuits by the use of special laminating techniques used to distribute the applied pressure on the thermoplastic encapsulant without concentration at the crossovers [207]. Optical loss tests have shown no significant attenuation due to microbends [207]. Note also that new fibers

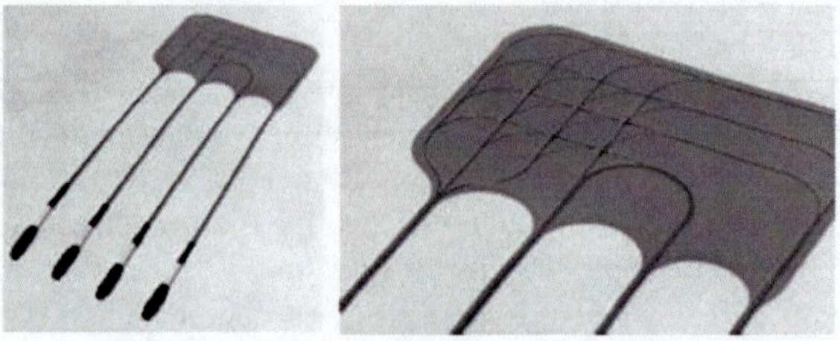

Fig. 6.36 The OptiFlex technology previously used for optical backplanes in optical networks

developed for fiber-to-the-home (FTTH) access networks, designed for sharper bends with low loss, can be used to accomplish more compact designs [208].

A method to increase bandwidth in rack-to-rack applications without the use of additional fiber paths is to use WDM [209]. Implementing this, however, in addition to requiring extra transceivers, adds the cost of multiplexers and demultiplexers. System advantages will come from examining novel ways to exploit the multiple wavelengths in a manner that simultaneously reduces cost and improves system performance. A promising technique is wavelength striping or using the multiple wavelengths in a parallel manner [210, 211]. In this case, the data packet is divided between the wavelengths reducing the time the packet spends traversing the network, thus reducing the overall latency and time-consuming network resources including switch fabrics. The wavelength striping technique is appropriate for a network where the wavelengths travel together from end node to end node, therefore it is more suitable to a short high-speed data link rather than a long-haul telecommunications network.

6.9.3 Shelf-to-shelf and Board-to-board

For optical interconnection from shelf-to-shelf or board-to-board fiber optic approaches similar to the ones used in rack-to-rack interconnections can be used. On the other hand, because of the shorter distances, FSOI as well as GWOI are applicable. Hence, there are two main classifications of board-to-board or shelf-to-shelf optical interconnects; these are the (a) rigid [212] or (b) flexible type [213]. The classification is based on the flexibility of the backplane. Figure 6.37 shows the basic approach of the optical backplane. Two cards are interconnected via the optical backplane. One connector on each of the cards is required to make the associated connections from the cards to the backplane.

A hybrid FSOI/GWOI for an optical backplane interconnecting different boards is shown in Fig. 6.38. Light from a VCSEL on the top circuit board is collimated by a ball microlens and is directed towards the optical backplane. The input lens focuses the optical beam, which after it is deflected by 90° by a mirror, is coupled into a waveguide (e.g., polymer). At the other end of the waveguide, a second mirror deflects the optical beam towards the output lens, which collimates the light. The collimated light is directed onto a ball microlens which focuses it onto

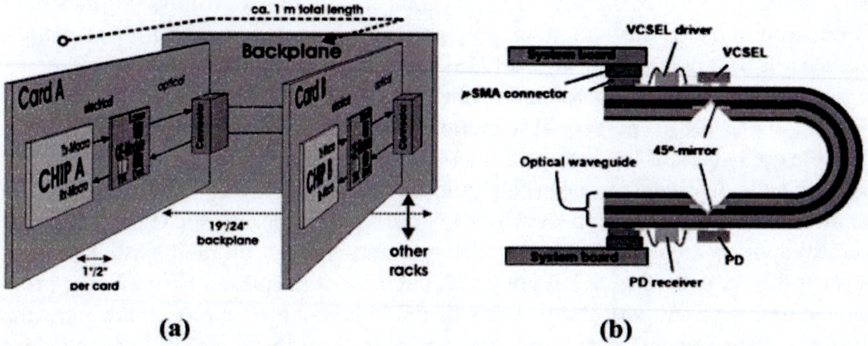

Fig. 6.37 Individual building blocks of a board-to-board technology platform for a wide range of applications are interconnected via **a** a rigid optical interconnect backplane (From [212], Fig. 1. Reproduced by permission of ©2010 IBM Research GmbH. Used by permission of Research GmbH.), **b** a flexible optical interconnect backplane (From [213], Fig. 5. Reproduced by permission of © 2009 The Institute of Electrical and Electronics Engineers)

Fig. 6.38 In the DaimlerChrysler's optical backplane, the beam from a laser diode passes through one set of lenses and reflects off a micromirror before reaching a polymer waveguide, then does the inverse before arriving at a photodiode and changing back into an electrical signal. A prototype operates at 1 Gb/s (From [217], page 34. Reproduced by permission of © 2002 The Institute of Electrical and Electronics Engineers)

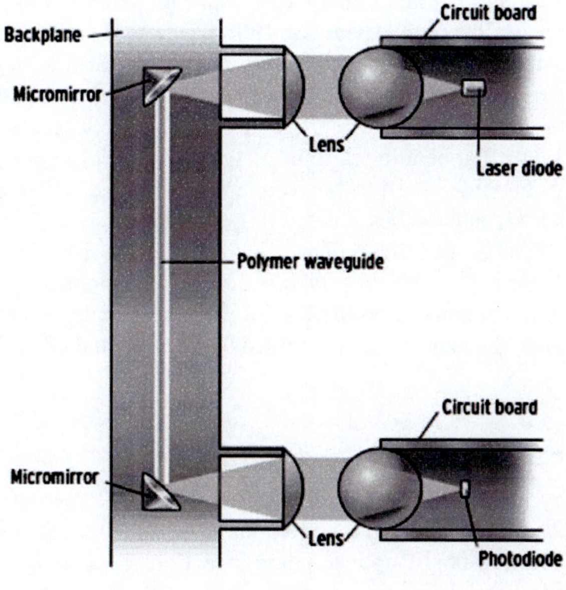

the photodiode of the second board [214]. Critical issues in this system are the low-loss polymer waveguides which can be fabricated from foil layers on arbitrary substrates (e.g., aluminum) by the hot embossing technique [215] or by laser writing so that the refractive index of the waveguides is increased in comparison to the surrounding material. For large-core multimode waveguides, e.g., 250 μm × 200 μm, with integrated mirrors and lenses, a high alignment tolerance in excess of ±500 μm for 1 dB loss has been achieved, which helps to reduce cost and increase system reliability [214]. A total of 10 Gbps transmission rates have been accomplished in 1 m long polymer waveguides and VCSELs as optical sources in backplane applications [216].

An optical flexible board interconnection subassembly was demonstrated by Hwang et al. [218]. The optical interconnect consists of an 850-nm VCSEL and a flexible optical board laminated with an upper polyimide layer, a multimode 1×8 splitter optical waveguide layer, and a lower polyimide layer. The light vertically emitted from a VCSEL is reflected by a 45°-mirror and then coupled to the input of an optical splitter waveguide. The VCSEL was flip-chipped on the board. The input light is then equally divided in power, through a 1×8 splitter, before being propagated to the output port. Parts of this flexible interconnect approach are shown in Fig. 6.39. This type of interconnection which is developed on a splitter-embedded flexible optical board has two main advantages. Unlike the case of interconnection based on silica-based optical waveguides, this approach provides bendable assemblies that can be easily applied to uneven structures. Moreover, it can improve integration and density since the optical and electrical components can be mounted on both sides of the flexible optical board; or, they can be fully embedded. Thus, the board-to-board optical interconnect not only interconnects two boards, but it can add an additional functionality, e.g., splitting, monitoring of signals etc. The twist loss of the flexible optical board on 180° twist was measured to be within 1.5 dB and a quasi-symmetric shape was observed. The edge optical waveguides are more mechanically stressed than the central optical waveguide. On the other hand, bending loss, when the flexible optical board was 180° folded, was measured at 0.13, 0.36, 0.7, and 1.38 dB at bending radii of 5, 4, 3, and 2 mm, respectively.

Flexible board-to-board interconnects (e.g., fiber-based Optiflex or waveguide-based) approaches are gaining popularity since this flexibility allows for easier assembly and interconnection between boards. In a rigid approach, the board position must be controlled and small deviations from the desired position may lead to misalignment problems. On the other hand, with the flexible designs any board misalignment can be addressed by the potential of the board to bend, twist, and flex.

6.9.4 Inter-chip and Intra-chip

As we described in the previous sections, both FSOI and GWOI can be applied for board-to-board optical interconnection. Each one has its advantages and disadvantages. In an FSOI, free space components are used to route optical signals generated by VCSELs from an optics board/chip to another optics board/chip with photodetectors as shown in Fig. 6.40. As we move towards shorter distances between chips on a board, GWOI approaches become more competitive compared to FSOI. This is because on one hand the FSOI techniques require more real estate to accommodate mounts and fixtures for the optical components; on the other hand propagation losses within the board waveguides are not as prohibiting as for longer distances.

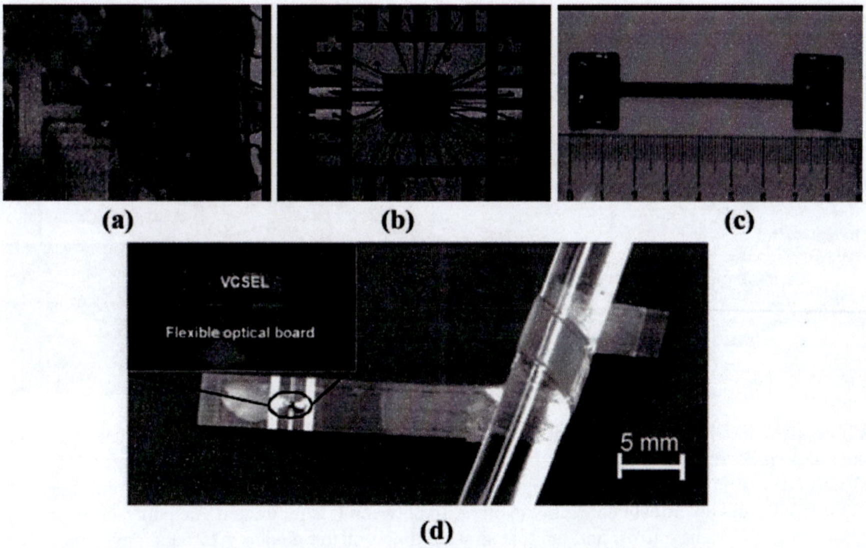

Fig. 6.39 Flexible optical interconnect board demonstrations **a** Flip-chip-bonded VCSEL on waveguide facet, **b** wire-bonded driver chip and **c** fabricated flexible interconnection module. (All a,b, and c are from [213], Fig. 6. Reproduced by permission of © 2010 The Institute of Electrical and Electronics Engineers.) **d** flexible optical interconnect can be used in tight spaces with tight bending radii (From [218], Fig. 4a. Reproduced by permission of © 2010 The Institute of Electrical and Electronics Engineers)

Different approaches for high-density on-board optical interconnects have been proposed and implemented [220–226]. Two-dimensional (2D) arrays with up to 540 optical transmitter and receiver elements have been demonstrated [227], high-speed driver and receiver circuits with low-power consumption have been designed [228], low-loss polymer materials with optical waveguides have been developed [229], and schemes that allow optical coupling between optoelectronic modules and waveguides on backplanes compatible with manufacturing processes are being pursued [230]. However, the challenge is to fulfill all these requirements together, and to develop simple packaging processes that permit the dense integration of high-speed components.

An approach often followed is the interconnection of chips, or opto-chips as often referred, by polymer waveguides integrated in printed circuit boards [231]. The IBM group has developed the Terabus technology designed to support high-density, high-speed, and low-cost data buses for longer lengths than electrical links. The Terabus board-level optical link is a highly integrated packaging approach, as schematically depicted in Fig. 6.41. It is based on a novel parallel optical module, or Optomodule, with sixteen transmitter and receiver channels. The Optomodule is a low-cost, low-profile module that can be directly surface-mounted to a circuit board using a conventional ball grid array (BGA) solder process. The printed circuit board incorporates a dense array of optical waveguides

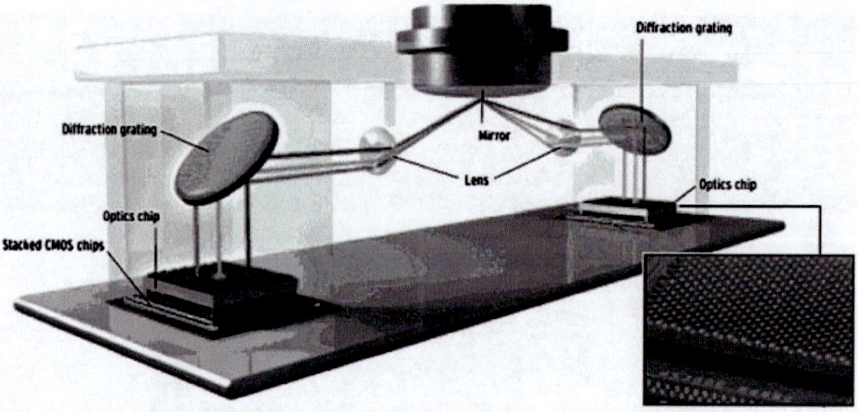

Fig. 6.40 An experimental module from the University of California, San Diego, just 2 cm high, connects stacks of CMOS chips. Each stack is topped with an optics chip consisting of 256 lasers (VCSELs) and photodiodes [219]. Light from the VCSELs makes a vertical exit from one stack [below, left] and a vertical entry into the other. In between it is redirected via a diffraction grating, lenses, an alignment mirror, and another grating. Each of the device's 256 channels operates at 1 Gb/s (From [217], Figure on page 35. Reproduced by permission of © 2002 The Institute of Electrical and Electronics Engineers)

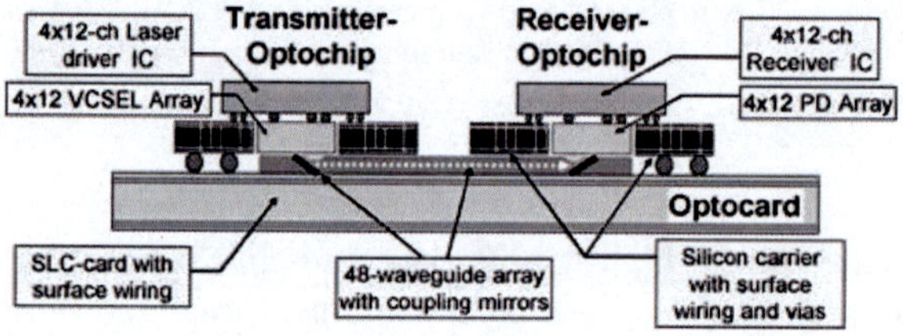

Fig. 6.41 Schematic view of the Terabus package composed of an Optocard with optical waveguides and transmitter and receiver Optochips (From [232], Fig. 1. Reproduced by permission of © 2006 The Institute of Electrical and Electronics Engineers)

with turning mirrors and lens arrays to form the Optocard. The full link corresponds to two Optomodules interconnected through the waveguides on the Optocard [232].

The Terabus technology offers the following advantages [232]:

1. high-speed, low-power CMOS analog amplifier circuits
2. efficient, high-speed 985-nm VCSEL and photodiode (PD) arrays
3. collimating lenses integrated into the backside of the substrate, illuminated/emitting VCSELs, and PDs
4. exclusive use of flip-chip packaging to minimize both the module footprint and associated packaging parasitics.

The Terabus uses 0.13-μm CMOS technology. The transceiver chip consists of 16 independent laser-diode driver (LDD) circuits and 16 receiver amplifier (Rx)

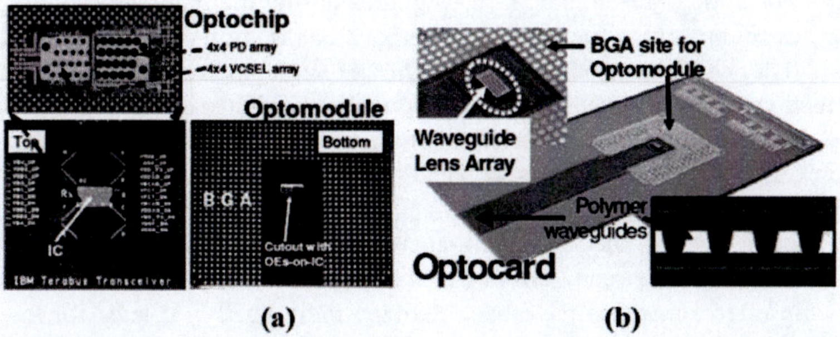

Fig. 6.42 **a** The Optochip (upper) consists of a CMOS IC with flip-chip attached VCSEL and PD arrays. The Optochip is then flip-chip attached to the organic carrier forming the Optomodule (lower). **b** Optocard with integrated polymer waveguides. It also incorporates 90° turning mirrors and collimating lens array. Inset section shows dense waveguide array (35 μm cores on 62.5 μm pitch) (From [233], Figs. 2 and 3. Reproduced by permission of © 2008 The Institute of Electrical and Electronics Engineers)

circuits arrayed in separate 4 × 4 blocks in the center of the chip. The OE devices are fabricated on a 250-μm × 350-μm pitch, but include a 62.5 μm offset between rows. The Optochip is flip-chip attached to the high density circuit card to form the Optomodule. The surface-laminar-circuitry (SLC^{TM}) substrate is 35-mm × 35-mm and includes a cutout in the center to accommodate the OE arrays and provide the optical input/output to/from the Optomodule shown in Fig. 6.42a. Transmission lines in the SLC connect the inputs (outputs) of the IC circuits to a surface probe pad that allow all channels to be characterized using high-speed electrical probes. The low-speed bias and control inputs are routed to BGA pads at the bottom of the SLC carrier. The Optocard is shown in Fig. 6.42b.

Recently, chip-to-chip optical interconnects using long-range surface plasmon polariton (LR-SPP) waveguides have been demonstrated [234, 235]. The approach features 2.5-cm-long gold strips embedded in a low-loss polymer cladding. An interconnect system based on the optimized waveguide with a 4-channel array is assembled with the arrayed optoelectronic chips [236]. This demonstration showed the feasibility of 10 Gbps (2.5 Gbps × 4 channels) signal transmission indicating that the LR-SPP waveguide is a potential transmission line for optical interconnection. The propagation loss for this approach is 6 dB/cm, but the demonstration shows that polymer-based Au LR-SPP optical waveguide is a potential transmission line for very short-range direct chip-to-chip optical interconnects.

An alternative approach is the use of silicon photonics. The potential of integration of electrical and optical functionality on the same chip and through the same processing steps are increasingly gaining popularity and silicon photonics can be the technology that will significantly reduce the cost. Intel Corporation recently announced the world's first silicon-based optical data connection with integrated lasers [237]. This photonic silicon link is composed of four optical channels, each running at 12.5 Gbps, which are combined onto a single fiber to transmit data up to 50 Gbps. Four hybrid silicon lasers, based on fusing a layer of

Indium Phosphide onto the silicon waveguides during the manufacturing process, are used as sources. Note that still the approach is based on optically multiplexing multiple (e.g., 4) lasers to accomplish the higher aggregate rate. The modulation is performed through integrated silicon modulators, which can have better performance than previously demonstrated direct modulation approaches. A passive multiplexer combines the four signals and their associated data onto a single fiber-optic cable, or waveguide. At the other end of the link, a demultiplexer and set of four photodetectors on the integrated receiver chip collect the light and convert it into an electrical data stream. All of the building blocks for the realization of such an on-chip interconnection have been realized individually (at least for lower bit rates) and/or combined with some of the other components. What remains to be demonstrated is high integration density of multiple of these components in a small form factor. The goal is to increase (a) the number of lasers multiplexed and (b) the bit rate for each modulator. For example, increasing the number of lasers on the chip from 4 to 25 will bring the available data rate to 312.5 Gbps and further increase of the modulation rate of each signal from 2.5 to 40 Gbps and provide 1 Tbps connectivity. This form factor must not increase the size of the chip considerably from today's size. One of the challenges is the difficulty of sharp bends and/or turns of the optical signal through waveguides. However, research is still ongoing and sharp bends of the optical path can be achieved through 45° integrated mirrors, or microring resonators.

Another practical constraint for the silicon photonics approach is the performance at elevated temperatures. The worst-case thermal environment occurs when the optical devices are located directly on the microprocessor. Both CMOS chips as well as photonic elements are sensitive to temperature variations and care is taken to ensure proper temperature control. At the same time, both CMOS chips and optical elements, and in particular lasers produce heat. Hence, integration of these two processes in the same chip brings some additional challenges in the heat dissipation of the entire chip. In addition to surviving at the maximum temperature, an optical device has to operate over a range of temperatures as the device heats and cools under normal operating conditions for a lifetime of the order of 10 years or more. Hence, it may be critical, as we move to full optical integration on the chip, to re-evaluate or redesign the chips based on the new technologies and around their advantages [238] and not to use current approaches. By separating processing, memory, interconnection, and other capabilities, a more efficient chip can be designed. This design flexibility will positively influence energy efficiency, systems performance, and the cost of development and ownership on a broad scale.

6.10 The Next Frontier: 100 Gbps and Beyond

We have already seen the great potential photonic approaches provide to the increase in speed of microprocessors and to overcome the electrical interconnection bottleneck. The race is on for achieving greater than 100 Gbps data rates at the inter-chip and intra-chip level. There have been numerous demonstrations of optical technologies suitable to be integrated on the board and placed close to the host adapter. Several papers are published every year in associated conferences and journals. Several consortia and projects have been actively addressing issues related to optimum device, module, and system architectures [239–242]. It is impossible to cover all these in just a chapter. We will again highlight the work of Intel and its collaborators [237] described in the previous section.

Another notable work is the one where IBM and its collaborators in [243] have developed a parallel optical transceiver module (Optochip) with 24 transmitters (operating at 12.5 Gbps each) plus 24 receivers achieving a total 300 Gbps bidirectional data rate [244]. The transceiver Optochip relies on silicon carrier technology to provide a high level of integration of the electrical and optical components onto a single substrate with high density interconnection. It consists of the Si carrier platform with 4 flip-chip attached components: two 24-channel 850 nm optoelectronic arrays (VCSELs and photodiodes) and two 24-channel CMOS ICs (receivers and laser drivers). 150-μm diameter "optical vias" are incorporated in the Si carrier at 48 locations of the VCSEL and PD array element in order to allow optical transmission through the silicon carrier. Lens arrays aligned to the optical vias can also be integrated into the Si carrier to collimate the optical inputs/outputs and facilitate optical coupling to/from the Optochip. The 24 receiver channels, characterized as full transmitter to receiver Optochip links, operate error-free up to 12.5 Gb/s. At 12.5 Gb/s, each complete link consumes only 11 mW/Gb/s.

NEC has developed a new form factor for 12-channel high-speed transmitters and receivers with 1 μm-range VCSELs to satisfy next-generation systems. All assembly parts to control the transmitters or receivers were integrated on 9-mm × 14-mm alumina substrates, and their optical receptacles were vertical to the motherboard. These structures, with which two-dimensional module arrays could be formed, were suitable for densely aligning modules near the logic LSIs. Error-free transmission was achieved with these structures at a data rate up to 20 Gbps, hence, achieving an aggregate rate of 240 Gbps. The power consumption for each link was less than 14.5 mW/Gb/s including waveform compensation functions [245].

6.11 Conclusions

In conclusion, this chapter gives an overview of optical interconnection systems. First, it defines the need and requirements for optical interconnects. Then, it presents the necessary components and processes as well as the different technologies and architectures involved in the design of optical interconnects. Compared to electrical solutions, optical interconnects can provide design simplification because of the absence of electromagnetic wave phenomena (e.g., impedance matching, electromagnetic interference/crosstalk, and inductance difficulties such as inductive voltage drops on pins and wires), connection distance independent performance, and frequency or bit rate independence. Depending on the type of the optical interconnect (e.g., FSOI or GWOI) 2D, 3D, and WDM architectures can be implemented, which can increase the channel count and throughput. Optical interconnects can also provide reduction of power dissipation and the potential for non-contact, parallel testing of chip operation.

As a first step, electrical and optical components are integrated in a hybrid fashion for board-level and off-board links. Low-port-count or lower-speed interconnects have been implemented. Several challenges need to be solved in order to enable electro-optical processor packages. The bandwidth requirements in computing systems demand the application of high-density optical transceivers with a large channel count. The enormous intra-chip bandwidth requirements and the limited real estate on the chips require the development of ultra-small (e.g., nano) electro-optical and optical components [246], as well as techniques that allow the optical signal to take sharp bends/turns so that folding the optical path can be used and hence result in a reduced chip size. Eventually, optical interconnect systems will be a combination of hybrid and monolithic approaches, thus taking full advantage of III–V-, Ge- and Si-based materials. Integration with CMOS circuits can provide low cost, integrated control, signal processing, and error correction. Simple and low-cost assembly of optical connectors and transceivers, as well as passive alignment schemes based on standard electronic assembly processes applied in today's electrical printed circuit board manufacturing would be instrumental in increasing yield and reducing cost. Yet, there are many challenges, including difficult issues such as packaging, cost, and reliability, as well as the need to move to potentially different microprocessor architectures that take advantage of both electronics and optics worlds.

Until recently, optics and electronics were considered competing technologies, nevertheless, it is becoming more and more acceptable that these two worlds can converge and successfully lead to solutions that are even beyond our imagination at this point in time.

References

1. Goodman JW, Leonberger FJ, Kung SC, Athale RA (1984) Optical interconnections for VLSI systems. Proceedings of the IEEE 72(7):850–866
2. Miller DAB (2000) Optical interconnects to silicon. IEEE J Sel Top Quantum Electron 6(6):1312–1317
3. Miller DAB (2010) Optical interconnects to electronic chips. Appl Opt 49(25):F59–F70
4. Moore GE (1965) Cramming more components onto integrated circuits. Electron Magazine 1(1):4
5. ftp://download.intel.com/museum/Moores_Law/VideoTranscripts/Excepts_A_Conversation_with_Gordon_Moore.pdf
6. Moussavi M (2006) Advanced conventional interconnects: state of the art, future trends, and limitations in optical interconnects. Pavesi L, Guillot G (eds) The silicon approach, Springer, Berlin

7. Chen RT, Choi C (2007) Optical interconnects. A Publication in the morgan & claypool publishers series on synthesis lectures on solid-state material and devices #2, (ed) Banerjee S, ISBN-10:1598290665
8. International technology roadmap for semiconductors 2009 edition—interconnect (2009). Available at http://www.itrs.net/Links/2009ITRS/2009Chapters_2009Tables/2009_Interconnect.pdf
9. Rana N, Bousman K, Shekhawat GS, Sirinakis G, Heuchling F, Welch J, Eisenbraun E, Geer R, Kaloyeros A (2002) Investigation of substrate selective covalent attachment for genetically engineered molecular interconnects. In: Proceedings of materials research society symposium, vol 728. San Francisco, CA
10. Benner AF, Ignatowski M, Kash JA, Kuchta DM, Ritter MB (2005) Exploitation of optical interconnects in future server architectures. IBM J Research and Development 49(4–5): 755–775
11. International technology roadmap for semiconductors (ITRS), assembly and packaging chapter, semiconductor industry association (2007). Available at http://www.public.itrs.net/Links/2009Winter/Presentations/Conference/AP_121609.pdf
12. Kuan-Neng C, Kobrinsky M, Barnett B, Reif R (2004) Comparison of conventional, 3D, optical, and RF interconnects for on-chip clock distribution. IEEE Trans Electron Devices 51(2):233–239
13. Kyung-Hoae K, Hoyeol C, Kapur P, Saraswat KC (2007) Performance comparisons between carbon nanotubes, optical, and Cu for future high-performance on-chip interconnect applications. IEEE Trans Electron Devices 54(12):3206–3215
14. Hoyeol C, Kapur P, Saraswat KC (2004) Power comparison between high-speed electrical and optical interconnects for interchip communication. IEEE/OSA J Light Technol 22(9):2021–2033
15. Cho H, Kapur P, Saraswat KC (2004) Power comparison between high-speed electrical and optical interconnects for inter-chip communication. In: Proceedings of IEEE international interconnect technical conference. San Francisco, CA, pp 116–118
16. Saraswat K, Cho H, Kapur P, Koo K-H (2008) Performance comparison between copper, carbon nanotube, and optical interconnects. In: Proceedings of IEEE international symposium on circuits and systems (ISCAS). Seattle, WA, pp 2781–2784
17. Dolfi DW (2003) Multi-channel optical interconnects for short-reach applications. In: Proceedings of the electronic components and technology conference (ECTC). New Orleans, LA
18. Levi FJ (2000) Optical interconnects in systems. Proc IEEE 88(6):750–757
19. Miller DAB (2005) Opportunities for optics to silicon chips. In: Proceedings of the IEEE lasers and electro-optics society annual meeting (IEEE/LEOS). Sydney, Australia, pp 202–203
20. Madamopoulos N (2010) Optical communications. 32 Centuries in the making. J Stud Res, The Grove School of Engineering, The City College of New York, City University of New York. Available at http://deepspace9.sci.ccny.cuny.edu/JSR/2010issue/pdfs/Innovations_Optical_Nicholas.pdf
21. Uzunoglu N (2008) Theoretical analysis of telecommunication through 'Friktories'. In: Paipetis SA (ed) History of mechanism and machine science, science and technology in Homeric epics. Springer Science and Business Media B.V., New York, pp 223–229
22. Rack to rack interconnection of the Los Alamos national laboratory and and IBM roadrunner first hybrid supercomputer. Available at www.lanl.gov/roadrunner/
23. Nelson BE, Keeler GA, Agarwal D, Helman NC, Miller DAB (2003) Wavelength division multiplexed optical interconnect using short pulses. IEEE J Select Top Quant Electron 9(2):486–491
24. Koyama F (2006) Recent advances of VCSEL photonics. IEEE/OSA J Light Technol 24(12):4502–413
25. Soda H, Iga K, Kitahara C, Suematsu Y (1979) GaInAsP/InP surface emitting injection lasers. Jpn J Appl Phys 18(12):2329–2330

26. Iga K (2000) Surface emitting laser-its birth and generation of new optoelectronics field. IEEE J Quant Electron 6(6):1201–1215
27. Liang D, Bowers JE (2010) Recent progress in lasers on silicon. Nat Photonics 4(8):511–517
28. Rong H, Jones R, Liu A, Cohen O, Hak D, Fang A, Paniccia M (2005) A continuous-wave Raman silicon laser. Nature 433(7027):725–728
29. Rong H, Xu S, Kuo YH, Sih V, Cohen O, Raday O, Paniccia M (2007) Low-threshold continuous-wave Raman silicon laser. Nat Photonics 1(4):232–237
30. Cheng S-L, Lu J, Shambat G, Yu HY, Saraswat K, Vuckovic J, Nishi Y (2009) Room temperature 1.6 μm electroluminescence from Ge light emitting diode on Si substrate. Opt Express 17(12):10019–10024
31. Liu J, Sun X, Camacho-Aguilera R, Kimerling LC, Michel J (2010) Ge-on-Si laser operating at room temperature. Opt Letters 35(5):679–681
32. Van Campenhout J, Rojo Romeo P, Regreny P, Seassal C, Van Thourhout D, Verstuyft S, Di Cioccio L, Fedeli J-M, Lagahe C, Baets R (2007) Electrically pumped InP-based microdisk lasers integrated with a nanophotonic silicon-on-insulator waveguide circuit. Opt Express 15(11):6744–6749
33. Liang D, Fiorentino M, Okumura T, Chang H-H, Spencer DT, Kuo Y-H, Fang A, Dai D, Beausoleil RG, Bowers JE (2009) Electrically-pumped compact hybrid silicon microring lasers for optical interconnects. Opt Express 17(22):20355–20364
34. Onomura A, Arai M, Kondo T, Matsutani A, Miyamoto T, Koyama F (2003) Densely integrated multiple-wavelength vertical-cavity surface-emitting laser array. Jpn J Appl Phys 42(5B):L529–L531
35. Gardner DS, Schrom G, Hazucha P, Paillet F, Karnik T, Borkar S, Saulters J, Owens J, Wetzel J (2006) Integrated on-chip inductors with magnetic films. In: Proceedings of international electron devices meeting (IEDM). San Francisco, CA, pp 1–4
36. Shi J-W, Chen Y-H, Liu T-M, Chan M-C, Gan K-G, Chiu Y-J, Bowers JE, Sun C-K (2002) Metal-semiconductor-metal traveling wave photodetectors. In: Proceedings of IEEE lasers and electro-optics society annual meeting (IEEE/LEOS), vol 2(2). Glasgow, Scotland, pp 445–446
37. Pinguet T et al (2008) Monolithically integrated high-speed CMOS photonic transceivers. In: Proceedings of IEEE international conference on group IV photonics. Sorrento, Italy, pp 362–364
38. Zhuang Y, Rejaei B, Boellaard E, Vroubel M, Burghartz JN (2003) Integrated solenoid inductors with patterned, sputter-deposited Cr/$Fe_{10}Co_{90}$/Cr ferromagnetic cores. IEEE Electron Device Lett 24(4):224–226
39. Huang Z, Kong N, Guo X, Liu M, Duan N, Beck AL, Banerjee SK, Campbell JC (2006) 21-GHz-bandwidth germanium-on silicon photodiode using thin SiGe buffer layers. IEEE J Select Topics Quant Electron 12(6):1450–1454
40. Hijioka K, Tanabe A, Amamiya Y, Hayashi Y, (2008) Crosstalk analysis method of 3D solenoid on-chip inductors for high-speed CMOS SoCs. In: Proceedings of international interconnect technology conference (IITC). Burlingame, CA, pp 186–188
41. Ohguro T, Sato N, Matsuo M, Kojima K, Momose HS, Ishimaru K, Ishiuchi H, (2004) Digest VLSI technology symposium. Honolulu, Hawaii, pp 220–221
42. Guo LH, Zhang QX, Lo GQ, Balasubramanian N, Kwong D-L (2005) High performance inductors on plastic substrate. IEEE Electron Dev Lett 26(9):619–621
43. Dahn, Hong C, Liu J, Giziewicz W, Beals M, Kimerling LC, Michel J, Chen J, Kärtner FX (2007) High performance, waveguide integrated Ge photodetectors. Opt Express 15(7):3916–3921
44. Yin T, Cohen R, Morse MM, Sarid G, Chetrit Y, Rubin D, Paniccia MJ (2007) 31 GHz Ge n-i-p waveguide photodetectors on silicon-on-insulator substrate. Opt Express 15(21): 13965–13971

45. Gupta A, Levitan SP, Selavo L, Chiarulli DM (2004) High-speed optoelectronics receivers in SiGe. In: Proceedings of international conference on VLSI design. Mumbai, India, pp 957–960
46. Kang Y, Morse M, Paniccia MJ, Zadka M, Saad Y, Sarid G, Pauchard A, Zaoui WS, Chen H-W, Daoxin D, Bowers JE, Liu H-D, Mcintosh DC, Zheng X, Campbell JC (2009) Monolithic Ge/Si avalanche photodiodes. In: Proceedings of IEEE international conference on group IV photonics, (GFP). San Francisco, CA, pp 25–27
47. Heck MJR, Chen H-W, Fang AW, Koch BR, Liang D, Park H, Sysak MN, Bowers JE (2010) Hybrid silicon photonics for optical interconnects. IEEE J Sel Topics Quant Electron 17(2):333–346
48. Okada K, Sugawara H, Ito H, Itoi K, Sato M, Abe H, Ito T, Masu K (2006) On-chip high-Q variable inductor using wafer-level chip-scale package technology. IEEE Trans on Electron Devices 53(9):2401–2406
49. Xu Q, Schmidt B, Pradhan S, Lipson M (2005) Micrometre-scale silicon electrooptic modulator. Nature 435(7040):325–327
50. Dong P, Liao S, Feng D, Liang H, Shafiiha R, Feng N-N, Li G, Zheng X, Krishnamoorthy AV, Asghari M (2010) Tunable high speed silicon microring modulator. In: Proceedings of IEEE/OSA conference lasers and electro-optics (CLEO)/quant electronics and laser science conference (QELS). San Jose, CA, pp 1–2
51. Fukano H, Yamanaka T, Tamura M, Kondo Y (2006) Very-low-driving-voltage electroabsorption modulators operating at 40 Gb/s. IEEE/OSA J Light Technol 24(5):2219–2224
52. Kuo Y-H, Chen H-W, Bowers JE (2008) High speed hybrid silicon evanescent electro-absorption modulator. Opt Express 16(13):9936–9941
53. Akiyama S, Itoh H, Takeuchi T, Kuramata A, Yamamoto T (2005) Wide-wavelength-band (30 nm) 10-Gb/s operation of InP-based Mach-Zehnder modulator with constant driving voltage of 2 Vpp. IEEE Photonics Technol Lett 17(7):1408–1410
54. Sysak MN, Anthes JO, Bowers JE, Raday O, Jones R (2008) Integration of hybrid silicon lasers and electroabsorption modulators. Opt Express 16(17):12478–12486
55. Chen H-W, Kuo Y-H, Bowers JE (2009) High speed silicon modulators. In: Proceedings of opto electronics and communications conference (OECC). Hong Kong, China, pp 1–2
56. Liu J, Beals M, Pomerene A, Bernardis S, Sun R, Jing Cheng, Kimerling LC, Michel J (2008) Waveguide-integrated, ultralow-energy GeSi electro-absorption modulators. Nat Photonics 2(7):433–437
57. Roth JE, Fidaner O, Schaevitz RK, Kuo Y-H, Kamins TI, Harris JS, Miller DAB (2007) Optical modulator on silicon employing germanium quantum wells. Opt Express 15(9):5851–5859
58. Kuo Y-H, Chen H-W, Bowers JE (2008) High speed hybrid silicon evanescent electroabsorption modulator. Opt Express 16(13):9936
59. Liu S, Liao L, Rubin D, Nguyen H, Ciftcioglu B, Chetrit Y, Izhaky N, Paniccia M (2007) High-speed optical modulation based on carrier depletion in a silicon waveguide. Opt Express 15(2):660–668
60. Marris-Morini D, Vivien L, Fédéli JM, Cassan E, Lyan P, Laval S (2008) Low loss and high speed silicon optical modulator based on a lateral carrier depletion structure. Opt Express 16(1):334–339
61. Pai CS, Bude MK, Frei M, Rogers SN, Jacobson DC, Merchant SM, Hui F, Liu R, Gregor RW (2001) Integrated W-silicide metal resistor for advanced CNOS technologies. In: Proceedings of the international interconnect technology conference (IITC). San Francisco, CA, pp 216–218
62. www.tycoelectronics.com
63. http://www.usconec.com
64. Turner EH, Stolen RH (1981) Fiber Faraday circulator or isolator. Opt Lett 6(7):322
65. Schulz PA (1989) Wavelength independent Faraday isolator. Appl Optics 28(20):4458
66. Microlens array parts 1: vocabulary. ISO 14880-1, 1, 2001

67. Wada O (1984) Ion-beam etching on InP and its applications to the fabrication of high radiance InGaAsP/InP light emitting diodes. J Electrochem Soc 131(10):2373–2380
68. Popovic ZD, Sprague RA, Conell GAN (1988) Technique for the monolithic fabrication of microlens arrays. Appl Optics 27(7):1281–1284
69. Hutley MC (1990) Optical techniques for the generation of microlens arrays. J Mod Optics 37(2):253–265
70. Daly D, Stevens RF, Hutley MC, Davies N (1990) The manufacture of microlenses by melting photoresist. Meas Sci Technol 1(8):759–766
71. Sinzinger S, Jahns J (2003) Microoptics, 2nd edn. Wiley, New York
72. Miyashita T (2005) The Microlens. In: Kawai S (ed) Handbook of optical interconnects. CRC-Taylor and Francis, Boca Raton, FL
73. Tomlinson WJ (1980) Application of GRIN-rod lenses in optical fiber communication systems. Appl Optics 19(7):1127–1138
74. Palais C (1980) Fiber coupling using graded-index rod lenses. Appl Optics 19(12): 2011–2018
75. Gilsdorf RW, Palais JC (1994) Single-mode fiber coupling efficiency with graded-index rod lenses. Appl Optics 33(16):3440–3445
76. Madamopoulos N, Riza NA (2000) Demonstration of an all-digital 7-bit 33-channel photonic delay line for phased array radars. Appl Optics 39(23):4168–4181
77. Ono Y (2003) Recent advances in diffractive optical elements: from diffractive optics towards periodic structure optics. Jpn J Optics 32(8):468–475
78. Kodate K (1997) Handbook of optical computing. Asakura, Tokyo
79. Keyworth BP, Corazza DJ, McMullin JN, Mabbott L (1997) Single-step fabrication of refractive microlens arrays. Appl Opt 36(10):2198–2202
80. http://www.gofoton.com/products/materials/micro_lenses.html
81. Chen C, Li M, Chang C, Chi G, Chang J, Cheng W, Yeh J, Wu C (2002) GaN diffractive microlens fabricated with gray-level mask. In: Magnusson R (ed) Diffractive optics and micro-optics, vol 75 of OSA trends in optics and photonics series. Optical Society of America, paper DMB3, USA
82. Zheng X, Kaman V, Yuan S, Xu Y, Jerphagnon O, Keating A, Anderson RC, Poulsen HN, Liu B, Sechrist JR, Pusarla C, Helkey R, Blumenthal DJ, Bowers JE (2003) Three-dimensional MEMS photonic cross-connect switch design and performance. J Sel Top Quant Electron 9(2):571–578
83. Bowers J (2007) Large scale micro-optic switches. In: Proceedings of the microoptics conference (MOC), Takamatsu, Japan
84. Wu MC, Solgaard O, Ford JE (2006) Optical MEMS for lightwave communication. IEEE/OSA J Light Technol 24(12):4433–4454
85. De Dobbelaere P, Falta K, Fan L, Gloeckner S, Patra S (2002) Digital MEMS for optical switching. IEEE Comm Mag 40(3):88–95
86. Helkey R, Adams S, Bowers J, Davis T, Jerphagnon O, Kaman V, Keating A, Liu B, Pusarla C, Xu D, Yuan S, Zheng X (2002) Design of large, MEMS-based photonic switches. Opt Photonics News 13(5):43–45
87. Wu MC (1997) Micromachining for optical and optoelectronic systems. Proc IEEE 85(11):1833–1856
88. Savastano L, Maier G, Pattavina A, Martinelli M (2005) Physical-parameter design in 2D MEMS optical switches. IEEE/OSA J Light Technol 23(10):3147–3155
89. Fan L, Gloeckner S, Dobblelaere PD, Patra S, Reiley D, King C, Yeh T, Gritters J, Gutierrez S, Loke Y, Harburn M, Chen R, Kruglick E, Wu M, Husain A (2002) Digital MEMS switch for planar photonic crossconnects. In: Proceedings of the IEEE/OSA optical fiber communication conference (OFC), paper TuO4, Anaheim, CA, pp 93–94
90. Dobbelaere PM, Gloeckner S, Patra SK, Fan L, King C, Falta K (2003) Design, manufacture and reliability of 2D MEMS optical switches. Proc SPIE Int Soc Opt Eng 4945(1):39–45

91. Hornbeck LJ (1996) Digital light processing and MEMS: reflecting the digital display needs of the networked society. In: Parriaux OM (ed) Proceedings of SPIE micro-optical technologies for measurement, sensors and microsystems, vol 2783, pp 2–13
92. Ford J, Aksyuk VA, Bishop DJ, Walker JA (1999) Wavelength add-drop switching using tilting micro-mirrors. IEEE/OSA J Light Technol 17(5):904–911
93. Bishop D, Giles CR, Austin G (2002) The Lucent lamda router: MEMS technology of the future here today. IEEE Comm Mag 40(3):75–79
94. Kim J, Nuzman CJ, Kumar B, Lieuwen DF, Kraus JS, Weiss A, Lichtenwalner CP, Papazian AR, Frahm RE, Basavanhally NR, Ramsey DA, Aksyuk VA, Pardo F, Simon ME, Lifton V, Chan HB, Haueis M, Gasparyan A, Shea HR, Arney S, Bolle CA, Kolodner PR, Ryf R, Neilson DT, Gates JV (2003) 1,100 × 1,100 port MEMS-based optical crossconnect with 4-dB maximum loss. IEEE Photonics Technol Lett 15(11):1537–1539
95. Bowers J (2009) Low power 3D MEMS optical switches. Optical MEMS and nanophotonics. In: Proceedings of IEEE lasers and electro-optics society annual meeting (IEEE/LEOS). Belek-Antalya, Turkey, pp 152–153
96. Savastano L, Maier G, Pattavina A, Martinelli M (2005) Physical-parameter design in 2D MEMS optical switches. IEEE/OSA J Light Technol 23(10):3147–3155
97. Yano M, Yamagishi F, Tsuda T (2005) Optical MEMS for photonic switching—compact and stable optical crossconnect switches for simple, fast, and flexible wavelength applications in recent photonic networks. IEEE J Sel Top Quant Electron 11(2):383–394
98. Noguchi K, Sakano T, Matsumoto T (1991) A rearrangeable multichannel free-space optical switch based on multistage network configuration. IEEE/OSA J Light Technol 9(12):1726–1732
99. Yamamoto T, Yamaguchi M, Hirabayashi K, Matsuo S, Amano C, Iwamura H, Kohama Y, Kurokawa T (1996) High-density digital free space photonic switches using micro-beam optical interconnections. IEEE Photonics Technol Lett 8(3):358–360
100. Chen H-W, Bowers JE (2010) Low-power, fast hybrid silicon switches for high-capacity optical networks. In: Proceedings of the photonics in switching (PS), paper PMC3. Monterey, CA
101. Chiao C, Dugan A (2001) The optical switching spectrum: a primer on wavelength switching technologies. Telecommunication. Available at http://www.telecoms-mag.com/issues/200105/tcs/theSkinny.html
102. Cheng DK, Liu Y, Sonek GJ (1995) Optical switch based on thermally-activated dye-doped biomolecular thin films. IEEE Photonics Technol Lett 7(4):366–369
103. Hardy S (1999) Liquid-crystal technology vies for switching applications. IEEE/OSA J Light Technol 16(13):44–46
104. Ware A (2000) New photonic-switching technology for all-optical networks. IEEE/OSA J Light Technol 17(3):92–98
105. Smith DA, Alessandro A, Baran JE, Fritz DJ, Jackel JL, Chakravarthy RS (1996) Multiwavelength performance of an apodized acousto-optic switch. IEEE/OSA J Light Technol 14(9):2044–2051
106. Chen H-W et al (2010) 25 Gb/s hybrid silicon switch using a capacitively loaded traveling wave electrode. Opt Express 18(2):1070–1075
107. Shin J, Ozturk C, Sakamoto S, Chiu YJ, Dagli N (2005) Novel T-rail electrodes for substrate removed low-voltage high-speed GaAs/AlGaAs electrooptic modulators. IEEE Trans Microwave Theory Tech 53(2):636–643
108. Akiyama S, Itoh H, Takeuchi T, Kuramata A, Yamamoto T (2005) Wide-wavelength-band (30 nm) 10 Gb/s operation of InP-based Mach-Zehnder modulator with constant driving voltage of 2 Vpp. IEEE Photonics Technol Lett 17(7):1408–1410
109. Little BE, Haus HA, Foresi JS, Kimerling LC, Ippen EP, Ripin RJ (1998) Wavelength switching and routing using absorption and resonance. IEEE Photonic Technol Lett 10(6):816–818
110. Vantrease D, Schreiber R, Monchiero M, McLaren M, Jouppi NP, Fiorentino M, Davis A, Binkert N, Beausoleil RG, Ahn JH (2008) Corona: system implications of emerging

nanophotonic technology. In: Proceedings of international symposium on computational architecture. Beijing, China, pp 153–164
111. Lee BG, Biberman A, Sherwood-Droz N, Poitras CB, Lipson M, Bergman K (2009) High-speed 2 × 2 switch for multi-wavelength silicon photonic networks on-chip. IEEE/OSA J Light Technol 27(14):2900–2907
112. Chan J, Hendry G, Biberman A, Bergman K (2010) Architectural exploration of chip-scale photonic interconnection network designs using physical-layer analysis. IEEE/OSA J Light Technol 28(9):1305–1315
113. Miller DAB (2009) Device requirements for optical interconnects to silicon chips. Proc IEEE 97(7):1166–1185
114. Ahn J, Fiorentino M, Beausoleil RG, Binkert N, Davis A, Fattal D, Jouppi NP, McLaren M, Santori CM, Schreiber RS, Spillane SM, Vantrease D, Xu Q (2009) Devices and architectures for photonic chipscale integration. Appl Phys 95(4):989–997
115. Dangel R, Beyeler R, Meier N, Lamprecht T, Horst F, Jubin D, Weiss J, Offrein BJ (2009) Optical interconnects for board level applications. In: Proceedings of the SPIE, 7219:721904-1–721904-7
116. Keyes RW (1970) Power dissipation in information processing. Science 168(3933):796–801
117. Gibbs HM, McCall SL, Venkatesan TNC, Gossard AC, Passner A, Wiegmann W (1979) Optical bistability in semiconductors. Appl Phys Lett 35(6):451–453
118. Miller DAB, Smith SD, Johnston A (1979) Optical bistability and signal amplification in a semiconductor crystal: application of new low power nonlinear effects in InSb. Appl Phys Lett 35(9):658–660
119. Smith PW (1982) On the physical limits of digital optical switching and logic elements. Bell Syst Tech J 61(8):1975–1993
120. Jenkins BK, Sawchuk AA, Strand TC, Forchheimer R, Soffer BH (1984) Sequential optical logic implementation. Appl Optics 23(19):3455–3464
121. Ranalli AR, Scott BA, Kondis JP (1999) Liquid crystal based wavelength selectable cross-connect. In: Proceedings European conference on optical communications (ECOC). Nice, France
122. Goodman JW, Leonberger FJ, Kung SC, Athale RA (1984) Optical interconnections for VLSI systems. Proc IEEE 72(7):850–866
123. Thienpont H, Debaes C, Baukens V, Ottevaere H, Vynck P, Tuteleers P, Verschaffelt G, Volckaerts B, Hermanne A, Hanney M (2000) Plastic microoptical interconnection modules for parallel free-space inter- and intra- MCM data communication. Proc IEEE 88(6):769–779
124. Kirk AG, Plant DV, Ayliffe MH, Châteauneuf M, Lacroix F (2003) Design rules for highly parallel free-space optical interconnects. IEEE J Sel Topics Quant Electron 9(2):531–547
125. Gibbs HM, McCall SL, Venkatesan TNC, Gossard AC, Passner A, Wiegmann W (1979) Optical bistability in semiconductors. Appl Phys Lett 35(6):451–453
126. Fisher AD, Lee JM (1986) The current status of two dimensional spatial light modulator technology. In: Proceedings SPIE meeting on optical and hybrid computing. Leesburg, VA, pp 352–371
127. Efron U, Braatz PO, Little MJ, Schwartz RN, Grinberg J (1983) Silicon liquid crystal light valve: status and issues. Opt Eng 22(6):682–686
128. Hornbeck LJ, Nelson WE (1988) Bistable deformable mirror device. In: Proceedings of OSA topical meeting on spatial light modulators and applications, paper ThB2. Washington, DC
129. Ross WE, Psaltis D, Anderson RH (1983) Two-dimensional magneto-optic spatial light modulator for signal processing. Opt Eng 22(4):485–490
130. Collings N, Crossland WA, Ayliffe PJ, Vass DG, Underwood I (1989) Evolutionary development of advanced liquid crystal spatial light modulators. Appl Optics 28(22):4740–4747
131. Streibl N, Brenner K-H, Huang A, Jahns J, Jewell J, Lohmann AW, Miller DAB, Murdocca M, Prise ME, Sizer T (1989) Digital optics. Proc IEEE 77(12):1954–1969

132. Jahns J, Huang A (1989) Planar integration of free—space optical components. Appl Optics 28(9):1602–1605
133. Prise ME, Craft NC, Downs MM, LaMarche RE, D'Asaro LA, Chirovsky LMF, Murdocca MJ (1991) Optical digital processor using arrays of symmetric self-electrooptic effect devices. Appl Optics 30(17):2287–2296
134. Craig RGA, Wherrett BS, Walker AC, Tooley FAP, Smith SD (1991) Optical cellular logic image processor: implementation and programming of a single channel digital optical circuit. Appl Optics 30(17):2297–2308
135. McCormick FB, Cloonan TJ, Tooley FAP, Lentine AL, Sasian JM, Brubaker JL, Morrison RL, Walker SL, Crisci RJ, Novotny RA, Hinterlong SJ, Hinton HS, Kerbis E (1993) Six-stage digital free-space optical switching network using symmetric self-electro-optic-effect devices. Appl Optics 32(26):5153–5171
136. Yamamoto T, Patel JS, Nakagami T (1992) A multi-channel free-space optical switch using liquid crystal polarization control devices. In: Proceedings SPIE topical meeting on photonic switching, Minsk, Belarus, pp 345–349
137. Noguchi K, Sakano T, Matsumoto T (1991) A rearrangeable multichannel free-space optical switch based on multistage network configuration. IEEE/OSA J Light Technol 9(12):1726–1732
138. Yamamoto T, Yamaguchi M, Hirabayashi K, Matsuo S, Amano C, Iwamura H, Kohama Y, Kurokawa T (1996) High-density digital free space photonic switches using micro-beam optical interconnections. IEEE Photonics Technol Lett 8(3):358–360
139. Iga K, Oikawa M, Misawa S, Banno J, Kokubun Y (1982) Stacked planar optics: an application of the planar micro-lens. Appl Optics 21(19):3456–3460
140. Singer W, Brenner K-H (1997) Stacked microoptical systems. In: Herzif HP (ed) Micro-optics: elements, systems and applications. Taylor & Francis, London
141. Jahns J, Huang A (1989) Planar integration of free space optical components. Appl Optics 28(9):1602–1605
142. Gruber M, Jahns J, Sinziger S (2000) Planar-integrated optical vector-matrix-multiplier. Appl Optics 39(29):5367–5373
143. Gruber M (2004) Multichip module with planar-integrated free-space optical vector-matrix-type interconnects. Appl Optics 43(2):463–470
144. Lin LY, Goldstein EL, Tkach RW (1998) Free-space micro machined optical switches with submillisecond switching time for large-scale optical crossconnects. IEEE Photonics Technol Lett 10(4):525–527
145. Yuan S, Madamopoulos N, Helkey R, Kaman V, Klingshirn J, Bowers J (2008) Fully integrated N × N MEMS wavelength selective switch with 100% colorless add-drop ports. In Proceedings of IEEE/OSA optical fiber communication conference (OFC), paper OWC3. San Diego, CA
146. Born M, Wolf E (1997) Principals of optics, 6th edn. Cambridge University Press, Cambridge
147. Hecht E (2002) Optics, 4th edn. Addison-Wesley, Reading
148. Goodman JW (1996) Introduction to Fourier Optics, 2nd edn. McGraw-Hill, Boston, MA
149. Lohmann AW (1991) Image formation of dilute arrays for optical information processing. Optics Commun 86(5):365–370
150. Tooley FAP (1996) Challenges in optically interconnecting electronics. IEEE J Sel Top Quant Electron 2(1):3–13
151. Hinton HS, Cloonan TJ, Mccormick FB, Lentine AL, Tooley FAP (1994) Free-space digital optical-systems. Proc IEEE 82(11):1632–1649
152. Walker AC, Desmulliez MPY, Forbes MG, Fancey SJ, Buller GS, Taghizadeh MR, Dines JAB, Stanley CR, Pennelli G, Boyd AR, Horan P, Byrne D, Hegarty J, Eitel S, Gauggel HP, Gulden KH, Gauthier A, Benabes P, Gutzwiller JL, Goetz M (1999) Design and construction of an optoelectronic crossbar switch containing a terabit per second free-space optical interconnect. IEEE J Sel Top Quant Electron 5(2):236–249

153. Berger C, Xiaoqing W, Ekman JT, Marchand PJ, Spaanenburg H, Wang MM, Kiamilev F, Esener SC (2001) Parallel processing demonstrator with plug-on-top free-space interconnect optics. Proc SPIE 4292:105–116
154. Neilson DT, Barrett CP (1996) Performance trade-offs for conventional lenses for free-space digital optics. Appl Optics 35(8):1240–1248
155. Hamanaka K (1991) Optical bus interconnection system using Selfoc lenses. Optics Lett 16(16):1222–1224
156. Rolston DR, Robertson B, Hinton HS, Plant DV (1996) Analysis of a microchannel interconnect based on the clustering of smart-pixel-device windows. Appl Optics 35(8):1220–1233
157. Chateauneuf M, Kirk AG (2003) Determination of the optimum cluster parameters in a clustered free-space optical interconnect. Appl Optics 42(29):5906–5917
158. Yuan S, Riza NA (1999) General formula for coupling-loss characterization of single-mode fiber collimators by use of gradient-index rod lenses. Appl Optics 38(15):214
159. Gilsdorf RW, Palais JC (1994) Single-mode fiber coupling efficiency with graded-index rod lenses. Appl Optics 33(16):3440–3445
160. Krishnamoorthy AV (2008) The intimate integration of photonics and electronics. In: Friberg AT, Dandliker R (eds) Advances in information optics and photonics. SPIE Press, Bellingham, pp 581–598
161. Soref R, Larenzo J (1986) All-silicon active and passive guided-wave components for $\lambda = 1.3$ and 1.6 μm. IEEE J Quant Electron 22(6):873–879
162. Soref R, Bennett B (1987) Electrooptical effects in silicon. IEEE J Quant Electron 23(1):123–129
163. Tang CK, Kewell AK, Reed GT, Rickman AG, Namavar F (1996) Development of a library of low-loss silicon-on-insulator optoelectronic devices. IEE Proc Optoelectron 143(5):312–315
164. Bestwick T (1998) ASOCVA silicon-based integrated optical manufacturing technology. In: Proceedings of IEEE electronic components and technology conference. Seattle, WA, pp 566–571
165. Pavesi L, Lockwood DJ (eds) (2004) Silicon photonics. Topics in applied physics 94. Springer, Germany
166. Kimerling L (2003) Silicon microphotonics: the next killer technology: towards the first silicon lasers, Pavesi (ed) Kluwer, Norwell, pp 465–476
167. Soref RA (1993) Silicon based optoelectronics. Proc IEEE 81(12):1687–1706
168. Canham LT (1990) Silicon quantum wire array fabrication by electrochemical and chemical dissolution of wafers. Appl Phys Lett 57(10):1046–1048
169. Tang CK, Reed GT (1995) Highly efficient optical-phase modulator in SOI waveguides. Electron Lett 31(6):451–452
170. Franzò G, Coffa S, Priolo F, Spinella C (1997) Mechanism and performance of forward and reverse bias electroluminescence at 1.54 μm from Er-doped Si diodes. J Appl Phys 81(6):2784–2793
171. Dainesi P, Kung A, Chabloz M, Lagos A, Fluckiger P, Ionescu A, Fazan P, Declerq M, Renaud P, Robert P (2000) CMOS compatible fully integrated Mach-Zehnder interferometer in SOI technology. IEEE Photonics Technol Lett 12(6):660–662
172. Pavesi L, Dal Negro L, Mazzoleni G, Franzo G, Priolo F (2000) Optical gain in silicon nanocrystals. Nature 408(6811):440–444
173. Ng WL, Lorenco MA, Gwilliam RM, Ledain S, Shao G, Homewood KP (2001) An efficient room-temperature silicon-based light-emitting diode. Nature 410(6825):192–194
174. Jalali B, Fathpour S (2006) Silicon photonics. IEEE/OSA J Light Technol 24(12): 4600–4615
175. Lipson M (2005) Guiding, modulating, and emitting light on Silicon-challenges and opportunities. IEEE/OSA J Light Technol 23(12):4222–4238

176. Reed GT, Png CE, Masanovic G, Lim ST, Chan SP, Headley W, Ang TW, Liu A, Paniccia M (2003) Silicon nanophotonic circuits. In: Proceedings of XLVII ETRAN conference. Herceg Novi, Serbia and Montenegro
177. Reed GT, Knights AP (2004) Silicon photonics: an introduction. Wiley Interscience, New York
178. Smit MK (1999) Advanced components for WDM. In: Proceedings of European conference on optical communications (ECOC), Nice, France, pp 98–99
179. Masanovic M, Val L, Burton JS, Skogen EJ, Coldren LA, Blumenthal DJ (2003) Monolithically integrated Mach-Zehnder interferometer wavelength converter and widely tunable laser in InP. IEEE Photonics Technol Lett 15(8):1117–1119
180. Rondeau G, Biblemont S, Decobert J, Post G (2000) Monolithic PIN-HEMT photoreceiver integration. In: Proceedings of international conference on indium phosphide and related materials. Williamsburg, VA, pp 333–336
181. Paniccia M, Morse M, Salib M (2004) Integrated photonics. In: Pavesi L, Lockwood DJ (eds) Silicon photonics, topics applied physics, vol 94. Springer, Germany
182. Goossen KW (1994) Integration of GaAs modulators to silicon IC's by monolithic and hybrid techniques. In: Proceedings of IEEE Sarnoff symposium. Princeton, NJ, pp 100–107
183. Almeida VR, Panepucci RR, Lipson M (2003) Nanotaper for compact mode conversion. OSA Opt Lett 28(15):1302–1304
184. Koch BR, Fang AW, Cohen O, Bowers JE (2007) Mode-locked silicon evanescent lasers. Opt Express 15(18):11225–11233
185. Fang AW, Koch BR, Gan K-G, Park H, Jones R, Cohen O, Paniccia MJ, Blumenthal DJ, Bowers JE (2008) A racetrack Mode-locked silicon evanescent laser. Opt Express 16(2):1393–1398
186. Vantrease D, Schreiber R, Monchiero M, McLaren M, Jouppi NP, Fiorentino M, Davis A, Binkert N, Beausoleil RG, Ahn JH (2008) Corona: system implications of emerging nanophotonic technology. In: Proceedings of international symposium on computer architecture. Beijing, China, pp 153–164
187. Tucker RS, Pei-Cheng Ku, Chang-Hasnain CJ (2005) Slow-light optical buffers: capabilities and fundamental limitations. IEEE/OSA J Light Technol 23(12):4046–4066
188. Henker R, Schneider T, Lauterbach K-U, Junker M, Ammann MJ, Schwarzbacher AT (2007) Slow and fast-light in optical fibers: an overview. In: Proceedings international students and young scientists workshop on photon and microsystems. Dresden, Germany, pp 32–37
189. Rawal S, Sinha RK (2009) Slow light based optical buffer with high delay bandwidth product in silicon-on-insulator photonic crystal waveguides. In: Proceedings of international conference on emerging trends in electronic and photonic devices and systems (ELECTRO). Varanasi, India, pp 580–582
190. Maxemchuk NF (1985) Regular and mesh topologies in local and metropolitan area networks. AT&T Tech J 64(7):1659–1686
191. Liboiron-Ladouceur O, Shacham A, Small BA, Lee BG, Wang H, Lai CP, Biberman A, Bergman K (2008) The data vortex optical packet switched interconnection network. IEEE/OSA J Light Technol 26(13):1777–1789
192. Batten C, Joshi A, Orcutt J, Khilo A, Moss B, Holzwarth CW, Popovic MA, Li H, Smith HI, Hoyt JL, Kartner FX, Ram RJ, Stojanovic V, Asanovic K (2009) Building many-core processor-to-DRAM networks with monolithic CMOS silicon photonics. IEEE Micro 29(4):8–21
193. Kirman N, Kirman M, Dokania RK, Martinez JF, Apsel AB, Watkins MA, Albonesi DH (2007) On-chip optical technology in future bus-based multicore designs. IEEE Micro 27(1):56–66
194. Shacham A, Bergman K, Carloni LP (2008) Photonic networks-on-chip for future generations of chip multiprocessors. IEEE Trans Comput 57(9):1246–1260

195. Chan J, Hendry G, Biberman A, Bergman K (2010) Architectural exploration of chip-scale photonic interconnection network designs using physical-layer analysis. IEEE/OSA J Light Technol 28(9):1305–1315
196. Alameh K, Yong Tak Lee (2009) Nanophotonics-based devices and applications. In: Proceedings of international symposium high-capacity optical networks and enabling technologies. Alexandria, Egypt, pp 91–95
197. Hattori HT, Jagadish C, Seassal C, Boutami S, BenBhakir B, Drouard E, Letartre X, Viktorovitch P (2006) Integrated photonic crystal devices for optical interconnect applications. In: Proceedings of international conference mathematical methods in electromagnetic theory. Moscow, Russia, pp 35–40
198. Barroso LA, Dean J, Holzle U (2003) Web search for a planet: the google cluster architecture. IEEE Micro 23(2):22–28
199. Madamopoulos N, Kaman V, Yuan S, Jerphagnon O, Helkey RJ, Bowers JE (2009) Applications of large-scale optical 3D-MEMS switches in fiber-based broadband-access networks. Photonic network communications 19(1):62–73
200. SFF-8436 specification for 14 QSFP + copper and optical modules, SFF committee review 3.5 9 July 2010. (ftp://ftp.seagate.com/pub/sff/SFF-8436.PDF)
201. Narasimha A, Analui B, Balmater E, Clark A, Gal T, Guckenberger D, Gutierrez S, Harrison M, Ingram R, Koumans R, Kucharski D, Leap K, Yi Liang, Mekis A, Mirsaidi S, Peterson M, Tan Pham, Pinguet T, Rines D, Sadagopan V, Sleboda TJ, Dan Song, Yanxin Wang, Welch B, Witzens J, Abdalla S, Gloeckner S, De Dobbelaere P (2008) A 40-Gb/s QSFP optoelectronic transceiver in a 0.13 μm CMOS silicon-on-insulator technology. In: Proceedings of IEEE/OSA optical fiber communication conference (OFC/NFOEC). San Diego, CA, pp 1–3
202. SFF-8642 specification for mini multilane series: shielded integrated connector, SFF committee, Rev 2.7, 26 Feb 2010. (ftp://ftp.seagate.com/pub/sff/SFF-8642.PDF)]
203. Mergeoptics, product brief: 12 channel 850 nm parallel active optical cable 120 Gbit/s CXP—3xQSFP ICD120GVP1420XYxx. www.mergeoptics.com
204. Holland WR, Burack JJ, Stawicki RP (1993) Optical fiber circuits. In: Proceedings of electronic components and technology conference. Orlando, FL, pp 711–717
205. Shahid MA, Roll RA, Shevchuk GJ (1994) Connectorized optical fiber circuits. In: Proceedings of electronic components and technology conference. Washington, DC, pp 981–985
206. Delbare W, Lauwers E, Qingsheng T, Vandewege J, Verbeke J (1993) Packaging for fiber in board electro-optical interconnections. In: Proceedings of IEEE lasers and electro-optics society annual meeting (IEEE/LEOS). San Jose, CA, 470–471
207. Nordin RA, Holland WR, Shahid MA (1995) Advanced optical interconnection technology in switching equipment. IEEE/OSA J Light Technol 13(6):987–994
208. Li M-J (2009) Bend-insensitive optical fibers for FTTH applications. In: Proceedings of the SPIE 7234-broadband access communication technologies III. Dingel BB, Jain R, Tsukamoto K (ed) 7234:72340B–72340B-9
209. Glick M (2008) Optical interconnects in next generation data centers; an end to end view. In: Proceedings of IEEE symposium on high performance interconnects. Stanford, CA, pp 178–181
210. Bergman K, Yeh C, Morookian J (2001) Advances in multichannel multigbytes/s bit-parallel WDM single fiber link. IEEE Trans Adv Packag 24(4):456–462
211. Lin T, Williams KA, Penty RV, Glick M, McAuley D (2006) Performance and scalability of a single-stage SOA switch for 10 × 10Gb/s wavelength striped packet routing. IEEE Photonics Technol Lett 18(5):691–693
212. http://www.zurich.ibm.com/sys/io_links/interconnects.html and http://www.zurich.ibm.com/sys/io_links/images_interconnects_fig1.html
213. Lee W-J, Hwang SH, Lim JW, Cho CH, Kim GW, Rho BS (2009) Optical interconnection module integrated on a flexible optical/electrical hybrid printed circuit board. In:

Proceedings of IEEE electronic components and technology conference (ECTC). San Diego, CA, pp 1802–1805
214. Moisel J, Guttmann J, Huber H-P, Krumpholz O, Rode M (2000) Optical back-planes with integrated polymer waveguides. Opt Eng 39(3):673–679
215. Lehmacher S, Neyer A (2000) Integration of polymer optical waveguides into printed circuit boards (PCB). Electron Lett 36(12):1052–1053
216. Mederer F, Michalzik R, Guttmann J, Huber H-P, Lunitz B, Moisel J, Wiedenmann D (2002) 10 Gbit/s data transmission with TO-packaged multimode GaAs VCSELs over 1 m long polymer waveguides for optical backplane applications. Opt Comm 206(4–6):309–312
217. Savage N (2002) Linking with Light. IEEE Spectrum 39(8):32–36
218. Hwang SH, Lim JW, Lee W-J, Kim GW, Cho CH, An JB, Jung KY, Cha KS, Rho BS (2010) Bendable and splitter-integrated optical subassembly based on a flexible optical board. IEEE Photonics Technol Lett 22(3):167–169
219. Guoqiang L, Dawei H, Yuceturk E, Mar PJ, Ozguz VH, Yue L, Esener SC (2002) Three-dimensional optoelectronic stacked processor by use of free-space optical interconnection and three-dimensional VLSI chip stacks. Appl Optics 41(2):348–360
220. Ishii Y, Koike S, Arai Y, Ando Y (2003) SMT-compatible large-tolerance 'OptoBump' interface for interchip optical interconnections. IEEE Trans Adv Packag 26(2):122–127
221. Cho S-Y, Seo S-W, Jokerst NM, Brooke MA (2004) Board-level optical interconnection and signal distribution using embedded thin-film optoelectronic devices. IEEE/OSA J Light Technol 22(9):2111–2118
222. Lukowicz P, Jahns J, Barbieri R, Benabes P, Bierhoff T, Gauthier A, Jarczynski M, Russell GA, Schrage J, Šullau W, Snowdon JF, Wirz M, Troster G (2003) Optoelectronic interconnection technology in the HOLMS system. IEEE J Sel Top Quant Electronics 9(2):624–635
223. Liu JJ, Kalavjian Z, Riely B, Chang W, Simonis GJ, Apsel A, Andreou A (2003) Multichannel ultrathin silicon-on-sapphire optical interconnects. IEEE J Sel Top Quant Electronics 9(2):380–386
224. Bona G-L, Offrein BJ, Bapst U, Berger C, Beyeler R, Budd R, Dangel R, Dellmann L, Horst F (2004) Characterization of parallel optical-interconnect waveguides integrated on a printed circuit board. Proc SPIE micro-optics, VCSELs, and photonic integration 5453:134–141
225. Kim J-S, Kim J-J (2004) Stacked polymeric multimode waveguide arrays for two-dimensional optical interconnects. IEEE/OSA J Light Technol 22(3):840–844
226. Hwang SH, Cho MH, Kang S-K, Park H-H, Cho HS, Kim S-H, Shin K-U, Ha S-W (2006) Passively assembled optical interconnection system based on an optical printed-circuit board. IEEE Photonics Technol Lett 18(5):652–654
227. Venditti MB, Laprise E, Faucher J, Laprise P-O, Lugo JEA, Plant DV (2003) Design and test of an optoelectronic-VLSI chip with 540- element receiver–transmitter arrays using differential optical signaling. IEEE J Sel Top Quant Electronics 9(2):361–379
228. Kromer C, Sialm G, Berger C, Morf T, Schmatz M, Ellinger F, Erni D, Bona G-L, Jackel H (2005) A 100 mW 4 × 10 Gb/s transceiver in 80 nm CMOS for high-density optical interconnects. In: Proceedings of IEEE international solid-state circuits conference. San Francisco, CA, pp 334–336
229. Yeniay R, Gao K, Takayama R, Gao A, Garito F (2004) Ultra-low-loss polymer waveguides. IEEE/OSA J Light Technol 22(1):154–158
230. Wang L, Wang X, Jiang W, Choi J, Bi H, Chen R (2005) 45 polymer-based total internal reflection coupling mirrors for fully embedded intraboard guided wave optical interconnects. Appl Phys Lett 87(14):3
231. Doany FE, Schow CL, Baks C, Budd R, Yin-Jung C, Pepeljugoski P, Schares L, Kuchta D, John R, Kash JA, Libsch F, Dangel R, Horst F, Offrein BJ (2007) 160-Gb/s bidirectional parallel optical transceiver module for board level interconnects using a single-chip CMOS IC. In: Proceedings of electronic components and technology conference (ECTC). Reno, NV, pp 1256–1261

232. Schares L, Kash JA, Doany FE, Schow CL, Schuster C, Kuchta DM, Pepeljugoski PK, Trewhella JM, Baks CW, John RA, Shan L, Kwark YH, Budd RA, Chiniwalla P, Libsch FR, Rosner J, Tsang CK, Patel CS, Schaub JD, Dangel R, Horst F, Offrein BJ, Kucharski D, Guckenberger D, Hegde S, Nyikal H, Chao-Kun L, Tandon A, Trott GR, Nystrom M, Bour DP, Tan, MRT, Dolfi DW (2006) Terabus: terabit/second-class card-level optical interconnect technologies. IEEE J Sel Top Quant Electronics 12(5):1032–1044
233. Doany FE, Schow CL, Budd R, Baks C, Kuchta DM, Pepeljugoski P, Kash JA, Libsch F, Dangel R, Horst F, Offrein BJ (2008) Chip-to-chip board-level optical data buses. In: Proceedings of IEEE/OSA optical fiber communication conference (OFC/NFOEC). San Diego, CA, pp 1–3
234. Nikolajsen T, Leosson K, Salakhutdinov I, BozheVolnyi SI (2003) Polymer-based surface-plasmon polariton stripe waveguides at telecommunication wavelengths. Appl Phys Lett 82(5):668–670
235. Charbonneau R, Scales C, Breukelaar I, Fafard S, Lahoud N, Mattiussi G, Berini P (2006) Passive integrated optics elements based on long-range surface plasmon polaritons. IEEE/OSA J Light Technol 24(1):477–494
236. Kim JT, Ju JJ, Park S, Kim M-S, Park SK, Lee M-H (2008) Chip-to-chip optical interconnect using gold long-range surface plasmon polariton waveguides. Opt Express 16(17):13133–13138
237. The 50G silicon photonics link-the world's first silicon-based optical data connection with integrated lasers. White paper, Intel labs (2010). http://techresearch.intel.com/spaw2/uploads/files/Intel_SiliconPhotonics50gLink_FINAL.pdf
238. Fehly D, Schlachetzhi A, Bakin A, Guttzeit A, Wehmann G (2001) Monolithic InGaAsP optoelectronic devices with silicon electronics. IEEE J Quant Electronics 37(10):1246–1252
239. Wong YM, Muehlner DJ, Faudskar CC, Fishteyn M, Gates JV, Anthony PJ, Cyr GJ, Choi J, Crow JD, Kuchta DM, Pepeljugoski PK, Stawiasz K, Nation W, Engebretsen D, Whitlock B, Morgan RA, Hibbs-Brenner MK, Lehman J, Walterson R, Kalweit E, Marta T (1996) OptoElectronic technology consortium (OETC) parallel optical data link: components, system applications, and simulation tools. In: Proceedings electronic components and technology conference. Orlando, FL, pp 269–278
240. Kodi AK, Louri A (2004) RAPID: reconfigurable and scalable all-photonic interconnect for distributed shared memory multiprocessors. IEEE/OSA J Light Technol 22(9):2101–2110
241. Minkenberg C, Abel F, Muller P, Raj Krishnamurthy, Gusat M, Hemenway BR (2005) Control path implementation for a low-latency optical HPC switch. In: Proceedings symposium high performance interconnects. Stanford, CA, pp 29–35
242. Chamberlain RD, Franklin MA, Baw Ch'ng Shi (2002) Gemini: an optical interconnection network for parallel processing. IEEE Trans Parallel Distrib Syst 13(10):1038–1055
243. Doany FE, Schow CL, Tsang CK, Ruiz N, Horton R, Kuchta DM, Patel CS, Knickerbocker JU, Kash JA (2008) 300 Gb/s 24-channel bidirectional Si carrier transceiver optochip for board-level interconnects. In: Proceedings of electronic components and technology conference (ECTC). Lake Buena Vista, FL, pp 238–243
244. Yashiki K, Hashimoto Y, Oda M, Suzuki N, Yamamoto K, Kurihara M, Sugimoto T, Hatakeyama H, Sakai J, Akagawa T, Narita K, Ono H, Fukatsu K, Tokutome K, Kouta H, Tsuji M, Ogura I, Noda A, Anan T, Suzuki Y, Kurata K (2010) 240-Gb/s on-board optical transmitters and receivers. In: Proceedings of IEEE/OSA optical fiber communication conference (OFC/NFOEC). San Diego, CA, pp 1–3
245. Vlasov Y, Green WMJ, Xia F (2008) High-throughput silicon nanophotonic wavelength-insensitive switch for on-chip optical networks. Nat Photonics 2(4):242–246

Chapter 7
Short Range (in-building) Systems and Networks: A Chance for Plastic Optical Fibers

María Angeles Losada and Javier Mateo

Abstract In this chapter the deployment of plastic optical fibers in short-range (in-building) networks is examined. The application is evaluated in the context of the problem of modeling optical power propagation through the fiber when it is deployed in peculiar environments such as those inside a house. A fast and versatile matrix method is developed to model optical power transmission throughout the fiber. This method is flexible in incorporating the effects of localized disturbances over transmission and compact in facilitating integration in commercial software packages.

7.1 Introduction to Plastic Optical Fibers

The first plastic optical fibers (POFs) were manufactured by DuPont in the late 1960s but they had huge attenuations due to impurities in the materials. When, in the middle 1970s, the attenuation was lowered to near the theoretical limit (125 dB/Km) there were already commercial glass fibers with a much lower attenuation and, most importantly, cheaper. At that point, the application of fibers in communications was mainly restricted to long-haul networks where POFs had no chance due to their restricted bandwidth and high attenuation, while the field of communications between computers was amply provided by copper wires. The digitalization of communications that started in the last decade of the past

M. A. Losada (✉) · J. Mateo
Photonic Technologies Goup, i3A
University of Zaragoza
Zaragoza, Spain
e-mail: alosada@unizar.es

J. Mateo
e-mail: jmateo@unizar.es

N. (Neo) Antoniades et al. (eds.), *WDM Systems and Networks*, Optical Networks,
DOI: 10.1007/978-1-4614-1093-5_7, © Springer Science+Business Media, LLC 2012

century has brought about the need of broadband links closer to private users. Short-range communication networks are increasingly demanding higher bitrates to answer the needs for distributing multimedia contents in automotive, industrial and domestic environments. It is in this context that POFs have demonstrated that they are a very suitable media as, at short distances, their relative low bandwidth and high attenuation are enough to cover the demands while they are easy to handle, resistant to curvatures and can be used with low-cost equipment [1].

Different polymers can be used to manufacture POFs, but the most frequently used material is polymethylmethacrylate (PMMA) with a refractive index of 1.492 which can bear temperatures up to 80°C. As in the case of glass fibers, the first POFs to be manufactured were the step index POFs (SI-POFs) and this type is still the most widely used today. Their standard size is generally 1 mm in diameter which is mainly referring to the core since the cladding is very thin, although there are commercial variants with smaller cores. The standard fiber numerical aperture (NA) is close to 0.5, but there are fibers with NAs down to 0.3. They have an attenuation profile with several minima in the visible range. In fact, the absolute minimum is at 565 nm but they are normally used at 650 nm where their attenuation is between 0.11 and 0.25 dB/m [2]. In standard conditions these fibers have bandwidths of 40 MHz for 100 m [3].

Fibers with more complex profiles and different materials have been developed in order to increase bandwidth and to reduce attenuation, but always at the cost of increasing complexity and prize and of reducing the amount of launched optical power. Double-step index profile polymer fibers (DSI-POFs) were developed to behave as low NA POFs selectively attenuating power propagating at higher angles. An extension of this idea is the multi-step index (MSI) profile which can be seen as a cheaper approximation to graded index profile fibers that restrict power propagation to a smaller area of the fiber. These fibers can achieve bandwidths ranging from 100 MHz-100m for DSI-POFs to 500 MHz-100m for MSI-POFs [4]. The Japanese company Ashahi Glass manufactures a small 250 microns core Graded Index POF (GI-POF) called Lucina™ [5] whose reported bandwidth reaches up to 2 GHz–100 m [6]. These fibers also show the lowest attenuation in POFs, less than 0.02 dB/m, as they are manufactured using CYTOP® [7], that is a material where all the hydrogen has been replaced with fluorine. There are also PMMA 1 mm core GI-POFs that try to reach a compromise between the advantages of launching ease and high bandwidth but they are limited by attenuation [8].

The multi-core fibers (MC-POF) offer another approach to increase bandwidth and decrease bending losses while preserving the advantages of the large size. There are different variants to the multi-core approach depending on the profile of each core. Their bandwidth can go up to 400 MHz–100 m but their cost is still high [9].

In the next sections, we will evaluate the application of POFs in domestic networks in the context of modeling the fiber behavior when it is deployed in environments where there are short-fiber segments with tight bends and linked by several connectors, such as those inside a house. First, in Sect. 7.2, different transmission media are compared to POFs, revealing that its many advantages place it in a competitive position relative to other more traditional propagation media. On the other hand, light propagation in POFs has some special characteristics that prevent the use of existing computer simulation models that are suitable for standard multimode fibers from describing their performance. The problems of modeling the transmission properties of these large fibers and of accounting for the impact of localized disturbances on them are addressed in Sect. 7.3, placing special emphasis on the approaches which are based on the diffusion equation. In Sect. 7.4, a matrix method to solve the power flow differential equation is described. This method can be applied to model not only the effects of propagation through the fiber but also those caused by bends, connectors and other non-uniformities as matrices. These matrices can be introduced as independent blocks into most commercial software packages when testing network designs prior to their actual deployment. Finally, the design of a prototype network that is used to distribute multimedia content and that is based on large core SI-POF is described in Sect. 7.5.

7.2 POFs in Home Networks

The demand for broadband services at home such as video-on-demand, online gaming and others raises the need for high-bandwidth services. Since people who require and buy the technology are not experts, one of the requirements is that installation and operation has to be easy and cheap, preferably the "do-it-yourself" type. Home networks offer a wide range of possible transmission media: coaxial cables, twisted copper pair cables, power lines and wireless among others. In general, a network that uses one of the above four media has been optimized for a particular set of services complicating the interoperability between devices. The ideal will be a single broadband network that provides an efficient solution for present and future devices [10].

One of the better candidates for a "do-it-yourself" home network would be Power Line Communications (PLC) because power outlets are already everywhere inside the house. Nevertheless, using PLC as the backbone will not provide enough bandwidth for some situations and has the major disadvantage of electrical background noise and dependence on the quality of the electrical infrastructure of each home which varies widely from construction to construction.

The new wireless standard 802.11n provides full connectivity for home applications and enough data rate to support real-time applications. However, in old houses with concrete walls, the signal coverage can be limited to a single room. In addition, even if theoretically there was enough bandwidth, frequent signal interruptions unacceptable for this kind of real-time applications have been reported [11].

Ethernet copper or fiber media are two technologies that require relatively high-installation expense in built homes. However, these approaches work well in new structured houses and are good solutions for the backbone of the house as they provide high bandwidth and reliability for data communications. The advantages of the POF over the RJ45 copper cables are not only physical but also aesthetical because in a home network a communication medium is required to introduce a

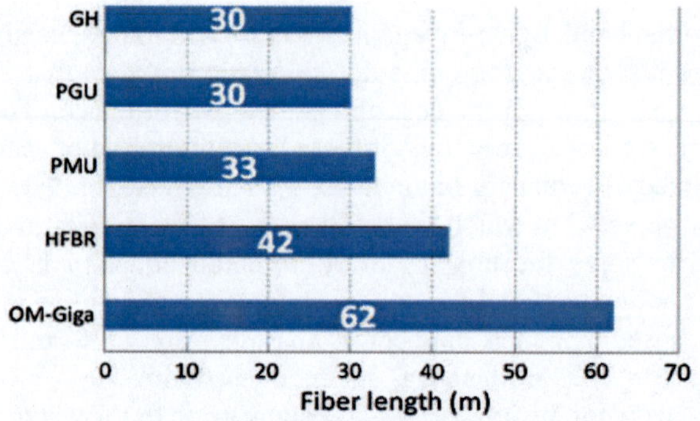

Fig. 7.1 The graph shows length limits for 1mm core diameter PMMA fibers when transmitting GbE data (1.25Gb/s, PRBS 27-1). A large core GI POF from Optimedia (OM GIGA) is compared to SI POFs from different manufacturers: PMU is a broadband fiber from Toray (PMU CD1002 22 E) with a relatively low numerical aperture (NA) of 0.33, HFB is the HFBR RUS500 from Hewlett Packard and GH is the ESKA PREMIER GH4001 from Mitsubishi, both with NA near 0.5

minimum environmental impact. Nowadays, RJ45 copper cable is very widely used in LANs but has a number of disadvantages when compared to POF: copper cable is thicker and less malleable than POF while small size and curvature resistance is imperative in this kind of network. Besides, electromagnetic radiation can interfere with the transmitted signal in copper cables, while electrical background noise does not affect optical wires. Therefore, optical fiber could be installed through the same conduits that provide electricity [10, 12].

POF combines the ease of installation benefit of copper networks with the transmission advantages of conventional glass fiber (GOF). It also overcomes the problems that can arise with GOF which is too brittle and more difficult to handle. The simplicity of installation of POF combined with the added flexibility of the POF cable (down to a radius of 25 mm) make it ideal for narrow cable shafts where installing POF can take shorter time than installing conventional thicker copper cable [13]. There are three more reasons to use POF as the network backbone. First, the PMMA fiber works best with visible light, mainly red, and thus hazard exposure can be easily avoided since now everyone can see it so they can avoid direct eye exposure. Second, it is impossible to get electrical shock using POF cables. Finally, the extensive use of optical fibers is steadily lowering the prices of connector and adapters needed for home installations.

Large core SI-POF PMMA fibers have large modal dispersion which limits their maximum supported data rates. However, a study of transmission performance of SI-POFs from different manufacturers based on bit error rate (BER) measurements has demonstrated that they are able to transmit Gigabit Ethernet at least up to 30 m [14]. This same study demonstrates that PMMA large core GI-POFs, such as OM-GIGA [8, 15], can be used for Gigabit Ethernet up to more than 60 m as shown in Fig. 7.1. These fibers keep most of the advantages of SI-POFs but it is difficult to find commercial optical transceivers working at 650 nm and gigabit rates. On the other hand, perfluorinated POFs are compatible with multimode GOF fiber and thus, can be used with commercial infrared transceivers [16]. Moreover, they can reach higher bit rates at longer link distances, but at the cost of losing some of the advantages of SI-POFs, as the lower hazard exposure is lost for infrared sources. In addition, these fibers are thinner than standard POFs and not as easy to connect as those with large cores.

Although there is still an uncertain future for POF's penetration into the European home, it is already widely used in other countries, mainly in Asia. For example, the GigaHouse Town implemented in a project designed by Keio University in Japan to demonstrate the viability and economic efficiency of a city with highly computerized homes to contribute in bringing the information technology to the average citizen [17]. Also, the Information Ministry of South Korea has a proposal for cabling homes using POF [18, 19]. In USA, the Plastic Optical Fiber Trade Organization (POFTO) is promoting industry awareness, acceptance and advancement of plastic optical fiber and component systems in the data communications markets and promoting the advancement and acceptance of standards incorporating POF [20]. In Europe, from the beginning of 2006, the POF-ALL project and its continuation, POF-PLUS, have gathered some European research centers and companies with the purpose of developing a low-cost solution based on POF to enable the delivery of broadband access to everyone [21, 22]. Over the last years, a considerable advance in the production of standard POF and commercially available devices has also been made by European companies such as Firecomms Ltd. [23], Homefibre Digital Network GmbH [24], Diemount GmbH [25] and Luceat S.p.A. [26] as main examples.

7.3 POF Modeling

7.3.1 Overview of Different Approaches

Similar to the conventional glass fibers, polymer optical fibers guide light on the basic principle of total internal reflection. There are, however, several aspects in POFs that prevent the direct use of the approaches traditionally used to describe light propagation in glass fibers. In the first place, POFs have large cores and high NAs and thus they propagate a much higher number of modes than the standard multimode glass fibers. For example, in a standard SI-POF of 1 mm core and 0.5 NA there are several millions of modes. POFs also have a much higher attenuation which impedes the approaches designed for weakly attenuating glass fibers. Another characteristic of POFs is the strong diffusion produced either by Rayleigh scattering of impurities in the bulk material or by Mie scattering produced by irregularities in the core-cladding interface [27]. This diffusion induces power transfer between neighboring modes which is translated as mode mixing, an effect that is much stronger than that present in multimode glass fibers. Bends, curvatures and other macroscopic strains also produce power diffusion to other further separated modes. All these diffusive effects intrinsic and extrinsic to the fiber have an impact on the fiber's modal dispersion which consequently determines its temporal response.

The strong diffusion in POFs however, does not guarantee a rapid achievement of steady conditions, which can only be reached after 75–100 m propagation [28]. Since the main use of current POFs is in short-range applications, we can assume that POF links will be mostly working in non-stationary conditions. In these distance ranges, the launching conditions have a strong impact on the fiber performance. Localized disturbances such as curvatures, bends and stress produced by external devices or manipulation have also a strong influence in the propagation properties [29, 30]. Their effect is not simply power loss but also power transfer from one mode to the others which are further away from their nearest neighbors. These changes in power distribution due to mode mixing alter the modes' relative temporal delays and have a direct relation to bandwidth [31]. Some disturbances, such as bends of very small radii, can even provoke permanent changes in the fiber modal behavior [32]. Therefore, it is important to devise a model that accounts for the behavior of the fiber in such conditions and that is able to integrate different disturbances to quantify their effects over propagation.

The different wave theory approaches to describe optical power propagation in fibers vary when it comes to the method they apply to find the solutions for the vector or scalar wave equation. Most of these approaches are based on approximations that are not applicable to POFs. The Wenzels, Kramer and Brillouin method (generally known as WKB) to solve the scalar wave equation is based on approximations that are close to a ray theory description of the power propagation and that can be applied to calculate efficiently the propagation times of many modes, allowing quick estimation of bandwidth [33, 34]. Some wave models incorporate a description of mode mixing [35, 36], such as the split-step algorithm that divides propagation into two parts: one that accounts for linear unmixed propagation and another that describes mode mixing using a matrix and which are calculated separately [37, 38].

In the ray tracing approach, a very large number of rays are generated at the fiber input, each following its own path along the fiber in order to draw a histogram at its output end. This histogram includes propagation time, attenuation, exit position and propagation angle. The obvious disadvantage of this procedure is that it requires large computational time, unless some approximations are made as in [39]. This method is flexible in studying fibers with different profiles [40–42] and even multi-core fibers [43]. This model can incorporate mode mixing as a statistical process where scattering centers change the power and angle of the incoming rays based on a probability distribution [44] and can also be useful in calculating bending losses [40, 45].

However, a more direct way to describe mode mixing is provided by the power flow differential equation. The models to describe optical power transmission based on this equation are the subject of the next subsection and of Sect. 7.4, where a matrix approach is described in more detail.

7.3.2 Power Flow Equation

The general diffusion equation has frequently been used to describe the evolution of the modal power distribution as it is transmitted throughout a multimode fiber, since Gloge published it in 1972 [46]. This approach has been restricted so far to fibers with step index profiles and large cores in order to guarantee that the number of modes transmitted inside a POF is high enough to assume that the propagation angle is a continuous variable. Thus, different modes are characterized by their propagation angle with respect to fiber axis θ. The optical power distribution, $P(\theta, z)$, is a function of fiber length z and of inner propagation angle θ. Its spatial evolution is then described by the following differential equation:

$$\frac{\partial P(\theta,z)}{\partial z} = -\alpha(\theta)P(\theta,z) + \frac{1}{\theta}\frac{\partial}{\partial\theta}\left(\theta \cdot d(\theta) \cdot \frac{\partial P(\theta,z)}{\partial\theta}\right), \tag{7.1}$$

where $\alpha(\theta)$ and $d(\theta)$ are the attenuation and diffusion functions, respectively.

The diffusion equation can be solved analytically only in some particular cases [46–48]. A more general approach can be obtained using numerical methods [49–52], but always imposing simplifying conditions over the attenuation or diffusion functions. In fact, all these works assumed that attenuation or diffusion or both have a constant value independent of the propagation angle. Another approach based on Gloge's power flow equation but without making assumptions about the angular diffusion and attenuation functions is described in [53]. There, attenuation and diffusion are described as general functions of the propagation angle and are estimated from radial profiles extracted from experimental far field patterns (FFPs). These functions are characteristic of the fiber type since they describe the fiber modal behavior. With this approach, the angular distribution of power can be obtained for different launching conditions and fiber lengths. However, as the power distribution does not include the time dependence, the frequency response and bandwidth cannot be obtained solving Eq. 7.1.

Following the procedure described by Gloge [54] the temporal dimension can be introduced as another variable of the optical power, $P(\theta, z, t)$, in Eq. 7.1. Thus, the derivative of optical power with respect to z results in two terms given the inter-dependence of z and t. If the relationship between light speed and trajectory given by ray theory is assumed, the following equation is obtained:

$$\frac{\partial P(\theta,z,t)}{\partial z} = -\alpha(\theta)P(\theta,z,t) - \frac{n}{c\,\cos\theta} \cdot \frac{\partial P(\theta,z,t)}{\partial t} + \frac{1}{\theta}\frac{\partial}{\partial\theta}\left(\theta \cdot d(\theta) \cdot \frac{\partial P(\theta,z,t)}{\partial\theta}\right). \tag{7.2}$$

Then, taking the Fourier transform on both sides of Eq. 7.2 and using the Fourier derivation property, the following simplified equation is derived:

$$\frac{\partial p(\theta,z,\omega)}{\partial z} = -\left(\alpha(\theta) + \frac{n}{c\ \cos\theta}\cdot j\omega\right)p(\theta,z,\omega) + \frac{1}{\theta}\frac{\partial}{\partial\theta}\left(\theta\cdot d(\theta)\cdot\frac{\partial p(\theta,z,\omega)}{\partial\theta}\right), \tag{7.3}$$

where $p(\theta, z, \omega)$ is the Fourier transform of $P(\theta, z, t)$. The solution of this equation can be found using different approaches as described in [55]. In the next subsection, we present a fast and robust matrix approach of the finite-difference method to solve the temporal generalization of the power flow equation in the frequency domain [56].

7.4 Matrix Approach to Solve the Generalized Power Flow Equation

The solution of the differential power flow equation can be approached using a finite-difference method [49]. Thus, if a forward difference is used for the first z derivative, and a first- and second-order central differences for the first and second angular derivatives, respectively, the power at angle θ and distance $z + \Delta z$ is obtained as the linear combination of the power at the same angle and the two adjacent angles $(\theta + \Delta\theta,\ \theta - \Delta\theta)$ for a distance z as shown in the following equation:

$$\begin{aligned} p(\theta,z+\Delta z,\omega) &= \left(1-\left(\alpha(\theta)+\frac{n}{c\cos\theta}\cdot j\omega\right)\Delta z\right)p(\theta,z,\omega) \\ &+\frac{\Delta z}{2\cdot\Delta\theta}\left(\frac{d(\theta)}{\theta}+d'(\theta)\right)(p(\theta+\Delta\theta,z,\omega)-p(\theta-\Delta\theta,z,\omega)) \\ &-\frac{2d(\theta)\Delta z}{\Delta\theta^2}p(\theta,z,\omega) \\ &+\frac{d(\theta)\Delta z}{\Delta\theta^2}(p(\theta+\Delta\theta,z,\omega)+p(\theta-\Delta\theta,z,\omega))\,. \end{aligned} \tag{7.4}$$

The dependence of the power at a given angle, θ, of the power at the adjacent ones: $\theta + \Delta\theta$ and $\theta - \Delta\theta$, suggests rewriting Eq. 7.4 using an algebraic formalism and introducing a matrix to describe power transmission through the fiber.

7.4.1 Propagation Matrix

Equation 7.4 can be expressed in a more compact representation in matrix form. In fact, the differential changes in the angular power distribution at each Δz step are given by a simple matrix product. Thus, given the angular power at an initial length z_1, the power distribution at a longer length z_2 can be calculated with the following matrix equation:

$$\mathbf{p}(z_2,\omega) = (\mathbf{A}(\omega) + \mathbf{D})^m\cdot\mathbf{p}(z_1,\omega), \tag{7.5}$$

In Eq. 7.5, **p** is a vector where each component k is the power at the discretized propagation angle $\theta = k \cdot \Delta\theta$, and $m = (z_2 - z_1)/\Delta z$ is an integer that can be found for any pair of lengths, $z_2 > z_1$ provided we choose a small Δz. **A** is a diagonal matrix that describes power propagation without diffusion. Its elements are obtained from Eq. 7.4 as:

$$A_{k,k}(\omega) \approx 1 - \Delta z \cdot \alpha(k \cdot \Delta\theta) - \Delta z \cdot \frac{n}{c\,\cos(k \cdot \Delta\theta)} \cdot j\omega\,. \tag{7.6}$$

Notice that **A** is the only frequency dependent term in Eq. 7.5. Iteration over the values of ω gives the complete spatial and temporal evolution of the optical power in the fiber. The complex values of $A_{k,k}(\omega)$ are obtained by sampling the angular frequency ω as required for a precise calculation of the inverse discrete Fourier transform of $p(\theta, z, \omega)$ to obtain $P(\theta, z, t)$.

The matrix **D** is a tri-diagonal matrix which accounts for diffusion along the fiber. Its elements for $k > 0$ are:

$$\begin{aligned} D_{k,k-1} &= \left(d(k \cdot \Delta\theta) - \frac{1}{2}\frac{d(k \cdot \Delta\theta)}{k} - \frac{1}{2} d'(k \cdot \Delta\theta)\Delta\theta \right) \frac{\Delta z}{\Delta\theta^2} \\ D_{k,k} &= -2d(k \cdot \Delta\theta)\frac{\Delta z}{\Delta\theta^2} \\ D_{k,k+1} &= \left(d(k \cdot \Delta\theta) + \frac{1}{2}\frac{d(k \cdot \Delta\theta)}{k} + \frac{1}{2} d'(k \cdot \Delta\theta)\Delta\theta \right) \frac{\Delta z}{\Delta\theta^2}. \end{aligned} \tag{7.7}$$

These matrix elements describe power diffusion through a differential fiber length representing the fraction of the power that flows out from a given angle and the power that drifts to this angle from the adjacent ones.

To obtain the boundary condition at $k = 0$, corresponding to $\theta = 0$, we use the approximation proposed in [50], and the fact that $P(\theta)$ is an even function of the angle, and we then obtain:

$$D_{0,0} = -4d(0)\frac{\Delta z}{\Delta\theta^2} \qquad D_{0,1} = 4d(0)\frac{\Delta z}{\Delta\theta^2}. \tag{7.8}$$

The other boundary condition at the maximum $k = N$ is given by imposing that there are no losses due to diffusion at this angle. Thus, the value of $D_{N,N-1}$ must compensate for the absence of the term $D_{N,N+1}$ in Eq. 7.7, such that the sum of terms is zero, resulting in the following:

$$D_{N,N-1} = 2d(N)\frac{\Delta z}{\Delta\theta^2} \qquad D_{N,N+1} = -2d(N)\frac{\Delta z}{\Delta\theta^2}. \tag{7.9}$$

Matrix $\mathbf{M}(\omega) = \mathbf{A}(\omega) + \mathbf{D}$ carries all space–time information concerning power propagation through the fiber and thus, gives a complete description of the fiber as a transmission system.

The key of this method is to take advantage of the sparse nature of the propagation matrix. Therefore, calculating multiple matrix powers is more efficient than performing the same number of iterations, particularly when using MatLab®.

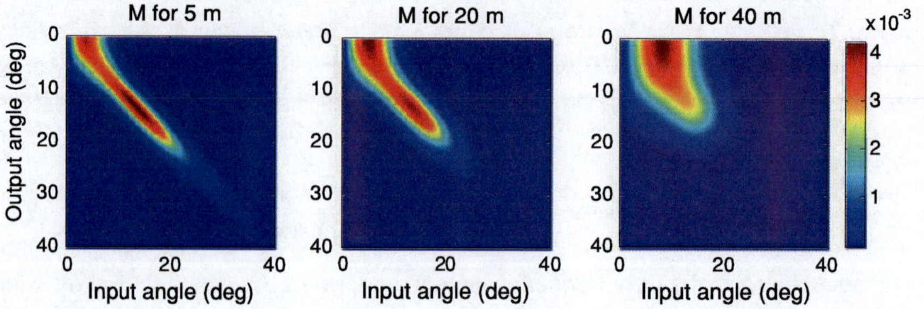

Fig. 7.2 Image representation of the propagation matrix for a SI-POF at three lengths

Even more, it is not necessary to re-calculate these matrices when changing the initial condition to obtain the output power distributions, as they only depend on the fiber diffusion and attenuation. The values of Δz and $\Delta\theta$ that are critical for convergence have been determined according to the required precision. In the calculations presented here the values used were $\Delta z = 0.001$ m and $\Delta\theta = 0.005$ rad producing accurate results. For these values, the execution time for a typical simulation of a 150 m fiber length is more than 70 times faster than the method used in [53], based on the MatLab® partial derivative equation solver.

Figure 7.2 shows the propagation matrix for a particular fiber (PGU from Toray) at three fiber lengths (5, 10 and 40 m), calculated from the attenuation and diffusion functions obtained in [53]. In this representation of the matrices, higher values are shown in deep red and lower in dark blue. As the matrices account for the effects of longer propagation lengths they acquire more high-value elements outside the neighborhood of the main diagonal indicating that power that was initially more confined is spreading through diffusion. Differential attenuation changes with increasing length are evident from the reduction of the values of the elements in the lower corner that correspond to the higher angles. The matrices have been individually normalized to their respective maxima. Otherwise, the image for 40 m will be hardly visible due to overall attenuation that gives the power loss with fiber length.

7.4.2 Space–Time Power Distribution

Once the initial condition in vector form is multiplied by the system matrix, it is possible to obtain the power distribution as a function of output angle and frequency $p(\theta, z, \omega)$ at a given length $z = L$ for a given input source. Its inverse Fourier transform gives the power temporal spread for each output angle at this length, $P(\theta, L, t)$ called space–time power distribution. In Fig. 7.3, this function is shown as a surface plot for a 150 m large core SI-POF [57] as an example. Power is represented on the vertical axis versus time and output angle on the horizontal

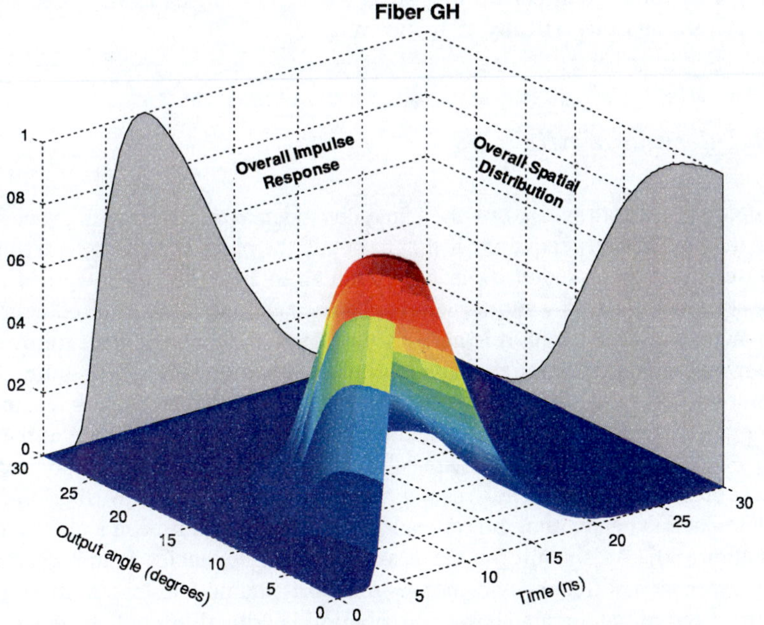

Fig. 7.3 The graph on the *upper left* is the image representation of the space–time power distribution at the output of 150 m of the PGU fiber. Below, the *lower leftmost* graph shows the overall pulse spread obtained as the integral of power over output angle. The power integral over time renders the radial profile of the FFP shown on the *upper rightmost* graph

plane. Time is shown in nanoseconds and output angle in degrees. Each vertical projection perpendicular to the axis representing the output angle gives the temporal pulse arriving at a given output angle. The sum of all these temporal functions yields the overall impulse response shown as a graph on the vertical plane perpendicular to the angular axis. The global frequency response $H(L, \omega)$ can be obtained as its Fourier transform.

The projections over the vertical plane perpendicular to the temporal axis are the radial profiles of the spatial power distribution at fixed times. The integrated power over time gives the radial profile of the FFP which is shown normalized on the plane perpendicular to the temporal axis.

The peak of the surface plot in Fig. 7.3 shows that optical power that exits the fiber over a cone up to 8° is concentrated over a relatively narrow time slot. Above this angular range, pulses have a wider time spread and their peaks increase with the output angle, but are reached at lower times than for a cosine prediction which indicates shorter delays than those that would be obtained in the absence of diffusion. In other words, diffusion improves the fiber transmission capability. In addition, it is clear that the power exiting the fiber at the highest angles is also that with the longest delays which suggest an easy way to improve fiber capacity using spatial filters to attenuate the tail at the higher angles as has been experimentally [58] and theoretically [59] shown.

7.4.3 Injection Matrix

An important aspect revealed by the triple dependence of power with propagation angle, length and time/frequency is that to obtain the pulse spread or the frequency response at one given length L, it is not enough to know the pulse spread or the frequency response at any shorter length. To compute the total acquired delay at a given angle and fiber length it is necessary to know the previous path followed by the power reaching that angle, which implies to know the space–time power distribution:$P(\theta, L_0 < L, t)$. Thus, to be able to calculate the frequency response at any length, it is necessary to know the angular power distribution right at the fiber input: $P(\theta, z = 0, t = 0)$ where there is no propagation acquired temporal delay. In fact, previous experimental results suggest that the input distribution has a strong impact on bandwidth, thus changing the balance of diffusion and differential attenuation [60]. As a result, on the basis of the discrepancies found when comparing experimental frequency responses for short- and middle-length fibers, it was found that the initial angular power distribution is critical to predict the temporal behavior [56]. The strong initial diffusion suffered by the optical power when it enters the fiber is introduced into the present framework as an independent effect by means of another matrix, which was called injection matrix **J**. The injection matrix is different for each fiber type and is estimated by measuring experimental radial profiles for short fibers obtained launching a He–Ne laser beam at different angles [56, 60].

Figure 7.4 shows the measurements for three PMMA POFs from different manufacturers as images. This representation of radial profiles versus input angle permits the compact visualization of the whole angular scan. In the images, each column represents the radial profile for the corresponding input angle as a function of the output angle on the vertical axis. The mirror images of the profiles, corresponding to negative output angles, are also shown for the sake of symmetry. The profiles for negative injection angles have also been measured and are not exactly the same as those obtained for the positive angles, revealing the deviations from the ideally symmetric response. The injection matrix is an average of the four quadrants shown in the images because its elements, as those of the propagation matrix shown in Fig. 7.2, represent power transfer only between non-negative angles.

The matrix product of the injection matrix and the source angular power distribution, $\mathbf{p}(z = 0_-)$, gives the vector describing the angular power distribution just after entering the fiber $\mathbf{p}(z = 0_+)$ as:

$$\mathbf{p}(z = 0_+) = \mathbf{J} \cdot \mathbf{p}(z = 0_+) \tag{7.10}$$

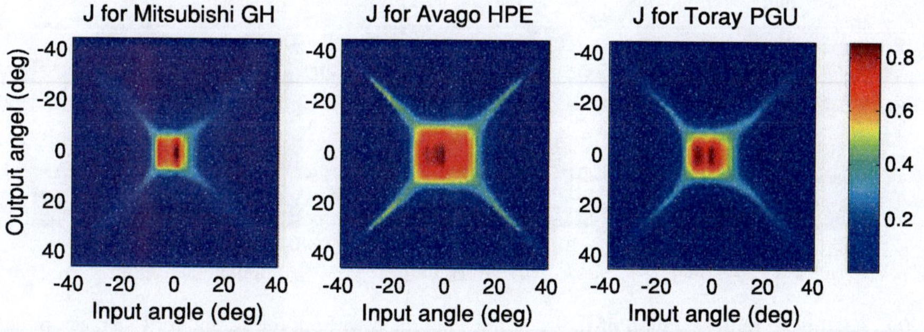

Fig. 7.4 Image representation of the injection matrix for the GH, HFB and PGU fibers (*left*, *center* and *right* graphs) obtained from inverse FFP measurements. Horizontal and vertical axes are input and output angles respectively

This equation, jointly with Eq. 7.5, provides the evolution of the space–time optical power distribution with fiber length from which angular power distribution, attenuation, bandwidth and pulse spreading can be derived. Model simulations reproduce experimental measurements, such as FFPs and frequency responses [53, 56], but can also be applied to obtain predictions of fiber properties where it is difficult or impractical to measure them. In addition, this method offers a flexible tool to study the impact of localized disturbances over transmission properties as will be shown in Sect. 7.4.4.

7.4.4 Modeling Spatially Localized Disturbances

Here, the potential of the matrix approach to incorporate localized disturbances is verified. These disturbances can be caused by different devices such as scramblers, tappers and connectors, or by curvatures and other kinds of strains. The effects of a device or defect, provided they have a linear behavior and can be characterized by a matrix, can be introduced into the model framework in a straightforward manner to reckon their impact over power propagation and to derive the changes imposed on spatial distribution or bandwidth.

In fact, an experimental method was devised to characterize a particular scrambler as a matrix that was then introduced into the model framework and successfully predicted experimental results of bandwidth versus fiber length [61]. This same approach can be adapted to characterize other devices or macroscopic effects with a linear behavior that admits a matrix representation. The procedure devised to obtain the characteristic matrix for a linear localized disturbance consists in isolating its effects from propagation effects by comparing experimental measurements obtained without the device and with the device near the fiber output end. These measurements were radial profiles extracted from output FFPs

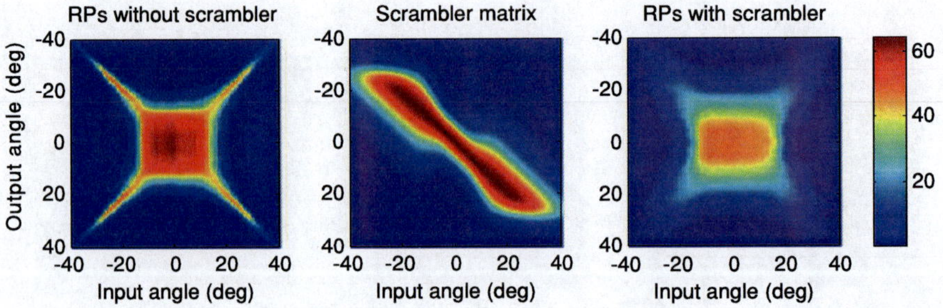

Fig. 7.5 Image representation of radial profiles obtained without the scrambler (*left*); scrambler characteristic matrix, S, (*middle*) and profiles obtained with the scrambler (*right*). Here, the scrambler matrix is represented also for negative angles to keep the same scaling for the three images

measured at different launching angles following a similar procedure as that devised to obtain the injection matrix.

The effects of the disturbance are modeled as a multiplicative matrix, **S**. Thus, given the input power profile as a column vector, $\mathbf{p}_i$, where each vector element is the power at a given angle, the output power vector, $\mathbf{p}_o$ can be obtained by the matrix product of $\mathbf{p}_i$, by the characteristic matrix **S**, or $\mathbf{p}_o = \mathbf{S}.\mathbf{p}_i$. The radial profiles for the whole angular scan can be arranged in a matrix form, where each column is a power vector whose horizontal index represents the input angle. The effects of the device over the whole data set can be directly calculated as the matrix product:

$$\mathbf{P_o} = \mathbf{S} \cdot \mathbf{P_i} \tag{7.11}$$

where $\mathbf{P}_o$ and $\mathbf{P}_i$ are matrices built as column power vector aggregates. Images in Fig. 7.5 show the radial profiles for a fiber before (on the left) and after (on the right) inserting a scrambler [61]. The scales are the same for the left and right images, so those measurements obtained with the scrambler are dimmer indicating power loss. Matrix **S**, which represents the characteristic of the scrambler, is shown in the center. In this representation, each column indicates the spread of the power in a given input angle due to the device. Each element in the column gives the relative power transferred to the angle indicated by the row index.

Once the matrix **S** is obtained to fit the experimental results, the effects of the device can be then introduced at any position in the fiber link. As an example, a device is inserted first right after the input end of the fiber and then, before its output end. Thus, the characteristic matrix has to be introduced as the right or left matrix product of **S** by the fiber propagation matrix, $\mathbf{M}(\omega)$, respectively, as described in the following equations:

$$\begin{aligned} &\textit{Device at input end,} \quad \mathbf{p}(z_2,\omega) = \mathbf{M}(\omega)^m \cdot \mathbf{S} \cdot \mathbf{J} \cdot \mathbf{p}(z_1,\omega) \\ &\textit{Device at output end,} \quad \mathbf{p}(z_2,\omega) = \mathbf{S} \cdot \mathbf{M}(\omega)^m \cdot \mathbf{J} \cdot \mathbf{p}(z_1,\omega) \end{aligned} \tag{7.12}$$

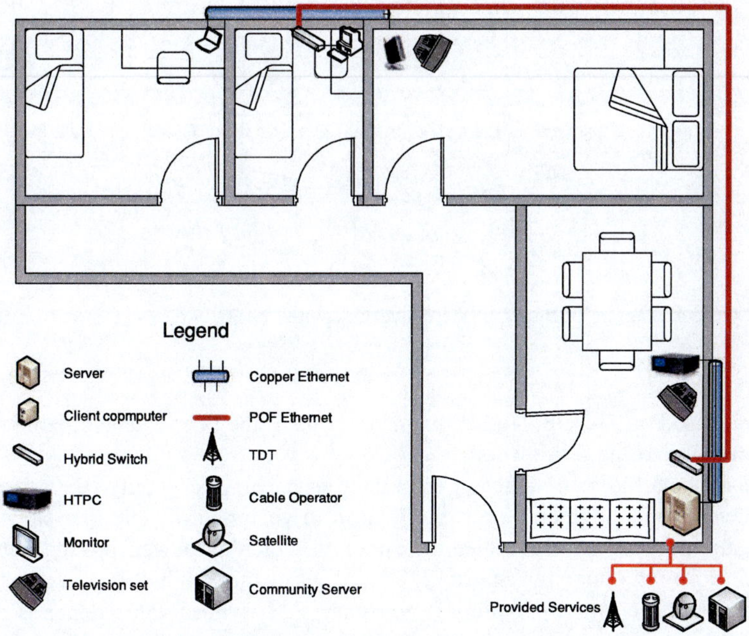

Fig. 7.6 Example of a domestic network, with the server in the living room from where it is connected to two other rooms by SI-POF

where **p** are the power vectors as defined in Eq. 7.5, **S** the localized disturbance characteristic matrix, **J** the injection matrix and **M** the fiber propagation matrix, all of which are described in the previous subsections. This approach was used to simulate the dependence of fiber bandwidth on the scrambler position. Simulations were in good agreement with experimental measurements showing how the scrambler near the detector increases the system bandwidth acting as a spatial filter while it degrades the performance if placed near the transmitter end of the fiber [31].

7.4.5 Network Simulation

Following the matrix approach described in the last subsections, an arbitrarily complicated link can be simulated if all its elements have been previously modeled. Next, Fig. 7.6 presents an example of the application of the matrix formalism to model a typical layout of a network which is based on POF. The server is placed at the living room where it is connected to the providers of different services. This server also acts as a client to provide TV and internet connections. Two of the three bedrooms in the house are connected to the server through a daisy chain

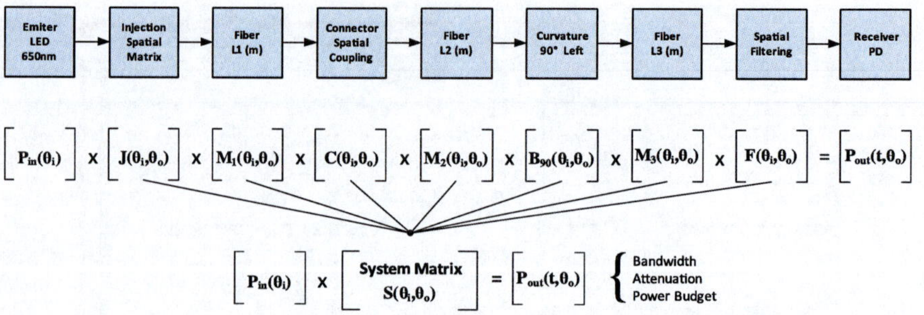

Fig. 7.7 Block diagram of the domestic network depicted in Fig. 7.6

network based on SI-POF. Notice that a 90° bend of the fiber is unavoidable to link the bedrooms to the living room.

The network has an optical source in the transceiver connected to the server that is modeled by its angular power distribution in vector form. The first step in the propagation model is to introduce the effect of launching power into the fiber that is modeled by the injection matrix **J**. The result of this product is a vector that is then multiplied by the fiber propagation matrix $\mathbf{M_1}$ whose value depends on the link distance L_1. After this first link, there is a connector that has an effect over the system losses and angular power distribution [62]. The connector effect is modeled by matrix **C**. After a second stretch of fiber of length L_2, modeled by $\mathbf{M_2}$, a 90° curvature is present whose effects are given by matrix $\mathbf{B_{90}}$. The last fiber segment L_3, is modeled by $\mathbf{M_3}$, and it ends in the receiver. Before the receiver a spatial filter modeled as a diagonal matrix has been placed to account for the angles that do not fit into the detector area [58, 59, 63].

In Fig. 7.7, a schematic of the domestic network depicted in Fig. 7.6 is shown with all the different system components represented as blocks. Each of these blocks is modeled by a vector or a matrix shown below. The whole system can then be modeled by a single matrix obtained as the product of all these matrices. Thus, the vector P_{out} yields all the information on output spatial and temporal distribution.

For a given source with a particular spatial distribution, the output angular power distribution, attenuation, bandwidth and pulse spreading can be derived. Moreover, it is easy to analyze the effects of changing the input source, the fiber type, the type of connector or the link lengths or the impact of the position of curvatures or strains. A study on those lines was performed to assess the effects over bandwidth of inserting a corrugated scrambler at the output of the fiber for different fiber lengths and input sources [61]. In addition, the blocks depicted in Fig. 7.7 can be easily included in simulation environments of commercial software tools that can model optical systems and thus can provide the BER and other estimates of system quality using the above realistic POF model. These software tools contain advanced models for single mode fibers (SMF) and various other optical components and even models for standard glass multimode fibers (MMF) but these are not able to describe large core POF behavior in non-stationary conditions as discussed above. An advantage of the above-proposed approach is that the optimization of a specific system case study does not imply re-calculating new matrix powers except if it involves a change in the fiber type. Instead, it only involves the change of several matrix products which are practically instantaneous, thus obtaining the final system response very quickly.

7.5 A Home Network Case Study

In this section, a POF-based network prototype is described. In the proposed network, all the home information sources converge within a single homogeneous open system. This prototype is based on standard equipment, devices and protocols allowing scalability and accommodation of future trends. Moreover, costs are kept as low as possible since this application is intended for the domestic market.

7.5.1 Network

The system is conceived as a hierarchical structure based on a server and several clients interconnected in a Fast Ethernet network. As soon as Gigabit Ethernet devices for POFs appear in the market, it will be upgraded to Gbit Ethernet. In this particular network the preferable topology would be either a daisy chain or a tree in order to reduce the installation cost and the aesthetical impact. In new premises with structured cabling, however, a centralized star from the server to the clients will be a better option. The backbone of the network is made of SI-POF operated at a visible wavelength, while the connection to the electronic devices is made by standard RJ45 copper cables.

The media conversion is made by low-cost transceivers or electro-optical switches. Although the principal transmission medium in this network is POF, a WiFi router has also been introduced to allow wireless connections. In addition, RJ45 copper cables have been used for some short connections because they represent the typical connection medium for consumer electronics and they are massively distributed in today homes. Thus, the performance of the three transmission media can be compared for audio, video and data signals.

The proposed system uses free open source software which are configured and adapted to meet the required objectives. There is a huge activity of many developing groups in this kind of programs around the world. Each month new versions appear with new and very interesting functionalities focused on the final user.

7.5.2 Server and Clients

Nowadays, the best cost-effective server can be implemented on a standard PC platform. In this design, all information sources converge on the server which is able to record simultaneously multiple live contents on-demand and can also follow a programable schedule. Moreover, the server stores films, music, pictures and live recordings to provide them later to each client on-demand. The technical requirements for a server-PC in such a system are lower than those of common desktop PCs. In addition, the server can also act as a client.

The clients can be multi-platform, depending on the individual user needs. In this prototype, desktop Personal Computers (PCs), set top boxes, laptops and pocket PCs were configured as different clients. The set top boxes are less versatile than PCs as the former are designed only for this particular multimedia application, while the latter have many other uses. Therefore, given that they both have good performance as clients and similar advantages, the use of a cheap box PC is recommended at any home dependence. Each client can display either the live-TV channels in real-time or any of the files stored on the server. Moreover, the users can easily program recordings from the client to be stored either on this same client or on the server. The client management is user-friendly and can be achieved through a single remote control.

7.5.3 Prototype Network Performance

Figure 7.8 shows several images of server and clients of the prototype network implemented in the Optical Communication Laboratory (University of Zaragoza). This network is presently working to demonstrate how all the home information sources can converge within a single open homogeneous system. The backbone of the network is SI-POF while the final end to the electronic devices is standard RJ45 copper cables. Although wireless is the best-positioned technology as the transmission medium for houses, it does not reach the effective bandwidth requirements due to wall-material problems and thus, its reliability remains in question. The ease of POF installation, which is its main advantage over the wired media, has been verified. In addition, when using POF, a connection can be easily checked because it operates at a visible wavelength. Also, the hand-made network assembly is rendered simpler as the visible light facilitates the connection of the fiber with its components. Therefore, POF and its components are at least as easy to install as copper cable. In addition, the flexibility of POF cable down to a small radius makes it ideal for narrow cable shafts where installing POF will be faster than installing conventional copper cable. Finally, it has been demonstrated that the system backbone is behaving correctly and that both the POF and the prototypes used in the implemented network (switches, converters and connectors) have shown a high reliability.

Fig. 7.8 Images of the server and several clients of the POF-based network installed at the Optical Communications Laboratory of the Engineering and Architecture School of University of Zaragoza

7.5.4 The Convergence Problem

The meaning of the term convergence depends on its context. From the content providers' point of view, convergence relates to the combination of audio, video and data signals over the same physical medium. This concept is what we simply call *triple play*. Internet technologies understand convergence as the transmission of TV content on the web. For the screen manufacturers this term means the interchangeability of PC monitors and TV screens.

The two major obstacles in the way of media convergence are interoperability and rights of management. The services and equipment providers do not have the initiative to provide open standards and systems that would allow users to feel a simple and cheap media experience. Accessing all the video and audio material stored in a hard disk is relatively simple; however it is conceptually complicated through a network. The interoperability problem does not only fall on content providers and device manufactures, but also on providers of interconnection devices [64].

The digital network interoperability problem is too complicated to be solved by a single company. The proprietary standards used in the actual market do not facilitate commercial solutions to implement a flexible home network. The new devices and standards are developed with the interest of the companies as a goal which does not necessarily tend to convergence. However, some progress has been made in order to homogenize standards. The most important is the Digital Living Network Alliance (DLNA^{TM}) that promotes the use of UpnP protocol for sharing media content across LAN and thus, standardizes formats. DLNA™ open system should be used for all the manufacturers to standardize the technology.

Convergence does not depend just on the technology, but also on the interest of the different players in the market place. In conclusion, the objectives of the electronics equipment manufacturers and of the service providers are far from those of the consumers. However, the present or near-future technologies will allow to satisfy the user needs at a minimum cost. There are a number of reliable transmission media and protocols widely known for video or voice-over-IP. Thus, the actual situation is ideal for the advent of domestic networks, where standard communication equipment is managed by the users to distribute multimedia contents homogeneously in their houses with quality and at a reasonable cost.

7.6 Conclusion

In this chapter the use of POF for in-house systems has been presented. Moreover, simulation of POF and a methodology for system engineering that includes POF is described. More specifically, the performance of a POF-based domestic network prototype for distributing multimedia contents has also been presented and verified, demonstrating how all the home information sources can converge within a single open homogeneous system. This system is conceived as a hierarchical structure based on a computer acting as the server and several clients interconnected in a network. The design of this type of network, where the links between active components are made by short fiber segments that frequently suffer tight bends, poses limitations which can be optimized if the layout of the system is analyzed prior the implementation. Therefore, a method is proposed based on matrix modeling of the different system components including fiber propagation that provides a complete description of the network performance and can be adapted to optical network simulation software packages.

References

1. Faller J (2006) POF for structured cabling in homes and small business. Tutorial of the POF World West, Santa Clara, CA
2. Ziemann O, Zamzow PE, Daum W (2008) POF handbook: optical short range transmission systems. Springer, Berlin, Germany
3. Zubia J, Arrue J (2001) Plastic optical fibers: an introduction to their technological processes and applications. Opt Fiber Technol 7(2):101–140
4. Mitsubishi Rayon Co LTD (2010) Eska™ specifications, from http://www.pofeska.com/pofeskae/product/02/index.html
5. AGC Asahi Glass (2003) Lucina specifications 2010, from http://www.lucina.jp/eg_lucina/indexeng.htm
6. Koike Y, Koike K (2011) Progress in low-loss and high-bandwidth plastic optical fibers. J Polym Sci Part B: Polym Phys 49(1):2–17

7. Murofushi M (1996) Low loss pefluorinated POF. In: Proceedings of the international conference on plastic optical fibers and applications (POF), Paris, France
8. Optimedia Inc Korea (2004) OM-giga specifications, from http://www.optimedia.co.kr/eng_optimedia_main_b_01.htm
9. Teshima S, Munekuni H (1998) Multi-core POF for high-speed data transmission. In: Proceedings of the international conference on plastic optical fibers and applications (POF), Berlin, Germany, pp 135–142
10. Ziemann O, Poisel H, Bloos M (2005) POF on the way home. In: Proceedings of the international conference on plastic optical fibers and applications (POF), Hong-Kong, China, pp 351–355
11. Mateo J, Oca A, Losada MA, Zubia J (2008) Domestic multimedia network based on POF. In: Proceedings of the international conference on plastic optical fibers and applications (POF), Santa Clara
12. Schelinski U, Frommhagen K, Scholles M (2004) A home networking infrastructure using POF based on IEEE 1394. In: Proceedings of the international conference on plastic optical fibers and applications (POF), Nuremberg, Germany, pp 59–66
13. Offenbeck B, Junger S, Tschekalinskij W, Weber N (2006) Possibilities and limitations of data communication over large core diameter PMMA POF. In: Proceedings of the international conference on plastic optical fibers and applications (POF), Seoul, Korea, pp 11–14
14. Lopez A, Heredia P, Mateo J, Losada M (2009) Comparative analysis of the transmission capabilities of large core plastic optical fibres. In: Proceedings of the international conference transparent optical networks-mediterranean winter conference (ICTON-MW), Angers, France, pp 1–4
15. Yoon K, Lee J, Kwon O, Woo S, Park S, Park C (2004) Fabrication of GI-POF by a multi-stage reaction method. In: Proceedings of the international conference on plastic optical fibers and applications (POF), Nuremberg, Germany
16. Watanabe Y, Tanaka C (2004) Current status of perfluorinated GI-POF and 2.5 Gbps data transmission over it. In: Proceedings of IEEE/OSA optical fiber communication conference, Los Angeles, CA, pp 12–13
17. Koike Y (2009) Status of high-speed plastic optical fibre towards giga house town. In: Proceedings of the international conference on plastic optical fibers and applications (POF), Sidney, Australia
18. Lee M, Kang S (1994) A FTTH network for integrated services of CATV, POTS and ISDN in Korea. Community networking integrated multimedia services to the home, In: Procedings of the 1st international workshop community networking integrated multimedia services to the home, San Francisco, CA, pp 261–264
19. Lee YK, Lee D (2003) Broadband access in Korea: experience and future perspective. IEEE Commun Mag 41(12):30–36
20. Plastic optical fiber trade organization—POFTO (2011), from http://www.pofto.org/home/
21. Cárdenas D, Nespola A, Camatel S, Abrate S, Gaudino R (2008) The rebirth of large-core plastic optical fibers: some recent results from the EU project "POF-ALL". In: Proceedings of IEEE/OSA optical fiber communication conference (OFC), San Diego, CA, pp 1–3
22. Okonkwo C, Tangdiongga E, Yang H, Visani D, Loquai S, Kruglov R, Charbonnier B, Ouzzif M, Greiss I, Ziemann O, Gaudino R, Koonen, AMJ (2011) Recent results from the EU POF-PLUS project: multi-gigabit transmission over 1 mm core diameter plastic optical fibers. IEEE/OSA J Lightwave Technol 29(2):186–193
23. Firecomms ltd. web page http://www.firecomms.com
24. Homefibre digital network GmbH web page http://homefibre.at
25. Diemount GmbH web page http://www.diemount.com
26. Luceat S.p.A. web page http://www.luceat.it
27. Bunge CA, Kruglov R, Poisel H (2006) Rayleigh and Mie scattering in polymer optical fibers. IEEE/OSA J Lightwave Technol 24(8):3137–3146

28. Djordjevich A, Savović S (2008) Coupling length as an algebraic function of the coupling coefficient in step-index plastic optical fibers. Opt Eng 12(47):125
29. Losada MA, Garcés I, Mateo J, Salinas I, Lou J, Zubía J (2002) Mode coupling contribution to radiation losses in curvatures for high and low numerical aperture plastic optical fibres. IEEE/OSA J Lightwave Technol 20(7):1160–1164
30. Fuster G, Kalymnios D (1998) Passive components in POF data. In: Proceedings of the international conference on plastic optical fibers and applications (POF), Berlin, Germany, pp 75–80
31. Kalymnios D (1999) Squeezing more bandwidth into high NA POF. In Proceedings of the international conference on plastic optical fibers and applications (POF), Chiba, Japan, pp 18–24
32. Losada MA, Mateo J, Garcés I, Zubía J, Casao JA, Pérez-Vela P (2004) Analysis of strained plastic optical fibres. IEEE Photon Technol Lett 16(6):1513–1515
33. Ishigure T, Endo H, Ohdoko K, Takahashi K, Koike Y (2005) Modal bandwidth enhancement in a plastic opticalfiber by W-refractive index profile. IEEE/OSA J Lightwave Technol 23(4):1754–1762
34. Ohdoko K, Ishigure T, Koike Y (2004) Propagating mode analysis and design of waveguide parameters of GI POF for very short-reach network use. IEEE Photon Technol Lett 17(1):79–81
35. Hardy A (2002) Exact derivation of the coupling coefficient in corrugated waveguides with arbitrary cross-section: application to optical fibers. IEEE J Quant Electron 31(3):505–511
36. Su L, Chiang KS, Lu C (2005) Microbend-induced mode coupling in a graded-index multimode fiber. Appl Opt 44(34):7394–7402
37. Appel't V, Zadorin A, Kruglov R (2005) Field transformation in a multimode optical waveguide with random irregularities of the film surface. Opt Spectrosc 99(4):611–619
38. Kruglov R, Appelt V, Zadorin A, Bunge C, Poisel H, Ziemann O (2006) Mode spectrum transformation in multimode fibres with rough surface. In: Proceedings of the international conference on plastic optical fibers and applications (POF), Seoul, Korea, pp 503–513
39. Aldabaldetreku G, Zubia J, Durana G, Arrue J (2007) Numerical implementation of the ray-tracing method in the propagation of light through multimode optical fibres. In: Bunge C-A, Poisel H (eds) POF modelling: theory, measurement and application. Books on Demand GmbH, Norderstedt, Germany, pp 25–48
40. Arrue J, Zubia J, Durana G, Mateo J (2001) Parameters affecting bending losses in graded-index polymer optical fibers. IEEE J Sel Top Quant Electron 7(5):836–844
41. Zubia J, Aldabaldetreku G, Durana G, Arrue J, Bunge CA, Poisel H (2004) Geometric optics analysis of multi-step index optical fibers. Fiber Integr Opt 23(3):121–156
42. Aldabaldetreku G, Durana G, Zubia J, Arrue J, Poisel H, Losada M (2005) Investigation and comparison of analytical, numerical, and experimentally measured coupling losses for multi-step index optical fibers. Opt Express 13(11):4012–4036
43. Berganza A, Aldabaldetreku G, Zubia J, Durana G (2010) Ray-tracing analysis of crosstalk in multi-core polymer optical fibers. Opt Express 18(21):22446–22461
44. Zubia J, Poisel H, Bunge C, Aldabaldetreku G, Arrue J (2002) POF modelling. In: Proceedings of the international conference on plastic optical fibers and applications (POF), Tokyo, Japan, pp 221–224
45. Durana G, Zubia J, Arrue J, Aldabaldetreku G, Mateo J (2003) Dependence of bending losses on cladding thickness in plastic optical fibers. Appl Opt 42(6):997–1002
46. Gloge D (1972) Optical power flow in multimode fibers. Bell Syst Tech J 51(8):1767–1783
47. Jiang G, Shi RF, Garito AF (1997) Mode coupling and equilibrium mode distribution conditions in plastic optical fibers. IEEE Photon Technol Lett 9(8):1028–1030
48. Zubia J, Durana G, Aldabaldetreku G, Arrue J, Losada MA, Lopez-Higuera M (2003) New method to calculate mode conversion coefficients in SI multimode optical fibers. IEEE/OSA J Lightwave Technol 21(3):776–781
49. Djordjevich A, Savovic S (2004) Numerical solution of the power flow equation in step-index plastic optical fibers. OSA J Opt Soc Am 21(8):1437–1442

50. Jeunhomme L, Fraise M, Pocholle JP (1976) Propagation model for long step-index optical fibers. Appl Opt 15(12):3040–3046
51. Rousseau M, Jeunhomme L (1977) Numerical solution of the coupled-power equation in step-index optical fibers. IEEE Trans Microw Theory Tech 25(7):577–585
52. Savovic S, Djordjevich A (2002) Optical power flow in plastic-clad silica fibers. Appl Opt 41(36):7588–7591
53. Mateo J, Losada MA, Garcés I, Zubia J (2006) Global characterization of optical power propagation in step-index plastic optical fibers. Opt Express 14(20):9028–9035
54. Gloge D (1973) Impulse response of clad optical multimode fibers. Bell SystTech J 52(6):801–816
55. Breyer F, Hanik N, Lee SCJ, Randel S (2007) Getting the impulse response of SI-POF by solving the time-dependent power-flow equation using the Crank-Nicholson scheme. In: Bunge CA, Poisel H (eds) POF modelling: theory, measurement and application. Verlag Books on Demand GmbH, Norderstedt, Germany
56. Mateo J, Losada MA, Zubia J (2009) Frequency response in step index plastic optical fibers obtained from the generalized power flow equation. Opt Express 17(4):2850–2860
57. Toray Industries Inc. (2011) Toray PGU specifications, from http://www.toray.co.jp/english/raytela/products/pro_b002a.html
58. Heredia P, Mateo J, Losada M (2007) Transmission capabilities of large core GI-POF based on BER measurements. In: Proceedings of the international conference on plastic optical fibers and applications (POF), Turin, Italy, pp 307–310
59. Losada MA, Mateo J, Martínez JJ, López A (2008) SI-POF frequency response obtained by solving the power flow equation. In: Proceedings of the international conference on plastic optical fibers and applications (POF), Santa Clara, pp 1–3
60. Losada MA, Mateo J, Serena L (2007) Analysis of propagation properties of step index plastic optical fibers at non-stationary conditions. In: Proceedings of the international conference on plastic optical fibers and applications (POF), Turin, Italy, pp 299–302
61. Losada MA, Mateo J, Martínez-Muro JJ (2011) Assessment of the impact of localized disturbances on SI-POF transmission using a matrix propagation model. J Opt 13(055406). doi: 10.1088/2040-8978/13/5/055406
62. Grivas E, Syvridis D, Friedrich G (2009) Influence of connectors on the performance of a VCSEL based standard step-index POF link. IEEE Photon Technol Lett 21(24):1888–1890
63. Grivas E, Raptis N, Syvridis D (2010) An optical mode filtering technique for the improvement of the large core SI-POF link performance. IEEE/OSA J Lightwave Technol 28(12):1796–1801
64. Cleevely D (2006) Convergence in the home. IET Commun Eng 4(3):10–15

Chapter 8
WDM Phase-Modulated Millimeter-Wave Fiber Systems

Xianbin Yu, Kamau Prince, Timothy B. Gibbon and Idelfonso T. Monroy

Abstract This chapter presents a computer simulation case study of two typical WDM phase-modulated millimeter-wave systems. The phase-modulated 60 GHz fiber multi-channel transmission systems employ single sideband (SSB) and double sideband subcarrier modulation (DSB-SC) schemes and present one of the latest research efforts in the rapidly emerging Radio-over-Fiber (RoF) application space for in-house access networks.

8.1 Introduction

The past 30 years have brought new opportunities for telecommunication networks related to incorporating radio frequency engineering and optoelectronics [1]. The development of next-generation communication systems have required innovative signal delivery schemes; as a result, the delivery of electrical signals over an optical fiber using the radio-over-fiber (RoF) technique has become one of the most important technologies for communication systems.

X. Yu (✉) · K. Prince · T. B. Gibbon · I. T. Monroy
Department of Photonics Engineering
Technical University of Denmark
Lyngby, Denmark
e-mail: xiyu@fotonik.dtu.dk

K. Prince
e-mail: kpri@fotonik.dtu.dk

T. B. Gibbon
e-mail: timothygibbon@yahoo.com

I. T. Monroy
e-mail: idtm@fotonik.dtu.dk

N. (Neo) Antoniades et al. (eds.), *WDM Systems and Networks*, Optical Networks,
DOI: 10.1007/978-1-4614-1093-5_8,

8.2 State of the Art in Radio Over Fiber Systems

8.2.1 60 GHz Wireless Fiber Systems

The recent advent of bandwidth demand increase for last-mile applications has caused a renewed interest in the use of wireless signals to cover last-meter access data distribution within the above space. The use of RoF schemes has proven invaluable in facilitating wireless microwave signal coverage within remote areas effectively delivering ubiquitous multi-gigabit wireless services through simple base stations and using the large bandwidth of the fiber optic backhaul. During the RoF technology implementations information is transported over an optical fiber by modulating the message-bearing radiofrequency (RF) signal onto an optical lightwave carrier. The RF signals are thus transported between the central station (CS) and base station (BS) nodes. We consider a scenario in which the CS acts as a centralized network controller, while BS nodes interface the optical network segment with the wireless coverage area. Most signal processing functions such as message coding, multiplexing, and modulation can therefore be centralized at the CS, which has a potential to dramatically reduce BS complexity and costs. Hence, RoF allows a micro-cellular network system with densely distributed antennas to be implemented [2].

The RoF technology concept is presented in Fig. 8.1 which essentially presents the part of the access networks in the last mile from the central office through the optical backhaul to the base stations. The combination of wireless and fiber optic technology enables the use of the ultra-broadband capacity of optical networks with the flexibility and mobility of wireless access networks. Moreover, adding the use of the optical distribution infrastructure to the wireless networks exploits the extremely low fiber transmission loss (0.2 dB/km at 1.55 μm) and immunity to electromagnetic interference of fiber to enable increased network performance.

For future broadband wireless applications, high RF carriers in the millimeter-wave band (30–300 GHz) will be required to meet the increasing demands of wireless end-users, to have instant access to high capacity information services, such as high-definition television (HDTV), image transfer, video conferencing, interactive gaming, and so on. Recently, the license-free 60 GHz band has been considered as a promising candidate to provide broadband services in next-generation wireless communications. The IEEE 802.15 group developed a millimeter-wave-based physical layer operating at 57–64 GHz for Wireless Personal Area Network (WPAN) in 2009 [3], and wireless high-definition (HD) is planning to use 60 GHz mm-wave to enable uncompressed digital transmission of full HD video, audio and data signal providing a wireless equivalent to High-Definition Multimedia Interface (HDMI) [4].

The transmission of 60 GHz signals is however limited to short-range free-space distances because of increased oxygen absorption in the fiber. Despite the atmospheric attenuation, the high frequencies (short wavelengths) allow small antennas which can enable the use of small portable receivers or possible

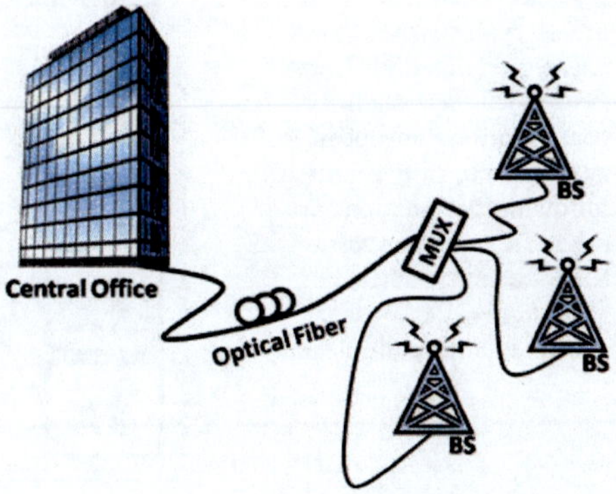

Fig. 8.1 The RoF technology concept. MUX—multiplexer, BS—base station

frequency reuse connected to a narrow beam width. Moreover, introducing RoF technology allows extending the possible reach of mm-wave wireless signals, which renders the combination of RoF and mm-wave radio frequencies one of the most important technologies for future communication systems [5].

Recent research has also shown that high-frequency electrical signal generation using optical methods is becoming more financially attractive than conventional electronic-based methods. Therefore, incorporating optical RF engineering and optoelectronics in the RoF technique is an important and promising technology [5–7].

In [8] the recent progress in the area of very high throughput wireless-over-fiber technologies and applications was presented and the focus, with the exception of a few papers like [9], was on the experimental side of research including testbeds. In [10] a new millimeter-wave generation scheme that takes advantage of the out-of-phase property between sidebands of a phase-modulated optical carrier was proposed as a novel idea. The work presented in this chapter focuses on a simulation case study performed by our group that uses all the above concepts.

8.2.2 Optical Phase Modulation for RoF Systems

Usually external modulation schemes, including intensity and phase, are employed to perform electro-optic conversions because these schemes have larger bandwidth (>100 GHz) than direct optical modulation [11]. However, in conventional intensity modulated RoF links, nonlinearities in electro-optic modulators severely limit the system performance and its application in future telecommunications.

Recent research shows that phase modulation (PM)-based electro-optic modulators can provide better linearity performance than intensity modulation in analog photonic links [12]. In particular, more than 12 $\mathrm{dB{\cdot}Hz^{2/3}}$ spur-free dynamic range (SFDR) improvement in an analog PM link can be obtained [13] using this approach. In addition, a phase modulator has the advantage of requiring no active

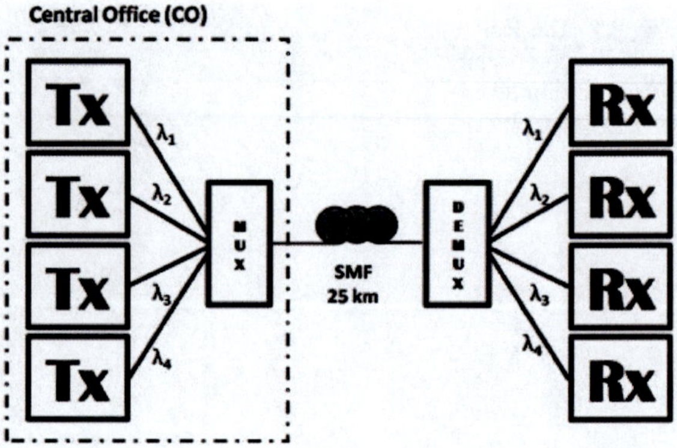

Fig. 8.2 A typical WDM access system where the phase-modulated 60 GHz system with a 4-channel configuration concept in RoF is simulated. In the figure the following designations are used: Tx = transmitter, Rx = receiver, and MUX/DEMUX = multiplexer/de-multiplexer

direct current (DC) bias and has 3 dB less optical power loss in the modulator compared to a quadrature biased intensity one. Furthermore, previous research in the area also indicated that for multichannel systems optical PM is less susceptible to crosstalk than its intensity modulation counterpart. All these advantages lead to obtaining increased signal-to-noise ratio benefits as well as higher RF signal gains [14–17]. Therefore, PM-based analog photonic systems are of great interest for fiber optical communication systems [15–20].

8.3 Design and Simulation of WDM Phase-Modulated Millimeter-Wave Fiber Systems: A Simulation Case Study

Based on the analysis above, it is clear that it can be of great interest to system performance and cost to create phase-modulated 60 GHz systems with 4-channel configurations. Suggested solutions should be based on simple and cost-effective receivers and should be able to provide efficient spectral usage, minimized crosstalk distortion and promising performance by enabling the longest possible wireless transmission range.

Based on our work, a proposed case study system overview is depicted in Fig. 8.2. Here, the performance obtained using two modulation solutions will be compared in simulation using the VPI Photonics™ simulation software and a single sideband (SSB) and a double-sideband suppressed-carrier (DSB-SC) system will be implemented. In order to demonstrate the feasibility of the proposed system setups, we distinguish and analyze the key elements of a representative system along with their schematics and parameters. These elements are: the transmitter, the multiplexer, the receivers, and finally the transmission line as described in Sects. 8.3.1–8.3.4 below.

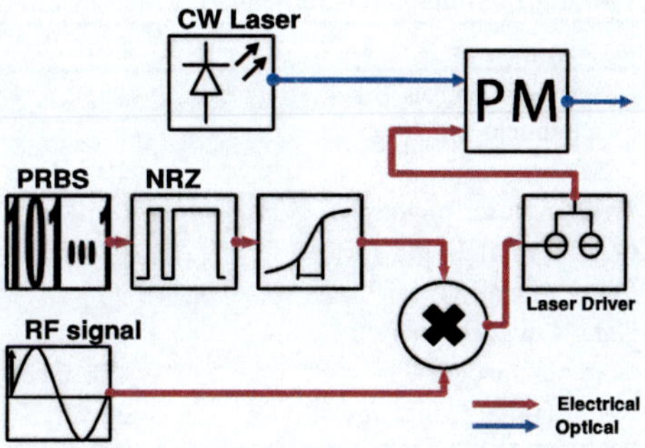

Fig. 8.3 The simulated phase-modulated RoF transmitter setup. In the figure, RF signal = radio frequency signal, PM = phase modulator and, CW Laser = a source of continuous light

8.3.1 Transmitter

The architecture of the transmitter is similar for all channels in both the proposed single sideband (SSB) and double sideband subcarrier (DSB-SC) systems and is presented in Fig. 8.3. The RoF signals are generated by mixing a 10 Gbit/s non-return-to-zero (NRZ) pseudorandom binary sequence (PRBS) with a word length of 2^7-1 with a millimeter-wave RF signal. This RF modulation format is the well-known amplitude-shift keying (ASK) modulation. Then the signals are shifted by a laser driver to achieve the required amplitude range at the phase modulator input. The amplitude-modulated RF signals are used for optical phase modulation of a continuous-wave (CW) lightwave from an external cavity laser with average optical power of 0 dBm. The laser linewidth is set to 10 MHz and phase deviation of the PM is set to 90°.

Detailed parameters for the SSB and DSB-SC transmitters are presented in Table 8.1 below. Note that 60 GHz is used in the SSB system, while 30 GHz is used in the DSB-SC system to optically generate 60 GHz signals.

8.3.2 Multiplexer

In order to combine the simulated optical channels which are to be launched into the transmission fiber, a multiplexing device is needed. In an actual designed system an arrayed waveguide grating (AWG) filter or an optical interleaver is used to perform this operation. Such multiplexer structures are shown in Table 8.2 for both the SSB as well as the DSB-SC systems (Table 8.3).

In the former (the SSB system), an input AWG is used to perform both multiplexing and filtering out the upper sideband. This filtering is done by shifting the channel AWG center frequency to a 25 GHz offset relative to the laser center frequency. Consequently, the transmitted optical signal consists of the original optical carrier and the first lower sideband at 60 GHz away from the optical carrier. Therefore the free spectral range (FSR) of the AWG can be narrowed to

Table 8.1 Simulated parameters for the transmitters of the SSB and DSB-SC case study systems

	SSB	DSB-SC
Laser wavelengths for 4 channels	1552.8, 1553.6, 1554.4, 1555.2 nm	1552.8, 1553.6, 1554.4, 1555.2 nm
Laser linewidth	10 MHz	10 MHz
Average laser power	0 dBm	0 dBm
Optical modulation format	PM	PM
Phase deviation	90°	90°
Data sequence bit rate	10 Gbit/s	10 Gbit/s
Channel spacing	100 GHz	100 GHz
Radio signal frequency	60 GHz	30 GHz
RF modulation format	ASK	ASK
Laser wavelength used in single channel configuration	1555.2 nm	1555.2 nm

Table 8.2 The simulated multiplexers for SSB and DSB-SC case study systems

	SSB	DSB-SC
Multiplexer structure	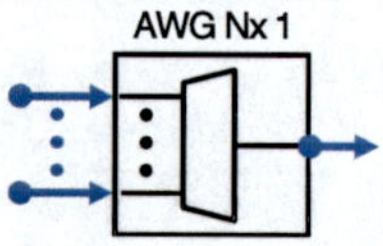	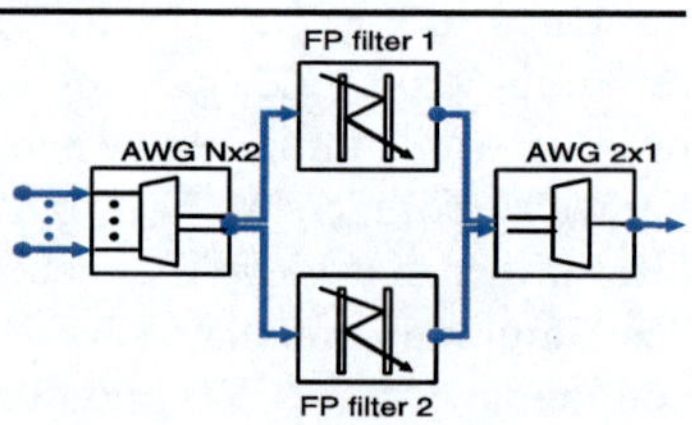
Center frequency	193.075 THz	FP1: 192.92 THz, FP2: 192.98 THz
Channel spacing	100 GHz	100 GHz
Output optical spectra	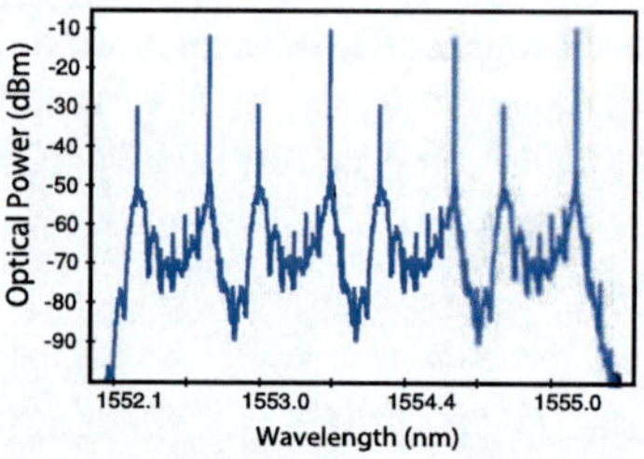	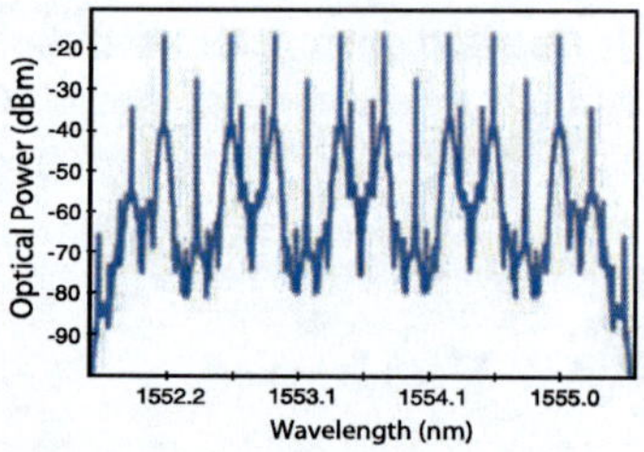

100 GHz due to the decrease in the occupied and desired spectrum. The optical spectra of all four of the WDM channels at the output of the AWG are also presented in Table 8.2.

In the latter (the DSB-SC system), an optical interleaver (OI) is designed as an *N*-input, 1-output device that filters and combines the first-order sidebands of the signals. The OI architecture is presented in Table 8.2. It is based on a pair of Fabry–Perot (FP) filters and two AWGs with (FSR) of 100 GHz. The *N*-input signals are then combined using the first AWG and sent into the parallel arms of the OI. The 4-channel optical spectra with 30 GHz modulation and transfer

Table 8.3 The simulated transmission line setup and its specifications

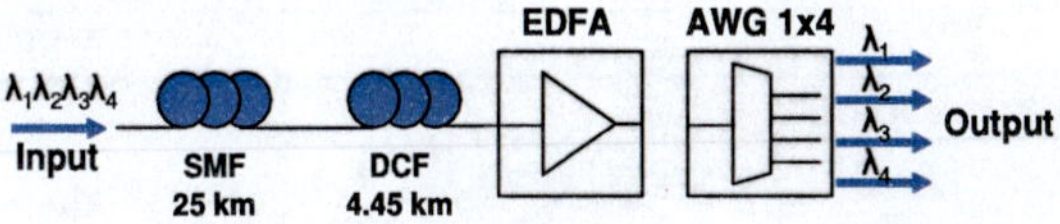

	Standard Single Mode Fiber (SMF)	Dispersion Compensating Fiber (DCF)
Length (L)	25 km	4.45 km
Dispersion (D)	16 ps/nm*km	–90 ps/nm*km
Attenuation	0.2 dB/km	0.2 dB/km
DEMUX	AWG 1x4	

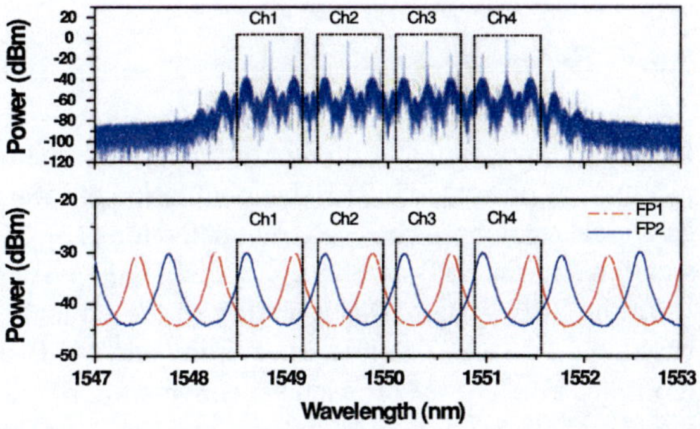

Fig. 8.4 The upper figure depicts the spectra of the four-channel input with 30 GHz modulation whereas the lower figure depicts the transfer functions of FP1 (*dashed*) and FP2 (*solid*)

functions mimicking that of the (FP) filters are displayed in Fig. 8.4. The FP filter 1 with the center frequency of 192.92 THz is used to filter out all upper sidebands, and the FP filter 2 with the center frequency of 192.98 THz filters out all lower sidebands. They are later combined using a 2 × 1 multiplexer (the second AWG). After the optical interleaver, the carrier suppression ratio related to first-order sidebands is larger than 10 dB, as seen in Table 8.2.

8.3.3 Transmission Line

The simulated transmission line is the same in both proposed systems, as shown in Table 8.3. The fiber link is composed of a 25 km single mode fiber (SMF) and dispersion matched 4.45 km dispersion compensation fiber (DCF). After the fiber transmission an Erbium-doped Fiber Amplifier (EDFA) with a gain of 10 dB is used to compensate for the fiber attenuation. The WDM optical mm-waves are then demultiplexed using a 1 × 4 AWG in order to separate the optical channels to be later demodulated at the receivers.

Table 8.4 The simulated receivers for the SSB and DSB-SC case study systems

	SSB	DSB-SC
Receiver structure	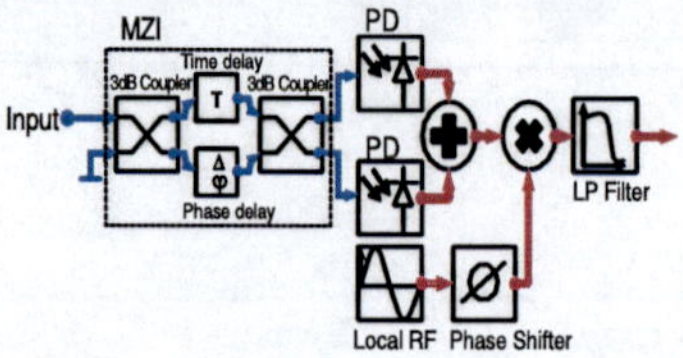	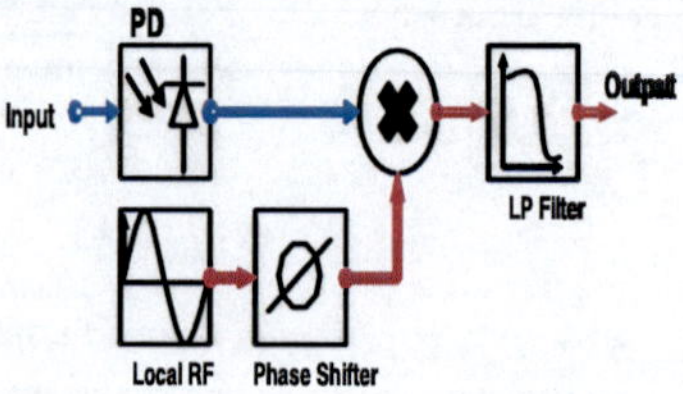

Table 8.5 The global parameters used in the simulations of the SSB and DSB-SC case study systems

Time Window	1,024/10e9 s
Bit Rate	10 Gbit/s
Sample Rate	1280 GSample/s

8.3.4 Receiver

It is well known that different modulation schemes demand different types of receivers as presented in [16]. Demodulation of information encoded on the phase of an optical carrier requires coherent detection [13]. The proposed receivers that were simulated in the SSB and DSB-CS case study systems are shown in Table 8.4.

In the SSB system, the reception of the optical phase-modulated RoF signal is based on a Mach–Zehnder interferometer (MZI) and a pair of balanced photodiodes (PD). The MZI enables conversion of phase-encoded information into intensity information [14]. It performs as a microwave photonic filter with its transfer function $H(\omega) \propto \sin^2(\omega\tau/2)$, where ω and τ are the RF frequency and time delay of the MZI respectively [13]. The time delay is set to $\tau = 8.33$ ps in the simulation, which equals that of half of the period of the 60 GHz RF signal.

In the DSB-SC system, the carrier suppression allows for a simpler receiver. The designed receiver is also presented in Table 8.4. The conversion from the optical domain into the electrical domain is performed by a high-speed photodiode. Further in the receiver, the incoming signals are mixed together with a phase-shifted local 60 GHz signal, to recover information carried on the sidebands. Then a low- pass filter is applied to remove the unwanted high-frequency components.

8.4 System Case Study: Simulated Performance and Discussions

8.4.1 Global Simulation Parameters

All the simulations for each system case study are done using the VPI-transmissionMaker^{TM} optical systems commercial tool. The global simulation parameters are presented in Table 8.5.

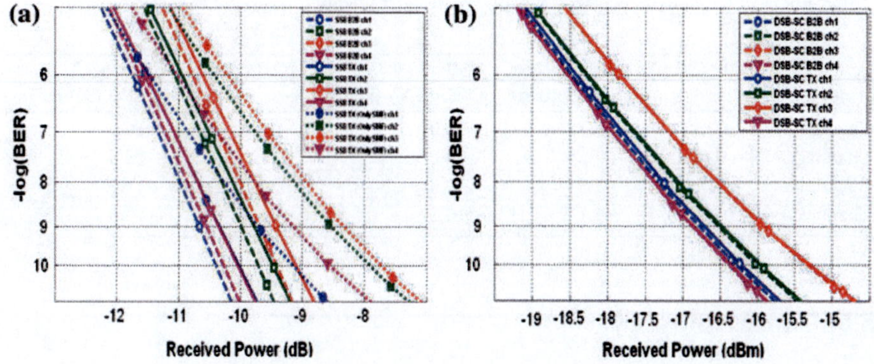

Fig. 8.5 The BER performance of **a** SSB, **b** DSB-SC systems. Transmission channel designated as TX

8.4.2 Simulation Results and Discussions

In order to compare the designed systems, we simulated the BER performances of both 4-channel systems. The results are presented in Fig. 8.5. All the simulated receiver sensitivities are listed in Table 8.6.

Figure 8.5(a) shows the BER performance for the SSB systems, including back-to-back (B2B), 25 km SMF transmission, and SMF + DCF fiber transmission for the transmitter side. We can see from Fig. 8.5(a) that the receiver sensitivity at a BER of 10^{-9} varies from −10.6 to −9.7 dBm for 4 channels in the B2B case, and from −10.3 to −9.4 dBm in the fiber transmission case. A transmission penalty of about 0.2–0.3 dB is thus clearly attributed to the SMF and DCF fibers. Moreover, the SSB modulation format is proven to be quite dispersion-tolerant. From Fig. 8.5, we note that the receiver sensitivity after SMF transmission without DCF varies from −9.7 to −8.3 dBm for 4 channels, and results in a penalty of about 0.7–1.3 dB compared to the transmission system with dispersion compensation. By analyzing 4 channels we also note that the performance of the middle channels (i.e. second and third channel) is worse than that of the other channels due to higher crosstalk introduced by the neighboring channels.

The DSB-SC system performance is correspondingly shown in Fig. 8.5(b). For a BER of 10^{-9}, the receiver sensitivity is shown to vary from –16.8 to –15.9 dB with negligible power penalty and crosstalk. Significantly the DSB-SC system presents the best BER performance from all the investigated systems. In addition to this, the SSB configuration presents promising yet slightly worse performance, enabling transmission without dispersion compensation. However, an almost 6.5 dB receiver sensitivity difference compared to the DSB-SC system is a significant power budget variance and can possibly affect the range of the wireless transmission. Table 8.6 presents the comparison summary of the receiver sensitivity of each channel as described above.

Table 8.6 Comparison of receiver sensitivity of each channel

	SSB			DSB-SC	
	B2B (dBm)	Only 25 km SMF (dBm)	SMF + DCF (TX) (dBm)	B2B (dBm)	SMF + DCF (TX) (dBm)
Channel 1	−10.6	−9.7	−10.3	−16.8	−16.7
Channel 2	−10.0	−8.5	−9.8	−16.5	−16.5
Channel 3	−9.7	−8.3	−9.4	−15.9	−15.9
Channel 4	−10.5	−9.1	−10.3	−16.8	−16.8

8.5 Conclusions

We have presented our results on the simulation of two system case studies involving phase-modulated 60 GHz fiber multi-channel transmission systems using SSB and DSB-SC modulation schemes. An optical interleaver was proposed and designed to implement WDM carrier suppression and optical generation at 60 GHz. Moreover, a simple receiver is needed to demodulate the fiber transmitted 60 GHz signals. Therefore, only 30 GHz RF signals are required at the transmitter side. The proposed WDM SSB system has a relatively simple multiplexer with higher RF frequency (60 GHz) in the transmitter and a complicated coherent receiver. Furthermore, by comparing the BER performances of the proposed SSB and DSB-SC, four-channel transmission systems, we can see that the DSB-SC system has about 6.5 dB better performance and less crosstalk. These advantageous features make the proposed DSB-SC system promising for further high-frequency wireless fiber communications.

The above work is a small sample of research performed along the lines of the emerging WDM phase-modulated millimeter-wave fiber systems which promise to revolutionize in-building access networks and systems and offer multi-gigabit per second bandwidth to bandwidth-hungry applications such as high-definition video, on-line gaming, and high date rate wireless communication.

Acknowledgments This work is supported by a Marie Curie International Incoming Fellowship within the 7th European Community Framework Program.

References

1. Capmany J, Novak D (2007) Microwave photonics combines two worlds. Nat Photonics 1(6):319–330
2. Mohamed N, Idrus SM, Mohammad AB (2008) Review on system architectures for the millimeter-wave generation techniques for RoF communication link. In: Proceedings of IEEE international RF and micorwave conference, Kuala Lumpur, Malaysia, pp 326–330
3. IEEE 802.15 Working Group for Wireless Personal Area Networks (WPANs) (2003) IEEE, http://www.ieee802.org/15/. Accessed 15 Feb 2009

4. WirelessHD®. WirelessHD® Consortium (2009) http://www.wirelesshd.org/. Accessed 15 Feb 2009
5. Fan S, Chien H-C, Hsueh Y-T, Chowdhury A, Yu J, Chang G-K (2009) Simultaneous transmission of wireless and wireline services using a single 60-GHz radio-over-fiber channel by coherent subcarrier modulation. IEEE Photonics Technol Lett 21(16):1127–1129
6. Zhu Z, Zheng X, Xu G, Guo Y, Zhang H (2009) A super-tripling technology used in radio over fiber systems for milti-service wireless signals within a millimeter-wave band. IEEE Photonics Technol Lett 21(20):1520–1522
7. Qi G, Yao J, Seregelyi J, Paquet S, Belise C (2005) Optical generation and distribution of continuously tunable millimeter-wave singnals using an optical phase modulator. IEEE/OSA J Lightwave Technol 23(9):2687–2695
8. Yu J, Chang G-K, Jia Z, Chowdhury A, Huang M-F, Chien H-C, Hsueh Y-T, Jian W, Liu C, Dong Z (2010) Cost-effective optical millimeter technologies and field demonstrations for very high throughput wireless-over-fiber access systems. IEEE/OSA J Lightwave Technol 28(16):2376–2397
9. Chowdhury P, Tornatore M, Sarkar S, Mukherjee B (2010) Building a green wireless-optical broadband access network (WOBAN). IEEE/OSA J Lightwave Technol 28(16):2230–2237
10. Chien H-C, Hsueh Y-T, Chowdhury A, Yu J, Chang G-K (2010) Optical millimeter-wave generation and transmission without carrier suppression for single- and multi-band wireless over fiber applications. IEEE/OSA J Lightwave Technol 28(16):2230–2237
11. Seeds J, Williams KJ (2006) Microwave photonics. IEEE/OSA J Lightwave Technol 24(12):4628–4641
12. Chou H-F, Ramaswamy A, Zibar D, Johansson LA, Bowers JE, Rodwell M, Coldren LA (2007) Highly linear coherent receiver with feedback. IEEE Photonics Technol Lett 19(12):940–942
13. Clark TR, Dennis ML (2007) Coherent optical phase-modulation link. IEEE Photonics Technol Lett 19(16):1206–1208
14. Urick VJ, Bucholtz F, Devgan PS, McKinney JD, Williams KJ (2007) Phase modulation with interferometric detection as an alternative to intensity modulation with direct detection for analog-photonic links. IEEE Trans Microw Theory Techniques 55(9):1978–1985
15. Kuri T, Kitayama K, Stöhr A, Ogawa Y (1999) Fiber-optic millimeter-wave downlink system using 60 GHz-band external modulation. IEEE/OSA J Lightwave Technol 17(5):799–806
16. Yu X, Jensen JB, Zibar D, Peucheret C, Monroy IT (2008) Converged wireless and wireline access system based on optical phase modulation for both radio-over-fiber and baseband signals. IEEE Photonics Technol Lett 20(21):1814–1816
17. Zibar D, Yu X, Peucheret C, Jeppesen P, Monroy IT (2009) Digital coherent receiver for phase-modulated radio-over-fiber optical links. IEEE Photonics Technol Lett 21(3):155–157
18. Yu J, Jia Z, Xu L, Chen L, Wang T, Chang G-K (2006) DWDM optical millimeter-wave generation for radio-over-fiber using an optical phase modulator and an optical interleaver. IEEE Photonics Technol Lett 18(13):1418–1420
19. Yu X, Gibbon TB, Monroy IT (2008) Bidirectional radio-over-fiber system with phase-modulation downlink and RF oscillator-free uplink using a reflective SOA. IEEE Photonics Technol Lett 20(24):2180–2182
20. Yu X, Jensen JB, Monroy IT (2008) Photonic implementation of 4-QAM/QPSK electrical modulation at millimeter-wave frequency. In: Proceedings of IEEE topical meeting on microwave photonics (MWP), Gold Coast, Queensland, Australia, p 27

Chapter 9
Fiber to the Home Through Passive Optical Networks

Alicia López, Noemí Merayo, Juan José Martínez and Patricia Fernández

Abstract This chapter presents a comprehensive overview of current deployed and novel proposed protocols and subsystem architectures for passive optical networks. Emphasis is given to link layer techniques that allow resource sharing emphasizing bandwidth utilization or so-called dynamic bandwidth allocation techniques. In the transport layer, methods for ONU remote-seeding and their limitations are described as they allow for cost reduction, which is of paramount importance in the access area.

9.1 Introduction

As fiber has established itself as the most suitable transmission medium for communication networks in the last two decades, passive optical networks have claimed their fair share of the market. In this area, the access, many architecture

---

A. López (✉) · J. J. Martínez
Photonic Technologies Group, i3A
University of Zaragoza
Zaragoza, Spain
e-mail: aliclope@unizar.es

J. J. Martínez
e-mail: jjosemar@unizar.es

N. Merayo · P. Fernández
Department of Signal Theory Communications and Telematic Engineering
University of Valladolid
Valladolid, Spain
e-mail: noemer@tel.uva.es

P. Fernández
e-mail: patfer@tel.uva.es

N. (Neo) Antoniades et al. (eds.), *WDM Systems and Networks*, Optical Networks,
DOI: 10.1007/978-1-4614-1093-5_9,

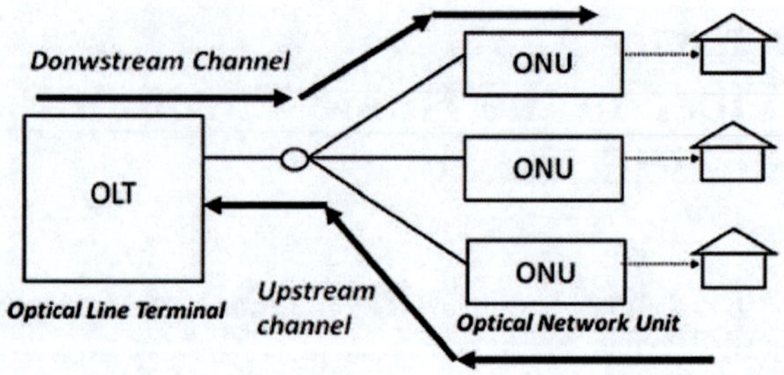

Fig. 9.1 Schematic diagram of a PON

alternatives have been proposed aimed at achieving the so-called fiber-to-the-home (FTTH) concept. The point-to-multipoint topology has been shown to be the most cost-saving option as it allows resources to be shared and thus involves a real and significant cost reduction. These architectures are basing their concept on the use of one trunk fiber from the location where centralized network management is accomplished (usually called the central office, CO) to a power splitter from which signals are redirected to the subscribers. In addition to fiber sharing, these networks also have advantages regarding the number of optical transceivers that are required. This network infrastructure is called Passive Optical access Network (PON) and entails a very attractive solution to deal with the well-known problem of communications in the first/last mile, as it can provide both high bandwidth and class of service differentiation.

This type of access network is mainly based on bidirectional communication between the optical line terminal (OLT) located inside the central office and several optical network units (ONUs) located inside or near the end subscribers [1], as Fig. 9.1 shows.

PONs do not require active components along the fiber infrastructure or in the outside plant, with the active components located and maintained at both ends of the network. PONs are typically based on a tree topology which covers distances up to 20–25 km. The network infrastructure is partially shared among users, as all of them use the same equipment at the OLT. This fact involves an important reduction in the operation and maintenance costs. Usually, a single wavelength is used for each of the two directions of communication (see Fig. 9.1), one for the downstream direction (from OLT to ONUs) and another for the upstream direction (from ONUs to OLT). Both wavelengths are multiplexed on the same fiber by means of wavelength division multiplexing (WDM). Optionally, some configurations use one additional wavelength for downstream video broadcasting.

9.1.1 PON Standards

The search for compatibility between equipment from different manufacturers has led to the creation of PON standards. Two standardization bodies have been active in the access area. On the one hand, the International Telecommunication Union (ITU) with help of the Full Service Access Network (FSAN) group has created BPON (Broadband PON, formerly APON, ATM PON) and GPON (Gigabit PON) standards. On the other hand, the Institute of Electrical and Electronics Engineers (IEEE) and particularly the Ethernet in the First Mile (EFM) group is responsible for EPON (Ethernet PON) standards.

9.1.1.1 APON Standard

The APON standard is based on the ATM protocol, defined in the ITU-T Recommendation G.983.1 (2005) [2]. Therefore, this standard uses fixed length cells of 53 bytes where downstream data are broadcasted towards all ONUs so that they can identify their own data.

The downstream APON frame consists of 54 data cells plus 2 PLOAM (physical layer operation, administration and maintenance) cells. Besides, the upstream frame consists of 53 data packets, each of them of 56 bytes and three remaining bytes used to synchronize with the OLT. Each frame includes one management slot called multiburst slot (MBS). As in the upstream direction all users share the same channel, the BPON standard defines the use of a TDMA (time division multiple access) protocol to avoid upstream collisions. Therefore, ONUs demand bandwidth by means of the MBS cells, whereas the OLT informs ONUs about the new allocated bandwidth using the PLOAM cells.

Regarding the transmission rates, the standard allows different bit rates such as 155 or 622 Mb/s symmetric or asymmetric configurations.

9.1.1.2 GPON Standard

The Gigabit PON standard is defined in the G.984.1 recommendation (2003) [3]. It supports several bit rates in both channels such as asymmetric or symmetric combinations, from 155 Mb/s to 2.5 Gb/s. It covers distances up to 25 km with splitting ratios of 1:32 and 1:64 (1:128 is planned).

The downstream frame format is a fixed frame of 125 μs, synchronized with a clock of 8 kHz. This frame periodicity assures the global synchronization of the whole system. Besides, the upstream frame has the same length and can contain information of several ONUs. Each upstream frame contains at least the Physical Layer Overhead upstream (PLOu) field. Besides the payload, it may also contain the physical layer operation and administration and management upstream (PLOAMu), the power leveling sequence upstream (PLSu), and the dynamic bandwidth report upstream (DBRu) sections. Furthermore, the GPON specification defines three ways in which one ONU can inform the OLT about its status: sending piggy-backed reports in the upstream DBRu field, using status indication bits in the PLOu field, or including an optional ONU report in the payload. On the other hand, the downstream frame consists of the physical control block downstream (PCBd) field, the ATM partition, and the GPON encapsulation method (GEM) partition. The OLT sends pointers in the PCBd field, each of them indicating the time at which each ONU starts and ends its upstream transmission. This performance allows that only one ONU can access the shared channel at the same time.

One important characteristic of the GPON standard is the GEM. This encapsulation method allows GPON networks to support all kind of traffic (native TDM, SONET, SDH, ATM, or Ethernet traffic) using a variant of SONET/SDH generic framing procedure (GFP). Furthermore, GPON permits the fragmentation of packets across multiple GEM frames, contrary to other standards such as EPON. This fragmentation process is useful to serve time- sensitive data as this encapsulated traffic is always sent at the beginning of each payload.

In the last years the work of the FSAN group has been focused on the specification of next-generation PON standards (NG-PON). In particular, XG-PON1 (Recommendation G.987, 2010) and XG-PON2 (not yet standardized) provide asymmetric 10-Gb/s downstream/2.5-Gb/s upstream and symmetric 10-Gb/s data rates, respectively [4].

9.1.1.3 Ethernet PON (EPON) Standard

The EPON standard has been standardized in the IEEE 802.3ah Task Force (2004) [5]. It was designed to support a symmetric bit rate of 1 Gb/s and to permit end-to-end distances up to 20 km with splitting ratios up to 1:32.

The EPON frame is a typical Ethernet frame with modifications in the preamble field. In particular, in this field the OLT inserts the logical link identifier (LLID) to locally identify each ONU inside the EPON.

In the downstream direction, the OLT broadcasts the packets to every ONU, which have to differentiate their own packets. To carry out this task, the medium access control (MAC) layer of the ONU extracts the LLID label of every Ethernet frame and checks if the ONU address corresponds to its ONU. On the other hand, in the upstream direction, each ONU sends its data according to a TDMA scheme. At the end of every transmission ONUs report their buffer status to demand bandwidth for the next cycle. To avoid upstream collisions, the OLT distributes the available bandwidth to every ONU and sends separate grant messages to inform them about the slot length for the next cycle and the time each of them is able to transmit. The EPON standard does not allow frame fragmentation, and hence the time slot length allocated to each ONU should match the packet lengths to fully use the total capacity of the allocated time slots.

From 2006 to 2009, the 10G-EPON standard (IEEE 802.3av) was developed as a promising solution for the bandwidth problem in future access networks [6].

The standard supports symmetric and asymmetric 10-Gb/s downstream and 1-Gb/s upstream data rates maintaining backward compatibility with already deployed EPON networks.

9.1.2 Objectives in PON Design

The increasing capacity demand at the user premises has motivated the introduction of highly efficient broadband solutions in the access area which, as mentioned before, provides a rather unique cost scenario. Therefore, PON design has been traditionally driven by the cost of network deployment and operation in terms of capital and operational expenditures (CapEx/OpEx), respectively. OpEx reduction is the most remarkable strength of PONs, while CapEx can pose a limitation to network deployment mainly due to the high fiber installation costs. Thus, when designing a PON it is important to ensure a long system life and/or ease of upgrading; in other words, the design should point at future-proof solutions with the main features of such solutions to include:

- High splitting ratio in order to serve the maximum number of subscribers and thus reduce OpEx per subscriber.
- High bandwidth in order to meet capacity demands; usually with a symmetric data rate.
- Single fiber access with bidirectional transmission to reduce CapEx.
- Long reach for increased network coverage.

Achievement of these objectives involves the introduction of novel techniques both in the link layer and the transport (physical) layer of the network. In this chapter we will deal with some of these techniques such as remote-seeding and dynamic bandwidth allocation mechanisms. Moreover, physical issues that are important for PON successful deployment will be analyzed and discussed as well.

9.2 PON Architectures

Already deployed PONs or new architecture proposals found in the literature differ mainly in both/either multiplexing domain and/or topology. In this section a comprehensive review of the alternative possible solutions is presented.

9.2.1 PON Topologies

In order to deploy a PON there are many topologies which can be applied. Some of the most commonly used topologies are summarized in the following and represented schematically in Fig. 9.2.

- *Tree topology*. This is a point-to-multipoint architecture in which all users share part of the infrastructure. This implies an important reduction in the implementation and maintenance costs in the access network. This architecture is the most widely used as most of the installed networks as well as the next-generation PON prototypes are based on this topology.

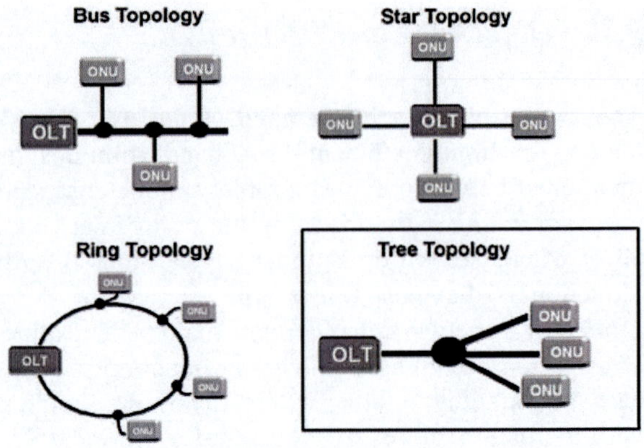

Fig. 9.2 Some typical topologies in Passive Optical Networks

- *Bus topology*. It provides a point-to-multipoint connectivity between the OLT and the ONUs. This architecture does not need splitters, so it allows a less centralized distribution and more distance coverage between the OLT and the ONUs. The main disadvantage of this topology is that faults in the bus may cause users to become disconnected from the network.
- *Star topology*. This topology supports a point-to-point connectivity between the OLT and the ONUs, as it deploys an optical fiber per user. This configuration provides very high bandwidth to end users keeping very low cost for the operation, administration, and maintenance (OAM) functions. However, this solution requires a large optical fiber deployment, which implies a very expensive implementation.
- *Ring topology*. This architecture offers the advantage of a point-to-multipoint connectivity between the OLT and the ONUs as well as possible inter-ONU connectivity. Moreover, in this type of architecture it is very easy to implement protection mechanisms since they are inherently present in any ring design. However, this topology shows important difficulties especially those related to the power budget. When the optical signal passes through several couplers, it becomes degraded and attenuated. Thus, the number of ONUs that can be connected to the ring PON is limited.

Although currently deployed PONs mainly focus on the tree topology, proposals found in the literature include architectures based on bus [7] as well as ring [8, 9] topologies. The latter ones offer the possibility of merging metro and access segments thus allowing communications among network end users. Furthermore, in these cases, PON infrastructure can be used as backbone for broadband wireless services.

9.2.2 Multiplexing Techniques in PON

Several multiplexing techniques can be introduced in a PON scenario. All of them aim at the efficient use of available resources (i.e. bandwidth and optical power).

They should take into account the ultimate design objective in PONs: to reach high splitting ratios and bandwidth offered to final users. Some of the most common multiplexing approaches are summarized next.

- *TDM* (*Time division multiplexing*). Multiplexing in the time domain involves intermittent, periodic transmission of ONUs connected to each OLT. Time is fragmented into slots which are assigned to ONUs to enable them for upstream transmission. Traditionally TDM slots are assigned in a fixed way leading to inefficient use of available bandwidth when ONU capacity requirements are not homogeneous. In this case, dynamic bandwidth allocation (DBA) mechanisms, which are thoroughly described in Sect. 9.3.1, must be introduced.
- *WDM* (*Wavelength division multiplexing*). Multiplexing in the wavelength domain is considered to be the ultimate solution for future-proof PONs. Using this technique, different wavelengths are dedicated to traffic transmission at individual ONUs or groups of ONUs. In such network designs, traditional power splitters are replaced by multiplex/demultiplex devices and appropriate transceivers must be placed at both OLT and ONUs. WDM-PONs use the huge fiber available bandwidth more efficiently than TDM-PONs. However, allocation of dedicated wavelengths to individual ONUs is usually not needed, being more common to assign wavelengths to groups of ONUs which share transmission capacity in a TDM fashion. Appropriate DBA algorithms for networks based on this multiplexing dimension are presented on Sect. 9.3.2.
- *OCDM* (*Optical code division multiplexing*). In orthogonal code division multiplexing, each ONU is assigned a unique code for data transmission. OLT and ONUs must be provided with data encoders/decoders, increasing the cost of networks. The utilized codes have special correlation properties so that different streams of encoded data can be transmitted simultaneously over the same wavelength. Limitations of this technique are related to the strict synchronization level required, which can be overcome by increasing the system complexity [10].
- *SCM* (*Subcarrier division multiplexing*). In subcarrier division multiplexing, radio-frequency (RF) carriers are assigned to each ONU, which are then modulated and transmitted over the same wavelength. At the link ends, separation and combination of data streams is carried out in the electrical domain. In order to multiplex high data rate streams, high electrical bandwidths are needed (i.e. high-speed and costly electrical components). This is one of the disadvantages of SCM, which together with the effect of optical beat interference (OBI) noise pose severe limitations to the application of this technique in PON [11, 12].

The described multiplexing techniques can also be combined in order to further increase the network splitting ratio and efficiently use fiber resources. Several hybrid approaches can be devised, the most common of which combines TDM and WDM because of its potential and simplicity [13–15].

WDM is present in every proposal for next-generation PONs, since it benefits from advantages such as high bandwidth, safety, longer transmission distance, and simplicity. It is foreseen as the preferred technology for future PONs operating at data rates of 10 Gb/s and beyond together with other optical access alternatives [16]. WDM-based PONs were first proposed in the early 1990s [17]. However, today already deployed networks massively use TDM, the main reason for which is the increased cost associated with WDM. Certainly, the increase in the number of wavelengths involves an increase in the number of transceivers at the OLT and the introduction of light sources at different and strictly tuned wavelengths at the ONUs. One possible solution in the short-term is the use of coarse WDM (CWDM) wavelength schemes with 20-nm channel spacings. CWDM components are relatively simple and do not require temperature control, which makes them much less costly than WDM components. Long-term solutions however lie in WDM or dense WDM (DWDM) schemes. In these cases great cost reduction can be accomplished by designing ONUs with wavelength-independent operation or so-called *colorless*.

9.2.3 PON Architectures with Colorless ONUs

Designing colorless ONUs has a number of advantages such as increased flexibility and scalability through the use of a uniform design for the whole network or even for similar networks. Moreover, it is possible to completely eliminate light sources at the subscriber ends or otherwise replace then by cost-effective components.

There are several approaches to achieve the desired colorless property, which can be classified into two main groups. On the one hand, *spectrum slicing* solutions [18–20] are based on the use of a broadband optical source in each ONU whose output is wavelength-filtered in order to obtain the appropriate upstream channels. On the other hand, *optical loopback* solutions [21–24] are based on remote feeding of ONUs with optical carriers that are generated at some point of the network, usually at the OLT.

In the optical loopback approach ONUs are usually referred to as *reflecting ONUs*. The main design options in this case are three: use of injection-locked Fabry-Perot lasers [21], use of reflective semiconductor optical amplifiers (RSOAs) [22], and use of reflective electroabsorption modulators (REAMs) in order to work with higher communication rates [23, 24]. In any of these options there are two ways for optical carrier provisioning or also called *remote seeding*. First, the OLT can transmit a dedicated wavelength together with the downstream channel so that ONUs perform upstream transmission over this dedicated wavelength. Secondly, upstream data can be transmitted over the same wavelength as downstream data, so that ONUs re-modulate the downstream optical carrier. Remote-seeding alternatives are schematically shown in Fig. 9.3.

The use of separate wavelengths for down- and up-stream requires many optical carriers to be generated at the OLT and so the number of transceivers is very high. Additionally, it can be considered to be a waste of fiber bandwidth. On the other hand, when downstream wavelength is reused for upstream transmission,

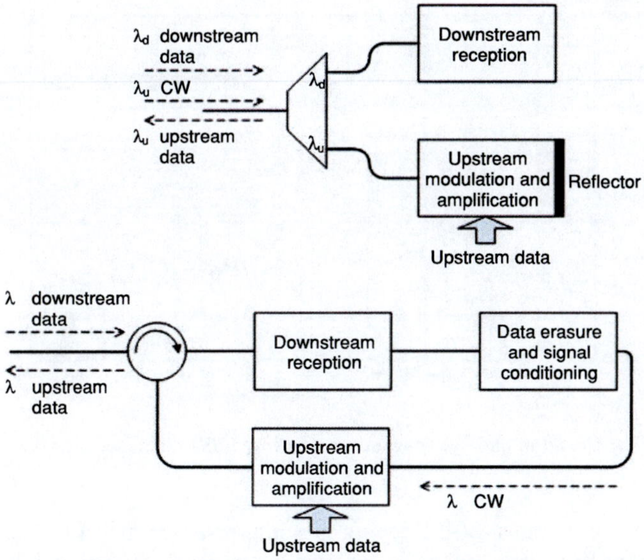

Fig. 9.3 Remote-seeding alternatives: separate wavelengths for down- and up-stream data (*up*), and reuse of downstream wavelength for upstream transmission (*down*)

bidirectional communication occurs over the same wavelength leading to physical impairments; these will be dealt with in Sect. 9.4. Additionally, in this latter case some mechanism must be implemented at the ONUs in order to avoid crosstalk between downstream and upstream data. Two main alternatives exist for this purpose, on the one hand, it is possible to erase downstream data upon reception at the ONU, which can then transmit upstream data by reusing the downstream wavelength [25]. The second solution requires no data erasing, but involves the use of orthogonal modulation formats in the bidirectional link [26–28]. Although the cost of implementing these modulation schemes might be too costly in an access scenario, the future advances in device integration will allow a steep reduction in the cost of a discrete element solution. Moreover, the possibility of using digital signal processing (DSP) to implement advanced modulations (OFDM, QPSK, M-QAM, etc.) represents a more drastic reduction in cost because it will allow substituting complex modulators and sources with low-cost electronic processors and direct modulated lasers.

The term *orthogonal modulations* encompasses such modulation formats that can coexist without mutual interference. Several combinations have been reported including differential phase-shift keying (DPSK)/intensity modulation (IM) [26] and frequency-shift keying (FSK)/IM [27, 28]. DPSK modulation requires the use of external devices and limits laser linewidth to very small values, which may be critical for the cost of the network [29]. On the other hand, many PON proposals based on FSK modulation make use of agile tunable lasers such as grating-assisted coupler with sampled reflector laser (GCSR) in order to be able to modulate

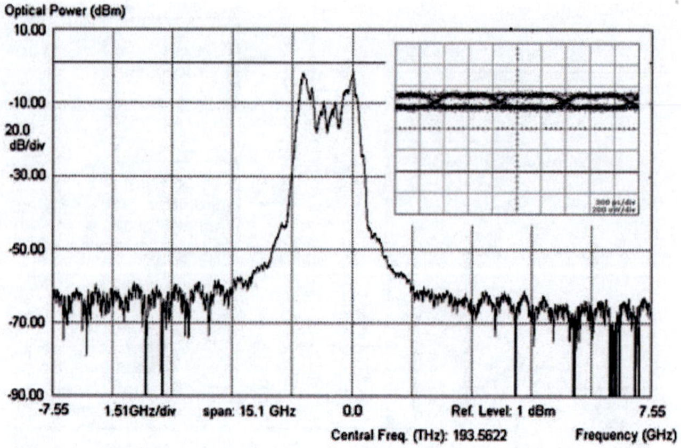

Fig. 9.4 Optical spectrum and eye diagram of a narrow FSK-modulated signal

relatively high data rates [30]. Keeping in mind, however, that PON scenarios are very cost-sensitive, the use of such tunable lasers is questionable and other modulation techniques should be considered. FSK can also be accomplished by directly modulating conventional lasers such as distributed feedback ones (DFBs) [31]. The main problem in this case is the residual intensity modulation (RIM), which makes downstream and upstream modulations to lose their orthogonality.

It is still not clear which is the winning solution; instead, it is reasonable to think that the suitability of the different alternatives will depend on the particular application. In particular, the use of QPSK modulation in combination with coherent detection has been demonstrated to be robust against dispersion in fiber [32]. Regarding FSK modulation, the use of GCSR lasers provides extraordinary flexibility to the network, since it allows easy modification of the operating wavelength. It can be also used to perform advanced techniques such as wavelength conversion [33]. However, proposals found in the literature report frequency deviations of the order of 20 GHz, which can limit the number of wavelengths and exhibit strong walk-off effect due to dispersion. Much smaller frequency deviations have also been proposed [28] by direct modulation of a DFB laser thanks to a physical phenomenon called *chirp* [34]. Figure 9.4 shows the optical spectrum and eye diagram of such an FSK-modulated signal (termed *narrow-FSK*) shown here with frequency deviation as small as 0.7 GHz. The main disadvantage of this solution is the appearance of residual intensity modulation (RIM), which is related to output power variations of the modulated laser and is visible in the eye diagram.

RIM is responsible for the loss of orthogonality between up- and downstream modulation formats and introduces a penalty in the upstream reception. The feasibility of such network design has been assessed in terms of bit error rate (BER) for both down- and upstream channels. Fig. 9.5 shows resulting curves obtained together with error-free received eye diagrams.

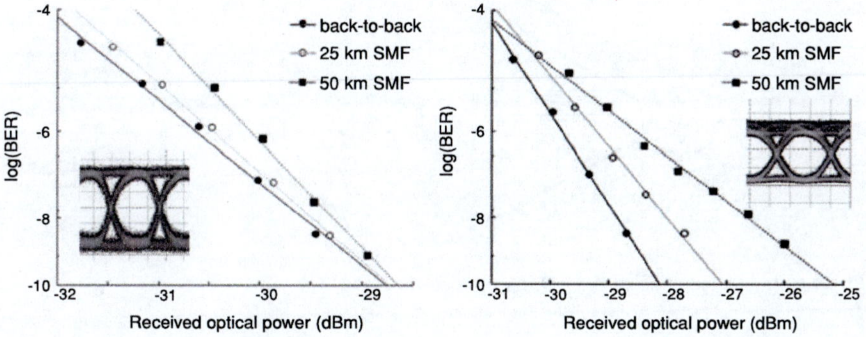

Fig. 9.5 Transmission performance for downstream narrow FSK-modulated data (*left*) and upstream IM-modulated data (*right*)

As results demonstrate, although non-strictly orthogonal, the combination of narrow-FSK and IM formats is suitable for PON architectures with colorless ONUs. Upstream transmission presents very low penalties for increasing fiber lengths due to the low impact of chromatic dispersion thanks to the narrow frequency spacing. Downstream transmission exhibits higher penalties which are mainly related to the introduction of noise by the reflective amplifier at the ONUs. The overall performance of the link is very satisfactory, which together with the simple design and cost-effective components that are needed make this design a powerful solution for future PONs.

9.3 Medium Access Control

The main problem in the link layer of PON networks occurs in the upstream direction (from ONUs to the OLT), as all users share the same wavelength, and a medium access control protocol (MAC) is necessary to avoid collisions between packets from different ONUs. There are a number of medium access control protocols which can be applied to PONs such as optical code division multiple access (OCDMA), subcarrier multiple access (SCMA), wavelength division multiple access (WDMA), and time division multiple access (TDMA).

Both OCDMA and SCMA are still not mature enough technologies to be applied in PONs. The main limitation of the former is the strict synchronization needed in the decoding stage, while the main problem of the latter is the presence of interference noise as stated in Sect. 9.2.2. On the other hand, WDMA is regarded as an expensive alternative for current access networks as it requires several wavelengths operating in the upstream, but it is a promising alternative for future PONs. Finally, TDMA is currently the most widespread control scheme in these networks as it allows a single upstream wavelength and it is relatively easy to implement.

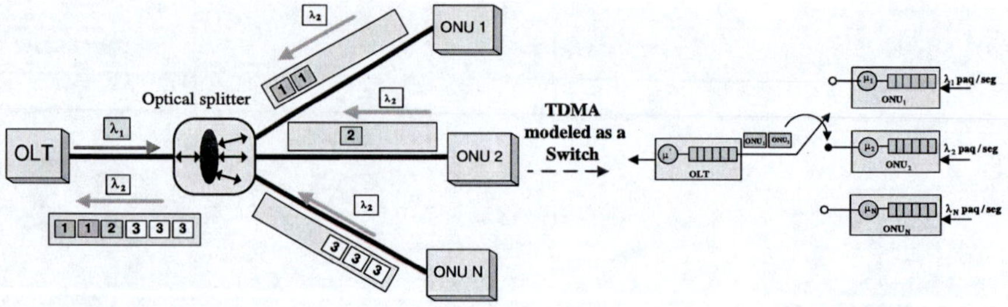

Fig. 9.6 Implementation of the TDMA protocol in PONs

In this section we will deal with access control mechanisms for PON architectures based on the tree topology, as it is the preferred in currently deployed networks. Medium access control for other network topologies has also been researched; an example of medium access control for a ring PON can be found in [35].

9.3.1 Time Division Multiple Access Protocol (TDMA) in PONs

In a TDMA scheme, time is divided into periodic cycles, and these cycles are divided into as many time slots as the number of ONUs which share the channel. Therefore, each slot is dedicated to one ONU and every cycle is organized in such a way that one slot transports packets from one ONU periodically, as it is shown in Fig. 9.6.

TDMA performance has been studied in some works [36], which model this scheme by means of a switch. When one specific ONU starts its transmission, it can be modeled as if the ONU is visited by a server, which stays inside the ONU taking out packets during its time slot. However, TDMA may be inefficient in the use of the bandwidth when the traffic nature is not homogeneous, as it always allocates the same bandwidth independently of the demand of each ONU. In fact, it is well known that real traffic is neither homogeneous nor continuous. To deal with this problem, algorithms which distribute the available bandwidth in a dynamic way have been proposed, called dynamic bandwidth allocation (DBA) algorithms. They adapt network capacity to the traffic conditions by changing the distribution of the bandwidth assigned to each ONU depending on the current requirements.

9.3.1.1 Dynamic Bandwidth Allocation (DBA) Algorithms in TDM-PONs

Since the Ethernet and the Gigabit PON standards are the most widespread deployed nowadays, some explanations of the channel control strategies are detailed for both standards.

The IEEE 802.3ah Task Force (EPON) developed the multipoint control protocol (MPCP), which controls the communication between the downstream and the upstream channels. Therefore, the implementation of a dynamic bandwidth assignment requires the use of the MPCP protocol to distribute the upstream bandwidth to each ONU. This protocol is implemented in the MAC layer located inside the OLT and uses five control messages. Among them, two control messages, called *Report* and *Gate*, arbitrate the communication between the OLT and the ONUs to assign bandwidth to each of them. In this protocol, the length of the upstream bandwidth units is not fixed, but dynamically calculated depending on the applied DBA algorithm. At the end of their time slot, ONUs send *Report* messages to the OLT to request bandwidth for the next cycle time. The OLT assigns bandwidth based on the demanded bandwidth of ONUs, and sends *Gate* control messages to inform ONUs about the new allocated bandwidth in the next cycle time. The time stamped into the *Gate* messages is used as a global time reference. Therefore, each ONU updates its local clock by means of the timestamp contained in each control message, so that global synchronization among ONUs and OLT is achieved. Moreover, each *Gate* message is able to support up to four transmission grants, which specify transmission length and transmission start time of four particular ONUs. The transmission start time of each ONU is expressed as an absolute timestamp according to the global synchronization of the system. Hence, each ONU sends packets during the assigned time slot according to its intra-ONU scheduler, which control the packet transmission from various local queues, each of them belonging to a different application.

On the other hand, the GPON standard defines three different methods to control the communications between both channels. In the first one, ONUs inform the OLT about their buffer state by sending control messages inside the DBRu field of the GPON frame. In the second method, ONUs use the bits of the physical layer overhead (PLOu) field of the upstream frame to send its next bandwidth demand. The last alternative consists of including an optional control message inside the upstream payload.

Independently of the implemented PON standard and apart from the standardized methods, DBA algorithms can be classified into offline (based on centralized management) and online (based on polling mechanisms) scheduling methods.

9.3.1.2 Online Scheduling Methods in DBA Algorithms

In the online strategies, also called polling strategies, the OLT assigns bandwidth to each ONU for the next cycle when it receives the individual control message of this ONU, which contains the next bandwidth demand. Therefore, the OLT permits each ONU to transmit just before the previous ONU finished its packet transmission. This operation allows an efficient upstream channel utilization [37–39], although the OLT does not consider the total demand of every ONU.

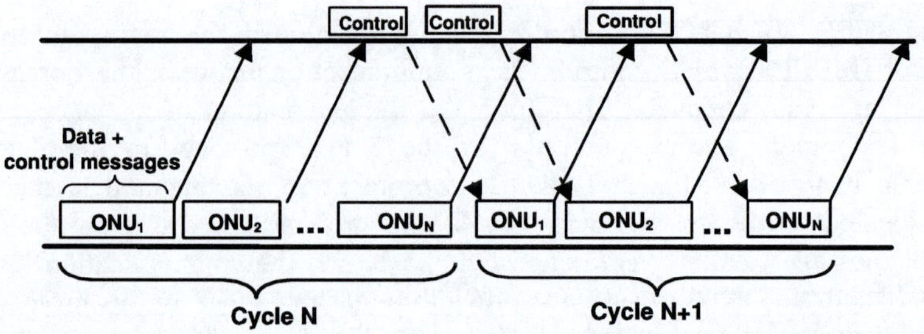

Fig. 9.7 Typical scheme of an online DBA method

In polling or online strategies, the OLT always knows the updated queue length (bytes) of every ONU and the distance to each of them. When the OLT receives a control message of one ONU with the demanded bandwidth for the next cycle, it immediately schedules its transmission for the next cycle. Then, the OLT sends a control message to this ONU to inform it about the transmission window for the next cycle (in bytes) and the instant of time this ONU is allowed to start the transmission. Once the ONU receives this message, it begins the transmission at that specified instant of time. At the end of its window, this ONU sends another different control message to the OLT with the current queue length, in order to demand bandwidth for the next cycle. By knowing queue lengths and distances to every ONU, the OLT is able to schedule transmission windows avoiding collisions. The general scheme of an online scheduling policy in PON networks is shown in Fig. 9.7.

Among online algorithms, one of the most well known is the interleaved polling with adaptive cycle time (IPACT) [38, 39]. This algorithm applies an adaptive cycle time approach in which ONUs are polled individually according to a round-robin strategy. However, under this scheme, if the OLT permits each ONU to empty its buffer at every cycle, the ONUs with high bandwidth demand will monopolize the upstream channel capacity. In order to avoid this situation the OLT should limit the maximum transmission window (MTW). Therefore, in IPACT several bandwidth allocation schemes are being studied including fixed, limited allocation, constant credit, the linear credit, and elastic credit schemes. From these schemes, the best results are achieved by using the limited service policy. Under this scheme, the OLT grants the required bandwidth to each ONU as long as the demand is lower than an imposed maximum bandwidth. When the demand is higher than this bandwidth, the OLT grants this maximum value. This operation makes the cycle time to be adaptive depending on the updated demand of each ONU.

The main disadvantage of polling schemes is that control messages sent by each ONU at the end of their transmission windows only report the current state of the queue lengths. Therefore, ONUs do not consider the traffic which arrives in the interval from transmission of control messages in cycle N to the time they are allowed to transmit in the next cycle $N + 1$. As a consequence, this traffic will not be reported until cycle $N + 1$, and it will not be sent out until at least cycle $N + 2$. As a result, this traffic suffers an additional delay of at least one cycle. In order to solve this problem, new algorithms based on IPACT have been proposed, such as the New-DBA algorithm [37]. This algorithm estimates the quantity of traffic which may arrive during this interval of time, and this estimation is included in the control message sent to the ONU in the next cycle time. In order to estimate traffic between consecutive cycles, the algorithm defines a parameter that is the control gain which determines the estimated traffic for the next cycle. Results show an improvement of the mean packet delay for low and medium loads, but the prediction makes the algorithm more complex.

In the DBA algorithm with multiple services (DBAM), a traffic prediction is also applied [40]. In order to allocate the extra bandwidth to one ONU_i for the next cycle $N + 1$, it takes into consideration the current bandwidth demand in cycle N. The assigned bandwidth for the next cycle is calculated by means of the updated bandwidth demand plus a prediction term based on the actual traffic received in the previous waiting period. The simulation study shows that DBAM improves the mean packet delay of real traffic when compared to centralized algorithms which apply fixed bandwidth allocation and limited bandwidth allocation policies.

Another algorithm is the so-called SLIding cycle time based DBA (SLICT), which is a polling algorithm that applies the limited allocation method [41]. In this scheme, the OLT allocates the required bandwidth to each ONU as long as the demand is lower than a maximum bandwidth imposed. When the demand is higher than this bandwidth, the OLT gives this latter maximum plus a proportional part of the remaining bandwidth which has not been used by previous ONUs in that cycle. This additional bandwidth is multiplied by a factor whose value is random between 0 and 1.

On the other hand, other polling schemes like the cycling polling-based DBA (BP-DBA) combine a pure TDMA scheme with a DBA method [42]. ONUs are periodically polled, but they are not polled at every cycle. When ONUs are not polled, they transmit following a fixed bandwidth allocation policy. The simulation study shows that BP-DBA achieves higher downstream throughput than IPACT, but it obtains higher mean packet delay and queue size.

Finally, Table 9.1 summarizes the main characteristics of the online DBA algorithms for PONs presented in this section.

9.3.1.3 Offline Scheduling Methods in DBA Algorithms

The offline algorithms are also called centralized algorithms (Fig. 9.8), as the OLT distributes the bandwidth for the next cycle once it receives the updated requirement of every ONU in one cycle. Consequently, the OLT makes the assignment based on knowledge of the global state of the network. The disadvantage of such operation is the waste of time that this waiting period involves. Besides, once the OLT has received every control message, the considered DBA algorithm is

Table 9.1 Main characteristics of online DBA algorithms in PONs

Algorithm	Policy	Properties
IPACT [38, 39]	Polling	• Each ONU can transmit just after the previous ONU transmits its last bit • There is no idle time between consecutive cycles • Better performance than centralized algorithms
New-DBA [37]	Polling	• Applies traffic prediction between consecutive transmissions • Adds the traffic estimation to the demanded bandwidth of ONUs for the next cycles • It is complex to implement when compared to traditional polling algorithms
DBAM [40]	Polling	• Applies traffic prediction between consecutive transmissions • Adds this traffic estimation to the demanded bandwidth for the next cycles • It is complex to implement when compare to other polling algorithms
SLICT [41]	Polling	• ONUs with high demand are scheduled more bandwidth than its maximum multiplied by a numeric factor • Has strong dependency on the multiplication factor
Cycling DBA [42]	Polling	• Consists of a periodic combination of fixed and dynamic bandwidth allocation policies • Improves the capacity of the downstream channel but achieves higher delay than traditional polling algorithms

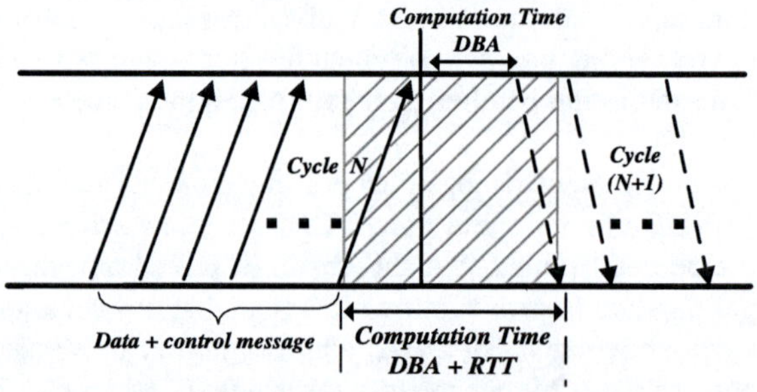

Fig. 9.8 Typical scheme of an offline DBA method

invoked and it needs some computation time to allocate the bandwidth and to generate the table of grants (*Gates*) for the next cycle time. This scheme increases even more the idle time when the upstream channel is not used. Algorithms such as dynamic minimum bandwidth (DMB) [43] and the DBA algorithm proposed in [44] (called here DBA-Choi) follow this allocation policy.

However, some other developed algorithms such as Enhanced DBA [45], apply a gate-ahead mechanism, in which the OLT sends some bandwidth allocation messages for the next cycle $N+1$ while it is receiving control messages from ONUs in the just current cycle N (see Fig. 9.9). In particular, the OLT applies an early bandwidth allocation scheme in which an ONU requesting a bandwidth

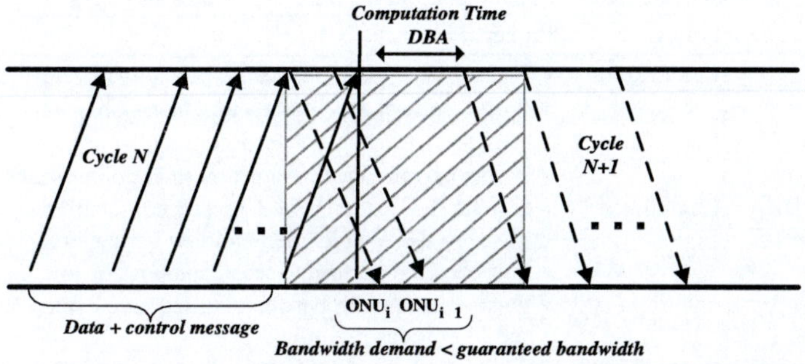

Fig. 9.9 Typical scheme of an offline DBA method with traffic prediction between consecutive cycles

lower than a determined minimum guaranteed bandwidth, can be scheduled at that moment without waiting for the arrival of every *Report*. If not, the ONU has to wait until every *Report* control message arrives. This mechanism results in an increase of the channel throughput and therefore a reduction in the waiting delay. Other algorithms such as the intra-ONU modified start-time fair queueing (M-SFQ) algorithm [46] follow the same bandwidth allocation policy at the OLT.

In order to improve the used bandwidth under high traffic loads, the algorithm n-DBA was proposed in [47]. In this algorithm, each time the OLT receives a control message from one ONU, it permits the ONU to instantaneously transmit data if its bandwidth demand is lower than its guaranteed bandwidth. At the same time, the remaining guaranteed bandwidth not used by the ONUs is added to the total non-used bandwidth in the present cycle. Depending on a variable of time updated by the arrival instant of the control message, each ONU is immediately permitted to transmit or not. Once the cycle ends, the remaining bandwidth of the current cycle is assigned to those ONUs with high bandwidth demand which have not been served before.

On the other hand, there exist centralized DBA algorithms which try to improve the bandwidth utilization by dynamically changing the order of the transmission of the ONUs at every cycle [48, 49]. In particular, the transmission order-based enhanced DBA (TOEDBA) algorithm [48] develops two different variants to predetermine the ONUs transmission order at every cycle. In the first option, called shortest request first (SRF), the transmission order starts with the ONU with the lowest bandwidth demand and it continues with the following ONUs in decreasing order. In the second option, called longest request first (LRF), the transmission order starts with the ONU with the highest bandwidth demand and then the other ONUs in an increasing order. On the other hand, in the algorithm dynamic polling order arrangement (DPOA) [49], the OLT fixes the transmission order for the next cycle $N + 1$ at the end of the cycle N. Hence, ONUs with high traffic load will transmit first in the next cycle time $N + 1$. This method benefits ONUs with heavy traffic load or ONUs with high demand of real-time traffic.

Table 9.2 Main characteristics of developed offline DBA algorithms in PONs

Algorithm	Policy	Properties
DBM [43]	Centralized	• Wasted bandwidth between consecutive cycles • Worse performance parameters than poling schemes
DBA-Choi [44]	Centralized	• Wasted bandwidth between consecutive cycles • Worse performance parameters than poling schemes
Enhanced DBA [45]	Centralized	• Permits the transmission between consecutive cycles of such ONUs which demand low bandwidth • Improves the performance parameters for low loads • Wasted bandwidth between consecutive cycles for medium and high loads
Intra-ONU M-FSQ [46]	Centralized	• Design and performance similar to enhanced DBA
n-DBA [47]	Centralized	• Permits the transmission of ONUs which comply with some premises • Improves the performance of the enhanced DBA and intra-ONU M-SFQ for every load • It is rather complex to implement
TOEDBA [48]	Centralized	• Changes the ONUs transmission between consecutive cycles • SRF (shortest request first): start ONUs with the lowest bandwidth demand. • LRF (longest request first): start ONUs with the highest bandwidth demand. • It chooses only a subset of ONUs to transmit at each cycle
DPOA [49]	Centralized	• Changes the ONUs transmission between consecutive cycles • ONUs with the highest bandwidth demand start first • It chooses only a subset of ONUs to transmit at each cycle • ONUs with high traffic load are likely to transmit in one cycle

Finally, Table 9.2 summarizes the main characteristics of the offline DBA algorithms for PONs presented in this section.

9.3.1.4 Quality of Service Strategies in DBA Algorithms

One of the most important challenges in PON networks is their capacity to offer quality of service (QoS). Network providers and operators have to offer a minimum guarantee to customers regarding their supported services. The available resources should be assigned according to the priority of certain kind of services and customers. Besides, it is essential that PONs efficiently provide both class of service and subscriber differentiation as current access networks deal with different subscriber profiles with very restrictive and heterogeneous traffic requirements.

9.3.1.5 Class of service (CoS) differentiation strategies in DBA algorithms

Dynamic bandwidth allocation algorithms should cope with class of service differentiation. Thus, it is necessary to define methods for categorizing traffic into classes of service (CoS). There are several schemes to classify traffic into different categories. In the priority queue method, used in some algorithms [39], packets belonging to different classes of services are inserted into its corresponding priority queue. All supported services share the same buffer in order to improve the performance of the highest priority traffic. Therefore, when the buffer is full, incoming packets with high priority are inserted inside the shared buffer if other packets with lower priority can be discarded. Furthermore, in other studies, packets are differentiated keeping separate queues of fixed length, one for each class of service [50]. Under this scheme, when queues are full, incoming packets will be dropped independently of their priority, as it is not permitted to replace packets with lower priority.

The priority and the separate queue methods, used for categorizing packets inside the queues, are combined with a scheduling scheme to transmit packets outside the queues. Some proposals apply the strict priority queue method defined in IEEE 802.1D, in which each ONU is equipped with a number of virtual queues equal to the number of supported services [50]. A scheduler located inside the ONU takes out packets from a queue only if queues with more priority are empty. However, during the time interval from the instant the ONU sends its control message at the end of its transmission until it starts to transmit in the next cycle, new high priority packets may arrive to that ONU. Thus, these priority packets which arrive in this interval will be sent out ahead of the lower priority packets reported in previous cycles. This performance leads to an increase in the mean packet delay and packet loss of the low priority traffic.

In order to reduce this problem other algorithms implement the fair nonstrict priority scheduling scheme [48], which works as the Strict Priority Queue, but in this case only packets reported in the last updated message can be sent out in the current cycle. Hence, if high priority packets arrive to one ONU just after the beginning of its current time slot, they have to wait until the reported low priority packets are transmitted.

In other packet scheduling strategies, the OLT allocates the exact bandwidth to each class of service in a centralized way, depending on its priority and its demand. In particular, in the DBA algorithm with multiple services (DBAM) [40], each class of service is set a maximum permitted bandwidth in every cycle. In this algorithm, the OLT allocates bandwidth firstly to the highest priority services, since they require bandwidth guarantees, and the remaining bandwidth is shared among the remaining low priority traffic depending on their demand. This bandwidth scheme is also followed by the algorithm service level agreement- algorithm (SLA-DBA) [50], the service scheduling scheme proposed in [44] and the cycling DBA algorithm [42].

Table 9.3 Main characteristics of developed Class of Service Differentiation Methods in PONs

Policy	Characteristics of the policy	Algorithms
Strict priority queue method	• Packets of one queue are sent out only if queues with higher priority are empty	• IPACT [38, 39]
Fair strict priority method	• Packets of each queue reported in the last updated message only can be sent out	• TOEDBA [48]
Exact bandwidth allocation to each class of service	• The OLT allocates the exact bandwidth to each service depending on the algorithm's policy	• SLA-DBA [50] • DBAM [40] • DBA-Choi [44] • Cycling DBA [42] • BP-DBA [51]

On the other hand, other algorithms, such as the burst-polling based delta dynamic bandwidth allocation (BP-DDBA) algorithm [51], distribute the available bandwidth to each class of service depending on fixed weights. Weighting factors are chosen according to the priority of each service. In fact, the proposal uses a different profile of weights depending on the ONU congestion.

Table 9.3 summarizes the main class of service differentiation strategies followed by DBA algorithms in PON to provide quality of service.

9.3.1.6 Customer Differentiation strategies in DBA Algorithms

Apart from class of service, DBA algorithms should differ between different user profiles, so they should provide customer differentiation. Typically, end users contract a service level agreement (SLA) with a provider, which contains technical specifications called service level specifications (SLSs). As a consequence, each SLA subscriber should be given different resources and DBA algorithms have to take into account the requirements for every SLA.

Some of the performed studies are focused on a fair bandwidth distribution among various service providers which offer the same services on the same upstream channel to different users (DUAL-SLA algorithm) [52]. However, most of the works take into account one unique service provider which offers multi-service levels according to subscriber requirements. As an example, the bandwidth guaranteed polling (BGP) method [53] divides ONUs into two sets of ONUs, the bandwidth guaranteed ONUs and best effort ONUs. Bandwidth guaranteed ONUs always receive the guaranteed bandwidth according to their SLA, whereas the other ONUs receive the remaining bandwidth. The upstream channel is divided into equal bandwidth units and the OLT keeps two entry tables, one for the guaranteed bandwidth ONUs and another for the best effort ONUs. Those bandwidth units which are not used by the guaranteed ONUs can be dynamically allocated to the best effort ONUs. If one bandwidth unit is not assigned to one guaranteed ONU, it will be offered to a non-guaranteed ONU in the order listed in the best effort table.

Table 9.4 Main characteristics of developed customer differentiation methods in PONs

Policy	Properties	Algorithms
Various service providers on the same channel	• Various service providers support the same services on the same upstream channel	• DUAL-SLA [52]
Upstream channel divided into bandwidth units	• Upstream channel divided into equal bandwidth units • Units are distributed among guaranteed and non-guaranteed ONUs	• BGP [53]
Based on static weights	• One service provider • Upstream channel divided among ONUs based on fixed weights according to their SLAs	• DMB [45] • Enhanced DBA [43]
Based on dynamic weights	• One service provider • Set of dynamic weights to distribute the upstream channel so that the highest priority traffic complies with the constrains of the mean packet delay for every subscriber	• DySLa [54] • DySGAB[55]

Another common way to provide client differentiation is to use a weighted factor assigned to each profile (SLA) in the PON. Consequently, the available bandwidth is allocated depending on these fixed weights. In such methods, each ONU is assigned a minimum guaranteed bandwidth based on its associated weight [43, 45]. The weighted factors are assigned depending on the priority of the SLA of each ONU.

On the other hand, other algorithms provide a suitable combination of class of service and customer differentiation. Most of the proposed DBA algorithms which provide customer differentiation do not control that the supported services comply with the standard constrains. However, this is a very important task since service providers have to ensure that high priority services comply with constraint parameters such as mean packet delay or packet loss rate. Following this philosophy, the algorithm dynamic service level agreement (DySLa) [54], provides both service and client differentiation but, unlike other previous works, it dynamically controls that each profile fulfill every class of service requirement. In particular, the algorithm ensures that the most sensitive services keep the mean packet delay below the maximum upper bound permitted for every priority profile. The same strategy is followed in dynamic service levels with guaranteed bandwidth (DySGAB) [55], which continuously auto-adapts the bandwidth assignment to fulfill the bandwidth requirements of all profiles, leading to an outstanding performance. This auto-adjustment, allows the network to evolve automatically to the stipulated values even if they change in real time.

Table 9.4 summarizes the main class of service differentiation strategies followed by DBA algorithms in PON to provide quality of service.

9.3.2 *Wavelength Division Multiple Access Protocol (WDMA) in PONs*

WDMA-PON architectures are viewed as a solution for the scalability problem of traditional PONs. In typical PONs, the available optical power limits the maximum number of ONUs connected to one OLT because only one wavelength is shared by all users. Additionally, the available bandwidth to each subscriber decreases with the number of subscribers. However, traditional PONs may be consciously upgraded adding multiple wavelengths over the same fiber infrastructure according to the bandwidth requirements of the access network. Therefore, only those nodes which need more capacity can be separately upgraded in a cost-effective way. The WDM-PON infrastructures as well as the WDM-DBA algorithms for such architectures are being strongly studied nowadays. However, no dominant architecture has emerged as of this writing. Therefore, the gradual WDM upgrade would be limited by technological costs and will be based on the necessity of service providers.

9.3.2.1 Dynamic Bandwidth Allocation (DBA) Algorithms in WDM-PONs

In hybrid WDM/TDM-PON architectures DBA algorithms should deal with both the bandwidth and the wavelength allocation. In this way, there are two approaches to make this double allocation: the separate wavelength and time assigment or the joined wavelength and time assignment. Most of recent studies consider this second allocation policy (joined wavelength and time assignment) as it allows multidimensional scheduling.

Regarding the dynamic wavelength assignment, there are multiple schemes, some of which have been extensively used in the transport network, such as fixed, random, least assigned, least loaded, or first fit. In all these, the OLT keeps track of the utilization of each wavelength and uses this information to decide the ONUs whose wavelength assignment will be changed. The fixed scheme makes the wavelength allocation very simple to implement but it lacks most of the statistical wavelength-domain multiplexing advantages. On the other hand, the random, the least assigned, and the least loaded methods tend to excessively overload certain wavelengths [56, 57]. Finally, the first-fit scheduling wavelength scheme, in which ONUs are able to transmit in the first free wavelength, leads to an efficient solution [58]. Most of the proposed algorithms follow this policy.

Regarding the dynamic bandwidth allocation schemes applied to WDM/TDM-PONs, algorithms can follow offline and online policies (as well as in traditional TDM-PONs) to distribute the available bandwidth in each wavelength. In this way, authors in [59] developed three different algorithms, called DWBA-1, DWBA-2, and DWBA-3, with different bandwidth distribution policies. The main factor which distinguishes their performance is the offline and online policy used by each of them. In the DWBA-1 algorithm, the OLT waits the arrival of every control message in one cycle to apply the allocation algorithm for the next cycle (offline policy). In contrast, in the DWBA-2 and DWBA-3 algorithms, the OLT permits that ONUs with low traffic demand can transmit before the reception of their individual control messages (online policy). In particular, in DWBA-2, the OLT immediately allows the transmission of the lightly loaded ONUs whose demanded bandwidth is lower than an imposed minimum bandwidth.

The algorithm proposed in [58], called WDM-IPACT, is an extension of the interleaved polling adaptive cycle time (IPACT) developed for EPONs. This algorithm applies the first-fit technique to dynamically select each wavelength channel. Hence, the algorithm permits each ONU to transmit in the first available upstream wavelength following a round-robin fashion discipline. In order to do that, the OLT dynamically calculates when each upstream channel becomes idle since it keeps track of the round-trip time of every ONU and its current allocated bandwidth on the different upstream channels.

The algorithm proposed in [57] developed an extension to the multi-point control protocol (MPCP) for WDM-PONs to provide dynamic bandwidth allocation. They implemented two scheduling paradigms for WDM-EPONs, one which follows an online discipline and another with an offline discipline. It is assumed that each ONU can transmit on all supported wavelengths in the WDM-EPON by means of a tunable transceiver with negligible tuning time. The simulations demonstrated that the online scheduling obtains lower delays than the offline scheduling, especially at medium and high ONU loads. This performance happens because the offline schemes allow an inter-scheduling cycle gap between consecutive cycles and ONUs are not able to transmit during them. The authors proposed a future approach to permit that the scheduler of the OLT could combine the online and the offline policies.

Other algorithms as the one presented in [60] support QoS in a differentiated service framework. The algorithm, called QoS-DBA-2, allows each ONU to simultaneously transmit in two channels, where each channel is dedicated to a different type of traffic. One wavelength channel is used for the highest priority traffic, and the other one for the medium and low priority traffic. This scheme may underutilize the channel capacity whether the high priority traffic is not enough to take advantage of its dedicated wavelength. To overcome this issue, the authors proposed a new algorithm which allows ONUs to transmit part of the low priority traffic on the high priority wavelength channel only if the minimum quality of service requirements of the high priority traffic is not affected.

Besides, the algorithm dynamic wavelength assignment to support multi-service level agreement (DyWaS-SLA) proposed in [61], simultaneously provides class of service and subscriber differentiation. The most important strength of the algorithm is that it ensures a minimum guaranteed bandwidth to every profile depending on the contracted SLA, by means of assigning a fixed weighted factor to each profile. Moreover, it uses the first-fit wavelength schemes and it leads to an efficient solution.

A summary of the main characteristics of the algorithms presented in this section regarding WDMA access protocol applied to WDM-EPON networks is shown in Table 9.5.

Table 9.5 Main characteristics of developed DBA algorithms in hybrid WDM/TDM-PONs

Strategies	Algorithms	Properties
Centralized-first-fit method	• DWBA-1 [59]	• The OLT waits for every Report to assign bandwidth for the next cycle • The excess bandwidth is distributed among the most overloaded ONUs at the end of the cycle
	• DWBA-2 [59]	• ONUs with low demand transmit before the reception of every Report • The excess bandwidth is distributed among the high loaded ONUs in a round-robin fashion discipline
	• DWBA-3 [59]	• ONUs with low demand transmit before the reception of every Report • It ensures a part of the excess bandwidth to every high loaded ONU depending on its bandwidth demand
Polling-first-fit method	• WDM-IPACT [58]	• The OLT schedules the transmission of every ONU just after receiving its Report message at every cycle • Supports class of service differentiation by means of the strict priority queue method
	• DyWas-SLA [61]	• The OLT schedules the transmission of every ONU just after receiving its Report message at every cycle • It ensures a guaranteed bandwidth by means of weighted factors
Separate wavelengths to each service	• QoS-DBA-2 [60]	• Each wavelength is dedicated to different classes of services depending on their priority

9.4 Transport Layer Issues in PON Networks

There are several transport layer issues that must be taken into account when dealing with PONs, as they can introduce serious restrictions to upstream and downstream data transmissions. These limitations are all related to the fact that the transmission media, i.e. the optical fiber, are shared among end users and thus both available optical power and bandwidth are shared resources. Because of the passive nature of PONs, physical degradations are even more pronounced, since they cannot be addressed by introducing active devices such as amplifiers.

In this section we will deal with fiber impairments derived from bidirectional transmission over single-mode fiber (SMF) including linear and non-linear effects. These degradations are harmful in already deployed TDM-PON architectures, but are especially deleterious in WDM-PONs that make use of several wavelengths. As a consequence, high optical power levels are present and out-of-band effects result in interchannel interference.

The excess of optical power injected into the fiber gives rise to effects which are negligible otherwise such as Rayleigh and Brillouin scattering. Particularly, if the colorless approach is adopted through an optical loopback technique as described in Sect. 9.2.1, optical power levels can be very high. This is because in such a case, ONUs are remotely-seeded by the OLT, which must provide the optical carrier(s) for upstream transmission. Thus, the seed signal(s) must have enough power in order to guarantee proper upstream reception considering the round-trip travel through the fiber link.

On the other hand, using more than one wavelength introduces the possibility that undesirable physical phenomena disrupt the transmission by means of inter-channel interference.

Some of the most common harmful effects are reflections, Rayleigh and Brillouin scattering, which are explained in the following sections. All of these effects scatter the injected optical power in the source counter-propagating direction (although Rayleigh distributes it in both directions). So, when we have a high-power source in the OLT, we also have strong backscattered powers at the OLT end, which can interfere with the proper reception of the upstream data, making it the most vulnerable channel of the communication. Therefore, at the end of the section below, some upstream transmission performance measurements are included in order to assess the degradations. As it will be shown, spectrally broadening the optical source is an effective technique to mitigate such degradations.

9.4.1 Reflections

Reflections present the most simple and intuitive degradations in bidirectional transmissions over fiber. They can appear both in the transmission ends and at intermediate points of the link. Generally reflections are caused by non-perfect connectors or splices and devices and can affect both downstream and upstream transmission [62, 63].

This effect involves losses, since part of the optical power transmitted in one direction returns in the opposite direction. More critically, the reflected signal can interfere with useful signals, increasing the number of errors and degrading the system throughput. As an example, the simplified diagram in Fig. 9.10 shows two possible reflections at an intermediate point in a remotely-seeded PON.

The OLT sends downstream data together with a continuous wave (CW) seed for the upstream data. As can be seen, when considering only single reflections, the upstream signal received at the OLT is affected by CW reflection and the down-stream seed received at the ONU is affected by the reflected upstream signal. In the depicted case, reflections and useful signals are transmitted at the same wave-length, and thus degradations are especially harmful since they cannot be addressed by filtering.

Reflection problems have been largely solved by the use of angled connectors (APC) and thus both loss in the propagating direction and interference in the counter-propagating direction can be neglected.

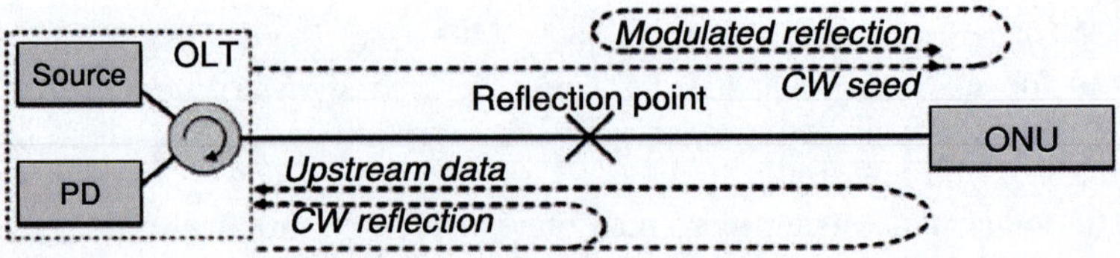

Fig. 9.10 Schematic diagram of two possible reflections at an intermediate point in a remotely-seeded PON

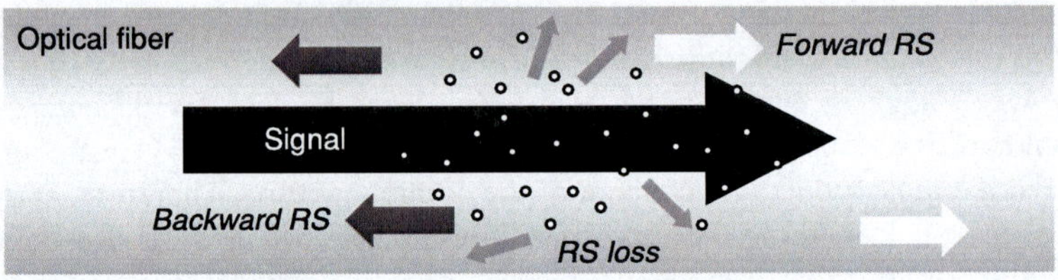

Fig. 9.11 Contributions to Rayleigh scattering: guided forward and backward RS and induced loss

9.4.2 Rayleigh Scattering

Rayleigh scattering (RS) is a linear phenomenon which light experiences when traversing any media, including optical fiber. It is caused by the non-homogeneities of the fiber refractive index due to defects in the manufacturing process. This phenomenon scatters light in every direction, which in fiber transmission implies that part of the optical power is transmitted forward, part is transmitted backward, and part is lost. Due to the nature of the effect, scattered power maintains the wavelength of the original signal and thus RS causes in-band interference [64, 65].

The sum of the three contributions to the Rayleigh scattering is depicted in Fig. 9.11. The forward contribution does not represent a problem, since it has an additive effect over the useful signal. On the other hand, the counter-propagating contribution involves one of the most serious transmission problems concerning bidirectional communication systems.

Rayleigh backscattering has been for long considered the most important degradation source in bidirectional optical transmission. Several techniques have been proposed in the literature to mitigate its effect. These in general involve spectrum broadening: using appropriate modulation formats [66] or by dithering the frequency of the optical source [67].

9.4.3 Brillouin Scattering

Stimulated Brillouin scattering (SBS) is a non-linear phenomenon caused by the interaction of the optical signal with acoustic vibrations of particles within the fiber. This kind of degradation arises when the optical power injected into the fiber exceeds a threshold value; in this case, scattering mostly occurs in the counter-propagating direction [68, 69]. It involves energy transfer so that the wavelength of the scattered light is slightly longer than that of the originating light. Frequency shift between original and scattered signal is lower than 20 GHz (about 11 GHz for wavelengths in the range of 1,550 nm) and the interaction bandwidth is as narrow as 20 MHz. Therefore, SBS particularly affects narrowband optical sources.

Being an out-of-band effect, SBS has been usually underestimated when assessing PON transmission performance. SBS degradation is negligible when dealing with traditional PON architectures, which do not require any optical loopback technique. This is because modulated sources exhibit bandwidths that are larger than the interaction bandwidth of this phenomenon. However, if we consider remotely-seeded PONs with colorless ONUs, the effect should be carefully studied, since high-power CW optical signals traverse the link and degrade upstream performance [70].

Figure 9.12 helps clarify this point: the back-reflected spectra for two levels of optical power injected into the fiber are plotted. The horizontal axis represents the frequency deviation relative to the frequency (wavelength) of the original signal. The optical source used in the measurements is a standard 1,545-nm DFB operated in CW mode.

Peaks observed at zero frequency deviation are caused by RS, while peaks observed at −11 GHz deviation are caused by SBS. As can be seen, back-reflected RS power smoothly increases when the optical power injected increases. On the other hand, SBS back-reflected power dramatically increases as more optical power is injected into the fiber because of the threshold effect of this phenomenon.

Figure 9.13 shows the eye diagram of the upstream data in a remotely-seeded PON. As can be seen, transmission performance is strongly impaired by the combination of back-scattering effects.

Next, some experimental measurements for the evaluation of the performance of remotely-seeded PON links are presented. Research in this area reveals that both Rayleigh and Brillouin Scattering should be taken into account and moreover the latter can be the more detrimental effect in such systems.

9.4.4 PON Upstream Performance Evaluation

Future WDM-remotely-seeded PONs will be strongly affected by reflections and backscattering phenomena. Upstream performance must be evaluated in order to quantify the degradations and implement the appropriate mechanisms to mitigate their effect. Reflections can be almost eliminated (reflected power more than 60 dB below signal power) by using angled connectors, but the effect of back-scattering phenomena remains an issue. In this section the influence of back-scattered optical power over upstream transmission is presented for a PON based on wavelength reuse such as the one described previously [29].

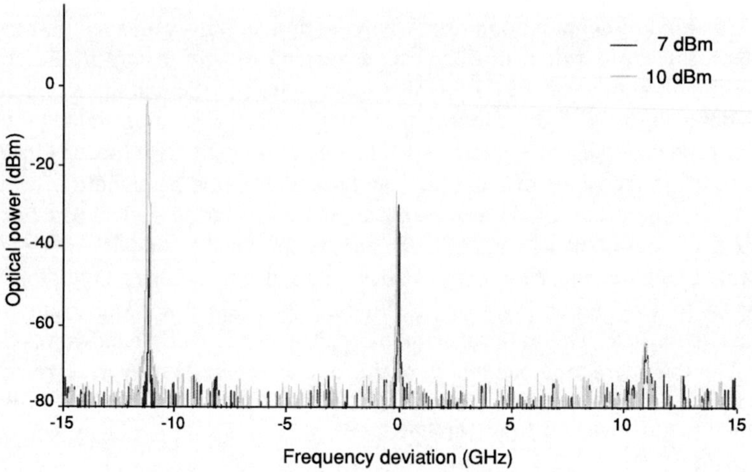

Fig. 9.12 Back-reflected optical spectrum (at OLT) for two different values of power injected into the fiber

Fig. 9.13 Upstream eye diagram affected by back-scattering phenomena

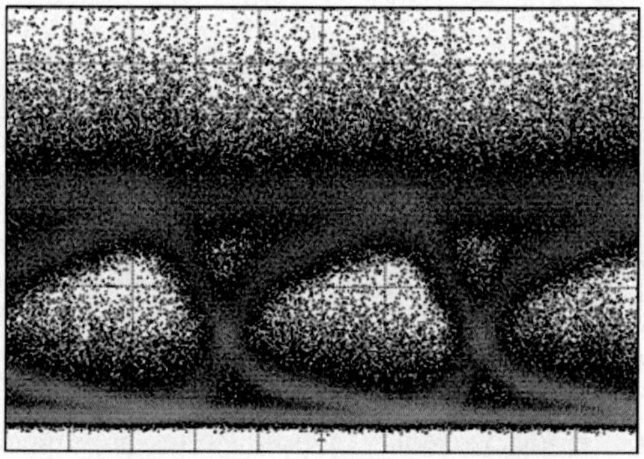

In order to assess the performance degradation in the presence of backscattering phenomena, two cases have been examined: optical carrier as generated by the source at the OLT and spectrally-broadened optical carrier. Spectral broadening is performed by applying a weak amplitude modulation over the bias current of the source. Increase of the injected optical power leads to an increase of the scattered powers (SBS located 11 GHz below reference frequency and RS located at the signal frequency). When the source linewidth is broadened, the SBS peak is reduced. This fact can be clearly seen in Fig. 9.14, which shows the back-scattered Rayleigh and Brillouin optical powers for different values of injected optical power and source broadening when 1.25-Gb/s data are transmitted.

As it can be observed, Rayleigh back-scattered power remains un-altered when spectral broadening is applied to the source. On the other hand, Brillouin back-scattered power can be efficiently reduced by this mitigation technique.

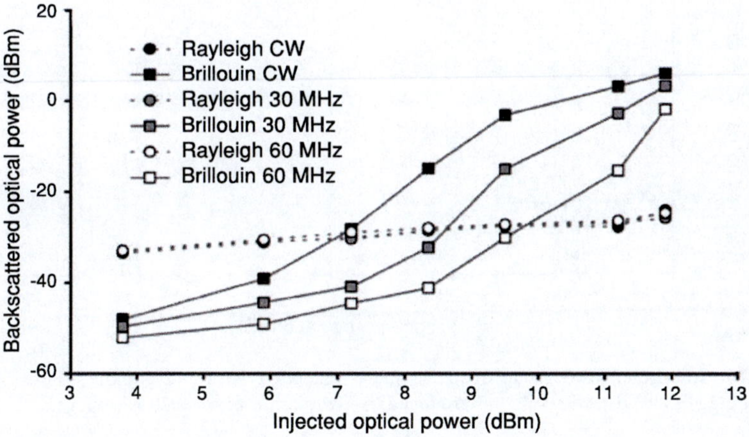

Fig. 9.14 RS and SBS back-scattered optical power for 1.25-Gb/s data transmission. (From [70], Fig. 3. Reproduced by permission of © 2010 Elsevier.)

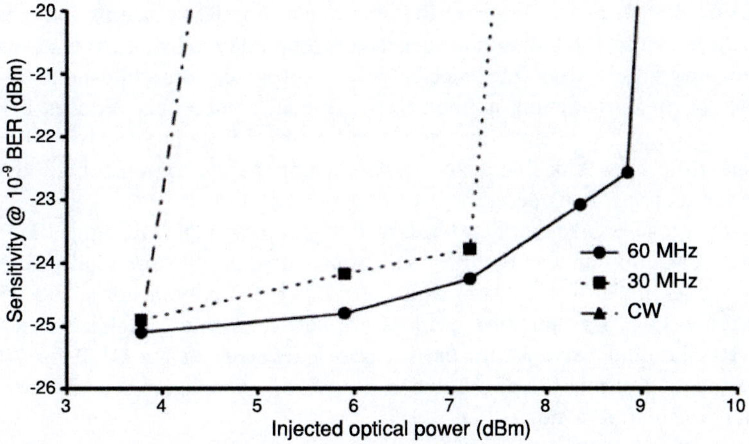

Fig. 9.15 Upstream BER performance for different spectral broadenings and 1.25-Gb/s data transmission (From [70], Fig. 9.8. Reproduced by permission of © 2010 Elsevier.)

The influence of back-scattering phenomena over upstream transmission can be measured in terms of BER of the signal received at the OLT end. Alternatively, it is possible to obtain the sensitivity of the link defined as the received optical power needed to obtain a particular BER value. Sensitivity at 10^{-9} BER for different injected powers is depicted in Fig. 9.15. Different configurations including no spectral broadening and 30-, 60-MHz broadenings are depicted.

As it can be seen, sensitivity curves for the different configurations exhibit the same behavior: smooth degradation in the first range of injected power and steep degradation when a threshold is reached. This steep degradation is related to

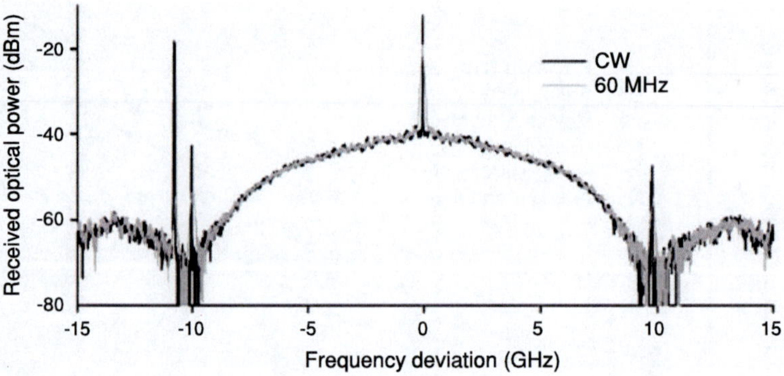

Fig. 9.16 Spectra of 10-Gb/s upstream signal at reception with and without source spectral broadening (From [70], Fig. 9.13. Reproduced by permission of © 2010 Elsevier.)

communication loss or total link failure. By source broadening, a shift of the threshold is introduced so that higher injected optical powers are possible.

Taking into account the results presented in Figs. 9.14 and 9.15 leads to the conclusion that there are two contributions degrading data transmission. The first one is responsible for the smooth sensitivity degradation, while the second one involves link failure when the injected power surpasses some threshold. The fact that the spectral broadening applied to the optical source only reduces Brillouin back-scattered power reveals that the cause for link failure is SBS. Assuming this, RS causes the slow link degradation, which remains unaffected for the different source broadenings considered.

Previous results revealed that back-scattering phenomena (RS and SBS) affect upstream transmission, being SBS the ultimate responsible for total link failure even when its power lies far from data spectra as is the previous case of 1.25-Gb/s transmission rate. The situation is more critical when future higher transmission rates (10-Gb/s and beyond) are used. Spectra received at the OLT for 10-Gb/s upstream data transmission are shown in Fig. 9.16, where again source broadening is either considered or not.

As the spectra show, Brillouin back-scattered power is located very close to the main lobe of data spectra and thus it is expected that its effect will be highly detrimental. In such case or when dealing with even higher transmission rates, spectral broadening or any other mitigation technique must be introduced in order to keep transmission performance within acceptable limits [70].

9.5 Industrial Deployments

It is expected that the number of worldwide FTTH/B subscribers will grow to 140 millions of users by 2014 [71]. In Europe, the growth rate for subscribers and homes/buildings passed picked up speed in 2010 – increasing to 22% and 21%, respectively, between June and December 2010. There were nearly 8.1 million FTTH/B subscribers and close to 33 million homes/buildings passed at the end of 2010 [72]. In March 2010, Deutsche Telekom unveiled a high-speed broadband plan, under which it will invest over 10 billion euros in the next three years in fiber optics, new mobile communications technologies and IT processes. It is projected that by 2012, there will be an FTTH subscriber base of 4 million covering 10% of all homes in Germany. On the other hand, British Telecom plans to cover two thirds of homes in the United Kingdom by 2015 based on FTTC network [73]. United Arab Emirates and Turkey are new entrants in the latest update of the Global FTTH Ranking. The United Arab Emirates made a jump start into the Global FTTH Ranking. They are now ranked fourth among all economies in FTTH market penetration, positioned ahead of all European and American economies and only behind South Korea, Japan and Hong Kong [74]. Asia will remain the leader of FTTH/B market, but the gap between the United States and Europe will be reduced. PON technology is nowadays the most preferable solution to provide FTTx all over the world. At the end of 2007 there were 29 millions of users connected to FTTx by means of PON deployments, but this number is likely to increase up to 100 millions of users by the end of 2012 [75]. This tendency means that the PON market will increase 15% per year from 2005 to 2012. North America will suffer the most aggressive deployment during this period, while the Asia-Pacific area will increase in a more stable way. However, the fastest PON deployments will happen in the west of Europe.

Regarding currently deployed networks, they are all characterized by tree topologies, with TDM being the most common multiplexing approach. Currently deployed TDM-PONs meet either GPON or EPON standard (see Sect. 9.1.1), with GPON solutions found in Europe and North America, while EPON is preferred in Asia.

In particular, EPON is one of the the most popular FTTH/B technologies in a worldwide basis. Of the 29 million of FTTH/B users at the end of 2008, 60% correspond to EPON deployments and 17% to GPON [71]. EPON is deployed mainly in the Asia-Pacific area. In Japan, which leads the FTTH market in the world, the service operator NTT (Nippon Telephone and Telegraph) had more than 10 million of connected users by EPON/GEPON technology at the end of 2008 [76], and it is expected that by the end of 2011 the number of subscribers will grow to 20 million [77]. In South Korea there were more than 5 millions of EPON/GEPON users with KT and SK Broadband operators [76] by the end of 2008. However, Japan and South Korea markets will be eclipsed by the new FFTx deployments in China with EPON as the dominant architecture there. In fact, it is expected that by the end of 2012 the number of users will grow up to 50 million [77]. Moreover, this success of EPON in China could influence both, the decrease of the EPON cost and the decisions of other emerging markets. For instance, eastern European countries such as Russia and Belarus are following the China example [78].

Regarding the United States, the main technology used is GPON, but it is also possible to find some EPON deployments. Some operators such as US SONET, Cable ONE, or CC Communications support triple play services in some states by means of EPON technology [79]. In contrast, Europe, whose FTTH/B leaders are Lithuania, Sweden, Norway, Slovenia and Slovakia [80], is deploying mainly GPON technology, whereas the EPON deployments are reduced to minor exploratory projects.

9.6 Conclusions

This chapter has reviewed some of the key protocols and techniques that will make future-proof Passive Optical Networks (PONs) a reality. In the link layer, a number of proposals for achieving dynamic bandwidth allocation among ONUs have been reported for both current TDM and also future hybrid WDM/TDM architectures. Online and offline methods have been classified and summarized with emphasis on the strengths and weaknesses of each method. Futhermore, one of the most important challenges in PONs is its capacity to offer quality of service (QoS). Network providers and operators have to offer minimum guarantees to customers regarding their supported services. The available resources should be properly assigned according to the priority of certain kind of services and customers, as current access networks are dealing with a multi-profile scenario, where a variety of users with different needs exist. Consequently, the main quality of service strategies in PONs infrastructures have been deeply described and summarized.

In the transport layer, the cost of the network infrastructure can be reduced by methods that avoid the use of an optical source at each ONU or in general permit a uniform design for all ONUs. Techniques for accomplishing ONU remote seeding have been outlined and their suitability has been discussed. Additionally, physical impairments in PON networks have been evaluated revealing an important limit to the maximum optical power in the network and thus the number of subscribers that the network can support. Finally, a description of current worldwide network deployments in the access area was presented.

References

1. Kramer G, Mukherjee B, Maislos A (2003) Ethernet passive optical networks. In: Dixit S (ed) Multiprotocol over DWDM: building the next generation optical internet. John Wiley & Sons, New York, pp 229–275
2. ITU-T standardization (2005) G.983.1: Broadband optical access systems based on passive optical networks (PON). Available at http://www.itu.int/rec/TREC-G.983.1-200501-I/en/
3. ITU-T standardization (2008) G.984.1: Gigabit-capable passive optical networks (GPON): General characteristics. Available at http://www.itu.int/rec/TREC-G.984.1-200803-P/en/
4. Effenberger FJ, Mukai H, Park S, Pfeiffer T (2009) Next-generation PON–part II: candidate systems for next-generation PON. IEEE Commun Mag 47(11):50–57
5. IEEE p802.3ah (2003) Media access control parameters, physical layers and management parameters for subscriber access networks. Available at http://www.ieee802.org/3/efm/

6. Tanaka K, Agata A, Horiuchi Y (2010) IEEE 802.3av 10G-EPON standardization and its research and development status. IEEE/OSA J Lightwave Technol 28(4):651–661
7. Bianco A, Cuda D, Finochietto JM, Neri F, Valcarenghi M (2008) WONDER: a PON over a folded bus. In: Proceedings of IEEE global communications conference (GLOBECOM), New Orleans, LA, pp 2600–2604
8. Prat J, Lázaro JA, Chanclou P, Soila R, Velanas P, Teixeira A, Tosi Beleffi G, Tomkos I, Kanonakis K (2009) Passive optical network for long-reach scalable and resilient access. In: Proceedings of international conference on telecommunications (ConTEL), Marrakech, Morocco, pp 271–275
9. Pathak B, Ummy MA, Madamopoulos N, Antoniades N, Ali MA, Dorsinville R (2010) Experimental demonstration of a distributed ring-based EPON architecture. Photon Netw Commun 19(1):55–61
10. Tancesvki L, Rush LA (2000) Impact of the beat noise on the performance of 2-D optical CDMA systems. IEEE Commun Lett 4(8):264–266
11. Mestgagh DJG (1995) Fundamentals of multiple access optical fiber networks. Artech House, New York
12. Soerensen S (2000) Optical beat noise suppression and power equalization in subcarrier multiple access passive optical networks by downstream feedback. IEEE/OSA J Lightwave Technol 18(10):104–107
13. Bock C, Prat J, Walker SD (2005) Hybrid WDM/TDM PON using the AWG FSR and featuring centralized light generation and dynamic bandwidth allocation. IEEE/OSA J Lightwave Technol 23(12):3981–3988
14. Kitayama K-I, Wang X, Wada N (2006) OCDMA over WDM PON—solution path to gigabit-symmetric FTTH. IEEE/OSA J Lightwave Technol 24(4):1654–1662
15. Baik J-S, Lee C-H (2006) Hybrid WDM/SCMA-PON using wavelength-locked Fabry–Pérot laser diodes. IEEE Photonics Technol Lett 18(15):1585–1587
16. Iannone PP, Reichmann KC (2010) Optical access beyond 10 Gb/s PON. In: Proceedings of European conference on optical communications (ECOC), paper Tu.3.B.1, Torino, Italy
17. Frigo NJ, Iannone PP, Magill PD, Darcie TE, Downs MM, Desai BN, Koren U, Koch TL, Dragone C, Presby HM, Bodeep GE (1994) A wavelength-division multiplexed passive optical network with cost-shared components. IEEE Photonics Technol Lett 6(11):1365–1367
18. Payoux F, Chanclou P, Moignard M, Brenot R (2005) Gigabit optical access using WDM PON based on spectrum slicing and reflective SOA. In: Proceedings of European conference on optical communications (ECOC), vol 3. Glasgow, UK, pp 455–456
19. Healey P, Townsend P, Ford C, Johnston L, Townley P, Lealman I, Rivers L, Perrin S, Moore R (2011) Spectral slicing WDM-PON using wavelength-seeded reflective SOAs. Electron Lett 37(19):1181–1182
20. Park SJ, Lee CH, Jeong KT, Park HJ, Ahn JG, Song KH (2004) Fiber-to-the-home services based on wavelength-division-multiplexing passive optical network. IEEE/OSA J Lightwave Technol 22(11):2582–2591
21. Choi KM, Baik JS, Lee CH (2006) Color-free operation of dense WDM-PON based on the wavelength-locked Fabry-Perot laser diodes injecting a low-noise BLS. IEEE Photonics Technol Lett 18(10):1167–1169
22. Park S-J, Choi Y-B, Oh J-M, Koo S-G, Lee D (2007) An evolution scenario of a broadband access network using R-SOA-Based WDM-PON technologies. IEEE/OSA J Lightwave Technol 25(11):3479–3487
23. Berrettini G, Meloni G, Giorgi L, Ponzini F, Cavaliere F, Ghiggino P, Poti L, Bogoni A (2009) Colorless WDM-PON performance improvement exploiting a service-ONU for multiwavelength distribution. In: Proceddings of IEEE/OSA optical fiber communications conference (OFC), paper OMN2, San Diego, CA
24. Schrenk B, Chatzi S, Bonada F, Lazaro JA, Klonidis D, Tomkos I, Prat J (2009) C + L band remote node for amplification in extended reach full-duplex 10 Gb/s WDM/TDM passive

optical networks RX RN. In: Proceedings of European conference on optical communications (ECOC), paper P6.19,Vienna, Austria
25. Takesue H, Sugie T (2003) Wavelength channel data rewrite using saturated SOA modulator for WDM networks with centralized light sources. IEEE/OSA J Lightwave Technol 21(11):2546–2556
26. Hung W, Chan CK, Chen LK, Tong F (2003) An optical network unit for WDM access networks with downstream DPSK and upstream remodulated OOK data using injection-locked FP laser. IEEE Photonics Technol Lett 15(10):1476–1478
27. Prat J, Polo V, Bock C, Arellano C, Vegas Olmos JJ (2005) Full-duplex single fiber transmission using FSK downstream and IM remote upstream modulations for fiber-to-the-home. IEEE Photonics Technol Lett 17(3):702–704
28. Martínez JJ, Garcés Gregorio JI, López Lucia A, Villafranca Velasco A, Aguado JC, Losada Binué MA (2008) Novel WDM-PON architecture based on a spectrally efficient IM-FSK scheme using DMLs and RSOAs. IEEE/OSA J Lightwave Technol 26(3):350–356
29. Vlachos K, Zhang J, Cheyns J, Sulur Chi N, Van Breusegem E, Tafur Monroy I, Jennen JGL, Holm-Nielsen PV, Peucheret C, O'Dowd R, Demeester P, Koonen AMJ (2003) An optical IM/FSK coding technique for the implementation of a label-controlled arrayed waveguide packet router. IEEE/OSA J Lightwave Technol 21(11):2617–2628
30. Vegas Olmos JJ, Tafur Monroy I, Koonen AMK (2003) High bit-rate combined FSK/IM modulated optical signal generation by using GCSR tunable laser sources. Opt Express 11(23):3136–3140
31. Vodhanel RS, Elrefaie AF, Iqbal MZ, Wagner RE, Gimlett JL, Tsuji S (1990) Performance of directly modulated DFB lasers in 10-Gb/s ASK, FSK, and DPSK lightwave systems. IEEE/OSA J Lightwave Technol 8(9):1379–1386
32. Savory SJ, Gavioli G, Killey RI, Bayvel P (2007) Transmission of 42.8Gbit/s polarization multiplexed NRZ-QPSK over 6400 km of standard fiber with no optical dispersion compensation. In Proceedings of IEEE/OSA optical fiber communications conference (OFC), paper OTuA1, Anaheim, CA
33. Lavrova OA, Rau L, Blumenthal DJ (2002) 10-Gb/s Agile wavelength conversion with nanosecond tuning times using a multisection widely tunable laser. IEEE/OSA J Lightwave Technol 20(4):712–717
34. Kobayashi S, Yamamoto Y, Ito M, Kimura T (1982) Direct frequency modulation in AlGaAs semiconductor lasers. IEEE J Quant Electron 18(4):582–595
35. Hossain ASMD, Dorsinville R, Ali MA, Shami A, Assi C (2006) Ring-based local access PON architecture for supporting private networking capability. J Opt Netw 5(1):26–39
36. Brunnel H (1986) Message delay in TDMA channels with contiguous output. IEEE Trans Commun 34(7):681–684
37. Byun HJ, Nho JM, Lim JT (2003) Dynamic bandwidth allocation algorithm in Ethernet passive optical networks. Electron Lett 39(13):1001–1002
38. Kramer G, Mukherjee B, Pesavento G (2002) Interleaved polling with adaptive cycle time (IPACT): a dynamic bandwidth distribution scheme in an optical access network. Photon Netw Commun 4(1):89–107
39. Kramer G, Mukherjee B, Ye Y, Dixit S, Hirth R (2002) Supporting differentiated classes of service in Ethernet passive optical networks. J Opt Netw 1(8):280–298
40. Luo Y, Ansari N (2005) Bandwidth allocation for multiservice access on EPONs. IEEE Commun Mag 43(2):16–21
41. Kim H, Park H, Kang C, Kim C, Yoo G (2005) Sliding cycle time-based mac protocol for service level agreeable Ethernet passive optical networks. In: Proceedings of IEEE international conference communications (ICC), pp 1848–1852, Seoul, Korea
42. Choi S-I (2004) Cycling polling-based dynamic bandwidth allocation for differentiated classes of service in Ethernet passive optical networks. Photon Netw Commun 7(1):87–96
43. Chang C-H, Kourtessis P, Senior JM (2006) GPON service level agreement based dynamic bandwidth assignment protocol. Electron Lett 42(20):1173–1174

44. Choi S-I, Huh J (2002) Dynamic bandwidth allocation algorithm for multimedia services over Ethernet PONs. ETRI Journal 24(6):465–468
45. Assi C, Ye Y, Dixit S, Ali MA (2003) Dynamic bandwidth allocation for quality-of-service over Ethernet PONs. IEEE J Select Areas Commun 21(9):1467–1477
46. Ghani N, Shami A, Assi C, Raja M (2004) Intra-ONU bandwidth scheduling in Ethernet passive optical networks. IEEE J Select Areas Commun 8(11):683–685
47. Zheng J (2006) Efficient bandwidth allocation algorithm for Ethernet passive optical networks. IEE Proc Commun 153(3):464–468
48. Sherif SR, Hadjiantonis A, Ellinas G, Assi C, Ali M (2004) A novel decentralized Ethernet-based PON access architecture for provisioning differentiated QoS. IEEE/OSA J Lightwave Technol 22(11):2483–2497
49. Ma M, Liu L, Cheng T-H (2005) Adaptive scheduling for differentiated services in an Ethernet passive optical network. IEEE J Select Areas Commun 4(10):661–670
50. Nowak D, Perry P, Murphy J (2004) A novel service level agreement based algorithm for differentiated services enabled Ethernet PONs. In: Procedings of IEEE international conference optical internet (IEICE), vol 1. Yokohama, Japan, pp 598–599
51. Yang YM, Ahny B, Nho J (2005) Supporting quality of service by using delta dynamic bandwidth allocations in Ethernet passive optical network. J Opt Netw 4(2):68–81
52. Banerjee A, Kramer G, Mukherjee B (2006) Fair sharing using dual service-level agreements to achieve open access in an Ethernet passive optical network (EPON). IEEE J Select Areas Commun 24(8):32–43
53. Ma M, Zhu Y, Cheng T.-H (2003) A bandwidth guaranteed polling MAC protocol for Ethernet passive optical networks. In: Proceedings of IEEE international conference on computer communications (INFOCOM), vol 1. San Francisco, CA, pp 122–131
54. Merayo N, Durán R-J, Fernández P, Lorenzo RM, de Miguel I, Abril EJ (2009) Bandwidth allocation algorithm based on automatic weight adaptation to provide client and service differentiation. Photon Netw Commun 1(1):119–128
55. Jiménez T, Merayo N, Fernández P, Durán RJ, Lorenzo M, Fernández N, de Miguel I, Abril EJ (2010) Self-adjustment bandwidth algorithm to ensure bandwidth levels in multi-profile LR-EPONs under heterogeneous traffic load. In: Proccedings of European conference on network and optical communications (NOC), vol 1. Algarve, Portugal, pp 1353–358
56. McGarry MP, Reisslein M, Maier M (2006) WDM Ethernet passive optical networks (EPONs). IEEE Commun Mag 44(2):15–22
57. McGarry MP, Reisslein M (2006) Bandwidth management for WDM EPONs. J Opt Netw 5(9):637–654
58. Kwong K.-H, Harle D, Andonovic I (2004) Dynamic bandwidth allocation algorithm for differentiated services over WDM EPONs. In: Proceedings of IEEE international conference on communication systems (ICCS) , vol 1. Singapore, pp 116–120
59. Dhaini AR, Assi CM, Maier M, Shami A (2007) Dynamic bandwidth allocation schemes in hybrid TDM/WDM passive optical networks. IEEE/OSA J Lightwave Technol 25(1):277–286
60. Merayo N, González R, de Miguel I, Jiménez T, Durán RJ, Fernández P, Aguado JC, Lorenzo RM, Abril EJ (2009) Hybrid dynamic bandwidth and wavelenght allocation algorithm to support multi-service level profile in a WDM-EPON. In: Proceedings of international conference AccessNet, vol 1. Hong Kong, China, pp 1–13
61. Dhani AR, Assi C, Shami A (2006) Quality of service in TDM/WDM Ethernet passive optical networks (EPONs). In: Proceedings of IEEE symposium on computers and communications (ISCC), vol 1. Pula-Cagliary, Italy, pp 621–626
62. Van Deventer MO (1993) Power penalties due to reflection and Rayleigh backscattering in a single frequency bidirectional coherent transmission system. IEEE Photonics Technol Lett 5(7):851–854
63. Fujiwara M, Kani MJ, Suzuki H, Iwatsuki K (2006) Impact of backreflection on upstream transmission in WDM single-fiber loopback access networks. IEEE/OSA J Lightwave Technol 24(2):740–746

64. Wood TH, Linke RA, Kasper BL, Carr EC (1998) Observation of coherent Rayleigh noise in single-source bidirectional optical fiber systems. IEEE/OSA J Lightwave Technol 6(2): 346–352
65. Gysel P, Staubli RK (1990) Spectral properties of Rayleigh backscattered light from single-mode fibers caused by a modulated probe signal. IEEE/OSA J Lightwave Technol 8(12):1792–1798
66. Chow CW, Talli G, Townsend PD (2007) Rayleigh noise reduction in 10-Gb/s DWDM-PONs by wavelength detuning and phase-modulation-induced spectral broadening. IEEE Photonics Technol Lett 19(6):423–425
67. Lázaro JA, Arrellano C, Polo V, Prat J (2007) Rayleigh scattering reduction by means of optical frequency dithering in passive optical networks with remotely seeded ONUs. IEEE Photonics Technol Lett 19(2):64–66
68. Agrawal GP (1989) Nonlinear fiber optics. Elsevier Science & Technology Books, New York
69. Waarts RG, Friesem AA, Lichtman E, Yaffe HH, Braun R-P (1990) Nonlinear effects in coherent multichannel transmission through optical fibers. Procceedings of the IEEE 78(8):1344–1368
70. Martínez JJ, Garcés I, López A, Villafranca A, Losada MA (2010) Analysis of the influence of backscattered optical power over bidirectional PON links. Elsevier Opt Commun 283(10):2243–2250
71. IDATE Consulting & Research (2009) World FTTx Markets. FTTx Market Report. Available at http://www.idate.org/en/Research-store
72. IDATE Consulting & Research (2011) Status of FTTH in Europe. Available at http://www.idate.org/en/News
73. Heshun L (2010) Fiber-optic broadband hits top gear. Huawei Communicate 57:17–20
74. FTTH Council Press Release (2011) Global FTTH Councils latest country ranking shows further momentum on all-fiber deployments. Available at http://www.ftthcouncil.org/en/newsroom
75. Hutcheson OL (2008) FTTx: current status and the future. IEEE Commun Mag 24(12):90–95
76. IDATE Consulting & Research (2009) FTTx: global operator rankings. Available at http://www.idate.org/en/Research-store
77. Hutcheson L (2009) FTTH/FTTB in Asia-Pacific. In: Proceedings of FTTH conference in Asia Pacific, Melbourne, Australia. Available at http://www.ftthcouncilap.org
78. Teknovus (2009) Teknovus Announces EPON FTTx in Russia and Belarus. Available at http://www.teknovus.com/News-Events/Press-Releases/2009/Teknovus-Announces-EPON-FTTx-in-Russia-and-Belarus
79. Zager M (2007) Independent telcos continue rolling out fiber. Broadband Properties, 72–87. Available at http://broadbandproperties.com
80. Press release of FTTH council (2009) Ranking of European FTTH penetration shows Scandinavia and smaller economies still ahead. FTTH Council Europe Annual Report 2010–2011. Available at http://www3.ftthcouncil.eu

Chapter 10
Coherent Optical Communication Systems

Ioannis Roudas

Abstract The rapid evolution of long-haul optical communications systems, witnessed in the last five years, is due to the gradual adoption of spectrally efficient, multilevel modulation formats, in conjunction with polarization division multiplexing (PDM) and coherent intradyne detection assisted by digital signal processing (DSP). The objective of this tutorial chapter is to briefly review the operating principles of state-of-the-art long-haul coherent optical communications systems. Due to limitations in space, it focuses mainly on coherent optical systems using quadrature phase-shift keying (QPSK) modulation.

10.1 Introduction

The commercialization in 2008 of the first 40 Gb/s coherent optical communications systems employing polarization division multiplexing (PDM) Quadrature phase-shift keying (QPSK) and intradyne detection assisted by digital signal processing (DSP) marked a major milestone in long-haul transmission [1, 2].

Coherent receivers were intensively studied in the eighties [3–7] because of their superiority to their direct-detection counterparts, mainly in terms of sensitivity and frequency selectivity. However, they were considered impractical at the time, due to their high cost and complexity, as well as their vulnerability to phase-noise and polarization rotations. The revived interest in coherent detection is largely due to the substitution of previously proposed analog electronic and

I. Roudas (✉)
Department of Electrical Engineering
University of Patras
26504 Rio, Greece
e-mail: roudas@ece.upatras.gr

N. (Neo) Antoniades et al. (eds.), *WDM Systems and Networks*, Optical Networks,
DOI: 10.1007/978-1-4614-1093-5_10,

optoelectronic modules (which were bulky, slow, expensive, and largely inefficient) in coherent optical receivers with relatively inexpensive, high-speed, application-specific integrated circuits (ASICs) (see recent surveys [8–11, 15] and the references therein). The latter enable adaptive electronic equalization of linear transmission impairments, i.e., chromatic dispersion, polarization mode dispersion and polarization-dependent loss, and to some extent, of fiber nonlinearities. They also allow for adaptive electronic compensation of imperfections of the analog optical transmitter and receiver front-ends, such as time skew of quadrature components and polarization tributaries, quadrature and polarization imbalance, etc. Finally, they perform all standard digital receiver functionalities such as digital clock recovery, intermediate frequency offset and phase-noise estimation, symbol decision, differential decoding, forward error correction, etc.

There is an emerging consensus among major system vendors that coherent optical PDM-QPSK communications systems are the most attractive candidates for 100 Gb/s Ethernet[1] transmission over existing terrestrial networks [2, 12, 16–19].

At the moment, several companies have announced the development of application-specific integrated circuits (ASICs) for DSP in coherent homodyne synchronous PDM-QPSK receivers operating at this symbol rate. Furthermore, recent field trials [21, 22] have demonstrated the practicality of long-haul 28 GBd coherent optical PDM-QPSK systems.

PDM M-ary Quadrature Amplitude Modulation (M-QAM) is actively investigated for use in next-generation long-haul terrestrial optical communications systems [12]. This modulation format is intended for either single carrier or multi-carrier systems using orthogonal frequency division multiplexing (OFDM), in order to achieve equivalent bit rates of the order of 400 Gb/s or even 1 Tb/s per wavelength channel [12]. PDM M-QAM allows for a nominal spectral efficiency of 2 M b/s/Hz. Recent hero experiments using coherent optical PDM M-QAM communication systems achieved several world records, most notably unprecedented aggregate WDM bit rates approaching 70 Tb/s [23, 24] and a spectral efficiency close to 12 b/s/ Hz [25].

Due to the proliferation of research studies on coherent optical PDM-QPSK and PDM M-QAM communication systems during the last five years, it is difficult to exhaustively cover all aspects of this topic here. The objective of this tutorial chapter is to briefly review the operating principles in long-haul PDM-QPSK coherent optical communications systems.

---

[1] Actually, it is necessary to use an effective bit rate of 112 Gb/s, which corresponds to a symbol rate of PDM-QPSK of 28 GBd, in order to achieve a net per channel 100 Gb/s data rate transmission. The reason is that one must take into account the overhead due to current forward error correction (FEC) (∼7%) and the Ethernet packet header (∼4%). We assume a WDM channel spacing of 50 GHz, which is compatible with the current ITU grid specifications and provides some margin for bandwidth narrowing due to the concatenation of several reconfigurable optical add-drop multiplexers (ROADMs). Then, the spectral efficiency of these systems is 2 b/s/Hz, i.e., half of the nominal spectral efficiency of PDM-QPSK [12–14].

The rest of the chapter is organized as follows: In Sect. 10.2, we initially present the digital M-PSK transmitter and receiver optimal architectures in a block diagram form. Next, we review the operating principle of coherent detection and describe different variants of coherent receivers. In Sect. 10.3, we describe the implementation of the functionalities of the optimal M-PSK transmitter and receiver using various photonic devices, i.e., a QM, a balanced receiver, a phase-diversity receiver with 90° hybrid, and a polarization-diversity receiver. In Sect. 10.4, we review the most prominent DSP algorithms. Finally, in Sect. 10.5, we develop an abstract model for the performance evaluation of an optical communication system using M-PSK modulation and synchronous homodyne detection. The details of the calculations are given in the Appendices.

10.2 Multilevel Differential Phase-Shift Keying

Multilevel phase-shift keying (M-PSK) is a type of digital modulation format whereby information is encoded into discrete changes $\Delta\phi_k$ of the phase of the carrier at time instants equal to multiples of the symbol period [26, 29]. Since phase changes are less affected by additive white Gaussian noise compared to amplitude changes, this modulation format exhibits higher sensitivity than amplitude shift keying (ASK).

10.2.1 Signal Representation

The M-PSK signal can be written as [26]

$$s(t) = \Re\{\tilde{s}(t)e^{j\omega_s t}\} \tag{10.1}$$

where $\Re\{.\}$denotes the real part and $\tilde{s}(t)$ is the complex envelope

$$\tilde{s}(t) = A\sum_{k=0}^{N-1} c_k g(t - kT) \tag{10.2}$$

where A is the carrier amplitude, N is the number of transmission symbols, T is the symbol period, $g(t)$is the symbol shape, and we defined the complex symbols

$$c_k \equiv e^{j\Delta\varphi_k} \tag{10.3}$$

In Eq. 10.3, the discrete phase changes $\Delta\phi_k$ take values in the set

$$\{2\pi(i-1)/M + \Phi\}_{i=1}^{M} \tag{10.4}$$

where Φ is an arbitrary initial phase.

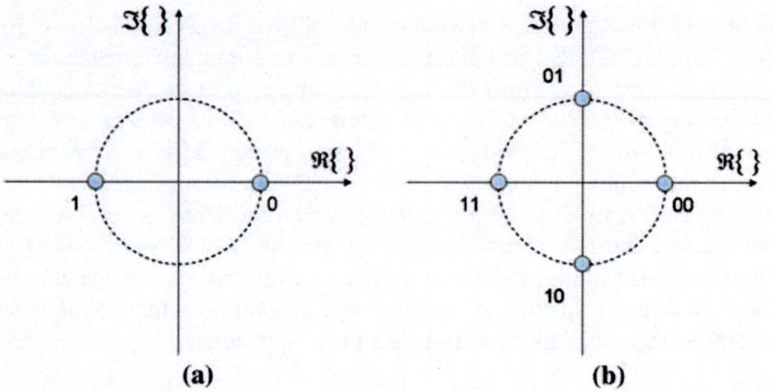

Fig. 10.1 Geometrical representation of M-PSK signal sets on the complex plane **a** $M = 2, \Phi = 0$; **b** $M = 4, \Phi = 0$

Substituting Eqs. 10.2, 10.3 into Eq. 10.1 and using trigonometric identities, we can express M-PSK modulation as the superposition of two carriers at the same frequency but with 90° phase difference carrying M-ary amplitude modulations (called *in-phase* and *quadrature components*)

$$s(t) = \underbrace{I(t)\cos\omega_s t}_{\substack{\text{in-phase}\\ \text{component}}} - \underbrace{Q(t)\sin\omega_s t}_{\substack{\text{quadrature}\\ \text{component}}} \tag{10.5}$$

where

$$\begin{aligned} I(t) &= A\sum_{k=0}^{N-1}\cos\Delta\phi_k g(t-kT) \\ Q(t) &= A\sum_{k=0}^{N-1}\sin\Delta\phi_k g(t-kT) \end{aligned} \tag{10.6}$$

From Eqs. 10.3, 10.4, we observe that symbols c_k can take M discrete complex values. A geometric representation of this set of M complex values is shown in Fig. 10.1. The symbols are represented on the complex plane as a constellation of equidistant points on a circle.

It is worth noting that random bits at the entrance of the transmitter must be mapped into the M discrete complex values that symbols c_k can take prior to transmission. In Fig. 10.1, words of $m = \log_2 M$bits are associated with different constellation points using Gray coding, e.g., words corresponding to adjacent constellation points differ by a single bit (see details below).

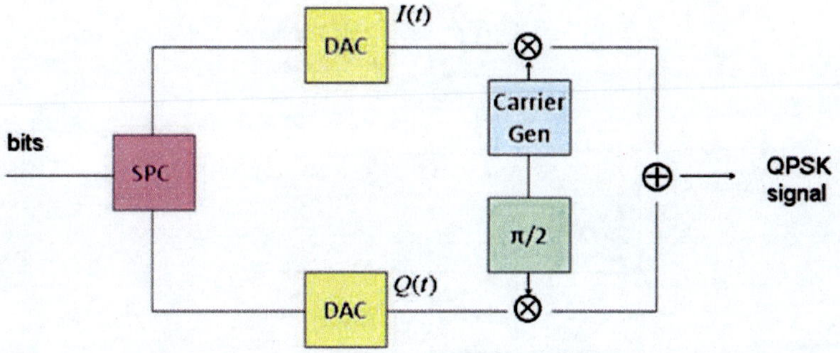

Fig. 10.2 QPSK transmitter (Symbols: SPC: Serial-to-parallel converter, DAC: Digital-to-analog converter, Carrier gen: Carrier generator, $\pi/2$: 90 deg. Phase shifter)

10.2.2 Transmitter and Receiver Architectures

From Eqs. 10.5, 10.6 it is straightforward to derive the block diagram of the QPSK transmitter. Figure 10.2 shows an example of implementation of an ideal QPSK transmitter for $\Phi = \pi/4$. The transmitter does not segment the input PRBS into words of two bits but instead, uses a serial-to-parallel converter to alternatively send bits to two binary pattern generators. These produce two baseband antipodal binary waveforms with instantaneous amplitude $\pm 1/\sqrt{2}$ at symbol rate $R = 1/T$, which modulate two CW carriers $A\cos\omega_s t$, $-A\sin\omega_s t$. Therefore, a signaling rate reduction by one-half is achieved. The two quadrature components are added and transmitted in the channel.

The optimal synchronous QPSK receiver structure is shown in Fig. 10.3. A synchronous receiver is equipped with a carrier phase recovery circuit, which computes the carrier phase change θ acquired during propagation. The received waveform is divided into two parts, each multiplied with $\cos(\omega_s t + \theta)$, $-\sin(\omega_s t + \theta)$, respectively, and filtered using two LPFs with impulse response $g(T - t)$ (matched filters). For ideal NRZ pulses, these LPFs can be ideally implemented as integrate and dump (I&D) filters (e.g., finite time integrators which perform a running average of duration T). In the absence of distortion, the waveforms at the output of the I&D filters are smoothed replicas of the baseband amplitude modulating waveforms $I(t), Q(t)$. These signals are sampled once per symbol, at the appropriate sampling instants, which are calculated using a symbol synchronization circuit, in order to minimize the error probability. Two binary decision devices, with thresholds equal to zero, are used to recover the bits. The decisions are made independently in the two receiver branches. Finally, the recovered binary sequences are combined into a single bit stream using a parallel-to-serial converter.

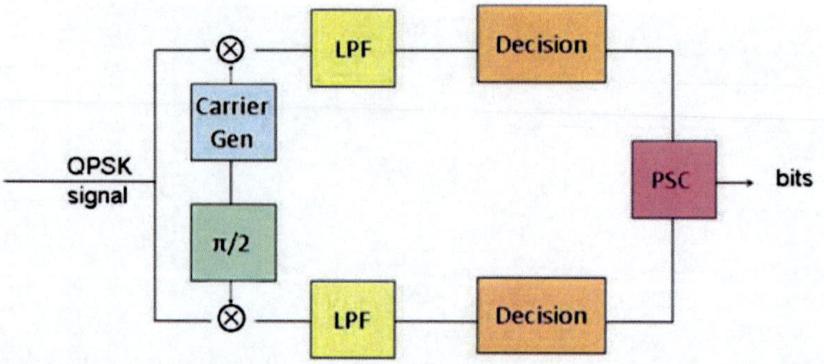

Fig. 10.3 Optimal synchronous QPSK receiver. (Symbols: Carrier gen: Carrier generator, $\pi/2$: 90 deg. Phase shifter, LPF: Lowpass filter, Decision: decision circuit, SPC: Serial-to-parallel converter)

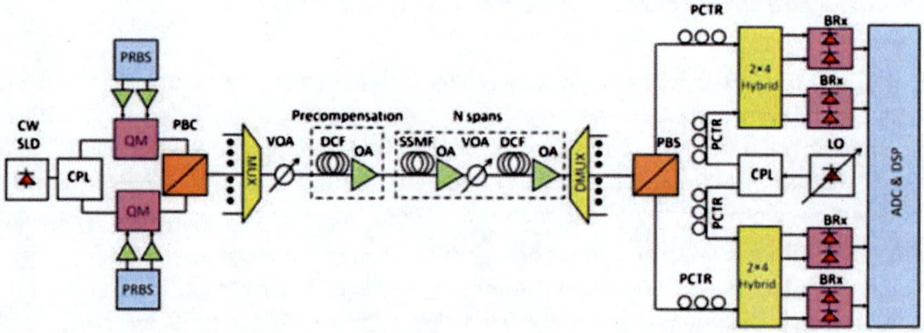

Fig. 10.4 Block diagram of a representative coherent optical PDM-QPSK system. (Symbols: PRBS: Pseudo-random bit sequence, SLD: Semiconductor laser diode, CPL: 3-dB coupler, QM: Quadrature modulator, PBC: Polarization beam combiner, MUX: Optical multiplexer, VOA: Variable optical attenuator, DCF: Dispersion compensating fiber, SSMF: Standard single-mode fiber, OA: Optical amplifier, DMUX: Optical demultiplexer, PBS: Polarization beam splitter, PCTR: Polarization controller, LO: Local oscillator, BRx: Balanced receivers, ADC: Analog-to-digital converter, DSP: Digital signal processing module)

10.2.3 Coherent Optical PDM QPSK System

The implementation of the aforementioned M-PSK transmitter and receiver structures for optical communications, in the case of PDM, is shown in Fig. 10.4.

More specifically, Fig. 10.4 shows the block diagram of a representative long-haul, PDM QPSK optical communications system with a polarization- and phase-diversity coherent optical receiver. At the transmitter, the optical signal from a CW semiconductor laser diode (SLD) is equally split and fed into two parallel quadrature modulators (QM). Two independent maximal-length pseudo-random bit sequences (PRBS) of period $2^n - 1$, at a bit rate R_b each, drive two QM. The two optical QPSK signals are superimposed with orthogonal SOPs, using a polarization beam combiner (PBC), to form a PDM-QPSK signal. The latter is wavelength division multiplexed (WDM) with additional channels carrying PDM-QPSK signal, using an optical multiplexer (MUX), and transmitted through N amplified spans composed of standard single-mode fiber (SSMF) and, possibly, dispersion compensating fiber (DCF). An additional dispersion pre-compensation module, composed of a DCF and a booster optical amplifier, might be included in the latter case.

The optical receiver front-end is composed of an optical demultiplexer (DMUX), acting as an optical bandpass filter (BPF), a polarization beam splitter (PBS), a laser diode, acting as a local oscillator (LO), two 2×4 90° optical hybrids, and four balanced photodetectors (BRx's). The x- and y-polarization components of the received optical signal and the local oscillator are separately combined and detected by two identical phase-diversity receivers composed of a 2×4 90° optical hybrid and two BRx's each, at the upper and lower polarization branches, respectively. The photocurrents at the output of the four balanced detectors are low-pass filtered (LPF), sampled at integer multiples of a fraction of the symbol period T_s, using an analog-to-digital converter (ADC), and fed to an application-specific integrated circuit (ASIC) for DSP (see below for details).

The aforementioned components of the M-PSK transmitter and receiver are explained in detail in the next section.

10.3 Optical Components

In this section, we describe the implementation of the functionalities of the optical M-PSK transmitter and receiver using various photonic devices, i.e., a QM, a balanced receiver, a phase-diversity receiver with 90° hybrid, and a polarization-diversity receiver.

10.3.1 Optical Transmitter: Quadrature Modulator

The QM is shown in Fig. 10.5 [28]. It is composed of a Mach–Zehnder interferometer which contains two push–pull Mach–Zehnder modulators [43], one in each arm, and a phase modulator at the lower arm that introduces a phase difference between the two arms.

The complex envelope of the modulated electric field at the output of the modulator $\tilde{E}_s(t)$can be written as a function of the unmodulated input electric field $\tilde{E}_{in}(t)$ (see Appendix A)

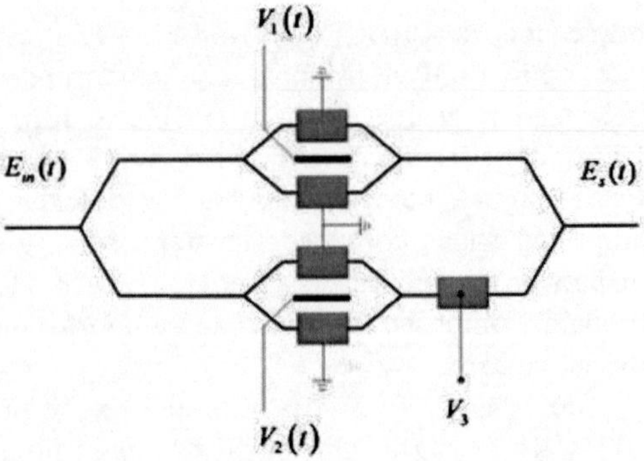

Fig. 10.5 Quadrature modulator

$$\tilde{E}_s(t) = \frac{j}{2} e^{-j\frac{\pi V_3}{2V_{\pi_3}}} \left\{ \sin\left[\frac{\pi V_1(t)}{2V_{\pi_1}}\right] - e^{-j\frac{\pi V_3}{2V_{\pi_3}}} \sin\left[\frac{\pi V_2(t)}{2V_{\pi_2}}\right] \right\} \tilde{E}_{in}(t) \tag{10.7}$$

where $V_1(t), V_2(t), V_3$ are the driving voltages and $V_{\pi_1}, V_{\pi_2}, V_{\pi_3}$ are the half-wave voltages of the two Mach–Zehnder modulators and the phase shifter, respectively.

The two Mach–Zehnder modulators operate with drive voltages that take values in the discrete sets

$$\begin{aligned} V_1 &\in \{-V_{\pi_1}, V_{\pi_1}\} \\ V_2 &\in \{-V_{\pi_2}, V_{\pi_2}\} \end{aligned} \tag{10.8}$$

The voltage of the phase shifter is set at $V_3 = V_{\pi_3}/2$, in order to introduce a phase difference of $\pi/2$ between the two arms.

It is assumed that that the input into the QM is an unmodulated optical wave whose complex envelope can be written as

$$\tilde{E}_{in}(t) = \sqrt{2P_s} e^{j\varphi_s} \tag{10.9}$$

where P_s is the average optical power and ϕ_s is the initial phase of the transmitted CW signal. In the previous formula, intensity and phase noises of the laser are neglected. Then, the complex envelope of the output electric field can be written as

$$\tilde{E}_s(t) = \sqrt{P_s} e^{j\varphi_s + j\varphi_k} \tag{10.10}$$

where $\varphi_k = \pm\pi/4, \pm 3\pi/4$.

10.3.2 Coherent Detection Fundamentals

The term *coherent* is used, in the context of optical communications, to refer to any technique employing nonlinear mixing between two optical waves on a semiconductor photodiode [4].[2] The carrier frequencies of the optical waves can be identical or different. In the former case, we have *coherent homodyne detection*. In the latter case, we have *coherent homodyne detection* (when the carrier frequency difference of the two waves is larger than or of the order of the symbol rate) or *coherent intradyne detection* (when the carrier frequency difference of the two waves is a fraction of the symbol rate).

The application of coherent homodyne detection for optical frequencies dates back to 1801, when Young proposed his now famous two-slit interference experiment as persuasive evidence of the wave nature of light [30]. In modern times, coherent heterodyne detection of electromagnetic waves has been used since the early days of radio communications. More specifically, heterodyne detection of radio waves was proposed by Fessenden [31], who also coined the term heterodyne from the Greek words 'heteros' (other) and 'dynamis' (force). Heterodyning gained immense popularity with the development of the superheterodyne receiver by E. H. Armstrong in 1921 [32]. Optical heterodyning was used for the first time by [33] in the visible part of the electromagnetic spectrum and by [34] in the infrared.

The earliest papers on coherent optical communication systems appeared in 1979, in Japanese, and in 1980, in English. The revived interest in coherent optical homodyne receivers in combination with advanced modulation formats started around 2004, e.g., see early articles [35–40].

10.3.2.1 Coherent Single-Ended Detection

The operating principle of coherent detection is explained in numerous textbooks, e.g., [41–42, 99]. Consider two traveling electromagnetic waves with carrier frequencies f_s andf_{lo}, respectively, from two independent laser sources, labeled the received signal and the local oscillator signal, respectively. The waves propagate in the same direction with identical states of polarization (SOP). Therefore, the electric fields of the two waves can be treated as scalars and they are denoted by $\vec{E}_s(t), \vec{E}_{lo}(t)$, respectively.

For simplicity, it is assumed that $E_s(t), E_{lo}(t)$ are both unmodulated (CW) sinusoidal signals

$$\begin{aligned} E_s(t) &= \sqrt{2P_s}\cos(\omega_s t + \varphi_s) \\ E_{lo}(t) &= \sqrt{2P_{lo}}\cos(\omega_{lo} t + \varphi_{lo}) \end{aligned} \tag{10.11}$$

---

[2] In contrast, in the digital communications literature the term *coherent* is used to refer to demodulation techniques in which the absolute phase of the incoming signal is tracked by the receiver. In optical communications, such receivers are called *synchronous*. In this report, we will be interested exclusively in coherent synchronous receivers.

where P_s, P_{lo} are the average optical powers, $\omega_s = 2\pi f_s, \omega_{lo} = 2\pi f_{lo}$ are the angular carrier frequencies, and φ_s, φ_{lo} are the initial phases of the received signal and the local oscillator signal, respectively. In the previous formulae, intensity and phase noises of the lasers are neglected.

The electric field of the combined signal impinging upon the photodiode, at a single detection point, can be written as the superposition of the electric fields of the received signal and the local oscillator

$$E_r(t) = E_s(t) + E_{lo}(t) \tag{10.12}$$

The photodiode is modeled as a square-law detector which responds to the square of the electric field

$$i(t) = R\left\langle E_s(t)^2 \right\rangle \tag{10.13}$$

where R is the responsivity of the photodiode and the angle brackets denote time averaging over an interval proportional to the response time of the photodiode.

By substituting Eqs. 10.12, 10.11 into Eq. 10.13 and using trigonometric identities, we obtain the following expression for the photocurrent in the absence of noise

$$\boxed{i(t) = \underbrace{R[P_s + P_{lo}]}_{\text{direct-detection term}} + \underbrace{2R\sqrt{P_s P_{lo}}\cos(\omega_{IF}t + \varphi_{IF})}_{\text{coherent-detection term}}} \tag{10.14}$$

where

$$\begin{aligned} \omega_{IF} &= 2\pi(f_s - f_{lo}) \\ \varphi_{IF} &= \varphi_s - \varphi_{lo} \end{aligned} \tag{10.15}$$

It is observed that Eq. 10.14 is the sum of three terms due to the direct-detection of the received signal and the local oscillator signal, and their mixing (coherent detection term), respectively. The latter preserves the information transferred by the amplitude, the frequency and the phase of the received signal. Therefore, this type of detection can be used in conjunction with amplitude, frequency or phase modulation formats. In addition, the amplitude of the coherent detection term depends on the power of the local oscillator, which can be made very large. This is the reason for the improved receiver sensitivity exhibited by coherent detection.

10.3.2.2 Balanced Receiver

An implementation of the coherent receiver with fiber-optic components is shown in Fig. 10.6 [44, 118]. This configuration uses a directional 3-dB coupler and two identical p-i-n photodiodes connected back-to-back. A received signal of average power P_s and carrier frequency f_s is combined with the signal of a local oscillator of average power P_{lo} and carrier frequency f_{lo} on a directional 3-dB coupler. It is

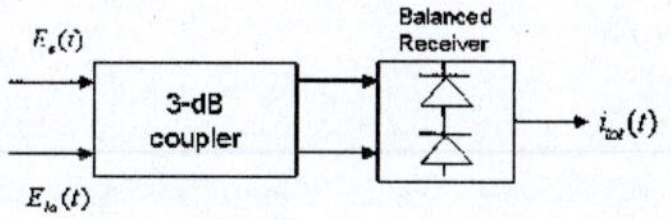

Fig. 10.6 Balanced receiver

assumed that the states of polarization of the signal and the local oscillator are identical. The electric fields of the signal and the local oscillator are combined with ±90° relative phase shift at the two outputs of an ideal, lossless, polarization independent, directional 3-dB coupler. The output fields then illuminate the pair of two identical p-i-n photodiodes and produce two photocurrents. The outputs of the photodiodes are then subtracted, resulting in a combined photocurrent $i_{tot}(t)$. Since this receiver uses two photodiodes instead of one, it is called a *balanced receiver*. Balanced receivers are particularly attractive for use in optical coherent communication systems because they detect all received signal power, eliminate direct-detection terms and cancel excess intensity noise due to the local oscillator [44, 118].

For CW signal and local oscillator, the input electric fields are given by Eq. 10.11 and the final expression of the photocurrent, neglecting a $\pi/2$ phase constant, is given by (see Appendix B)

$$\boxed{i_{tot}(t) = \underbrace{2R\sqrt{P_s P_{lo}}\cos(\omega_{IF}t + \varphi_{IF})}_{\text{coherent - detection term}}} \tag{10.16}$$

where R is the photodiode responsivity, ω_{IF} is the intermediate angular frequency defined as $\omega_{IF} = 2\pi(f_s - f_{lo})$ and $\varphi_{IF} = \varphi_s - \varphi_{lo}$.

10.3.2.3 Phase-Diversity Receiver

Phase-diversity receivers were initially proposed as a means of achieving homodyne detection with increased phase-noise tolerance, compared to homodyne detection with optical phase locking of the local oscillator [46]. They are based on multi-port couplers (called *optical hybrids*), which combine the signal and the local oscillator (LO) fields, introducing various phase shifts between the two, followed by multiple photodiodes to detect the combined optical signals. In the current context, phase-diversity receivers with 90° optical hybrids are used to separately recover the in-phase and quadrature components of the M-PSK signal downshifted in the baseband [47]. The phase noise is subsequently removed using a feed-forward phase estimation technique (see below).

Figure 10.7 shows an example of implementation of a phase-diversity receiver composed of a 4 × 4 90° optical hybrid and two balanced receivers (only the two-input ports are shown in the schematic). This configuration combines the best features of both phase-diversity and balanced receivers: it recovers the in-phase

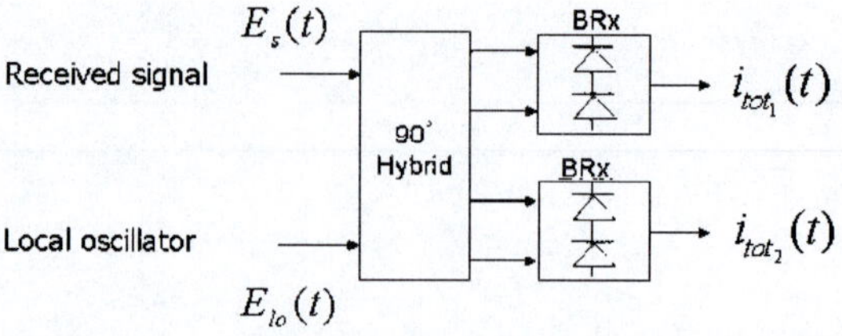

Fig. 10.7 Phase-diversity receiver using a 4 × 4 90° optical hybrid and two balanced receivers

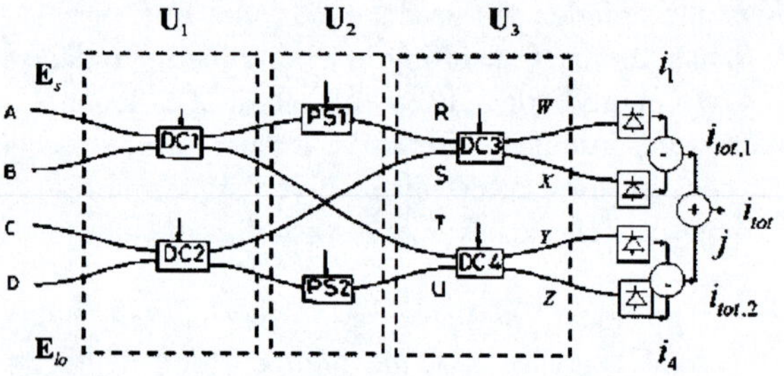

Fig. 10.8 Implementation of the 90° 4 × 4 hybrid. (Abbreviations: DC = direct current, PS = Phase shifter). (Symbols: $\mathbf{E}_s$=Received signal electric field vector, $\mathbf{E}_{lo}$=local oscillator electric field vector, $i_1 - i_4$: photocurrents at the four photodiodes, $i_{tot_{1,2}}$= phtocurrents at the output of the balanced receivers, i_{tot}=total complex photocurrent) (From [48], Fig. 10.2. Reproduced by permission of © 1989 The Institute of Electrical and Electronics Engineers.)

and quadrature components, while using all the available optical power of the received signal and the local oscillator, rejecting the direct-detection terms and the intensity noise of the local oscillator.

A hardware implementation of a 4 × 4 90° optical hybrid is shown in Fig. 10.8 and is analyzed in Appendix C [48]. A simpler but suboptimal implementation of a phase-diversity receiver proposed by [46, 117] uses a 2 × 2 90° optical hybrid (Fig. 10.9) and two matched photodiodes (not connected back-to-back, as in the case of a balanced receiver) and is analyzed in Appendix D.

As shown in Appendix C, for CW signal and local oscillator, the input electric fields are given by Eq. 10.11 and the final expression of the photocurrent, using the arrangement of Fig. 10.8, is given by

$$\boxed{\begin{aligned} i_{tot_1}(t) &= \underbrace{R\sqrt{P_s P_{lo}}\cos(\omega_{IF}t + \varphi_{IF})}_{\text{in-phase component}} \\ i_{tot_1}(t) &= \underbrace{R\sqrt{P_s P_{lo}}\sin(\omega_{IF}t + \varphi_{IF})}_{\text{quadrature component}} \end{aligned}} \tag{10.17}$$

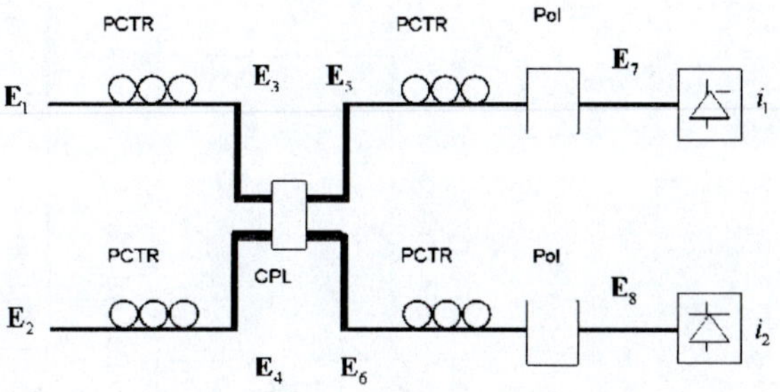

Fig. 10.9 Implementation of the 90° 2 × 2 hybrid. (Symbols: $\mathbf{E}_{1-8}$=Electric field vectors, $i_1 - i_2$: photocurrents at the two photodiodes)

10.3.2.4 Polarization- and Phase-Diversity Receiver

So far, it was assumed that the SOPs of the signal and the local oscillator were identical. In reality, the SOP of the received signal is unknown and changes over time. In order to avoid outages, we use the polarization-diversity receiver configuration of Fig. 10.10. Using two polarization beam splitters (PBS), the signal and the local oscillator are analyzed in *x, y* linear polarizations, which are detected separately using two phase-diversity receivers.

As shown in Appendix E, for CW signal and local oscillator, the input electric fields are given by Eq. 10.11 and the final expression of the photocurrents in the upper branch are given by:

$$\boxed{\begin{aligned} i_{tot_{1,x}}(t) &= \underbrace{R\sqrt{P_{s,x}P_{lo}/2}\cos(\omega_{IF}t+\varphi_{IF}+\varphi_x)}_{\text{in-phase component}} \\ i_{tot_{2,x}}(t) &= \underbrace{R\sqrt{P_{s,x}P_{lo}/2}\sin(\omega_{IF}t+\varphi_{IF}+\varphi_x)}_{\text{quadrature component}} \end{aligned}} \tag{10.18}$$

where $P_{s,x}$ is the power of the optical signal in the *x*-polarization and φ_x is a phase angle arising from the polarization mismatch between the signal and the local oscillator in the upper branch.

Similarly, the final expressions of the photocurrents in the lower branch are given by:

$$\boxed{\begin{aligned} i_{tot_{1,y}}(t) &= \underbrace{R\sqrt{P_{s,y}P_{lo}/2}\cos(\omega_{IF}t+\varphi_{IF}+\varphi_y)}_{\text{in-phase component}} \\ i_{tot_{2,y}}(t) &= \underbrace{R\sqrt{P_{s,y}P_{lo}/2}\sin(\omega_{IF}t+\varphi_{IF}+\varphi_y)}_{\text{quadrature component}} \end{aligned}} \tag{10.19}$$

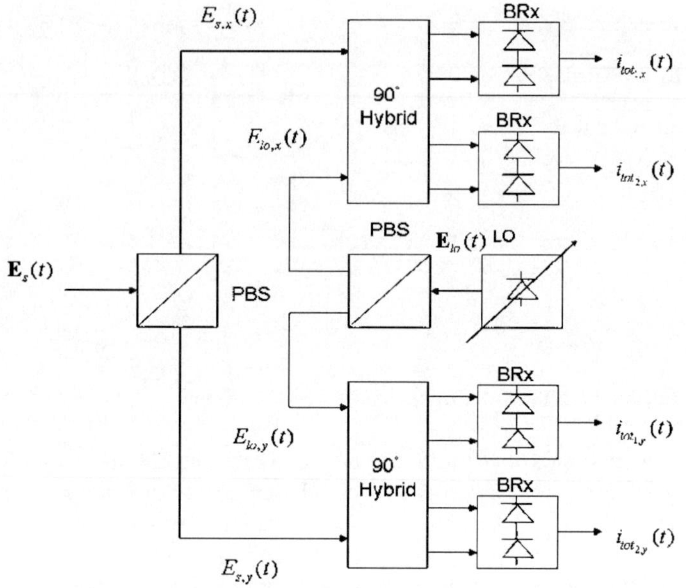

Fig. 10.10 Implementation of the polarization-diversity receiver

where $P_{s,y}$ is the power of the optical signal in the x-polarization and φ_y is a phase angle arising from the polarization mismatch between the signal and the local oscillator in the lower branch.

By combining the photocurrents at the outputs of the balanced receivers in the upper branch of the polarization-diversity receiver with 90° phase shift we obtain

$$\boxed{i_{tot_x}(t) = i_{tot_{1,x}}(t) + j i_{tot_{2,x}}(t) = R\sqrt{P_{s,x}P_{lo}/2}e^{j(\omega_{IF}t+\varphi_{IF}+\varphi_x)}} \tag{10.20}$$

Similarly, by combining the photocurrents at the outputs of the balanced receivers in the lower branch of the polarization-diversity receiver with 90° phase shift, we obtain

$$\boxed{i_{tot_y}(t) = i_{tot_{1,y}}(t) + j i_{tot_{2,y}}(t) = R\sqrt{P_{s,y}P_{lo}/2}e^{j(\omega_{IF}t+\varphi_{IF}+\varphi_y)}} \tag{10.21}$$

Finally, the DSP unit samples the photocurrents with a sampling period T_s and forms the complex array

$$\mathbf{I}_{tot}(nT_s) = \begin{pmatrix} i_{tot,x} \\ i_{tot,y} \end{pmatrix} = R\sqrt{P_s P_{lo}/2}e^{j(\omega_{IF}nT_s+\varphi_{IF})}|e_s(nT_s)\rangle \tag{10.22}$$

where $|e_s(nT_s)\rangle$ is a normalized Jones vector which represents the received signal SOP at the instant nT_s.

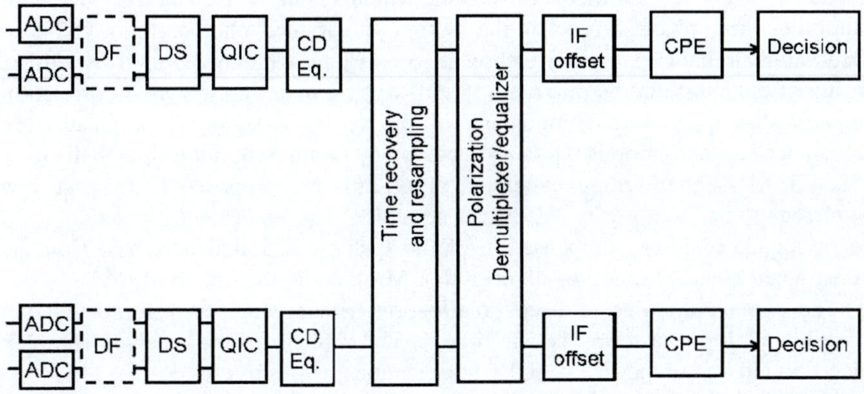

Fig. 10.11 DSP ASIC architecture (Symbols: ADC: analog-to-digital converter, DF: digital filter, QIC.: Quadrature imbalance estimation and compensation, CD eq.: fixed chromatic dispersion equalizer, IF offset: IF offset estimation and compensation, CPE: carrier phase estimation)

10.4 DSP Functionalities

In this section, we briefly review the DSP functionalities used in coherent optical intradyne PDM-QPSK receivers [50–53].

There is no unanimous agreement in the optical communications community regarding the optimal breakdown of DSP tasks, the order of the DSP modules, and the hardware implementation of different DSP algorithms. Figure 10.11 shows the most prominent view of the DSP ASIC's architecture [50–53].

Initially, the analog signals are converted to digital signals using analog-to-digital converters (ADCs) [54]. Conceptually, ADCs are composed of a sampler, which operates at a frequency that is not, in general, an integer multiple of the symbol rate, followed by a quantizer [26]. The amplitude of the signal samples at the output of the quantizer takes 2^k discrete values, where k denotes the number of ADC's resolution bits. Typically, state-of-the-art 56 GSa/s ADCs have 6–8 b resolution [54]. Current ADCs operate at a sampling frequency above the Nyquist rate, in order to achieve optimal performance, at the expense of hardware complexity and power consumption. For instance, employing two samples per symbol decreases timing errors and enables the use of robust fractionally spaced linear equalizers and digital timing recovery algorithms [51].

After the ADC, digital filtering (DF) might be performed in order to reject out-of-band signal frequency components and ASE noise [55]. The signal deskew (DS) module compensates for timing errors due to optical path length differences between the quadratures of each polarization tributary [56]. Then, quadrature imbalance, occurring at each phase-diversity receiver, is estimated and corrected (QIC). Quadrature imbalance is due to imperfections of the optical hybrid and balanced receiver photodiode mismatch, which result in DC offsets, and both amplitude and phase errors in the output quadrature photocurrents [57–59]. Quadrature imbalance can be estimated by various techniques, e.g., by ellipse fitting of the constellation diagram [58, 60] or by computing the cross-correlation between the quadrature components of the received signal [61]. Quadrature imbalance compensation is typically achieved by Gram-Schmidt orthogonalization [58, 60, 61]. Other orthogonalization algorithms are proposed in [51]. A low implementation complexity feedback structure can be used in practice [53]. In principle, adaptive compensation of the quadrature imbalance, based on the constrained constant modulus algorithm (CMA), could also be used [69].

The two quadratures of each polarization tributary are then combined, via complex addition, to form discrete-time, scaled replicas of the received complex electric field vectors at the x- and y-polarizations, respectively.

Performance-wise, electronic chromatic dispersion post-compensation in coherent PDM-QPSK optical communications systems is superior to in-line optical compensation using dispersion compensating fibers (DCFs) [52]. Another advantage of electronic chromatic dispersion post-compensation over DCFs is that, in reconfigurable networks, the link length changes dynamically. The taps of electronic equalizers can be easily adjusted to accommodate for these changes [9]. In addition, in practice, the dispersion parameters of the optical fibers cannot be fully matched by DCFs, whereas the opposite is true for electronic equalizers [9]. Furthermore, the capital expenditure for implementing electronic equalization functionalities in the ASIC DSP chips can be relatively insignificant, when the latter are produced in large numbers. Finally, electronic equalization can compensate simultaneously for fiber nonlinearities by using back-propagation, e.g., based on the Manakov equation [62–68].

Electronic chromatic dispersion compensation is performed using a linear equalizer [50–53]. The transfer function of the linear equalizer is ideally the inverse transfer function of the optical fiber, the latter being modeled as a parabolic-phase, all-pass filter [42]. In practice, it can be implemented in time-domain by using a finite impulse response (FIR) filter with fixed coefficients [70, 71]. For large amounts of accumulated chromatic dispersion, a frequency domain implementation [76] is more computationally efficient than its time-domain counterpart [50, 52]. Several alternative approaches for reducing hardware complexity have been recently proposed (see [72–75, 50–53] and the references therein).

In experiments using real-time digital sampling oscilloscopes, the ADC sampling rate might not coincide with an integer multiple of the symbol rate. Therefore, the signals must be resampled to obtain an integer number of samples per bit. In a commercially available coherent receiver specifically designed for a specific symbol rate, such an operation might only be required for fine tuning. Signal resampling is a two-stage process that requires a clock-recovery scheme and an interpolation scheme. Typical non-data-aided digital clock-recovery algorithms from digital communications can be used [77] but their performance might be severely impaired, due to fiber transmission effects [78]. Several algorithms are specifically proposed for coherent optical intradyne PDM-QPSK receivers that are robust to fiber transmission effects [79–82, 84]. Resampling can be achieved using either polynomial interpolation [27] or a fractional delay filter [85].

For polarization demultiplexing, as well as for removing signal distortion due to intersymbol interference arising from residual chromatic dispersion, polarization mode dispersion, polarization-dependent loss, and deterministic fiber nonlinearities, a blind adaptive two-input two-output (TITO) equalizer is used [50–53]. The equalizer attempts to counteract the channel effects by forming a linear superposition of the photocurrents. It is composed of four complex transversal filters with impulse responses w_{kl}, $k, l = 1, 2$, which are connected in a butterfly structure. For updating the filter coefficients, several variants of the constant modulus algorithm (CMA) [86, 87] can be used, possibly in conjunction with the decision-directed least-mean squares (DD-LMS) algorithm [87]. CMA-based blind adaptive equalizers are popular, due to their low computational complexity and their robustness in the presence of intermediate frequency (IF) offsets and laser phase noise. The second feature allows for decoupling between polarization demultiplexing and carrier frequency/phase recovery, so the latter two impairments can be addressed by separate DSP modules. A disadvantage of CMA-based modules is their possible erroneous convergence to the same PDM channel. The introduction of constraints on the CMA filter coefficients has been proposed for the correction of this defect, both for polarization demultiplexing only [88, 89], as well as for joint polarization demultiplexing/adaptive equalization [90, 91].

After polarization demultiplexing, the complex envelopes of the electric fields of the PDM-QPSK tributaries are recovered separately, at the upper and lower branches of the ASIC. The non-zero IF offset, due to the carrier frequency difference between the transmitter and the local oscillator lasers is estimated and removed. Several algorithms have been proposed for this purpose [92–95, 122].

A proliferation of algorithms has been proposed for carrier phase estimation (see for example [96, 97] and the references therein). The most commonly used is a feed-forward scheme initially proposed for burst digital transmission [123]. The algorithm can be modified to compensate for phase distortion due to interchannel cross-phase modulation as well [103, 124]. Its operating principle and its performance are explained in detail below.

10.4.1 Feed-Forward QPSK Carrier Phase Estimation Algorithm

The complex photocurrent at the input of the feed-forward carrier phase estimation module is proportional to the complex envelope of the received signal

$$x_k = A_k e^{j\phi_k} + n_{1_k} + jn_{2_k} \tag{10.23}$$

where A_k is the instantaneous amplitude, φ_k is the instantaneous phase due to modulation, $\phi_k = \phi_s(kT_s) - \phi_{lo}(kT_s)$ is the total phase noise, and $n_{1,k} = n_1(kT_s), n_{2,k} = n_2(kT_s)$ are two independent additive Gaussian noises due to the combined action of ASE, shot and thermal noises.

The feed-forward phase-noise estimation algorithm uses N successive samples of x_k to produce an estimate of the average phase noise $\hat{\phi}$ during the N-symbol interval. To eliminate the phase modulation, the samples x_k are raised to the fourth power and added together. The receiver then uses the argument of the sum to estimate the phase noise, based on the expression

$$\hat{\phi} = \frac{1}{4} \arg\left(\sum_{k=1}^{N} x_k^4 \right) \tag{10.24}$$

Since the arg{.} function yields values in the interval $(-\pi, \pi)$, $\hat{\phi}$ is wrapped in the interval $(-\pi/4, \pi/4)$, whereas the actual phase noise is unbounded. Several phase unwrapping algorithms have been proposed in the optical communications literature, e.g., [96, 97]. Unsuccessful phase unwrapping can cause cycle slips [96, 97], which, in turn, can lead to catastrophic error bursts. Cycle slips can be mitigated by differential encoding and decoding (see below), at the expense of a slight increase of the bit error probability.

10.5 Other Receiver Functionalities

10.5.1 Gray Coding

At the M-PSK transmitter, an m-tuple of bits, where$m = \log_2 M$, is mapped into one of M distinct phase values. It is desirable to optimally associate the M phases with the binary m-tuples, in order to minimize the bit error probability. Since it is more likely that the noise added by the optical components will cause adjacent phase errors, mapping is performed using Gray coding [27, 83, 100–102], i.e., adjacent symbols in phase space differ only by one significant digit. With this code, an adjacent symbol error causes only one bit to be incorrect.

For illustration, Table 10.1(a)–(c) shows a two-, four-, and eight-level reflected Gray code [100]. It is observed that any two successive triplets in the list differ in exactly one position. The construction of the reflected Gray code is performed iteratively as follows: During the first iteration, $M = 2$, the Gray code set consists of the words 0 and 1 [Table 10.1(a)]. During the second iteration, $M = 4$, a set of new Gray code words is obtained by reflecting the set of words obtained during the first iteration about a horizontal line, then prefixing every element of the original set by zeros and its reflection by ones [Table 10.1(b)]. The same procedure is repeated during the third iteration, $M = 8$ [Table 10.1(c)], and so on.

Table 10.1 Two, four and eight-level Gray code (a), (b), (c)

Gray code word	Decimal no (n)	Phase change $\Delta\varphi_k = 2\pi_n/M$
(a)		
0	0	0
1	1	π
(b)		
00	0	0
01	1	$\pi/2$
11	2	π
10	3	$3\pi/2$
(c)		
000	0	0
001	1	$\pi/8$
011	2	$\pi/4$
010	3	$3\pi/8$
110	4	$\pi/2$
111	5	$5\pi/8$
101	6	$3\pi/4$
100	7	$7\pi/8$

10.5.2 Differential Coding

In coherent synchronous M-PSK receivers, there is a phase ambiguity due to the algorithm used in the phase-noise estimation [26, 29]. The phase ambiguity arises as follows: as described above, in order to estimate the carrier phase change due to phase noise, the received M-PSK signal is first raised to the M-th power to remove phase modulation. The argument of the resulting signal is divided by M to provide the carrier phase estimate. This process introduces a phase ambiguity of π/M. This is due to the fact that the M-PSK constellation is π/M rotation invariant. To eliminate this constant phase error due to the phase ambiguity, the information is encoded, not in absolute phases, but in the phase differences between two successive symbols. The resulting M-PSK signal is said to be differentially encoded (multilevel differentially encoded phase-shift keying, M-DEC-PSK).[3] A differential decoder is added at the output of the receiver in Fig. 10.3, in order to recover the original bit sequence. The decoder forms the differences between two successive symbols. Then, the constant phase error, due to the phase ambiguity in the carrier recovery, is eliminated.

[3] Without loss of generality, here we will adopt the term M-PSK instead of the more proper M-DEC-PSK.

Formally, we can mathematically describe the differential encoding process as follows:

At the transmitter, at the output of the mapper we obtain the complex symbols

$$c_k \equiv e^{j\Delta\varphi_k} \tag{10.25}$$

The differential encoder multiplies each consecutive symbol with the previously sent one. At the output of the differential encoder, we obtain the complex symbols with unit magnitude and argument equal to the sum between the present input and the previous total phase

$$d_k = e^{j\varphi_k} = d_{k-1}c_k = e^{j(\Delta\varphi_k+\varphi_{k-1})} \tag{10.26}$$

with initial condition $d_{-1} = e^{j\Phi}$, where Φ is a random initial phase.

At the output of the receiver, we recover the estimated differentially encoded symbols $\hat{d}_n$which, in the ideal case, are equal to the transmitted ones d_n.

In order to form an estimate $\hat{c}_n$of the original symbols c_n, the differential decoder uses the recursive function

$$\hat{c}_n = \hat{d}_n\hat{d}^*_{n-1} = e^{j(\Delta\varphi_n+\varphi_{n-1})}e^{-j\varphi_{n-1}} = e^{j\Delta\varphi_n} \tag{10.27}$$

where it is assumed that the initial value is $\hat{d}_{-1} = 1$.

10.6 Error Probability Evaluation

Error probability is the most appropriate design criterion for digital communications systems, since it is uniquely related to system capacity [26]. Different analytical expressions for the error probability of coherent optical QPSK systems in the presence of additive Gaussian noise and Gaussian phase noise, with various degrees of accuracy, are derived by different authors, e.g., [9, 36, 99, 104–106].

In the most rudimentary case, in the absence of intersymbol interference (ISI) and phase noise, in the exclusive presence of additive Gaussian noise, the bit error probability of a coherent optical QPSK system, assuming Gray coding, can be analytically calculated [26]

$$P_{e|b} = \frac{1}{2}\mathrm{erfc}\left(\sqrt{\frac{\rho_s}{2}}\right) \tag{10.28}$$

where $\mathrm{erfc}(z)$ is the complementary error function defined as [26]

$$\mathrm{erfc}(z) = \frac{2}{\sqrt{\pi}}\int_z^{\infty} e^{-t^2}dt \tag{10.29}$$

In Eq. 10.28, ρ_s denotes the electronic SNR before the decision circuit, which is defined as [26]

$$\rho_s = \frac{A^2}{2\sigma^2} \tag{10.30}$$

where A is the amplitude of the complex envelope of the QPSK signal and σ^2 is the variance of each quadrature additive Gaussian noise component, due to the combined action of amplified spontaneous emission (ASE), shot and thermal noises.

If the additive Gaussian noise is exclusively due to amplified spontaneous emission (ASE) noise, the electronic symbol SNR is related to the optical signal-to-noise ratio (OSNR) by the expression (adapted from [20])

$$\text{OSNR} = \frac{pB_{eq}\rho_s}{2\Delta v_{res}}, \tag{10.31}$$

where $p = 1$ for single polarization channel transmission and $p = 2$ for PDM. In Eq. 10.31, Δv_{res} is the resolution bandwidth for the measurement of the ASE noise from the optical amplifiers and B_{eq} is the equivalent noise bandwidth [29] of the equivalent-baseband, aggregate transfer function of the transmission channel and the coherent optical receiver.

In the general case, in the presence of ISI, additive Gaussian noise, and Gaussian phase noise, we can use a computationally-efficient, semi-analytical method for the evaluation of the error probability of coherent optical PDM-QPSK systems.

According to the deterministic semi-analytical method [107], the distortion of the signal due to transmission impairments and the electronic DSP functionalities is computed by simulation in the absence of noise, the noise statistics at the input of the decision circuit are calculated analytically, and the average bit error probability is estimated using an analytical formula. More specifically, we evaluate the average symbol error probability using the arithmetic mean

$$\bar{P}_{e|s} = \frac{1}{N_{symbols}} \sum_{k=1}^{N_{symbols}} P_{e|s_k} \tag{10.32}$$

where $N_{symbols}$ denotes the number of simulated symbols and $P_{e|s_k}$ is the conditional error probability for the kth symbol.

We can transfer all constellation points to the upper right quadrant of the complex plane using appropriate rotations (i.e., multiples of $\pi/2$). Then, we can calculate the conditional error probability for the kth symbol by numerical integration using the formula (adapted from [36] with slight modifications)

$$P_{e|s_k} = \int_{-\infty}^{\infty} \text{erfc}\left[\sqrt{\rho_{s_k}}\cos(\theta_k + \varphi_k)\right] p_{\varphi_k}(\varphi_k)d\varphi_k + \int_{-\infty}^{\infty} \text{erfc}\left[\sqrt{\rho_{s_k}}\sin(\theta_k + \varphi_k)\right] p_{\varphi_k}(\varphi_k)d\varphi_k \quad (10.33)$$

In Eq. 10.33, ρ_{s_k} is the instantaneous electronic symbol SNR before the decision circuit, defined as $\rho_{s|k} = A_k^2/(2\sigma^2)$ and $p_{\varphi_k}(\varphi_k)$ is the pdf of the residual phase noise, which is considered Gaussian to a first order approximation [97], with variance equal to (adapted from [106] with slight modifications, see Appendix F)

$$\sigma_{\varphi_k}^2 = Dt_k + \frac{DT}{3} - 2\frac{Dt_k}{T}\left(T - \frac{t_k}{2}\right) + \frac{\sigma^2}{2B_{eq}T} \quad (10.34)$$

In Eq. 10.34, $D = 2\pi(\Delta\nu_s + \Delta\nu_{lo})$, where $\Delta\nu_s$, $\Delta\nu_{lo}$ is the 3-dB spectral linewidth of the lasers, T is the estimation (block) interval, t_k is the sampling time within the estimation interval, and B_{eq} is the equivalent-baseband, aggregate equivalent noise bandwidth of the coherent optical receiver. For a given system topology and assuming that the coherent homodyne receiver's adaptive filter coefficients are known, it is straightforward to analytically calculate the filtered additive Gaussian noise standard deviation σ, based on formulas from noise theory [26]. Finally, the average bit error probability, for Gray coding, is approximately $\bar{P}_{e|b} = 0.5\bar{P}_{e|s}$.

In order to take into account the ISI arising from m adjacent symbols on each side of the symbol we want to detect, we simulate all possible combinations of $2m+1$ symbols [107]. Since each QPSK symbol takes four possible values, one needs to simulate, in principle, 4^{2m+1} quaternary symbols. This can be achieved by various methods [98, 107–109]. Here, we use two independent de Bruijn pseudo-random bit sequences, generated by different polynomials of periods equal to 2^{4m+3}, at the inputs of the QM.

In practice, the design of optical communications systems is based on the effective Q factor, which is related to the average bit error probability by the (arbitrary) equation [42]

$$\bar{P}_{e|b} = \frac{1}{2}\text{erfc}\left(\frac{\bar{Q}}{\sqrt{2}}\right) \quad (10.35)$$

The corresponding effective Q factor is calculated by inverting (10.35)

$$\bar{Q} = \sqrt{2}\text{erfc}(2\bar{P}_{e|b}) \quad (10.36)$$

10.7 Future Trends

Commercially available long-haul terrestrial coherent optical PDM-QPSK systems will be mostly used for transmitting 112 Gb/s per wavelength over 1500–2000 km of standard single-mode fiber (SSMF) and several reconfigurable optical add/drop multiplexers (ROADMs), using the current 50-GHz ITU WDM grid [12].

However, given the anticipated rate of future network traffic growth [110], the capacity of current long-haul terrestrial systems has to rapidly evolve to satisfy future bandwidth demands. Current research focuses on the development of coherent optical systems operating at 400 Gb/s or even 1 Tb/s per wavelength channel until 2020.

To achieve these data rates, a continuous increase in spectral efficiency is necessary, given the restrictions imposed by the ADC sampling rate and resolution, the required WDM channel spacing, and the optical amplifier bandwidth. In principle, higher spectral efficiencies can be achieved by using M-QAM, in single or multi-carrier coherent optical systems. However, as the size of the QAM constellation grows, there are severe limitations due to the increase in OSNR, the impact of nonlinearities, and the stringent requirements in laser linewidth imposed by laser phase noise. Winzer [12] explores the advantages and disadvantages of different alternative solutions.

Low-attenuation, large effective area optical fibers [111, 112], electronic compensation of fiber nonlinearities [62–68] and stronger forward error correction (FEC) codes [113], are some of the key enabling technologies that might influence the design of future coherent optical communications systems.

As a longer term solution for a drastic capacity increase, mode division multiplexing [114, 125] over few-mode fibers [115, 116] or multicore fibers [120] is also actively investigated.

Acknowledgments The author would like to thank Dr. T. Vgenis and Dr. N. Mantzoukis, Department of Electrical and Computer Engineering, University of Patras, Greece, for contributions to different aspects of this work, as well as Mr. X. Zhu, Corning Inc., and Prof. J.C. Cartledge, Queen's University for stimulating discussions.

Appendix A: Quadrature Modulator

In this Appendix, we derive a simplified mathematical model for the electro-optic QM in $LiNbO_3$.

In the subsequent analysis, the following notations are used [49]: Dirac's ket vectors denote Jones vectors, and boldface letters denote electric field vectors or scattering matrices (the distinction must be clear from context). In addition, analytic signals and complex envelopes are used (see definitions [26], Chap. 4) and are denoted by a hat and a tilde, respectively. These representations lead to elegant and concise mathematical expressions.

Using low pass equivalent formulation, the vector of the electric field of the transmitted optical signal can be written as

$$\mathbf{E}_s(t) = \Re\{\hat{\mathbf{E}}_s(t)\} \tag{10.37}$$

where $\Re\{\}$ denotes real part and $\hat{\mathbf{E}}_s(t)$ is the analytic electric field vector

$$\hat{\mathbf{E}}_s(t) = \tilde{\mathbf{E}}_s(t)e^{j\omega_s t} \tag{10.38}$$

In the above relationship, ω_s is the carrier angular frequency and $\tilde{\mathbf{E}}_s(t)$ is the complex envelope of the electric field

$$\tilde{\mathbf{E}}_s(t) = \tilde{E}_s(t)|e_s\rangle \tag{10.39}$$

where $|e_s\rangle$ is the normalized Jones vector that denotes the state of polarization (SOP) of the transmitted signal and $\tilde{E}_s(t)$ is the scalar complex envelope of the transmitted optical signal. For monochromatic electric fields, $\tilde{\mathbf{E}}_s(t)$corresponds to the electric field phasor.

As a starting point, we derive the transfer function of a Mach–Zehnder modulator, in terms of scattering matrices. A scattering matrix defines the relationship between the input and output analytic electric fields [119].

The scattering matrix of a lossless, polarization independent, directional 3-dB coupler is written [44, 45, 119]

$$\mathbf{S} = \frac{1}{\sqrt{2}}\begin{pmatrix} 1 & j \\ j & 1 \end{pmatrix} \tag{10.40}$$

The scattering matrix of the two parallel branches of a push–pull Mach–Zehnder modulator is written

$$\mathbf{D} = \begin{pmatrix} e^{\frac{j\phi}{2}} & 0 \\ 0 & e^{-\frac{j\phi}{2}} \end{pmatrix} \tag{10.41}$$

In the previous relationship, ϕ is the sum of a phase shift due to the propagation φ_0 and a phase shift due to the voltage-dependent refractive index (Pockels effect) [45]

$$\phi = \varphi_0 - \pi\frac{V}{V_\pi} \tag{10.42}$$

where

$$\begin{aligned} \varphi_0 &= \frac{2\pi}{\lambda_0}nL \\ V_\pi &= \frac{\lambda_0 d}{rn^3L} \end{aligned} \tag{10.43}$$

where λ_0 is the free-space wavelength of the input optical beam, L is the length of the device, d is the distance between the electrodes, n is the effective refractive index in the absence of voltage, and $\mathfrak{r}$ is the Pockels electro-optic coefficient. The constant V_π is called half-wave voltage. The birefringence of $LiNbO_3$ is neglected.

The analytic electric fields at the output ports of the Mach–Zehnder modulator are

$$\begin{pmatrix} \hat{\mathbf{E}}_{o,1} \\ \hat{\mathbf{E}}_{o,2} \end{pmatrix} = \mathbf{SDS} \begin{pmatrix} \hat{\mathbf{E}}_{i,1} \\ 0 \end{pmatrix} = j \begin{pmatrix} \sin\frac{\phi}{2} \\ \cos\frac{\phi}{2} \end{pmatrix} \hat{\mathbf{E}}_{i,1} \tag{10.44}$$

where $\hat{\mathbf{E}}_{i,1}$ is the analytic electric field at one of the input ports of the Mach–Zehnder modulator.

Discarding the second output, the final result is

$$\hat{\mathbf{E}}_{o,1} = j \sin\frac{\phi}{2} \hat{\mathbf{E}}_{i,1} \tag{10.45}$$

For the analysis of the QM, we define the scattering matrices for the Mach–Zehnder modulators and the phase shifter as

$$\mathbf{M} = \begin{pmatrix} j\sin\frac{\phi_1}{2} & 0 \\ 0 & j\sin\frac{\phi_2}{2} \end{pmatrix}$$
$$\boldsymbol{\Phi} = \begin{pmatrix} e^{\frac{j\phi_3}{2}} & 0 \\ 0 & e^{-\frac{j\phi_3}{2}} \end{pmatrix} \tag{10.46}$$

The analytic electric fields at the output ports of the QM are

$$\begin{pmatrix} \hat{\mathbf{E}}_{o,1} \\ \hat{\mathbf{E}}_{o,2} \end{pmatrix} = \mathbf{S\Phi MS} \begin{pmatrix} \hat{\mathbf{E}}_{i,1} \\ 0 \end{pmatrix} \tag{10.47}$$

It is straightforward to show that, neglecting the phase shift due to the propagation φ_0,

$$\hat{\mathbf{E}}_{o,1} = \frac{j}{2} e^{-j\frac{\pi V_3}{2V_{\pi 3}}} \left\{ \sin\left[\frac{\pi V_1(t)}{2V_{\pi_1}}\right] - e^{-j\frac{\pi V_3}{2V_{\pi 3}}} \sin\left[\frac{\pi V_2(t)}{2V_{\pi_2}}\right] \right\} \hat{\mathbf{E}}_{i,1} \tag{10.48}$$

Since the polarizations are preserved we can write the following final relationship for the scalar complex envelopes

$$\tilde{E}_s(t) = \frac{j}{2} e^{-j\frac{\pi V_3}{2V_{\pi 3}}} \left\{ \sin\left[\frac{\pi V_1(t)}{2V_{\pi_1}}\right] - e^{-j\frac{\pi V_3}{2V_{\pi 3}}} \sin\left[\frac{\pi V_2(t)}{2V_{\pi_2}}\right] \right\} \tilde{E}_{in}(t) \tag{10.49}$$

Appendix B: Balanced Receiver

As discussed in Appendix A, the scattering matrix of a lossless, polarization independent, directional 3-dB coupler is written [45, 119]

$$\mathbf{S} = \frac{1}{\sqrt{2}}\begin{pmatrix} 1 & j \\ j & 1 \end{pmatrix} \tag{10.50}$$

The analytic electric fields at the output ports of the 3-dB coupler are

$$\begin{pmatrix} \hat{\mathbf{E}}_{o,1} \\ \hat{\mathbf{E}}_{o,2} \end{pmatrix} = \mathbf{S}\begin{pmatrix} \hat{\mathbf{E}}_{s} \\ \hat{\mathbf{E}}_{lo} \end{pmatrix} = \frac{1}{\sqrt{2}}\begin{pmatrix} \hat{\mathbf{E}}_{s} + j\hat{\mathbf{E}}_{lo} \\ j\hat{\mathbf{E}}_{s} + \hat{\mathbf{E}}_{lo} \end{pmatrix} \tag{10.51}$$

The photocurrent at the output of the two photodiodes is given by

$$i_k = \frac{R_k}{2}\hat{\mathbf{E}}_k^{\dagger}\hat{\mathbf{E}}_k \tag{10.52}$$

where R_k is the responsivity of the photodiodes and dagger denotes the adjoint matrix.

If the received analytic electric field $\hat{\mathbf{E}}_k$ is the sum of two analytic electric fields $\hat{\mathbf{E}}_k = \hat{\mathbf{E}}_{k,1} + \hat{\mathbf{E}}_{k,2}$, the photocurrent can be written

$$i_k = \frac{R_k}{2}\left[\hat{\mathbf{E}}_{k,1}^{\dagger} + \hat{\mathbf{E}}_{k,2}^{\dagger}\right]\left[\hat{\mathbf{E}}_{k,1} + \hat{\mathbf{E}}_{k,2}\right] = \frac{R_k}{2}\left\{\left\|\hat{\mathbf{E}}_{k,1}\right\|^2 + \left\|\hat{\mathbf{E}}_{k,2}\right\|^2 + 2\Re\left[\hat{\mathbf{E}}_{k,2}^{\dagger}\hat{\mathbf{E}}_{k,1}\right]\right\} \tag{10.53}$$

By substitution of Eq. 10.51 into Eq. 10.53 we obtain

$$\begin{aligned} i_1 &= \frac{R_1}{4}\left\{\left\|\hat{\mathbf{E}}_{s}\right\|^2 + \left\|\hat{\mathbf{E}}_{lo}\right\|^2 + 2\Im\left[\hat{\mathbf{E}}_{lo}^{\dagger}\hat{\mathbf{E}}_{s}\right]\right\} \\ i_2 &= \frac{R_2}{4}\left\{\left\|\hat{\mathbf{E}}_{s}\right\|^2 + \left\|\hat{\mathbf{E}}_{lo}\right\|^2 - 2\Im\left[\hat{\mathbf{E}}_{lo}^{\dagger}\hat{\mathbf{E}}_{s}\right]\right\} \end{aligned} \tag{10.54}$$

where $\Im\{.\}$ denotes the imaginary part.

In the case $R_1 = R_2 = R$, we obtain

$$i_{tot} = i_1 - i_2 = R\Im\left[\hat{\mathbf{E}}_{lo}^{\dagger}\hat{\mathbf{E}}_{s}\right] \tag{10.55}$$

Appendix C: Analysis of a 4 × 4 90° Hybrid

The architecture of the four input/four output port optical hybrid is shown in Fig. 10.12 [48]. The received signal is fed into input A and split by an ideal,

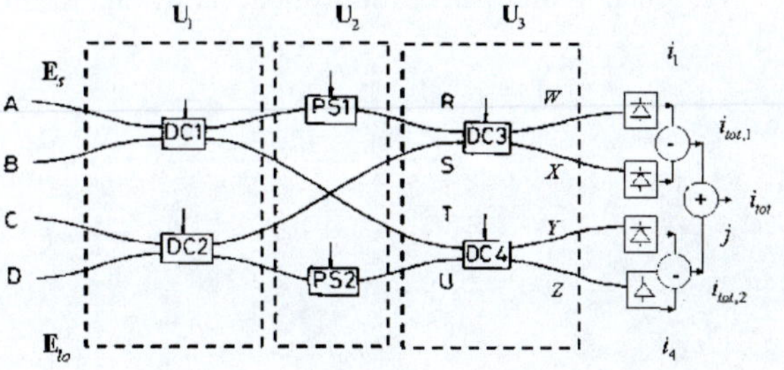

Fig. 10.12 Implementation of the 90° 4 × 4 hybrid

lossless, polarization-independent 3-dB coupler (DC 1). The relative phase between the signals at the ports R and T can be tuned by a phase shifter (PS 1) to $2k\pi$. Similarly, the local oscillator signal is fed into input D and split by a second ideal, lossless, polarization-independent 3-dB coupler (DC 2). The phase between the local oscillator signal at the ports S and U is adjusted to 90° by a second phase shifter (PS 2). The split and phase shifted signals from the transmitter and the local oscillator are combined with four different phase shifts using two additional ideal, lossless, polarization-independent 3-dB couplers (DC 3, DC 4). The resulting optical signals at the output ports W, X, Y, and Z are detected by two balanced receivers. It will be shown that the total photocurrents at the output of the balanced receiversi_{tot_1}, i_{tot_2} correspond to coherent beating terms with a 90° phase difference between them.

The relationship that links the analytic electric fields at the input and output ports is:

$$\begin{pmatrix} \hat{\mathbf{E}}_W \\ \hat{\mathbf{E}}_X \\ \hat{\mathbf{E}}_Y \\ \hat{\mathbf{E}}_Z \end{pmatrix} = \mathbf{U}_3\mathbf{U}_2\mathbf{U}_1 \begin{pmatrix} \hat{\mathbf{E}}_A \\ \hat{\mathbf{E}}_B \\ \hat{\mathbf{E}}_C \\ \hat{\mathbf{E}}_D \end{pmatrix} = \mathbf{U}_3\mathbf{U}_2\mathbf{U}_1 \begin{pmatrix} \hat{\mathbf{E}}_s \\ 0 \\ 0 \\ \hat{\mathbf{E}}_{lo} \end{pmatrix} \tag{10.56}$$

where $\hat{\mathbf{E}}_s, \hat{\mathbf{E}}_{lo}$ are the analytic electric fields of the signal and the local oscillator, respectively, omitting the time dependence to alleviate the formalism, and the matrices $\mathbf{U}_1, \mathbf{U}_2, \mathbf{U}_3$, that describe the transfer function of the respective broken rectangles in Fig. 10.12, are given by:

$$\mathbf{U}_1 = \mathbf{U}_3 = \frac{1}{\sqrt{2}} \begin{pmatrix} 1 & j & 0 & 0 \\ j & 1 & 0 & 0 \\ 0 & 0 & 1 & j \\ 0 & 0 & j & 1 \end{pmatrix} \tag{10.57}$$

$$\mathbf{U}_2 = \begin{pmatrix} 1 & 0 & 0 & 0 \\ 0 & 0 & 1 & 0 \\ 0 & 1 & 0 & 0 \\ 0 & 0 & 0 & j \end{pmatrix} \tag{10.58}$$

Performing the matrix multiplication in Eq. 10.56, it is straightforward to show that

$$\begin{aligned}\hat{\mathbf{E}}_W &= \frac{\hat{\mathbf{E}}_s - \hat{\mathbf{E}}_{lo}}{2}\\ \hat{\mathbf{E}}_X &= \frac{j}{2}\left(\hat{\mathbf{E}}_s + \hat{\mathbf{E}}_{lo}\right)\\ \hat{\mathbf{E}}_Y &= \frac{j}{2}\left(\hat{\mathbf{E}}_s + j\hat{\mathbf{E}}_{lo}\right)\\ \hat{\mathbf{E}}_Z &= \frac{j}{2}\left(j\hat{\mathbf{E}}_s + \hat{\mathbf{E}}_{lo}\right)\end{aligned} \tag{10.59}$$

The photocurrent at the output of the four photodiodes is given by

$$i_k = \frac{R_k}{2}\hat{\mathbf{E}}_k^{\dagger}\hat{\mathbf{E}}_k \tag{10.60}$$

where R is the responsivity of the photodiodes and dagger denotes the adjoint matrix.

If the received analytic electric field $\hat{\mathbf{E}}_k$ is the sum of two analytic electric fields $\hat{\mathbf{E}}_k = \hat{\mathbf{E}}_{k,1} + \hat{\mathbf{E}}_{k,2}$, the photocurrent can be written

$$i_k = \frac{R_k}{2}\left[\hat{\mathbf{E}}_{k,1}^{\dagger} + \hat{\mathbf{E}}_{k,2}^{\dagger}\right]\left[\hat{\mathbf{E}}_{k,1} + \hat{\mathbf{E}}_{k,2}\right] = \frac{R_k}{2}\left\{\left\|\hat{\mathbf{E}}_{k,1}\right\|^2 + \left\|\hat{\mathbf{E}}_{k,2}\right\|^2 + 2\Re\left[\hat{\mathbf{E}}_{k,2}^{\dagger}\hat{\mathbf{E}}_{k,1}\right]\right\} \tag{10.61}$$

where $\Re\{.\}$ denotes the real part.

By substitution of Eq. 10.59 into Eq. 10.61 we obtain

$$\begin{aligned}i_1 &= \frac{R_1}{8}\left\{\left\|\hat{\mathbf{E}}_s\right\|^2 + \left\|\hat{\mathbf{E}}_{lo}\right\|^2 - 2\Re\left[\hat{\mathbf{E}}_{lo}^{\dagger}\hat{\mathbf{E}}_s\right]\right\}\\ i_2 &= \frac{R_2}{8}\left\{\left\|\hat{\mathbf{E}}_s\right\|^2 + \left\|\hat{\mathbf{E}}_{lo}\right\|^2 + 2\Re\left[\hat{\mathbf{E}}_{lo}^{\dagger}\hat{\mathbf{E}}_s\right]\right\}\\ i_3 &= \frac{R_3}{8}\left\{\left\|\hat{\mathbf{E}}_s\right\|^2 + \left\|\hat{\mathbf{E}}_{lo}\right\|^2 + 2\Im\left[\hat{\mathbf{E}}_{lo}^{\dagger}\hat{\mathbf{E}}_s\right]\right\}\\ i_4 &= \frac{R_4}{8}\left\{\left\|\hat{\mathbf{E}}_s\right\|^2 + \left\|\hat{\mathbf{E}}_{lo}\right\|^2 - 2\Im\left[\hat{\mathbf{E}}_{lo}^{\dagger}\hat{\mathbf{E}}_s\right]\right\}\end{aligned} \tag{10.62}$$

In the case $R_1 = R_2 = R_3 = R_4 = R$ we obtain

$$\begin{aligned}i_{tot,1} &= i_2 - i_1 = \frac{R}{2}\Re\left[\hat{\mathbf{E}}_{lo}^{\dagger}\hat{\mathbf{E}}_s\right]\\ i_{tot,2} &= i_3 - i_4 = \frac{R}{2}\Im\left[\hat{\mathbf{E}}_{lo}^{\dagger}\hat{\mathbf{E}}_s\right]\end{aligned} \tag{10.63}$$

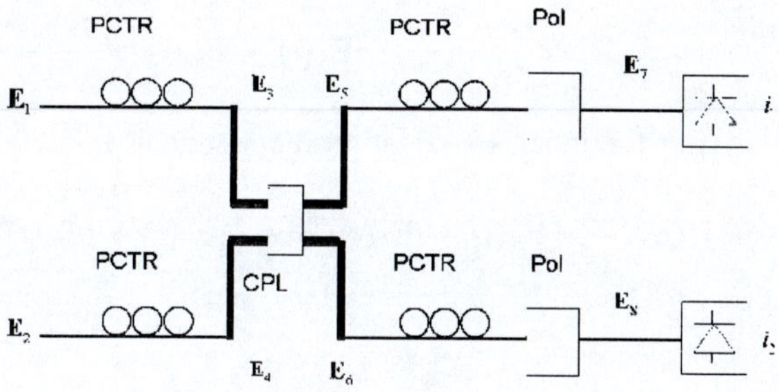

Fig. 10.13 Implementation of the 90° 2 × 2 hybrid

At the DSP we can form the complex photocurrent

$$i_{tot} = i_{tot,1} + ji_{tot,2} = \frac{R}{2}\hat{\mathbf{E}}_{lo}^{\dagger}\hat{\mathbf{E}}_s \tag{10.64}$$

Appendix D: Analysis of a 2 × 2 90° Hybrid

A 2 × 2 90° hybrid was proposed by [46, 117] (Fig. 10.13). The hybrid is composed of four polarization controllers (PCTR), an ideal, lossless, polarization-independent 3-dB coupler (CPL), and two fiber polarizers (Pol). The two-input polarization controllers change the state of polarization of the input signals to linear 45° and right-circular, respectively. The two-output polarization controllers change the principal axes of the fiber polarizers so that they select the $|x\rangle, |y\rangle$ polarization components of the optical signals at the output ports of the 3-dB coupler. The signals impingent to the photodiodes are proportional to $\hat{E}_1 + \hat{E}_2$ and $\hat{E}_1 + j\hat{E}_2$, respectively, where $\hat{E}_1, \hat{E}_2$ are the analytic signals corresponding to the input electric fields. The two-output photocurrents are calculated in this appendix and are shown to be in quadrature. In addition, it is shown that the detection of the two quadratures can be achieved by several different settings of the polarization controllers.

Simplified Model [46, 117]

After the two-input polarization controllers, the electric fields of the signal and the local oscillator can be written as

$$\hat{\mathbf{E}}_3(t) = \hat{E}_r(t)\frac{|x\rangle + |y\rangle}{\sqrt{2}} \tag{10.65}$$

$$\hat{\mathbf{E}}_4(t) = \hat{E}_{lo}(t)\frac{|x\rangle + j|y\rangle}{\sqrt{2}} \tag{10.66}$$

After the ideal, polarization-independent, lossless 3-dB coupler

$$\hat{\mathbf{E}}_5(t) = \frac{1}{\sqrt{2}}\left[\hat{\mathbf{E}}_3(t) + j\hat{\mathbf{E}}_4(t)\right] = \frac{1}{2}\left\{\left[\hat{E}_r(t) + j\hat{E}_{lo}(t)\right]|x\rangle + \left[\hat{E}_r(t) - \hat{E}_{lo}(t)\right]|y\rangle\right\} \tag{10.67}$$

$$\hat{\mathbf{E}}_6(t) = \frac{1}{\sqrt{2}}\left[j\hat{\mathbf{E}}_3(t) + \hat{\mathbf{E}}_4(t)\right] = \frac{1}{2}\left\{\left[j\hat{E}_r(t) + \hat{E}_{lo}(t)\right]|x\rangle + j\left[\hat{E}_r(t) + \hat{E}_{lo}(t)\right]|y\rangle\right\} \tag{10.68}$$

After the fiber polarizers

$$\hat{\mathbf{E}}_7(t) = \frac{1}{2}\left[\hat{E}_r(t) + j\hat{E}_{lo}(t)\right]|x\rangle \tag{10.69}$$

$$\hat{\mathbf{E}}_8(t) = \frac{j}{2}\left[\hat{E}_r(t) + \hat{E}_{lo}(t)\right]|y\rangle \tag{10.70}$$

At the photodiodes

$$\begin{aligned} i_1 &= \frac{R_1}{2}\hat{\mathbf{E}}_7^{\dagger}(t)\hat{\mathbf{E}}_7(t) = \frac{R_1}{8}\left|\hat{E}_r(t) + j\hat{E}_{lo}(t)\right|^2 \\ &= \frac{R_1}{8}\left\{\left|\hat{E}_r(t)\right|^2 + \left|\hat{E}_{lo}(t)\right|^2 + 2\Im\left[\hat{E}_r(t)\hat{E}_{lo}^*(t)\right]\right\} \\ i_2 &= \frac{R_2}{2}\hat{\mathbf{E}}_8^{\dagger}(t)\hat{\mathbf{E}}_8(t) = \frac{R_2}{8}\left|\hat{E}_r(t) + \hat{E}_{lo}(t)\right|^2 \\ &= \frac{R_2}{8}\left\{\left|\hat{E}_r(t)\right|^2 + \left|\hat{E}_{lo}(t)\right|^2 + 2\Re\left[\hat{E}_r(t)\hat{E}_{lo}^*(t)\right]\right\} \end{aligned} \tag{10.71}$$

Final expressions for laser offset and unmatched photodiodes

$$i_1 = \frac{R_1}{4}\left\{P_r + P_{lo} + 2\sqrt{P_r P_{lo}}\sum_k g(t - kT)\sin[\omega_{IF}t + \phi_k + \theta(t)]\right\} \tag{10.72}$$

$$i_2 = \frac{R_2}{4}\left\{P_r + P_{lo} + 2\sqrt{P_r P_{lo}}\sum_k g(t - kT)\cos[\omega_{IF}t + \phi_k + \theta(t)]\right\} \tag{10.73}$$

Generalized Model

The vector of the electric field of the received signal can be written as

$$\hat{\mathbf{E}}_1(t) = \hat{E}_r(t)\left|e_r'\right\rangle \tag{10.74}$$

The vector of the electric field of the local oscillator can be written as

$$\hat{\mathbf{E}}_2(t) = \hat{E}_{lo}(t)\left|e_{lo}'\right\rangle \tag{10.75}$$

After the polarization controllers

$$\hat{\mathbf{E}}_3(t) = \hat{E}_r(t)\left|e_r\right\rangle \tag{10.76}$$

$$\hat{\mathbf{E}}_4(t) = \hat{E}_{lo}(t)\left|e_{lo}\right\rangle \tag{10.77}$$

After the ideal, polarization-independent, lossless 3-dB coupler

$$\hat{\mathbf{E}}_5(t) = \frac{1}{\sqrt{2}}\left[\hat{\mathbf{E}}_3(t) + j\hat{\mathbf{E}}_4(t)\right] = \frac{1}{2}\left[\hat{E}_r(t)\left|e_r\right\rangle + j\hat{E}_{lo}(t)\left|e_{lo}\right\rangle\right] \tag{10.78}$$

$$\hat{\mathbf{E}}_6(t) = \frac{1}{\sqrt{2}}\left[j\hat{\mathbf{E}}_3(t) + \hat{\mathbf{E}}_4(t)\right] = \frac{1}{2}\left[j\hat{E}_r(t)\left|e_r\right\rangle + \hat{E}_{lo}(t)\left|e_{lo}\right\rangle\right] \tag{10.79}$$

After the fiber polarizers

$$\hat{\mathbf{E}}_7(t) = \frac{1}{2}\left[\hat{E}_r(t)\langle p_1 \mid e_r\rangle + j\hat{E}_{lo}(t)\langle p_1 \mid e_{lo}\rangle\right]\left|p_1\right\rangle \tag{10.80}$$

$$\hat{\mathbf{E}}_8(t) = \frac{1}{2}\left[j\hat{E}_r(t)\langle p_2 \mid e_r\rangle + \hat{E}_{lo}(t)\langle p_2 \mid e_{lo}\rangle\right]\left|p_2\right\rangle \tag{10.81}$$

We define

$$\begin{aligned}\langle p_1 \mid e_r\rangle &= \alpha_1 e^{i\zeta_1}\\ \langle p_1 \mid e_{lo}\rangle &= \beta_1 e^{i\xi_1}\\ \langle p_2 \mid e_r\rangle &= \alpha_2 e^{i\zeta_2}\\ \langle p_2 \mid e_{lo}\rangle &= \beta_2 e^{i\xi_2}\end{aligned} \tag{10.82}$$

At the photodiodes

$$i_1 = \frac{R_1}{2}\hat{\mathbf{E}}_7^{\dagger}(t)\hat{\mathbf{E}}_7(t) = \frac{R_1}{8}\left\{\alpha_1^2\left|\hat{E}_r(t)\right|^2 + \beta_1^2\left|\hat{E}_{lo}(t)\right|^2 + 2\alpha_1\beta_1\Im\left[\hat{E}_r(t)\hat{E}_{lo}^*(t)e^{i(\zeta_1-\xi_1)}\right]\right\} \tag{10.83}$$

$$i_2 = \frac{R_2}{2}\hat{\mathbf{E}}_8^{\dagger}(t)\hat{\mathbf{E}}_8(t) = \frac{R_2}{8}\left\{\alpha_2^2\left|\hat{E}_r(t)\right|^2 + \beta_2^2\left|\hat{E}_{lo}(t)\right|^2 - 2\alpha_2\beta_2\Im\left[\hat{E}_r(t)\hat{E}_{lo}^*(t)e^{i(\zeta_2-\xi_2)}\right]\right\}$$

Final expressions for laser offset and unmatched photodiodes

$$i_1 = \frac{R_1}{4}\left\{\alpha_1^2 P_r + \beta_1^2 P_{lo} + 2\alpha_1\beta_1\sqrt{P_r P_{lo}}\sum_k g(t-kT)\sin[\omega_{IF}t + \phi_k + \theta(t) + \zeta_1 - \xi_1]\right\} \tag{10.84}$$

$$i_2 = \frac{R_2}{4}\left\{\alpha_2^2 P_r + \beta_2^2 P_{lo} - 2\alpha_2\beta_2\sqrt{P_r P_{lo}}\sum_k g(t-kT)\sin[\omega_{IF}t + \phi_k + \theta(t) + \zeta_2 - \xi_2]\right\} \tag{10.85}$$

For ideal hybrid operation

$$(\zeta_2 - \xi_2) - (\zeta_1 - \xi_1) = (2k+1)\frac{\pi}{2} \tag{10.86}$$

Appendix E: Model of Polarization and Phase Diversity Receiver

In the upper branch, we receive

$$\begin{aligned} \hat{\mathbf{E}}_{s,x}(t) &= \hat{E}_s(t)\langle x \mid e_s(t)\rangle|x\rangle \\ \hat{\mathbf{E}}_{lo,x}(t) &= \frac{1}{\sqrt{2}}\hat{E}_{lo}(t)|x\rangle \end{aligned} \tag{10.87}$$

In the lower branch, we receive

$$\begin{aligned} \hat{\mathbf{E}}_{s,y}(t) &= \hat{E}_s(t)\langle y \mid e_s(t)\rangle|y\rangle \\ \hat{\mathbf{E}}_{lo,y}(t) &= \frac{1}{\sqrt{2}}\hat{E}_{lo}(t)|y\rangle \end{aligned} \tag{10.88}$$

The total photocurrent in the upper branch, in the absence of noise, is calculated by substituting Eq. 10.87 into Eq. 10.64

$$i_{tot,x} = \frac{R}{2\sqrt{2}}\hat{\mathbf{E}}_{lo}^{\dagger}\hat{\mathbf{E}}_{s,x} = \frac{R}{2\sqrt{2}}\hat{E}_{lo}^{*}\hat{E}_s\langle x \mid e_s(t)\rangle \tag{10.89}$$

Similarly, the total photocurrent in the lower branch, in the absence of noise, is calculated by substituting Eq. 10.88 into Eq. 10.64

$$i_{tot,y} = \frac{R}{2\sqrt{2}}\hat{\mathbf{E}}_{lo}^{\dagger}\hat{\mathbf{E}}_{s,y} = \frac{R}{2\sqrt{2}}\hat{E}_{lo}^{*}\hat{E}_s\langle y \mid e_s(t)\rangle \tag{10.90}$$

The DSP forms the following array, using the photocurrent of the two branches

$$\mathbf{I}_{tot} = \begin{pmatrix} i_{tot,x} \\ i_{tot,y} \end{pmatrix} = \frac{R}{2\sqrt{2}}\hat{E}_{lo}^{*}\hat{E}_s|e_s(t)\rangle \tag{10.91}$$

We observe that we are able to recover the full information content of the signal, regardless of the received signal SOP. However, there is a reduction in the photocurrent amplitude by $1/\sqrt{2}$.

Appendix F: Residual Phase-Noise Variance

The analytical calculation of the residual laser phase-noise variance, after the carrier phase estimation circuit proposed by [123], was done by [106]. In this Appendix, we rederive Eq. 10.34 from first principles, by approximating the phase-noise estimation averaging over discrete received signal samples as a continuous linear filtering process.

F.1 Laser Phase-Noise Properties

First, we review the basic properties of laser phase noise.

Phase noise is due to the spontaneous emission of semiconductor lasers. It can be modeled as a Wiener-Levy process, where the instantaneous phase is written as the integral of the random instantaneous angular frequency deviation $\dot{\phi}(t)$ [41]

$$\phi(t) = \int_0^t \dot{\phi}(t')dt' \tag{10.92}$$

The instantaneous angular frequency deviation $\dot{\phi}(t)$ can be interpreted as the difference between the instantaneous laser angular frequency $\omega_s(t)$ and the laser nominal angular carrier frequency ω_0, i.e.,

$$\dot{\phi}(t) = \omega_s(t) - \omega_0 \tag{10.93}$$

A common approximation is to assume that the instantaneous angular frequency deviation $\dot{\phi}(t)$ is just a zero-mean white Gaussian noise with psd

$$S_{\dot{\phi}}(\omega) = D \tag{10.94}$$

where D is a constant, which is called *phase diffusion coefficient*. It is linked to the 3-dB spectral linewidth of the laser. Since we are interested here for the total phase noise due to the transmitter laser and the local oscillator

$$D = 2\pi(\Delta\nu_s + \Delta\nu_{lo}), \tag{10.95}$$

where $\Delta\nu_s$, $\Delta\nu_{lo}$ is the 3-dB spectral linewidth of the lasers, respectively.

The autocorrelation function of the instantaneous angular frequency deviation is given by

$$R_{\dot{\phi}}(\tau) = \int_{-\infty}^{\infty} S_{\dot{\phi}}(\omega)e^{j\omega\tau}df = D\int_{-\infty}^{\infty} e^{j\omega\tau}df = D\delta(\tau) \tag{10.96}$$

where $\delta(t)$ is the Dirac delta function.

From the above relationship, for $\tau = t_1 - t_2$

$$R_{\dot{\phi}}(t_1 - t_2) = E\left\{\dot{\phi}(t_1)\dot{\phi}(t_2)\right\} = D\delta(t_1 - t_2) \tag{10.97}$$

The mean of the instantaneous angular frequency deviation is zero

$$\mu_{\dot{\phi}} = E\left\{\dot{\phi}(t)\right\} = 0 \tag{10.98}$$

The variance of the instantaneous angular frequency deviation is given by

$$\sigma_{\dot{\phi}}^2 = R_{\dot{\phi}}(0) = D\delta(0) \tag{10.99}$$

F.2 Mean and Variance of the Phase Noise

To determine the mean, we take the expected value of both sides of Eq. 10.92

$$\mu_{\phi} = E\{\phi(t)\} = \int_0^t E\left\{\dot{\phi}(t')\right\}dt' = 0 \tag{10.100}$$

To determine the variance, we use the definition

$$\begin{aligned}\sigma_{\phi}^2 = E\{\phi^2(t)\} &= \int_0^t\int_0^t E\left\{\dot{\phi}(t_1)\dot{\phi}(t_2)\right\}dt_1dt_2 \\ &= D\int_0^t\int_0^t \delta(t_1 - t_2)dt_1dt_2 = D\int_0^t dt_2 = Dt\end{aligned} \tag{10.101}$$

The cross-correlation function is

$$
\begin{aligned}
E\{\phi(t_1)\phi(t_2)\} &= \int_0^{t_1}\int_0^{t_2} E\left\{\dot{\phi}(t_1')\dot{\phi}(t_2')\right\}dt_1'dt_2' \\
&= D\int_0^{t_1}\int_0^{t_2} \delta\left(t_1' - t_2'\right)dt_1'dt_2' = D\min(t_1, t_2)
\end{aligned}
\tag{10.102}
$$

This agrees with the relationship Eq. 11.22 of [121], which is derived in a different manner.

F.3 Operation of the Feed-Forward Phase-Noise Estimation Circuit

The feed-forward phase-noise estimation circuit estimates the average of the received signal phase over a time interval T.

$$
\bar{\phi} = \frac{1}{T}\int_0^T \phi(t)dt \tag{10.103}
$$

To determine the mean, we take the expected value of both sides of Eq. 10.92

$$
\mu_{\bar{\phi}} = \frac{1}{T}\int_0^T E\{\phi(t)\}dt = \mu_\phi = 0 \tag{10.104}
$$

To determine the variance, we use the definition

$$
\begin{aligned}
\sigma_{\bar{\phi}}^2 &= \frac{1}{T^2}\int_0^T\int_0^T E\{\phi(t_1)\phi(t_2)\}dt_1 dt_2 \\
&= \frac{D}{T^2}\int_0^T\int_0^T \min(t_1, t_2)dt_1 dt_2
\end{aligned}
\tag{10.105}
$$

To evaluate the integral, we break it into a sum of two integrals, which correspond to the cases where $t_1 > t_2$ and $t_1 < t_2$

$$\sigma_{\bar{\phi}}^2 = \frac{D}{T^2}\left[\int_0^T\left(\int_0^{t_2} t_1 dt_1\right)dt_2 + \int_0^T\left(\int_0^{t_1} t_2 dt_2\right)dt_1\right] \\ = \frac{D}{T^2}\int_0^T t_2^2 dt_2 = \frac{DT}{3} \tag{10.106}$$

The averaging of the phasor of the unmodulated electric field corrupted by phase noise, in the case of an integrate and dump filter, i.e., an LPF whose impulse response is a rectangular pulse of duration T, was studied by [126]. In the realm of small phase noise, an analytical expression for the moment generating function of $\bar{\phi}$ is given by relation (6) in [126]. Double differentiation of (6) in [126] yields the same result like Eq. 10.105.

F.4 Operation of the Decision Circuit

The decision circuit makes a decision, which is corrupted by the remainder of phase noise

$$\theta(t) \cong \phi(t) - \bar{\phi} \tag{10.107}$$

To determine the mean, we take the expected value of both sides of Eq. 10.107

$$\mu_\theta = E\{\theta(t)\} = \mu_\phi - \mu_{\bar{\phi}} = 0 \tag{10.108}$$

To determine the variance, we use the definition

$$\sigma_\theta^2 = E\left\{\left[\phi(t) - \bar{\phi}\right]^2\right\} = E\left\{\phi^2(t) - 2\bar{\phi}\phi(t) + \bar{\phi}^2\right\} \\ = \sigma_\phi^2 + \sigma_{\bar{\phi}}^2 - 2E\left\{\bar{\phi}\phi(t)\right\} \tag{10.109}$$

The variances $\sigma_\phi^2, \sigma_{\bar{\phi}}^2$ are given by Eq. 10.101 and Eq. 10.105, respectively. We just have to evaluate the last term in Eq. 10.109.

$$E\left\{\bar{\phi}\phi(t)\right\} = \frac{1}{T}\int_0^T E\{\phi(t)\phi(t_1)\}dt_1 = \frac{D}{T}\int_0^T \min(t, t_1)dt_1 \tag{10.110}$$

Obviously, $0 \leq t \leq T$. As above, to evaluate the integral, we break it into a sum of two integrals which correspond to the cases $t_1 < t$ and $t_1 > t$

$$E\left\{\bar{\phi}\phi(t)\right\} = \frac{D}{T}\left(\int_0^t t_1 dt_1 + t\int_t^T dt_1\right) = \frac{Dt}{T}\left(T - \frac{t}{2}\right) \tag{10.111}$$

In conclusion, the variance of the remainder of phase noise is given by

$$\sigma_\theta^2 = Dt + \frac{DT}{3} - 2\frac{Dt}{T}\left(T - \frac{t}{2}\right) \tag{10.112}$$

It is observed that the variance of the remainder of phase noise depends on the time instant in the time interval $[0, T]$. The maximum value occurs at the edges $t = 0$ and $t = T$ of the time interval and is equal to

$$\max\left(\sigma_\theta^2\right) = \frac{DT}{3} \tag{10.113}$$

The minimum value of error occurs in the middle $t = T/2$ of the time interval and is equal to

$$\min\left(\sigma_\theta^2\right) = \frac{DT}{12} \tag{10.114}$$

F.5 Receiver Photocurrent in the Presence of Additive Noise

For the k−th sample at the receiver, we define the auxiliary random variable

$$x_k = A_k e^{j\phi_k} + n_{1_k} + jn_{2_k} \tag{10.115}$$

where $n_{1_k}, n_{2_k}$are zero-mean white Gaussian processes with psd

$$S_{n_1}(f) = S_{n_2}(f) = \frac{N_0}{2} \tag{10.116}$$

Their autocorrelation functions are

$$R_{n_1}(t_1 - t_2) = R_{n_2}(t_1 - t_2) = \frac{N_0}{2}\delta(t_1 - t_2) \tag{10.117}$$

F.6 Small-Noise Approximation

Assuming negligible ISI, we set $A_k = A$. We factor out $Ae^{j\phi_k}$

$$x_k = A_k e^{j\phi_k}\left[\left(1 + n'_{1_k}\right) + jn'_{2_k}\right] \tag{10.118}$$

where we defined two new additive Gaussian noises n'_{1_k}, n'_{2_k} with zero mean and variance σ^2/A^2.

For small values of additive noise

$$x_k \cong Ae^{j\phi_k}\left(1+jn'_{2_k}\right) = Ae^{j(\phi_k+\xi_k)} = Ae^{j\psi_k} \tag{10.119}$$

where we defined

$$\begin{aligned}\psi_k &= \phi_k + \xi_k \\ \xi_k &= \tan^{-1} n'_{1_k} \cong n'_{2_k}\end{aligned} \tag{10.120}$$

The new phase variable ξ_k has zero mean $\mu_\xi = 0$ and variance equal to

$$\sigma_\xi^2 = \sigma^2/A^2 \tag{10.121}$$

The new phase variable ψ_k has zero mean $\mu_\psi = 0$ and variance equal to

$$\sigma_\psi^2 = \sigma_\phi^2 + \sigma_\xi^2 \tag{10.122}$$

F.7 Operation of the Feed-Forward Phase-Noise Estimation Circuit

The feed-forward phase-noise estimation circuit estimates the average of the received signal phase over a time interval T.

$$\bar{\psi} = \frac{1}{T}\int_0^T \psi(t)dt = \frac{1}{T}\int_0^T [\phi(t) + \xi(t)]dt = \bar{\phi} + \bar{\xi} \tag{10.123}$$

To determine the mean, we take the expected value of both sides of Eq. 10.123

$$\mu_{\bar{\psi}} = \mu_{\bar{\phi}} + \mu_{\bar{\xi}} = 0 \tag{10.124}$$

To determine the variance, we use the property

$$\sigma_{\bar{\psi}}^2 = \sigma_{\bar{\phi}}^2 + \sigma_{\bar{\xi}}^2 \tag{10.125}$$

To determine the variance $\sigma_{\bar{\xi}}^2$, we use the definition

$$\begin{aligned}\sigma_{\bar{\xi}}^2 &= \frac{1}{T^2}\int_0^T\int_0^T E\{\xi(t_1)\xi(t_2)\}dt_1dt_2 \\ &= \frac{N_0'}{2T^2}\int_0^T\int_0^T \delta(t_1-t_2)dt_1dt_2 = \frac{N_0'}{2T} = N_0'B_e\end{aligned} \tag{10.126}$$

where we defined the equivalent electronic noise bandwidth as [29]

$$B_e = \frac{1}{2}\frac{\int_{-\infty}^{\infty} |H_e(f)|^2 df}{|H_e(0)|^2} \tag{10.127}$$

where $H_e(f)$ is the transfer function of the filter.

In our case, the feed-forward phase-noise estimation circuit can be thought of as an integrate and dump filter with impulse response

$$h_e(t) = \begin{cases} \frac{1}{T} & 0 \le t \le T \\ 0 & \text{elsewhere} \end{cases} \tag{10.128}$$

Taking the inverse Fourier transform, we calculate the transfer function of the integrate and dump filter

$$H_e(f) = \int_{-\infty}^{\infty} h_e(t) e^{-j\omega t} dt = \frac{\sin \pi f T}{\pi f T} e^{-j\omega T/2} \tag{10.129}$$

From the above relationship, we notice that $|H_e(0)| = 1$ and using Parseval's theorem, we can analytically calculate the value of the equivalent electronic noise bandwidth

$$B_e = \frac{1}{2} \int_{-\infty}^{\infty} |H_e(f)|^2 df = \frac{1}{2} \int_{-\infty}^{\infty} h_e(t)^2 dt = \frac{1}{2T} \tag{10.130}$$

It is worth noting that we do not take into account the correlation of the additive Gaussian noise samples due to the optical filter!

F.8 Operation of the Decision Circuit

The decision circuit makes a decision, which is corrupted by the remainder of phase noise and the additive white Gaussian noise

$$\varphi(t) = \phi(t) - \bar{\psi} = \theta(t) - \bar{\xi} \tag{10.131}$$

To determine the mean, we take the expected value of both sides of Eq. 10.107

$$\mu_\varphi = E\{\varphi(t)\} = \mu_\theta - \mu_{\bar{\xi}} = 0 \tag{10.132}$$

To determine the variance, we use the definition

$$\begin{aligned} \sigma_\varphi^2 =& E\left\{ \left[\theta(t) - \bar{\xi}\right]^2 \right\} = E\{\theta^2(t) - 2\bar{\xi}\theta(t) + \bar{\xi}^2\} \\ =& \sigma_\theta^2 + \sigma_{\bar{\xi}}^2 - 2E\{\bar{\xi}\theta(t)\} \end{aligned} \tag{10.133}$$

The final result is

$$\sigma_{\varphi_k}^2 = Dt_k + \frac{DT}{3} - 2\frac{Dt_k}{T}\left(T - \frac{t_k}{2}\right) + \frac{\sigma^2}{2B_{eq}T} \tag{10.134}$$

References

1. Sun H, Wu K-T, Roberts K (2008) Real-time measurements of a 40 Gb/s coherent system. Opt Exp 16(2):873–879
2. Roberts K, O'Sullivan M, Wu K-T, Sun H, Awadalla A, Krause DJ, Laperle C (2009) Performance of dual-polarization QPSK for optical transport systems. IEEE/OSA J Lightwave Technol 27(16):3546–3559
3. Okoshi T (1987) Recent advances in coherent optical fiber communication systems. IEEE/OSA J Lightwave Technol 5(1):44–52
4. Linke RA, Gnauck AH (1988) High-capacity coherent lightwave systems. IEEE/OSA J Lightwave Technol 6(11):1750–1769
5. Okoshi T, Kikuchi K (1988) Coherent optical fiber communications. Kluwer, Norwell
6. Betti S, de Marchis G, Iannone E (1995) Coherent optical communication systems. Wiley, New York
7. Ryu S (1995) Coherent lightwave communication systems. Artech House, Boston
8. Kikuchi K (2008) Coherent optical communication systems. In: Kaminow IP, Li T, Willner AE (eds.) Optical fiber telecommunications. vol VB, Academic, San Diego, CA, pp 95–129 (Chapter 3)
9. Ip E, Lau APT, Barros DFJ, Kahn JM (2008) Coherent detection in optical fiber systems. Opt Exp 16(2):753–791
10. Charlet G (2008) Coherent detection associated with digital signal processing for fiber optics communication. Comptes Rendus Physique 9(9–10):1012–1030
11. Li G (2009) Recent advances in coherent optical communication. Adv Opt Photon 1(2):279–307
12. Winzer P (2010) Beyond 100G Ethernet. IEEE Commun Mag 48(7):26–30
13. Kahn JM, Ho K-P (2004) Spectral efficiency limits and modulation/detection techniques for DWDM systems. IEEE J Select Topics Quant Electron 10(2):259–272
14. Winzer PJ, Essiambre R.-J (2008) Advanced optical modulation formats. In: Kaminow IP, Li T, Willner AE (eds.) Optical fiber telecommunications. vol. VB, Academic, San Diego, CA, pp 23–93 (Chapter 2)
15. Yu J, Zhou X (2009) Multilevel modulations and digital coherent detection. Opt Fiber Technol 15(3):197–208
16. Van den Borne D, Sleiffer V, Alfiad MS, Jansen SL, Wuth T (2009) POLMUX-QPSK modulation and coherent detection: The challenge of long-haul 100G transmission. In: Proceedings of European conference on optical communications (ECOC), paper 3.4.1, Vienna, Austria
17. Faure J.-P, Lavigne B, Bresson C, Bertran-Pardo O, Colomer AC, Canto R (2010) 40G and 100G deployment on 10G infrastructure: market overview and trends, coherent versus conventional technology. In: Proceedings of IEEE/OSA optical fiber communication conference (OFC/NFOEC), paper OThE3, San Diego, CA
18. Saunders R, Traverso M, Schmidt T, Malouin C (2010) Economics of 100 Gb/s transport. In: Proceedings of IEEE/OSA optical fiber communication conference (OFC/NFOEC), paper NMB2, San Diego, CA

19. Camera M (2010) 100GbE optical transport, appropriate modulation formats, and impact on deployed transport networks. In: Proceedings of IEEE/OSA optical fiber communication conference (OFC/NFOEC), paper NME3, San Diego, CA
20. Essiambre R-J, Kramer G, Winzer PJ, Foschini GJ, Goebel B (2010) Capacity limits of optical fiber networks. IEEE/OSA J Lightwave Technol 28(4):662–701
21. Ruhl F, Sorbello L, Evans D, Diaconescu L, Doucet D, Fasken D, Nimiczeck M, Sitch J, O'Sullivan M, Belanger M (2010) 2038 km and four 50 GHz OADM/filters transmission field trial of 115.2 Gb/s coherent CoFDM modem in the Telstra network. In: Proceedings of IEEE/OSA optical fiber communication conference (OFC/NFOEC), paper OThE3, San Diego, CA
22. Birk M, Gerard P, Curto R, Nelson LE, Zhou X, Magill P, Schmidt TJ, Malouin C, Zhang B, Ibragimov E, Khatana S, Glavanovic M, Lofland R, Marcoccia R, Saunders R, Nicholl G, Nowell M, Forghieri F (2010) Coherent 100 Gb/s PM-QPSK field trial. IEEE Commun Mag 48(7):52–60
23. Sano A, Masuda H, Kobayashi T, Fujiwara M, Horikoshi K, Yoshida E, Miyamoto Y, Matsui M, Mizoguchi M, Yamazaki H, Sakamaki Y, Ishii H (2010) 69.1-Tb/s (432 × 171-Gb/s) C- and extended L-band transmission over 240 km using PDM-16-QAM modulation and digital coherent detection. In: Proceedings of IEEE/OSA optical fiber communication conference (OFC/NFOEC), paper PDPB7, San Diego, CA
24. Zhou X, Yu J, Huang M.-F, Shao Y, Wang T, Nelson L, Magill P, Birk M, Borel PI, Peckham DW, Lingle R (2010) 64-Tb/s (640 × 107-Gb/s) PDM-36QAM transmission over 320 km using both pre- and post-transmission digital equalization. In: Proceedings of IEEE/OSA optical fiber communication conference (OFC/NFOEC), paper PDPB9, San Diego, CA
25. Nakazawa M, Okamoto S, Omiya T, Kasai K, Yoshida M (2010) 256 QAM (64 Gbit/s) coherent optical transmission over 160 km with an optical bandwidth of 5.4 GHz. In: Proceedings of IEEE/OSA optical fiber communication conference (OFC/NFOEC), paper OMJ5, San Diego, CA
26. Proakis J (2000) Digital communications, 4th edn. McGraw-Hill, New York, pp 177–178
27. Press WH, Teukolsky SA, Vetterling WT, Flannery BP (1992) Numerical recipes in C. Cambridge University Press, Cambridge, UK
28. Griffin RA, Carter AC (2002) Optical differential quadrature phase-shift key (oQPSK) for high capacity optical transmission. In: Proceedings of IEEE/OSA optical fiber communication conference (OFC), paper WX6, Anaheim, CA
29. Benedetto S, Biglieri E, Castellani V (1987) Digital transmission theory. Prentice-Hall, Englewood Cliffs, NJ, pp 213–220
30. Serway R (1990) Physics for scientists and engineers. Saunders, Philadelphia
31. Fessenden RA (1902) Wireless signaling. U.S. Patent No. 706,740
32. Armstrong EH (1921) A new system of short wave amplification. Proc IRE 9(1):3–11
33. Forrester AT, Gudmundsen RA, Johnson PO (1961) Photoelectric mixing as a spectroscopic tool. J Opt Soc America 51(3):253–259
34. Teich MC, Keyes RJ, Kingston RH (1966) Optimum heterodyne detection at 10.6 pm in photoconductive Ge:Cu. Appl Phys Lett 9(10):357–360
35. Taylor MG (2004) Coherent detection method using DSP for demodulation of signal and subsequent equalization of propagation impairments. IEEE Photon Technol Lett 16(2):674–676
36. Noé R (2005) Phase noise tolerant synchronous QPSK/BPSK baseband-type intradyne receiver concept with feedforward carrier recovery. IEEE/OSA J Lightwave Technol 23(2):802–808
37. Noé R (2005) PLL-free synchronous QPSK polarization multiplex/diversity receiver concept with digital I&Q baseband processing. IEEE Photon Technol Lett 17(4):887–889
38. Ly-Gagnon D, Tsukamoto S, Katoh K, Kikuchi K (2006) Coherent detection of optical quadrature phase shift keying signals with carrier phase estimation. IEEE/OSA J Lightwave Technol 24(1):12–21

39. Kikuchi K (2006) Phase-diversity homodyne detection of multilevel optical modulation with digital carrier phase estimation. IEEE J Select Topics Quant Electron 12(4):563–570
40. Kazovsky LG, Kalogerakis G, Shaw WT (2006) Homodyne phase-shift-keying systems: past challenges and future opportunities. IEEE/OSA J Lightwave Technol 24(12):4876–4884
41. Kazovsky L, Benedetto S, Willner A (1996) Optical fiber communication system. Artech House, Boston, MA Chapter 4
42. Agrawal GP (2002) Fiber-optic communication systems, 3rd edn. Wiley, New York Chapter 10
43. Wenke G, Klimmek M (1996) Considerations on the α-factor of nonideal, external optical Mach–Zehnder modulators. J Opt Commun 17(2):42–48
44. Abbas GL, Chan VWS, Yee TK (1985) A dual-detector optical heterodyne receiver for local oscillator noise suppression. IEEE/OSA J Lightwave Technol 3(5):1110–1122
45. Saleh BEA, Teich MC (1991) Fundamentals of photonics. Wiley, New York Chapter 7
46. Kazovsky LG (1989) Phase- and polarization-diversity coherent optical techniques. IEEE/OSA J Lightwave Technol 7(2):279–292
47. Derr F (1992) Coherent optical QPSK intradyne system: concept and digital receiver realization. IEEE/OSA J Lightwave Technol 10(9):1290–1296
48. Hoffman D, Heidrich H, Wenke G, Langenhorst R, Dietrich E (1989) Integrated optics eight-port 90 hybrid on $LiNbO_3$. IEEE/OSA J LightwaveTechnol 7(5):794–798
49. Gordon JP, Kogelnik H (2000) PMD fundamentals: polarization mode dispersion in optical fibers. PNAS 97(9):4541–4550
50. Kuschnerov M, Hauske FN, Piyawanno K, Spinnler B, Alfiad MS, Napoli A, Lankl B (2009) DSP for coherent single-carrier receivers. IEEE/OSA J Lightwave Technol 27(16):3614–3622
51. Savory SJ (2010) Digital coherent optical receivers: algorithms and subsystems. IEEE J Select Topics Quant Electron 16(5):1164–1179
52. Spinnler B (2010) Equalizer design and complexity for digital coherent receivers. IEEE J Select Topics Quant Electron 16(5):1180–1192
53. Leven A, Kaneda N, Corteselli S (2010) Real-time implementation of digital signal processing for coherent optical digital communication systems. IEEE J Select Topics Quant Electron 16(5):1227–1234
54. Dedic I (2010) 56 GS/s ADC: Enabling 100GbE. In: Proceedings of IEEE/OSA optical fiber communication conference (OFC/NFOEC), paper OThT6, San Diego, CA
55. Winzer PJ, Gnauck AH, Doerr CR, Magarini M, Buhl LL (2010) Spectrally efficient long-haul optical networking using 112-Gb/s polarization-multiplexed 16-QAM. IEEE/OSA J Lightwave Technol 28(4):547–556
56. Tanimura T, Oda S, Tanaka T, Hoshida T, Tao Z, Rasmussen J (2009) A simple digital skew compensator for coherent receiver. In: Proceedings of European conference on optical communications (ECOC), paper 7.3.2, Vienna, Austria
57. Koc U, Leven A, Chen Y, Kaneda N (2006) Digital coherent quadrature phase-shift-keying (QPSK). In: Proceedings of IEEE/OSA optical fiber communication conference (OFC/NFOEC), paper OThI1, Anaheim, CA
58. Roudas I, Sauer M, Hurley J, MauroY, Raghavan S (2007) Compensation of coherent DQPSK receiver imperfections. In: Proceedings of IEEE LEOS summer topicals, paper MA3.4, Portland, OR
59. Petrou CS, Roudas I, Raptis L (2008) Impact of transmitter and receiver imperfections on the performance of coherent optical QPSK communication systems. In: Proceedings of the conference on lasers and electro-optics (CLEO), paper CThJJ1, San Jose, CA
60. Petrou CS, Vgenis A, Kiourti A, Roudas I, Hurley J, Sauer M, Downie J, Mauro Y, Raghavan S (2008) Impact of transmitter and receiver imperfections on the performance of coherent optical QPSK communication systems. In: Proceedings of IEEE/LEOS annual meeting, paper TuFF3, Newport Beach, CA
61. Fatadin I, Savory S, Ives D (2008) Compensation of quadrature imbalance in an optical QPSK coherent receiver. IEEE Photon Technol Lett 20(20):1733–1735

62. Ip E, Kahn JM (2008) Compensation of dispersion and nonlinear impairments using digital backpropagation. IEEE/OSA J Lightwave Technol 26(20):3416–3425
63. Mateo E, Li G (2009) Compensation of interchannel nonlinearities using enhanced coupled equations for digital backward propagation. Appl Opt 48(25):F6–F10
64. Goldfarb G, Li G (2009) Efficient backward-propagation using wavelet-based filtering for fiber backward-propagation. Opt Exp 17(11):8815–8821
65. Ip E, Kahn JM (2010) Fiber impairment compensation using coherent detection and digital signal processing. IEEE/OSA J Lightwave Technol 28(4):502–519
66. Ip E (2010) Nonlinear compensation using backpropagation for polarization-multiplexed transmission. IEEE/OSA J Lightwave Technol 28(6):939–951
67. Millar DS, Makovejs S, Behrens C, Hellerbrand S, Killey RI, Bayvel P, Savory SJ (2010) Mitigation of fiber nonlinearity using a digital coherent receiver. IEEE J Select Topics Quant Elec 16(5):1217–1226
68. Yaman F, Li G (2010) Nonlinear impairment compensation for polarization-division multiplexed WDM transmission using digital backward propagation. IEEE Photon J 2(5):816–832
69. Petrou CS, Vgenis A, Roudas I, Raptis L (2009) Quadrature imbalance compensation for PDM QPSK coherent optical systems. IEEE Photon Technol Lett 21(24):1876–1888
70. Savory SJ, Gavioli G, Killey RI, Bayvel P (2007) Electronic compensation of chromatic dispersion using a digital coherent receiver. Opt Exp 15(5):2120–2126
71. Savory SJ (2008) Digital filters for coherent optical receivers. Opt Exp 16(2):804–817
72. Goldfarb G, Li G (2007) Chromatic dispersion compensation using digital IIR filtering with coherent detection. IEEE Photon Technol Lett 19(13):969–971
73. Geyer, JC, Fludger CRS, Duthel T, Schulien C, Schmauss B (2010) Efficient frequency domain chromatic dispersion compensation in a coherent Polmux QPSK-receiver. In: Proceedings of IEEE/OSA optical fiber communication conference (OFC/NFOEC), paper OWV5, San Diego, CA
74. Geyer JC, Bisplinghoff A, Duthel T, Fludger CRS, Schulien C, Schmauss B (2010) Optimization of the chromatic dispersion equalizer of a 43 Gb/s realtime coherent receiver. In: Proceedings of IEEE/OSA optical fiber communication conference (OFC/NFOEC), paper OWV8, San Diego, CA
75. Khan FA, Agrell E, Karlsson M, (2010) Electronic dispersion compensation by Hadamard transformation. In: Proceedings of IEEE/OSA optical fiber communication conference (OFC/NFOEC), per OWV4, San Diego, CA
76. Kudo R, Kobayashi T, Ishihara K, Takatori Y, Sano A, Miyamoto Y (2009) Coherent optical single carrier transmission using overlap frequency domain equalization for long-haul optical systems. IEEE/OSA J Lightwave Technol 27(16):3721–3728
77. Meyr H, Moeneclaey M, Fechtel SA (1997) Digital communication receivers: synchronization, channel estimation, and signal processing. Wiley, New York
78. Zibar D, Bianciotto A, Wang Z, Napoli A, Spinnler B (2009) Analysis and dimensioning of fully digital clock recovery for 112 Gb/s coherent polmux QPSK systems. In: Proceedings of European conference on optical communications (ECOC), paper 7.3.4, Vienna, Austria
79. Fludger C, Duthel T, van den Borne D, Schulien C, Schmidt E-D, Wuth T, Geyer J, De Man E, Khoe G-D, de Waardt H (2008) Coherent equalization and POLMUX-RZ-DQPSK for robust 100-GE transmission. IEEE/OSA J Lightwave Technol 26(1):64–72
80. Kuschnerov M, Hauske FN, Gourdon E, Piyawanno K, Lankl B, Spinnler B (2008) Digital timing recovery for coherent fiber optic systems. In: Proceedings of IEEE/OSA optical fiber communication conference (OFC/NFOEC), paper JThA63, San Diego, CA
81. Tanimura T, Hoshida T, Oda S, Nakashima H, Yuki M, Tao Z, Liu L, Rasmussen JC (2008) Digital clock recovery algorithm for optical coherent receivers operating independent of laser frequency offset. In: Proceedings of European conference on optical communications (ECOC), Mo.3.D.2, Brussels, Belgium
82. Chang SH, Chung HS, Kim K (2008) Digital non-data-aided symbol synchronization in optical coherent intradyne reception. Opt Exp 16(19):15 097–15 103

83. Adachi F, Ohno K (1991) BER performance of QDPSK with postdetection diversity reception in mobile radio channels. IEEE Trans Vehic Technology 40(1):237–249
84. Louchet H, Kuzmin K, Richter A (2009) FFT-based digital clock recovery for coherent transmission systems with multilevel modulation formats. In: Proceedings of European conference on optical communications (ECOC), paper 7.3.5, Vienna, Austria
85. Laakso TI, Valimaki V, Karjalainen M, Laine UK (1996) Splitting the unit delay. IEEE Signal Process Mag 13(1):30–60
86. Godard D (1980) Self-recovering equalization and carrier tracking in two dimensional data communication systems. IEEE Trans Commun 28(11):1867–1875
87. Haykin S (1996) Adaptive filter theory, 3rd edn. Prentice Hall, Englewood Cliffs, NJ
88. Kikuchi K (2008) Polarization-demultiplexing algorithm in the digital coherent receiver. In: Proceedings of IEEE LEOS summer topicals, paper MC2.2, Acapulco, Mexico
89. Roudas I, Vgenis A, Petrou CS, Toumpakaris D, Hurley J, Sauer M, Downie J, Mauro Y, Raghavan S (2010) Optimal polarization demultiplexing for coherent optical communications systems. IEEE/OSA J Lightwave Technol 28(7):1121–1134
90. Vgenis A, Petrou CS, Papadias CB, Roudas I, Raptis L (2010) Nonsingular constant modulus equalizer for PDM-QPSK coherent optical receivers. IEEE Photonics Technol Lett 22(1):45–47
91. Xie C, Chandrasekhar S (2010) Two-stage constant modulus algorithm equalizer for singularity free operation and optical performance monitoring in optical coherent receiver. In: Proceedings of IEEE/OSA optical fiber communication conference (OFC/NFOEC), paper OMK3, San Diego, CA
92. Morelli M, Mengali U (1998) Feedforward frequency estimation for PSK: a tutorial review. European Trans Telecommun 2(2):103–116
93. Leven A, Kaneda N, Koc U-V, Chen Y-K (2007) Frequency estimation in intradyne reception. IEEE Photon Technol Lett 19(6):366–368
94. Li L, Tao Z, Oda S, Hoshida T, Rasmussen J (2008) Wide-range, accurate and simple digital frequency offset compensator for optical coherent receivers. In: Proceedings of IEEE/OSA optical fiber communication conference (OFC/NFOEC), paper OWT4, San Diego, CA
95. Hoffmann S, Bhandare S, Pfau T, Adamczyk O, Wördehoff C, Peveling R, Porrmann M, Noé R (2008) Frequency and phase-estimation for coherent QPSK transmission with unlocked DFB lasers. IEEE Photon Technol Lett 20(18):1569–1571
96. Ip E, Kahn J (2007) Feedforward carrier recovery for coherent optical communications. IEEE/OSA J LightwaveTechnol 25(9):2675–2692
97. Taylor MG (2009) Phase estimation methods for optical coherent detection using digital signal processing. IEEE/OSA J Lightwave Technol 27(7):901–914
98. Grellier E, Antona J.-C, Bigo S (2010) Are multilevel pseudorandom sequences really needed to emulate highly dispersive optical transmission systems?. In: Proceedings of European conference on optical communications (ECOC), paper We.6.A.1, Torino, Italy
99. Ho K-P (2005) Phase-modulated optical communication systems. Springer, New York
100. Lucky RW, Salz J, Weldon EJ (1968) Principles of data communication. McGraw-Hill, New York 13–15, 247–248
101. Weber WJ (1978) Differential encoding for multiple amplitude and phase shift keying systems. IEEE Trans Commun 26(3):385–391
102. Ma Y, Zhang QT (2002) Accurate evaluation for MDPSK with noncoherent diversity. IEEE Trans Commun 50(7):1189–1200
103. Tao Z, Li L, Liu L, Yan W, Nakashima H, Tanimura T, Oda S, Hoshida T, Rasmussen JC (2010) Improvements to digital carrier phase recovery algorithm for high-performance optical coherent receivers. IEEE J Select Topics Quant Electron 16(5):1201–1209
104. Bononi A, Bertolini M, Serena P, Bellotti G (2009) Cross-phase modulation induced by OOK channels on higher-rate DQPSK and coherent QPSK channels. IEEE/OSA J Lightwave Technol 27(18):3974–3983
105. Vanin E, Jacobsen G (2010) Analytical estimation of laser phase noise induced BER floor in coherent receiver with digital signal processing. Opt Exp 18(5):4246–4259

106. Goldfarb G, Li G (2006) BER estimation of QPSK homodyne detection with carrier phase estimation using digital signal processing. Opt Exp 14(18):8043–8053
107. Jeruchim M (1984) Techniques for estimating the bit error rate in the simulation of digital communication systems. IEEE J Select Areas Commun 2(1):153–170
108. van den Borne D, Gottwald E, Khoe GD, de Waardt H (2007) Bit pattern dependence in optical DQPSK modulation. Electron Lett 43(22):1223–1225
109. Spinnler B, Xie C (2007) Performance assessment of DQPSK using pseudo-random quaternary sequences. In: Proceedings of European conference on optical communications (ECOC), paper 2.3.6, Berlin, Germany
110. Tkach RW (2010) Scaling optical communications for the next decade and beyond. Bell Labs Tech J 14(4):3–10
111. Chen R, O'Sullivan M, Ward C, Asselin S, Belanger M (2010) Next generation transmission fiber for coherent systems. In: Proceedings of IEEE/OSA optical fiber communication conference (OFC/NFOEC), paper OTuI1, San Diego, CA
112. Charlet G (2010) Fiber characteristics for next-generation ultra-long-haul transmission systems. In: Proceedings of European conference on optical communications (ECOC), paper We.8.F.1, Torino, Italy
113. Smith BP, Kschischang FR (2010) Future prospects for FEC in fiber-optic communications. IEEE J Select Topics Quant Electron 16(5):1245–1257
114. Stuart HR (2000) Dispersive multiplexing in multimode optical fiber. Science 289(5477):281–283
115. Koebele C, Salsi M, Sperti D, Tran P, Brindel P, Mardoyan H, Bigo S, Boutin A, Verluise F, Sillard P, Astruc M, Provost L, Cerou F, Charlet G (2011) Two mode transmission at 2x100 Gb/s, over 40 km-long prototype few-mode fiber, using LCOS-based programmable mode multiplexer and demultiplexer. Opt Exp 19(17):16593–16600
116. Randel S, Ryf R, Sierra A, Winzer PJ, Gnauck AH, Bolle CA, EssiambreR-J, Peckham DW, McCurdy A, Lingle R (2011) 6×56-Gb/s mode-division multiplexed transmission over 33-km few-mode fiber enabled by 6×6 MIMO equalization. Opt Exp 19(17):16697–16707
117. Kazovsky LG, Curtis L, Young WC, Cheung NK (1987) An all fiber 90° optical hybrid for coherent communications. Appl Opt 26(3):437–439
118. Alexander SB (1987) Design of wide-band optical heterodyne balanced mixer receivers. IEEE/OSA J LightwaveTechnology 5(4):523–537
119. Green PE (1993) Fiber optic networks. Prentice Hall, Englewood Cliffs, NJ
120. Zhu B, Taunay TF, Fishteyn M, Liu X, Chandrasekhar S, Yan MF, Fini JM, Monberg EM, Dimarcello FV (2010) 112-Tb/s space-division multiplexed DWDM transmission with 14-b/s/Hz aggregate spectral efficiency over a 76.8-km seven-core fiber. Opt Exp 19(17): 16665–16671
121. Papoulis A (1991) Probability, random variables, and stochastic processes, 3rd edn. McGraw-Hill, New York
122. Nakashima H, Tanimura T, Hoshida T, Oda S, Rasmussen J, Li L, Tao Z, Ishii Y, Shiota K, Sugitani K, Adachi H (2008) Novel wide-range frequency offset compensator demonstrated with real-time digital coherent receiver. In: Proceedings of European conference on optical communications (ECOC), paper Mo3.D.4, Brussels, Belgium
123. Viterbi A, Viterbi A (1983) Nonlinear estimation of PSK-modulated carrier phase with application to burst digital transmission. IEEE Trans on Inform Theory 29(4):543–551
124. Kuschnerov M, Chouayakh M, Piyawanno K, Spinnler B, de Man E, Kainzmaier P, Alfiad MS, Napoli A, Lankl B (2010) Data-aided versus blind single-carrier coherent receivers. IEEE Photon J 2(3):387–403
125. Shah AR, Hsu RCJ, Tarighat A, Sayed AH, Jalali B (2005) Coherent optical MIMO (COMIMO). IEEE/OSA J Lightwave Technol 23(8):2410–2419
126. Foschini GJ, Vannucci G (1998) Characterizing filtered light waves corrupted by phase noise. IEEE Trans Inform Theory 34(6):1437–1448

Chapter 11
Design Process for Terrestrial and Undersea DWDM Network Upgrades

Sergey Burtsev

Abstract The design process for DWDM upgrades of terrestrial and undersea systems is presented in the form of realistic industry scenarios. Fiber plant quality, margins for repairs and fiber aging, and upgrade capacity target are taken into account. Estimation of repeater characteristics, management of multipath interference and nonlinearities as well as the Raman amplification aspects in the various designs are also analyzed.

11.1 Introduction

Building green-field undersea and terrestrial fiber-optic Dense Wavelength Division Multiplexed (DWDM) systems is expensive for carriers. The goal of capital cost reduction forces carriers down the path of legacy systems upgrades whenever an upgrade provides such network capacity increase that is sufficient for network growth. The goal of this chapter is to cover the design process of such DWDM upgrades of legacy fiber-optic systems.

Our work focuses on the following three DWDM upgrade applications: (1) undersea repeatered systems with the emphasis on regional size systems; (2) unrepeatered (UR) systems (undersea and terrestrial), and finally (3) terrestrial multi-span long-haul systems. Here, we use a broad definition for an upgrade. The definition includes not only the case of undersea repeatered and UR systems where the only part that is replaced during the upgrade is submarine line terminal equipment

S. Burtsev (✉)
Xtera Communications Inc
500 W Bethany Dr
Allen, TX 75013, USA
e-mail: sergey.burtsev@xtera.com

N. (Neo) Antoniades et al. (eds.), *WDM Systems and Networks*, Optical Networks,
DOI: 10.1007/978-1-4614-1093-5_11,

(SLTE) and no modification is performed on the submerged plant, but, also, the case of terrestrial multi-span long-haul upgrades, when both new terminal as well as in-line active equipment are added, while the legacy fiber plant is preserved.

To provide a quality DWDM network upgrade design a system engineer has to address various issues that include:

(1) Customization of the network design process for a specific DWDM application, which is a result of the fact that no single design process can fit all applications
(2) Customization of the DWDM network design process for a specific network that needs to be upgraded
(3) Accuracy of the design capacity predictions and a study of the major factors that affect such accuracy

To address the above, we select specific DWDM applications that help the researcher look into these issues from various distinct angles. We start with areas and subtopics that are common for all DWDM system types, namely:

- Fiber plant quality that is related to:
  - In-line fiber type
  - Fiber loss data
  - Fiber chromatic dispersion measurement data
  - Fiber polarization mode dispersion (PMD) data
- Margins for repair/aging which are required by the customer and
- Upgrade capacity target

We also look at the main issues specific for a particular type of DWDM system, like:

- Estimating repeater characteristics for legacy undersea repeatered systems
- Managing multi-path interference (MPI) and fiber nonlinearities in UR systems
- Raman aspects of a design for UR systems and terrestrial multi-span systems

In all the above, the key question that pertains to any DWDM fiber-optic system design is how much upgrade capacity can be guaranteed by a given system vendor. Capacity design results based on the models that will be presented below are as accurate as the underlying link data is accurate and comprehensive. Quite often in the real world, a system designer has to deal with the fact that link data might be either incomplete and/or inaccurate.

The purpose of this chapter is to provide a real-life discussion of existing system upgrade strategies and scenarios and so discussion of issues affecting the quality of upgrade design predictions such as the accuracy of design tools used in the process, laboratory test-bed data that drive such upgrades, and field trial data for feasibility studies of such upgrades are beyond the scope of this work. We do not delve into the above areas because they are usually very complex and carrier-specific. In addition there are usually critical issues that arise such as: the validation process of the design tools usually lags behind the sales department drive to bid on various types of links, including the stressful links with small margins, the type that design tools are not validated for yet. Validation of design predictions could be done with the experimental data, both test-bed data and field trial data; however, getting laboratory data involves quite a bit of expense mainly related to the building of a test-bed which can be prohibitive. In addition, test-bed implementation (emulation) of a network may lack important details that are specific to the system that is planned for an upgrade. Finally, the arrangement of a field trial on a system which is scheduled for an upgrade is also expensive and difficult as it requires specialized test equipment and resolution of various logistical issues, including an arrangement of test time with a carrier.

The focus of this chapter is on the optical transport part of any given fiber-optic system upgrade design. In Sects. 11.2 and 11.3 the subject of upgrades for repeatered and un-repeatered undersea DWM links is presented. In Sect. 11.3 upgrades for terrestrial long-haul DWDM links are discussed. In this chapter, issues such as link availability, protection, network management, supervisory, and other related subjects, and forming the technical part of an upgrade bid, aren't analyzed.

11.2 Repeatered Undersea DWDM links

Undersea fiber-optic repeatered systems have been deployed for transoceanic and regional markets since 1995. Four equipment generations have been developed and deployed since then, starting with Gen-1 system, a single channel (2.5 Gbs or 5 Gbs) all the way to Gen-3- and Gen-4-systems supporting in excess of hundred 10 Gbs channels at 25 GHz channel spacings.

Table 11.1 summarizes the major characteristics of the legacy undersea repeatered system for the Gen-1, Gen-2, and Gen-3 systems. Characteristics such as line rate, type, in-line fiber, repeater information, compensating fiber type, and others are presented. With each generation the system capacity increased due to several factors most important of which were the broader EDFA-based repeater bandwidth and the higher repeater output power that supported more channels, plus transponder enhancements due to higher line rates and better Forward-Error-Correction (FEC) technologies.

Table 11.2 below summarizes the major characteristics of line fibers used in the above Gen-1, Gen-2, and Gen-3 systems. These fibers were commercial brands developed with dispersion characteristics tailored for different applications. Dispersion shifted fiber (DSF), standard single mode fiber (SSMF), and non-zero dispersion shifted fiber (NZ-DSF) which is similar in characteristics to Corning Inc.'s LEAF^{TM} fiber were introduced and discussed in Chap. 2. Submarine LEAF (subLEAF) and submarine SMF-LS (subLS) are Corning Inc. fibers tailored to the submarine systems as shown in Tables 11.1, 11.2. Finally pure silica core fiber (PSCF) is applied for both longer wavelengths and shorter wavelengths (C- and L-bands).

Table 11.1 Major characteristics of legacy undersea repeatered systems since the deployment of the first undersea fiber-optic repeatered systems in 1995. Fibers used: standard single mode fiber (SSMF), Dispersion shifted fiber (DSF), pure silica core fiber (PSCF), and finally submarine LEAF (subLEAF) and submarine SMF-LS (subLS) are Corning Inc. fibers tailored to the submarine systems

	GEN-1	GEN-2	GEN-3
Start of commercial deployment	1995	1998	2001
Line Rate (Gbs)	**5**	10	10
Type	Single Channel	WDM	DWDM
Line Fiber	DSF	subLS	subLEAF/subLS
Dispersion compensating Fiber Fiber	SSMF	SSMF or PSCF	SSMF or PSCF
Output Repeater Power (dBm) (Typical)	4	6–8	12–15
Repeater Noise Figure (dB) (Typical)	6	6	4.5
Original Bandwidth (nm) (Typical)	2–3 nm	7–8 nm	22–27 nm
Effective Upgrade Bandwidth (nm)	7–8 nm	15–20 nm	22–27 nm

Table 11.2 Major characteristics of line fibers used in undersea repeatered Gen-1, Gen-2 and Gen-3 systems

Characteristic	Unit	DSF	Submarine SMF-LS™	Submarine LEAF™	SSMF	PSCF
ITU Standard document		G.653	G.655	G.655	G.652	G.654
Loss at 1550 nm	dB/km	0.21	0.21	0.21	0.2	0.18
Chromatic dispersion at 1550 nm (Typical)	ps/(nm km)	−0.7	−3	−4	+17	+19
Chromatic dispersion slope at 1550 nm (Typical)	ps/(nm^2 km)	0.07	0.07	0.12	0.06	0.06
Fiber effective area (Typical)	μm^2	50	50	70	80	75

Gen-4 systems were omitted from the upgrade discussion in this chapter mainly because only few Gen-4 systems have been installed so far and they were not widely upgraded as systems of prior generations. A very thorough description of the past evolution as well as the current state-of-the-art in this area is presented in [1].

During the upgrade phase of an undersea fiber-optic repeatered system no change is done to the submerged plant which is comprised of: the fiber-optical cable, undersea repeaters, and power feed equipment (PFE) that provide electrical power in the form of stable direct current (DC) to the undersea repeaters [1]. The only part that is changed is the submarine line terminal equipment (SLTE). SLTE equipment consist of transponders (TXP) or muxponders (MXP), and equipment, used to prime the signal before it is sent into or received from the undersea line optical MUX/DEMUX optics, the dispersion compensating units (based on fiber or fiber Bragg gratings (FBG)), and the optical amplifiers that need to amplify the already attenuated optical signal.

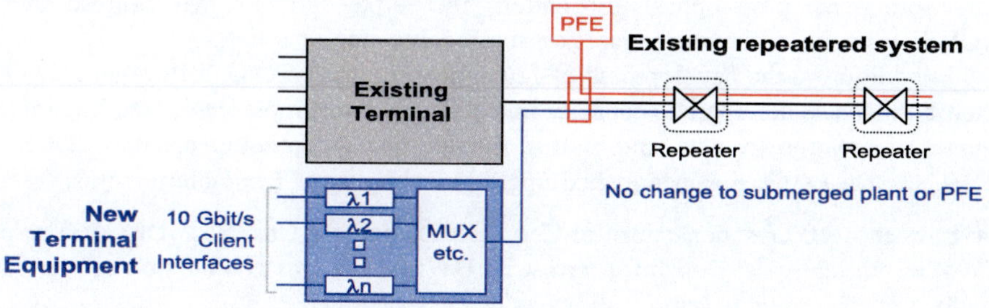

Fig. 11.1 Example of a dark fiber upgrade

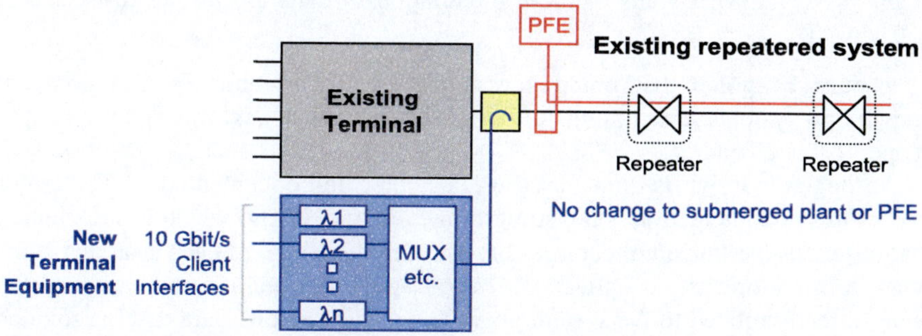

Fig. 11.2 Example of an overlay upgrade

There are two different types of upgrades that a carrier might request: (1) the dark fiber upgrade is when the upgrade SLTE uses the fiber pair in the undersea cable that is not used by a legacy system as shown in Fig. 11.1 or when the old SLTE is replaced altogether with a new one. Thus, the word "dark" means only that the fiber pair is not lit (no data channels), but still there are active repeaters supporting the fiber; (2) the overlay upgrade is when legacy SLTE and legacy channels are maintained and new upgrade channels from a new SLTE are coupled together with the legacy channels by an overlay coupler as shown in Fig. 11.2.

The major target of an upgrade is to significantly increase system capacity. This goal is technologically achieved by using several ways. For example, GEN-1 systems were implemented originally with TXPs/MXPs without FEC at all, while Gen-2 systems used classic FEC (Reed-Solomon) only. Nowadays, upgraded MXPs/TXPs have modern built-in FEC chips that provide enhanced and even ultra-FEC coding gains leading to significant system margin increases and an increase from the initial bandwidth (BW) to an effective upgrade BW (see Table 11.1). An effective upgrade BW increase depends on the specifics of the legacy link and the upgrade equipment. Increasing the link bandwidth involves accurate pre-emphasis tuning because the upgrade channels might operate in the part of the repeater bandwidth that is not designed for normal operation. In this case, one needs pre-emphasis to "flatten" the output optical signal-to-noise ratio (OSNR) for the different channels on the receive side of a link.

In addition to the development of FEC chips used in MXPs/TXPs other factors help to increase the upgrade capacity in a system, for example: increased line rate, novel modulation formats, and built-in tunable dispersion compensators (TDC).

There are two major established upgrade technologies available now:

- Low-end: 10 Gbs non-return-to-Zero (NRZ) and 40 Gbs NRZ-DPSK. These were originally developed for terrestrial DWDM systems and later on converted for undersea regional applications as discussed in [2]
- High-end: 10 Gbs return-to-zero differential phase shift keying (RZ-DPSK) and 20 Gbs RZ-DPSK, developed specifically for the upgrades of transoceanic systems [3]. Details of these modulation formats are further discussed in Chap. 10

It must be noted that brand new high-end 40 Gbs and 100 Gbs coherent technology systems are coming to the market now, but this topic is beyond the scope of this chapter.

An upgrade design is done based on customer link documentation. The major part of this documentation is the straight line diagram (SLD) which is a document that describes the link architecture. This includes the length of each span and some other information such as optical cable type, splice information and anything that is not directly related to the system upgrade design. The link data that is provided by the customer usually contains the SLD part, but other important parts of the repeater information such as output total power and bandwidth or fiber data might be missing. This adds uncertainty to the design predictions.

For a given link an upgrade design has to meet the following transmission performance targets. Some of the design targets are generic for DWDM deployments, like end-of-life (EOL) maximum capacity and corresponding EOL system margins. EOL maximum capacity represents a guaranteed maximum upgrade capacity estimate that accounts for repairs due to cable cuts incurred over the system life-time and an associated OSNR degradation. Changing the link span loss and/or the repair margins estimates will have a direct consequence on the link performance and will ultimately affect the maximum end-of-life (EOL) capacity of the system.

Other performance targets specific to undersea repeatered deployments are:

- Minimization of channel loaders
  - Loaders are usually nothing but continuous wave (CW) lasers used to manage signal power budget by draining the power from undersea repeaters which are always working in the saturated regime. In the saturated regime the power per channel (in dBm) is computed according to the algebraic expression:

$$P_{ch}(dBm) = P_{repeater}(dBm) - 10LOG_{10}(Number\_of\_Channels) \tag{11.1}$$

An increase in the number of system channels leads to a corresponding decrease in the channel power. The need for loaders is there because the install capacity during the initial upgrade could be significantly less than the EOL capacity which leads to an excessive signal power and corresponding nonlinear penalties in the absence of loaders
If a customer does not want to pay for the loaders, these are usually provided for free by the system vendor

- For an overlay upgrade, there is an additional target of minimizing the impact of upgrade channels as well as legacy channels on each other. This goal is usually achieved by providing large enough guard bands between upgrade and legacy channels.

As a typical example of a low-cost 10 Gbs DWDM dark fiber upgrade let us look at the 1,000 km-long Gen-1 system which had an original 1×5 Gbs EOL capacity. For simplicity, assume that all repeater spans are equal and the link is composed of twelve undersea repeaters and thirteen spans of 83 km each. For simplicity assume that the transponders use enhanced FEC with a pre-FEC Q target of 9.3 dB at a Bit Error Rate (BER) of 10^{-13}. Note that throughout this chapter we follow a traditional convention measuring the system parameter Q, which is introduced in various parts of this book, in decibel units and defining Q as:

$$Q = 20 * LOG_{10}(Q_{linear}) \tag{11.2}$$

where Q_{linear} is directly related to the major measurable system performance characteristic BER as described in [4]. The relation between Q and BER is given as:

$$Q_{linear} = \sqrt{2}\ erfinv(1 - 2BER) \tag{11.3}$$

where erfinv denotes the inverse error function.

In the above example, the original system capacity (pre-upgrade) is provided by a single 5 Gbs channel at 1558 nm. A dark fiber upgrade could be achieved by either using dark fiber or by using lit fiber and removing the legacy 5 Gbs channel. During the upgrade process, a single fiber pair can be equipped with 12 channels at 10 Gbs capacity each and 50 GHz channel spacing. Such an upgrade will result in a 24 factor increase of the initial design capacity. So in the above example a high upgrade figure-of-merit with very low-cost DWDM equipment was achieved as a result of two factors: (1) the initial system (Gen-1) had a very low capacity; (2) the system link length was rather short. An even higher upgrade capacity, for example 5×40 Gbs, for the same 1,000 km link is obtained by using 40 Gbs NRZ-DPSK muxponders spaced at 100 GHz (0.8 nm). Such upgrade results in a 40 factor increase in the network capacity.

In undersea applications the standard performance summary, called power budget table, is the main figure-of-merit that is computed by a system designer. For illustration purposes we present an example of such a table in Table 11.3 below for the case of our 12×10 Gbs upgrade example above. The table provides segment margins (see for example line 8) for EOL capacity for two different cases: a

Table 11.3 A power budget example for the Gen-1 dark fiber upgrade case study. The system has a 120 Gbs total capacity (12 × 10 Gbs per channel at 50 GHz (0.8 nm) channel spacing)

Line	Parameter	BOL		EOL		Comments
0	Calculated OSNR	16.3	dB/0.1 nm	14.5	dB/0.1 nm	The Optical Signal to Noise Ratio (OSNR) at the end of the link due to the sum of ASE s-sp and sp—sp noises from all the submerged amplifiers, taking into account all various span losses. Calculated.
1	Mean Q value (from calculated OSNR)	15.0	dB	13.5	dB	Calculated from OSNR, based on the back-2-back lab/manufacturing data for a typical transponder.
1.1	Propagation impairments	1.0	dB	1.0	dB	Propagation penalties due to fiber nonlinearities: Self Phase Modulation, Cross Phase Modulation, Four Wave Mixing, Stimulated Brillouin and Raman Scattering, and dispersion penalties. Calculated and/or estimated based on field trial results.
1.2	Gain Flatness impairments	0.2	dB	0.2	dB	Penalty due to non-flat repeater gain curve. Estimated based on customer/field data.
1.3	Non-optimal optical pre-emphasis impairment	0.2	dB	0.2	dB	
1.4	Wavelength tolerance impairment	0.1	dB	0.1	dB	Due to non-ideal channel filtering in MUX/DEMUX optics. The penalty is small due to wavelength lockers used in signal lasers of modern transponders. Estimated.
1.5–7	Mean polarisation penalties (PMD + PPDG + PDL)	0.2	dB	0.2	dB	Together with the penalty in the line 2 present total penalty due to PMD/PDL/PDG. Estimated based on field data.
1.8	Supervisory impairment	0.1	dB	0.1	dB	The supervisory signal providing line monitoring and communication with wet plant elements uses a low frequency over-modulation of the signal for the case of active supervisory. Estimated based on an experimental data.

(continued)

Table 11.3 (continued)

Line	Parameter	BOL		EOL		Comments
1.9	Manufacturing and environmental impairment	1.0	dB	1.0	dB	Due to transponder manufacturing distribution plus changes in the environmental conditions.
2	Time varying system performance (5 sigma rule)	0.4	dB	0.4	dB	See comments for the line 1.5–7
3	Line Q Value (line 1—lines 1. 1 line 2)	11.8	dB	10.3	dB	
4	Specified TTE Q value (back to back)	25.0	dB	25.0	dB	
5	Segment Q value	11.6	dB	10.1	dB	
5.1	BER corresponding to segment Q without EFEC	7.2E–05		6.9E–04		Pre-EFEC BER
5.2	BER corresponding to segment Q with EFEC	<1E–13		<1E–13		Post-EFEC BER
5.3	Effective segment Q value with EFEC	>17.3	dB	>17.3	dB	Post-EFEC Q
6	Q limit beforeEFEC correction (1E 13)	9.3	dB	9.3	dB	Pre-EFEC limit Q necessary to get post-EFEC BER value better than the theshold (in this case 10–13)
7	Repairs, ageing and pump failures	1.5	dB	0.0	dB	Difference between line 5 BOL- and EOL-column Q-values
8	Segment margin	2.3	dB	0.8	dB	Line 5—Line 6. For EOL case ITU standard G.977 recommends 1 dB.
9	Unallocated supplier margin	0.0	dB	0.0	dB	Margin for unallocated penalty.
10	Commissioning limit	11.6	dB			Line 6 + line 8

beginning-of-life (BOL) case and an end-of-life (EOL) case that account for repair margins. In addition it provides itemized performance penalties. Essentially this is a power budget approach that outlines the various effects in the system and starting from a performance target subtracts all the above effects in the form of a Q-penalty. More details about the calculation of the penalty of each effect can be found in the ITU G.977 standard.

The power budget of Table 11.3 shows both BOL and EOL cases. The former refers to the performance of the system at the time of the upgrade and the latter is normally derived by using the following undersea industry typical assumptions with regard to the repair span loss margins:

- There are two types of repairs of cable cuts (mostly caused by the fishing industry): shallow water (depth < 1000 m) and deep water (depth > 1000 m) repairs.
  - Each repair in deep water adds 4 dB of loss to the span and the repair frequency rule is one repair per 1000 km of deep water cable over the system lifetime
  - A shallow water repair adds 0.5 dB of loss to the span and the repair frequency rule is one repair every 25 km over the system lifetime
- Aging increases the fiber attenuation by up to 0.004 dB/km over the system lifetime

To verify whether the standard repair margins are adequate for a specific link upgrade design it helps to have a repair report for a given link so that one could evaluate whether the impact of the old repairs is accounted for in the BOL upgrade case and whether the EOL repair budget is adequate to account for future repairs, assuming that in the future the cable cuts happen with the same frequency as in the past. It must be noted that the way EOL margins are set in this specific budget exercise they are specific for undersea DWDM applications. In terrestrial DWDM applications EOL margins are accounted for differently. This is described in Sect. 11.3 below.

Various issues related to the lack of accurate and/or comprehensive link data make it imperative to perform field trials on the system which is scheduled for an upgrade. Although field trials are expensive, require specialized test equipment, and involve various logistical issues, including an arrangement of test time with a carrier, field trials are still very essential for the validation of the design results for the upgrade capacity.

11.3 Unrepeatered DWDM links

For undersea and terrestrial links with a length under 500 km unrepeatered (UR) technology could be used instead of the repeatered one discussed above. In UR applications passive optical cable only connects terminal stations; while all active

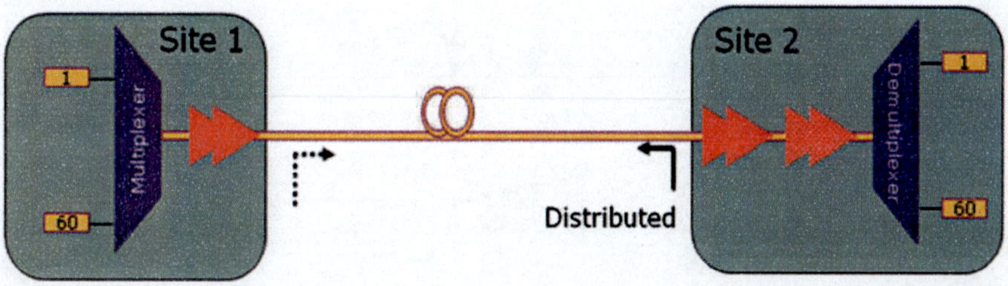

Fig. 11.3 Schematic of a typical UR link. Passive optical cable connects terminal stations

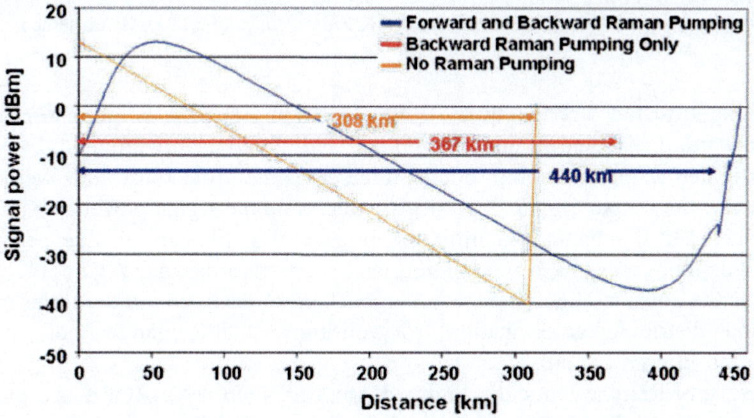

Fig. 11.4 Amplification comparison for three different cases: no Raman pumping (EDFA amplification only) with the shortest reach, backward distributed Raman pumping with the medium reach, and, finally, forward and backward distributed Raman pumping with the longest reach. For all three cases the output OSNR is 13 dB/0.1 nm and the maximum signal power is +13dBm/channel

equipment is located in the terminal stations as shown in Fig. 11.3. The elimination of undersea repeaters helps to cut link cost significantly and makes UR technology appealing to the carriers. An excellent description of the development history as well as the state-of-the-art for this technological area can be found in Chap. 6 of [1].

There are several ways to increase UR link reach and capacity. The major ones are listed below:

- The use of ultra-low loss fibers like pure silica core fiber (PSCF) with the typical loss as low as 0.17 dB/km, instead of the SSMF fiber which has higher loss typically around 0.20 dB/km.
- The use of Raman optical distributed amplifiers as discussed in [5–7]. Raman amplifiers can pump an optical cable in both forward and backward directions. In Fig. 11.4 an amplification comparison is done for three different cases and

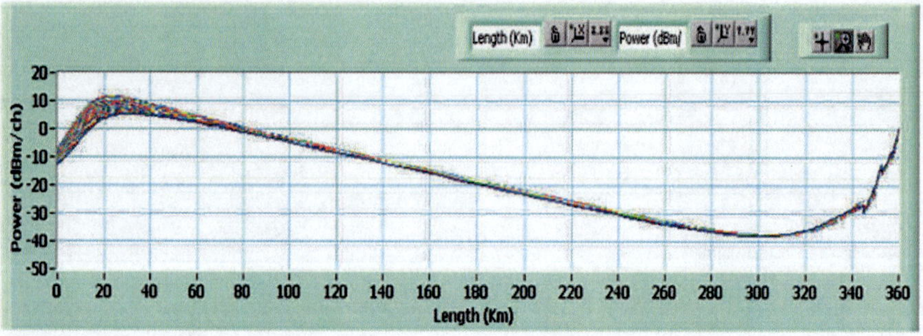

Fig. 11.5 Signal power profile for the typical UR system of Fig. 11.3. Signal power (dBm/channel) vs. cable length (in km) for a PSCF link length of 345 km is presented

compared to the link length: no Raman pumping (EDFA amplification only), backward distributed Raman pumping, and, finally, forward and backward distributed Raman pumping. For all three cases the same constraints were used, namely: the output OSNR is 13 dB/0.1 nm while the signal power maximum is set at +13dBm/channel to limit the impact of nonlinearities like self-phase modulation, cross-phase modulation, and four-wave mixing. Figure 11.4 illustrates the considerable increase in single channel reach due to the large gain of Raman distributed amplification. It it well-known that Raman technology could be used in both C- and L-bands.

- The use of Remote Optically Pumped Amplifiers (ROPA): ROPA is a piece of Erbium-doped fiber inserted into the line fiber about 100 km from the receiving end. Pumping of ROPA is done remotely and so Raman distributed amplification (in this case the backward one) pumps the line and also the ROPA. Note, that the ROPA uses a very small part of Raman pump power, while the majority of it is lost in pumping the line. Despite the presence of ROPA this configuration is considered an un-repeatered one since ROPA does not require any electrical components for its operation.
- Increasing the line rate from 10 Gbs per channel to 40 Gbs per channel. This will involve the cost of upgrading the line equipment.

The latest developments in UR applications can be found in [5, 6].

The typical power distribution in an UR link with forward and backward Raman amplifiers is also provided in Figs. 11.5 and 11.6 for a typical DWDM configuration as discussed in [8]. In Fig. 11.5 the signal power profile for a 345 km long PSCF fiber span with distributed forward (transmit-side) and backward (receive side) Raman pumping is presented. This corresponds to the typical equipment schematic that is shown in Fig. 11.3 above. Additional amplification, provided by lumped Raman amplifiers on the receiver side is necessary in order to increase the signal power within the tolerance range of the transponders.

In Fig. 11.6 the typical signal power profile for a 402 km long PSCF fiber span with ROPA is presented.

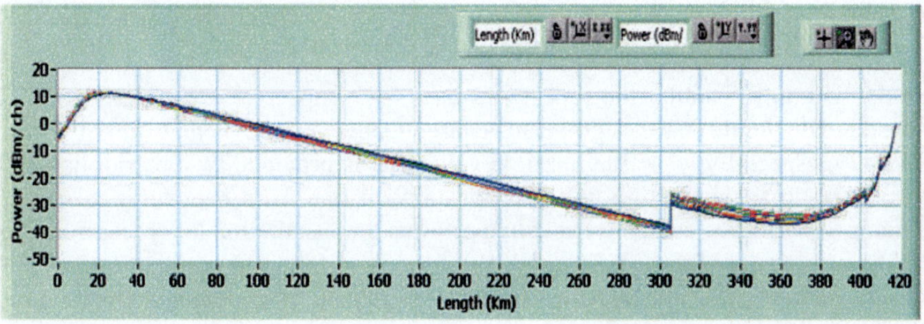

Fig. 11.6 Signal power profile with ROPA located at 97 km from the receiver side of the typical system of Fig. 11.3. Signal power (dBm/channel) vs. cable length (in km) for a PSCF link length of 402 km is presented

For an UR link upgrade design the usual design issues of a DWDM repeatered system typically are:

- The output OSNR across the supported bandwidth and the output gain ripple: Unlike in undersea repeatered links, in the UR link case the computation of the link gain is quite complicated. The designer has to account for its various Raman gain components such as pump-to-signal, pump-to-pump, and signal-to-signal components as discussed in [7]. To achieve a flat output OSNR across the supported bandwidth the system designer has to apply significant signal pre-emphasis at the transmit-side.
- The link dispersion budget and fiber nonlinearities

Standard design parts are accompanied by extra design consideration factors like the necessity to manage multi-path interference (MPI) [7] and Stimulated Brillion Scattering (SBS). Both MPI and SBS are typical effects in an UR link design that can be significant factors due to the large Raman gain within an UR span, especially at the beginning/end of a link as demonstrated in Figs. 11.4 to 11.6. On the other hand these are not an issue in a typical undersea repeatered system where MPI- and SBS-related penalties are small.

The single most significant factor affecting all aspects of an UR link upgrade design is the span fiber loss. The system designer simply cannot trust completely the link data that are provided by a customer for an upgrade analysis.

Below we list the main reasons why such data have proven unreliable time and time again:

- Accurate loss data which were taken during the original system deployment could be outdated by the time the upgrade takes place. This could be, for example, due to the optical cable damages that occurred since the original cable deployments
- There could be measuring discrepancies as high as several dBs between different loss data sets for the same links that are provided by a carrier
- Sometimes loss data are not available at all

- Taking loss data for links with ROPA presents additional difficulties due to the nature of the amplification process

For UR links, unlike the repeatered ones, the impact of a link loss error of several dBs could have significant consequences in the actual link upgrade capacity, especially when the link loss value that is provided by a carrier happens to underestimate the actual link loss. To prevent this from happening, a system vendor (or a service company) usually performs the installation loss measurements before an upgrade is started. To get the best loss data for a link, one has to combine results of two different loss measurements—optical time domain reflectometry (OTDR) [9] and power [10].

Each of these measurement methods has limitations and errors and none of them is able to provide accurate loss data for all the link components such as the optical fiber, connectors, splices, fiber defects, and ROPA. The errors are removed by comparing and/or stitching together power and OTDR measurement results.

In an OTDR measurement a fiber is probed with a short laser pulse (from one end only unlike for span loss measurements where one needs access to both ends) [9]. OTDR measurements are done in both link directions and at several wavelengths, and have the following advantages:

- They provide accurate measurements of fiber loss (this is the slope in an OTDR trace [8]) in both directions
- Even in the opposite direction they are beneficial in identifying whether there is or not an optical isolator preceding the ROPA

The main disadvantages of using OTDR measurements are:

- Limited dynamic range of about 40 dB or so for a typical OTDR test-set. Dynamic range provides information on the maximum fiber loss value that can be measured by the OTDR module. As a result, the middle parts of long (>300 km) links might not be able to be measured accurately by OTDR and have to be complemented by power measurements [10].
- Dead zones which in a fiber measurement will vary mainly due to the presence of reflections and the corresponding overload of an OTDR receiver (as explained in more detail in [9])

On the other hand, power loss measurements which are typically done in both link directions and at several wavelengths have the main advantage that they provide an independent reference for fiber loss that is used to debug and tune up the OTDR loss data. On the other hand, the main disadvantage of using power loss measurements is that it allows for more variability of loss results due to human error and/or poor source calibration compared with the OTDR which tends to be more accurate.

Continuing on the upgrade engineering process, after the current link loss data is obtained, the designer has to account for a repair loss margin. For UR links the repair loss margin is either provided by a carrier or is estimated based on the guidelines that were outlined in Sect. 11.1 above for undersea repeatered links.

Table 11.4 Simplified UR link design summary for a 345 km-long system with no ROPA and the use of enhanced FEC as described in Sect. 11.2 above

BOL link loss (dB)	Loss margin (dB)	EOL link loss (dB)	BW (nm)	Line rate (Gbs/ch)	Capacity	EOL OSNR (dB/0.1 nm) worst channel
58.9	5.8	64.7	1567–1592	10	60 × 10 Gbs 50 GHz channel spacing	12

Unlike with link loss data, the uncertainty in the reliability of the chromatic dispersion data and the related impact on the upgrade capacity is rather small for the UR links under discussion. The main reasons why this is the case are:

- In typical UR links an optical cable usually contains PSCF and SSMF fibers. These fibers have very well-known and tight manufacturing chromatic dispersion distributions
- Modern dispersion compensating fibers developed for PSCF and SSMF fibers provide excellent wide-band dispersion compensation in both C- and L-bands. As a result, errors are rather small in dispersion budgets done for typical PSCF, SSMF, and corresponding dispersion compensating fibers
- Even when errors in the dispersion budget design occur, the system designer can supply several trimming dispersion compensating units as options during the actual deployment and then, when the final system is ready for deployment or upgrade in the field, choose the one dispersion compensating unit that provides the best BER results during commissioning tests

Although, in most of the cases as described above, the uncertainty in the chromatic dispersion link budget is very small, practically it is never exactly zero. Upgrade problems are rare but they might arise, for example, when a customer has an NZDSF-based UR link whose dispersion type/data was lost over the years since the original deployment. In such case a trial-and-error design process might be required.

After all link information is accounted for, a system designer provides a performance summary for an UR link upgrade in a form similar to the one presented in Table 11.3. Table 11.4 presents a short form design summary for a typical 10 Gbps per channel DWDM UR link (without ROPA) upgrade. The corresponding signal power profile follows the one presented in Fig. 11.4. In Table 11.4 to compute the loss margin we assumed a 0.5 dB repair loss for every 30 km of the link over its entire lifetime.

11.4 Terrestrial Long-Haul DWDM links

The architecture of any undersea DWDM system or network could be usually reduced to a point-to-point link. Unlike that, terrestrial DWDM systems and networks represent mostly a combination of sites exchanging traffic between themselves and any fiber-optic links connecting the sites. Rich connectivity in terrestrial DWDM networks is supported by both fixed optical add/drop multiplexers (OADM) and reconfigurable add/drop multiplexers (ROADMs) [11]. In this section we focus on long-haul systems where the longest optical path exceeds 600 km (see network example in the Table 11.5).

Unlike in an undersea repeatered system upgrade where only SLTEs are upgraded while maintaining all the same in-line equipment (wet plant), in terrestrial upgrades a vendor is given the existing fiber plant only. At the predetermined sites, which are provided by a carrier, a system vendor installs terminal equipment (at the end points of a network) used to generate and receive traffic. A system vendor also installs the following in-line equipment: (1) optical add/drop (fixed and reconfigurable) modules that serve to add and/or drop a certain number of wavelengths and to amplify optical signals as well as to compensate for their losses; (2) in-line amplifiers (ILA) whose only role is amplification of optical signals weakened during propagation in an optical cable. The derived upgrade design requirements then specify all network sites plus EOL capacity from any site A to any site B.

The starting point in the design process by a system designer is to evaluate the existing fiber plant quality that is destined for an upgrade, as there could be multiple issues with the optical cable. First, an existing optical cable could be damaged in some spans which usually translates into extra fiber losses. This might increase these losses beyond or even well beyond the standard values creating problems. The cause of the optical cable damages could be multiple: defects in optical cable manufacturing that passed through manufacturing testing undetected, fiber handling errors during original installation, damages incurred since original cable deployment, and others. If several fiber pairs are offered by a carrier to choose from then the system designer has to identify for the deployment the best available. The availability of alternative fiber pairs is more likely to happen for terrestrial upgrades, because unlike undersea cable that carries only few fiber pairs, terrestrial optical cable fiber counts could be as high as a few hundreds. If alternative fiber pairs are not available, the next step of the design process is to decide which damaged cable sections have to be replaced and which sections could stay. Replacement of the damaged optical cable translates into a significant extra expense and time delays for the customer, a situation that every customer always tries to avoid. The usage for upgrades of an all-Raman system with its extra system margin, relative to EDFA-based systems, helps to support the extra losses without an optical cable change, while keeping sufficient system upgrade capacity. For example, the NuWave CXR all-Raman system supports up to 240 × 10 Gbs/40 Gbs per channel capacity for long-haul networks [12].

During the second step in the design process the designer has to look into the fiber polarization mode dispersion (PMD) effect. Legacy fibers manufactured and deployed in mid-1990s and earlier usually have high PMD. This is because these fibers were manufactured before fiber spinning was added to the fiber drawing process in the mid-1990s that resulted in lower fiber PMD. It must be noted here that PMD of legacy DSF fiber can be as high as 1 ps/sqrt(km) and that it can

Table 11.5 A realistic terrestrial 900 km-long link case study that is due for an upgrade

Sites *U*	Site Type	Span Length (km)	BOL Span Loss (dB/km)	EOL Span Loss margin (dB)	EOL Span Loss (dB/km)
1	TERM 1				
2	OADM	73	16.1	2.0	18.1
3	OADM	5	1.1	2.0	3.1
4	ILA	100	22.0	2.0	24.0
5	ILA	94	20.7	2.0	22.7
6	OADM	54	11.9	2.0	13.9
7	OADM	29	6.4	2.0	8.4
8	ILA	110	24.2	2.0	26.2
9	ILA	71	15.6	2.0	17.6
10	OADM	86	18.9	2.0	20.9
11	OADM	27	5.9	2.0	7.9
12	ILA	112	24.6	2.0	26.6
13	ILA	55	12.1	2.0	14.1
14	TERM 2	86	18.9	2.0	20.9

impact the design process even for systems with channels at 10 Gbs line rate. Moderate PMD of the order of 0.1 ps/sqrt(km), which is typical for optical cables manufactured since the mid-1990s, has negligible impact on a link upgrade design for a 10 Gbs line rate, while still might be an important factor for a 40 Gbs line rate. Clearly in all the above the question that needs to be addressed is whether or not to use polarization mode dispersion compensators (PMDC).

Finally, upgrade design capacity results depend on the legacy fiber type; for example a network design that uses SSMF fiber is different from that which uses NZDSF fiber. Compensation of chromatic dispersion and dispersion slope using dispersion compensating fibers developed for SSMF has proven to be better than that for NZDSF fibers.

A typical terrestrial multi-span DWDM link, similar to the one deployed in the past by Xtera Communications [12], is described in Table 11.5. In this realistic case study an Xtera NuWave CXR all-Raman system [12] is used. Link design was done with the maximum EOL capacity of 240 × 10 Gbps and a 50 GHz channel spacing. Such capacity is made possible by using both C- and L-optical bands.

Table 11.5 provides all the required link information that is essential for link design: the node sites, the type of equipment per site, the span length, the span losses, and the repair margins. The system begins with two terminal sites TERM1 and TERM2 which are labeled in Table 11.5. Each TERM-site contains terminal equipment consisting of 10 Gbps transponders (TXP), and equipment used to prime the signal before it is sent into or received from the line. These equipment are the optical MUX/DEMUX optics, dispersion compensating units, and the optical amplifiers that are needed in order to amplify the attenuated optical signal.

In addition to the above, there are two different in-line network elements: 1) in-line amplifiers (ILA) and 2) fixed optical add/drop multiplexers (OADM). Finally, in the above 900 km-long link, span lengths vary in the range from 5 km to

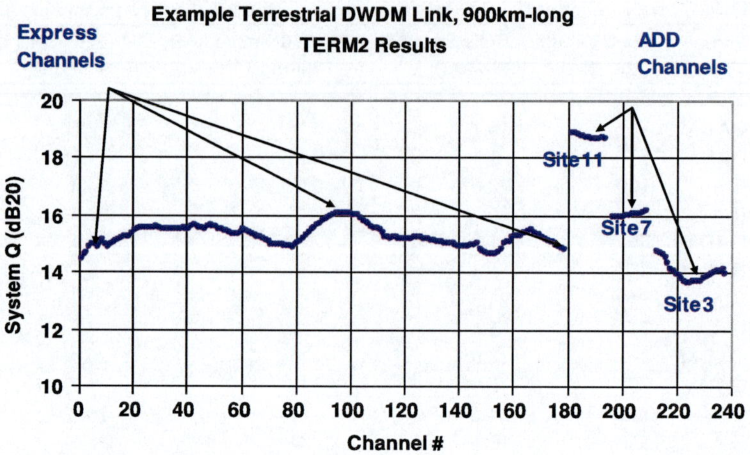

Fig. 11.7 Q-value for the TERM 2 system performance for the 10 Gbs/channel forward direction system

112 km. In this particular case study, the 5 km span shown in Table 11.5 was not concatenated with an adjacent span to illustrate the fact that a customer might request an OADM at that site.

For simplicity, in this case study results are provided only for channels in the forward direction as follows: a) for the express channels propagating from TERM1 to TERM2; b) for ADD channels, added at the OADM sites: site 3 (26 channels total added), site 7 (13 channels total added), and site 11 (13 channels total added) that propagate from the corresponding OADM site to TERM2.

Figure 11.7 presents link design results in the form of Q-values per channel at TERM2. Channel 1 corresponds to the edge channel on the red side of the spectrum (L-band), while channel 240 corresponds to the edge channel on the blue side of the spectrum (C-band). Express and ADD channels are indicated as well. The minimum pre-EFEC Q is assumed to be 13.6 dB (for channels added at the OADM site 3). This provides a 4.2 dBQ margin with enhanced FEC (EFEC) transponders having a pre-EFEC Q target of 9.4 dBQ guaranteeing a BER of 10^{-15} after EFEC. It should be clear to the reader then that all express and added channels are error-free after EFEC. All the corresponding OSNR (dB/in 0.1 nm) values are provided in the accompanying Fig. 11.8.

Figure 11.9 presents the Q-values for all channels that are dropped at the sites 2, 6, and 10. In this case, the minimum pre-EFEC Q is 16.4 dB (for channels dropped at the OADM site 10), which provides a 7.0 dBQ margin with EFEC transponders having a pre-EFEC Q target of 9.4 dBQ at a BER of 10^{-15}. Thus, it should be again clear that all dropped channels are error-free after EFEC.

From the above, it is clear that there are differences between the design process for a terrestrial DWDM upgrade and its counterpart for an undersea repeatered

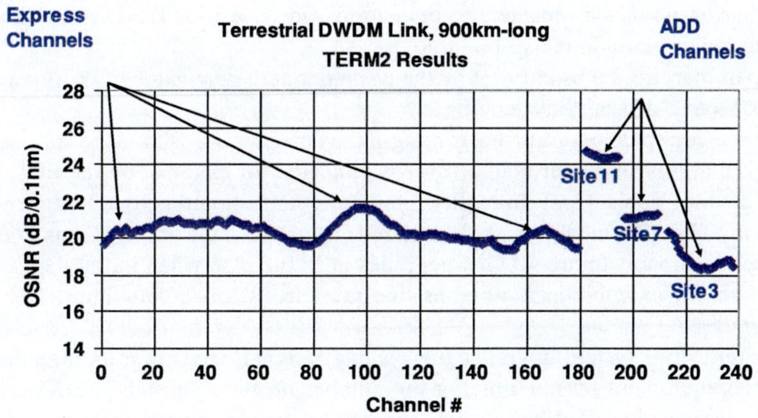

Fig. 11.8 OSNR for the TERM 2 system performance for the 10 Gbs/channel forward direction system

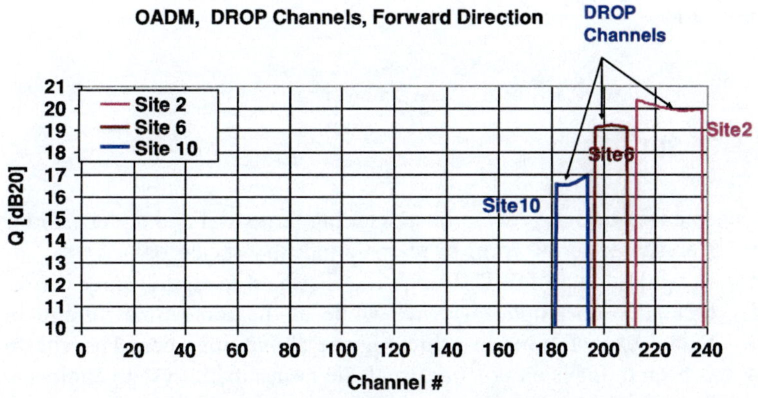

Fig. 11.9 Q-value system performance at OADM sites for DROP channels and a 10 Gbps per channel system in the forward direction

DWDM system upgrade. A given terrestrial link upgrade design has to address the following issues which are specific to the nature of this system:

- Terrestrial DWDM networks have rich connectivity. A system designer needs to provide capacity for both the express (going from terminal to terminal) channels, as well as the add/drop channels. Undersea DWDM links are mostly point-to-point.
- In a terrestrial system the system designer needs to consider link designs in both the forward and the backward directions, since the links are usually asymmetric

- Unlike in undersea repeatered applications, in terrestrial DWDM applications span length variations can be large
- Repair margins are usually set by the customer at the flat value of 2 dB per span, irrespective of span/link lengths.
- For a given span loss plus repair margin the designer needs to make sure that the EOL margins are better than 0dBQ (as indicated for example on the EOL-value in line 8 of Table 11.3). In reality the requirement should be much higher since the initial deployment, which for example might be done at a 10 Gbps line rate, needs to support future 40 Gbs-upgrades or in the case when initial deployment is done for example at a 40 Gbps line rate, needs to support future 100 Gbps upgrades
- For terrestrial system upgrades the system designer needs to minimize the cost of the equipment by maximizing the number of glass-through sites. These are sites without installed active line equipment, system margins permitting. In a glass-through site two consecutive spans are connected by a short few meters long passive optical cable (jumper), which results in span concatenation and an increase in a span length before amplification is actually done.
- Finally the designer needs to minimize the number of different types of in-line equipment used, which is helpful for both manufacturing and spare minimization.

11.5 Conclusions

In this chapter the design process for upgrading terrestrial and undersea long-haul systems is presented in the form of realistic industry scenarios.

We discussed in detail DWDM upgrades that help to significantly increase the capacity of legacy fiber-optic systems, while at the same time minimizing the carrier's capital expenditures in achieving the above upgrades. The goal of this chapter has been to focus on practical, real-life issues that a system engineer out in the field has to deal with in order to provide the quality DWDM network upgrade design needed on existing systems. Clearly the goal is to try to maximize an upgraded capacity and minimize cost.

To illustrate these issues and their resolution from different angles, three different DWDM upgraded applications were chosen: (1) upgrades of undersea repeatered systems with the emphasis on regional size systems; (2) unrepeatered (UR) systems (undersea and terrestrial); and (3) terrestrial multi-span long-haul systems. The chapter covers common design issues for all these applications, and looks into issues that are specific for each application.

Acknowledgments The author is grateful to his present and former collogues at Xtera Communications: D. Chang, S. Edirisinghe, G. Grosso, D. Krishnappa, W. Pelouch, A. Puc, A. Robinson, J. Schwartz, N. Taylor, and W. Wong.

References

1. Chesnoy J (2002) Undersea fiber communication systems. Academic Press, New York
2. Paling J, Mertz P, Grubb S (2010) Application of photonic integrated circuit based DWDM transmission equipment on legacy repeatered submarine cable systems. In: Proceedings of suboptic industry conference, Yokohama, Japan
3. Schwartz J, Edirisinghe S, Wong W, Acebo R, Palinski G (2010) Optical layer concatenation of long-span amplified systems. In: Proceedings of suboptic industry conference, Yokohama, Japan
4. Agrawal GP (1997) Fiber-optic communication systems. p 172. Wiley, New York
5. Bissessur H, Brylski I, Mongardien D, Bousselet P (2010) Unrepeatered systems: state of the art. In: Proceedings of suboptic industry conference, Yokohama, Japan
6. Perrier P, Février H, Lyons J (2010) Advances in repeaterless system architectures and their applications to Festoon cable systems. In: Proceedings of suboptic industry conference, Yokohama, Japan
7. Islam M (ed) (2003) Raman amplifiers for telecommunications. Springer Inc, New York
8. Puc A, Chang DI, Grosso G, Ellison J, Oberland R, Burtsev S, Perrier PA, Pelouch W, Webb S (2008) L-band unrepeatered WDM experiment over 451 km using all-Raman amplification and ROPA. In: Proceedings of IEEE/LEOS annual meeting, paper WH5,Newport Beach, CA
9. Derikson D (1998) (ed) Fiber optic test and measurement. Chapter 11, Prentice Hall, Englewood Cliffs
10. Derikson D (1998) (ed) Fiber optic test and measurement. Chapter 13, Prentice Hall, Englewood Cliffs
11. Kaminow IP, Li T, Willner AE (eds) (2008) Optical fiber telecommunications V, B: systems and networks. Chapter 8, Academic, New York
12. Xtera Communications Inc. http://xtera.com

Part III
Logical Layer

Chapter 12
Impairment-Aware Optical Networking: A Survey

Siamak Azodolmolky, Marianna Angelou, Ioannis Tomkos, Tania Panayiotou, Georgios Ellinas and Neophytos (Neo) Antoniades

Abstract In optical networks the effect of physical layer impairments can play an important role in the routing and wavelength assignment decisions that are taken in the control layer. Furthermore, as networks evolve to support more bandwidth-intensive applications, and as rich multimedia and real-time services become more popular, next-generation networks are expected to support traffic that will be heterogeneous in nature with both unicast and multicast applications. In this chapter we investigate the problems of routing and wavelength assignment in transparent optical

---

S. Azodolmolky (✉) · M. Angelou · I. Tomkos
Athens Information Technology
19.5 Km Markopoulo Avenue
19002 Peania, Attiki, Greece
e-mail: sazo@ait.edu.gr

M. Angelou
e-mail: mang@ait.edu.gr

I. Tomkos
e-mail: itom@ait.edu.gr

T. Panayiotou · G. Ellinas (✉)
Department of Electrical and Computer Engineering
University of Cyprus
1678 Nicosia, Cyprus
e-mail: gellinas@ucy.ac.cy

T. Panayiotou
e-mail: Panayiotou.tania@ucy.ac.cy

N. (Neo) Antoniades
Department of Engineering Science and Physics
College of Staten Island/The City University of New York
Staten Island, NY 10314, USA
e-mail: neo.antoniades@csi.cuny.edu

N. (Neo) Antoniades et al. (eds.), *WDM Systems and Networks*, Optical Networks,
DOI: 10.1007/978-1-4614-1093-5_12,

networks that support unicast and multicast applications, while taking into consideration the physical layer impairments during the provisioning of each application.

12.1 Introduction

Optical networking has undergone various changes and enhancements during the past few years and the trend clearly draws an evolution path towards lower cost, both in terms of capital and operational expenditures (CAPEX and OPEX) and higher capacity. In addition to cost reduction, there are major concerns regarding the energy consumption, heat dissipation, and physical space requirements. This evolution trend has been governed by progress in networking capabilities (e.g., higher line rates, advanced and robust modulation formats, more wavelengths per fiber, and future multi-core fibers) and emerging applications (e.g., high definition 3D gaming, ultra high definition video streams, telepresence, etc.). Providing more capacity in a cost-effective manner was the focus of optical network evolution. Considering the optical transmission systems, this evolution can be mapped to denser wavelength division multiplexing (WDM) transmission systems (i.e., systems with 80–160 wavelengths per fiber, or emerging multi-core fibers) operating at higher line rates (e.g., 40–100 Gbps), and switching equipments with multiple granularities (i.e., wavelength and sub-wavelength switching) [1]. To address the demands of emerging *dynamic applications* efficiently, static and high-capacity transmission pipes are no longer sufficient. Thus, a configurable and dynamic optical layer along with a corresponding control plane, which is able to serve dynamic requests, is the expected outcome of the aforementioned evolution trend. However, any proposal should address the requirements in a cost-effective way.

As depicted in Fig. 12.1, optical network architectures are evolving from traditional opaque networks toward all-optical (i.e., transparent) networks. The optical signal carrying traffic undergoes an optical-electronic-optical (OEO) conversion at every switching (or routing) node in the opaque networks. The transmission reach of the optical signal without any regeneration and considering the practical and economical factors is limited (e.g., 2000–2500 km) [2]. Signal regeneration is required to re-amplify, re-shape, and re-time the optical signal (also known as 3R). The regeneration process simply improves the quality of the signal. The OEO conversion removes the impact of physical impairments and enables the optical signal to reach long distances. However, this process (i.e., OEO conversions) is quite expensive due to several factors such as the number of required regenerators in the network and the line rate and modulation format dependency of the OEO conversion process. Electronic regeneration has its own scalability issues related to cost, power consumption, heat dissipation, and space requirements. Network architects and designers have to consider more OEO equipment as the size of the network increases, which consequently adds extra cost, power consumption, heat dissipation, required space, and operation and maintenance costs.

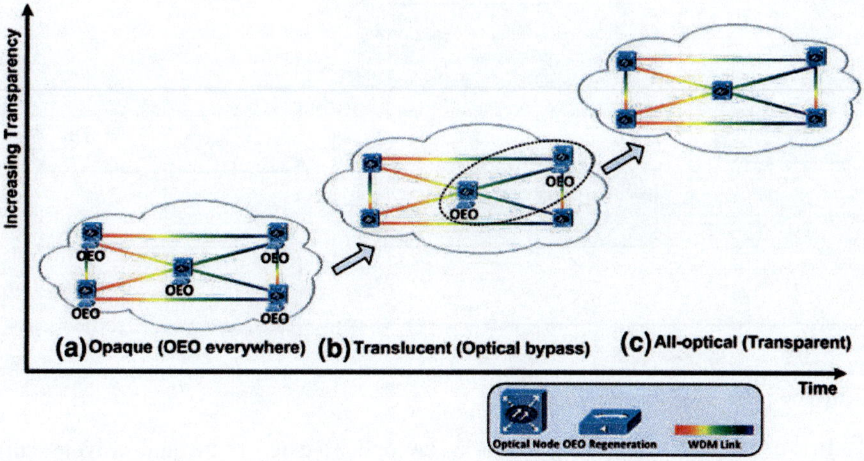

Fig. 12.1 Evolution of optical networks (From [14], Figure 1. Reproduced by permission of © 2009 Elsevier)

Different proposals are suggested to address these issues. Sparsely placed electrical or optical regenerators is one such approach [3]. Regeneration can be performed completely in the optical domain [4, 5], however electrical regeneration (i.e., OEO conversion) is still the most economic, reliable, and mature technique. All-optical regeneration is still an area of active research on many fronts and is not mature enough for practical deployments. Intermediate optical network architectures, which are identified as translucent [6] (or optical-bypass [7]) networks, fill the gap of practical and mature all-optical regeneration. Translucent network architectures have been proposed as a compromise between opaque and all-optical networks. In translucent networks, a set of sparsely but strategically selected regenerators is deployed in the network to maintain the acceptable level of quality of transmission (QoT) of signals. This solution in fact eliminates most of the OEO nodes and allows a signal to remain in the optical domain for much of its path from source to destination. Keeping the signal in the optical domain brings a significant cost reduction due to the elimination of OEO (i.e., electronic processing) equipment, since optical technology can operate on a spectrum of wavelengths at once and can also operate on wavelengths independent of their line rate and/or modulation format [8]. This removal of OEO equipment also paves the way for lower energy consumption, heat dissipation, and space requirements. Translucent (optical-bypass) core WDM networks using reconfigurable optical add/drop multiplexers (ROADMs) and tunable lasers appear to be on the road toward widespread deployment and could evolve to all-optical mesh networks based on optical cross-connects (OXCs) in the near future. No matter which architecture is considered, these architectures aim at providing the required infrastructure for end-to-end connection establishment with acceptable QoT.

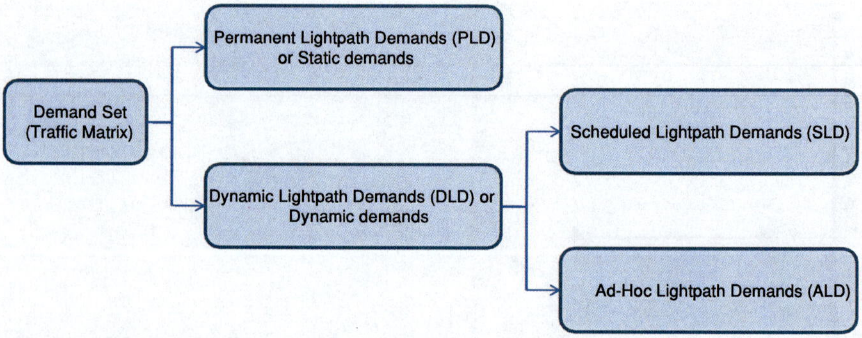

Fig. 12.2 Demand set (traffic matrix) categories (From [14], Figure 2. Reproduced by permission of © 2009 Elsevier)

In optical networks, a *lightpath* is an optical path (route and wavelength) established between a pair of source–destination nodes. The demand set (or traffic matrix) in the network is a set of lightpath requests that must be established in the network. Each "demand" represents an individual lightpath establishment request. In the framework of network planning, some demands are permanent and are defined as Permanent Lightpath Demands (PLD) or static demands. Other demands are referred to as Dynamic Lightpath Demands (DLD) or dynamic demands for short, in which demand requests have a finite lifetime (i.e., start and end of connection) [9]. In this context there exist two variants of DLD (see Fig. 12.2):

- Scheduled Lightpath Demands (SLD): The activation time and lifetime of these demands are known in advance. For instance, provisioning of layer 1 Virtual Private Networks (L1 VPNs) falls under this category. Since the lifetime of SLDs is known a priori the network planning process can consider them as a whole.
- Ad Hoc Lightpath Demands (ALD): This category of demands is characterized by the fact that their arrival time and their lifetime are not known in advance. These two parameters (i.e., arrival time and duration) may be modeled in general by two random processes.

If the allocated wavelength for a given lightpath remains the same across all fiber links that it traverses, then the Routing and Wavelength Assignment (RWA) [10–12] is said to satisfy the wavelength continuity constraint. Considering the wavelength continuity constraint, each lightpath is created by allocating a single wavelength throughout its route. However, if a switching node is equipped with a wavelength converter, then the wavelength continuity constraint does not apply and the routing problem is reduced to normal routing in circuit-switched networks, where the only limiting factor is the number of available wavelengths (channels) on each link.

Previous studies [10] have already investigated the RWA problem. The RWA problem is known to be NP-complete. Most of these works assume an all-optical network, in which all intermediate regenerations (i.e., OEO conversions) are eliminated. Furthermore, in most RWA approaches the optical layer is assumed as a perfect transmission system, and therefore all outcomes of the RWA process are considered as feasible solutions with respect to the QoT requirements. However, the actual QoT of the lightpaths may be unacceptable. Therefore, the incorporation of physical layer impairments in the planning and operation of all-optical/translucent networks has recently received considerable attention from the research community. The physical layer impairments (PLIs) are either considered as constraints for the RWA decisions (i.e., physical layer impairment constrained) or the impact of physical layer impairments are integrated in the RWA decisions (i.e., physical layer impairment aware). It is possible to find alternate routes and/or wavelengths considering the PLIs in the latter case, while in the former approach, the routing decisions are constrained by the PLIs. However, for simplicity, in the rest of this chapter we use the generic impairment aware (i.e., IA-RWA) term to refer to both approaches. Given the limited reach of optics, some intermediate regeneration is required in long-haul backbone networks. Therefore, the IA-RWA problem will be inevitably coupled with the regenerator placement problem [13], in which the network designers plan and design translucent (or optical bypass) networks with optimum number of regeneration sites and/or regenerators per site for a given network topology and demand sets (traffic matrix). The regenerator placement problem is also known to be NP-complete [10, 13].

In order to report the current proposals for IA-RWA optical networking, more than one hundred papers were collected. Considering 28 different metrics, these works were reviewed and ranked [14]. The result of this study is compiled in this chapter and includes IA-RWA algorithms and their algorithmic approaches, classification of IA-RWA proposals, protection and resilience consideration, performance metrics of IA-RWA algorithms, impairment aware control plane extensions, and impairment-aware muticasting architectures and RWA techniques.

12.2 IA-RWA Algorithms

In this section we present the algorithmic approaches for solving the IA-RWA problem and the classification of the different IA-RWA proposals in the investigated papers. This section also includes a description of the performance metrics and the methods adopted to evaluate the different proposals.

12.2.1 Algorithmic Approach

In general, the algorithmic approach for the IA-RWA problem can be categorized either as a combinatorial approach, which searches for an optimal solution or as a sequential approach based on some heuristic or meta-heuristic algorithms, which usually give a suboptimal solution.

The classic RWA problem (i.e., without PLIs constraints/awareness) is NP-complete [10] and thus its optimal solution cannot be found in polynomial time using any known algorithm. The IA-RWA problem introduces additional complexity to the RWA problem since it involves a number of physical layer-related constraints. To eliminate these obstacles the RWA problem can be decomposed into two sub-problems and solved separately: (a) a routing (R) problem, i.e., choice of a suitable route from source to destination, and (b) a wavelength assignment (WA) problem, which finds an available wavelength for the selected route. When the routing and wavelength assignment steps are considered individually and separately, each one (i.e., routing and wavelength assignment) can be further split into: (1) search and (2) selection process. The former one concerns the search for a set of candidate routes/wavelengths, and the latter is a decision function to select the solution from the candidate set.

Although such a decomposition of the RWA problem does not guarantee its optimal solution, yet the computation time can be reduced considerably. Further simplification of the IA-RWA problem can be achieved with the application of a heuristic (or meta-heuristic) algorithm, which can be used to solve any of the routing (R) and/or wavelength assignment (WA) subproblems.

12.2.1.1 Heuristics

Most of the IA-RWA algorithms reported in the literature are based on simple heuristics, since the IA-RWA problem involves additional physical-layer constraints, which in most cases are verified using complex analytical models of physical impairments.

A variety of heuristic algorithms can be found in the literature regarding the routing subproblem and, in general, are based on the shortest path (SP) routing algorithm (e.g., Dijkstra's shortest path algorithm). Among these algorithms, two classes of algorithms can be distinguished: (1) single-path routing algorithms and (2) multi-path routing algorithms. The multi-path routing algorithms are also known as k-shortest path (k-SP) algorithms. In any SP-based algorithm there is a cost (or weight) parameter assigned to each network link. This link cost should have additive properties and it is used by the algorithm to find a path of a minimum overall cost. This link cost may simply represent a hop count, link length, or link state (i.e., link usage, congestion, or physical layer impairment-related information).

A number of IA-RWA algorithms follow the minimum hop SP approach with respect to the single-path routing as reported in [13, 15–22]. The simplest metric, which has additive property and considers the physical layer impairments, is the

---

[1] The physical layer impairments (including four-wave mixing) are presented in detail in Chap. 2.

physical link length [20]. Other IA-RWA proposals utilize the link cost that is a function of the residual dispersion parameter [18], the four-wave mixing (FWM)[1] crosstalk [23], the Q-factor [20], or the noise variance [24]. In these proposals, both linear and nonlinear impairments [25] are considered by the variance of impairments (i.e., noise variance). Since noise variances are additive hop by hop, this solution may be preferable.

Some IA-RWA algorithms modify the SP routing algorithm. For example, the Minimum Coincidence and Distance (MINCOD) algorithm [26] computes the paths that minimize the path distance and the number of shared links, while the least-congested algorithm [19] aims at network load balancing. An impairment constrained-based SP algorithm, which considers the physical impairments due to FWM crosstalk and network resources was proposed in [23]. Furthermore, an enhanced Bellman-Ford shortest path algorithm that deals with constraint-based and multi-objective lightpath provisioning was proposed in [27].

IA-RWA algorithms which exploit multi-path routing algorithms, operate on a set of pre-computed alternative paths. This class of multi-path algorithms can be also referred to as *k*-SP, since these alternative paths are usually the shortest paths. In some proposals the set of candidate paths is restricted to disjoint paths [28]. In the multi-path routing, similar to the single-path routing algorithms, the cost metric can be related to the hop count as reported in [9, 29–32] or can be PLI-aware. The distance metric (i.e., the link length) is the simplest PLI-aware link cost [28, 33]. Other physical impairments, such as Polarization Mode Dispersion (PMD), Amplifier Spontaneous Emission (ASE) noise, Crosstalk (XT), Chromatic Dispersion (CD), and Filter Concatenation (FC), can also be adopted to represent the link cost [34]. Also, some proposals consider a Q-factor penalty as a link cost as described in Sect. 12.7 of this chapter. These costs can be measured using performance monitoring devices [35] or can be calculated analytically either as the worst case Q-factor penalties [28] or taking into account as linear [36] and non-linear impairments [37]. Information about regenerating modules, the number of available and total wavelengths, and the link length can also be combined in a complex link cost formulation [38].

Once candidate routes are found, the selection of an appropriate route is performed either sequentially or in parallel. In the former approach a sequence of re-attempts is performed until the first available route that satisfies a certain performance requirement is found [30, 32, 39]. In the latter, the route is chosen that is the most suitable one according to a given selection criteria (i.e., selection function) [29, 30]. The Breath-First Search (BFS) algorithm [30, 40] can be exploited for searching among multiple alternative routes. In order to explore all possible solutions, the BFS algorithm tries to examine all nodes of the network in a systematic way.

The wavelength assignment procedure operates on a set of candidate wavelengths that are available on previously selected routes. The set of candidate wavelengths may be ordered according to a given policy or unordered. Wavelength ordering was suggested in [41] as a technique to select the wavelength with the minimum number of adjacent channels (i.e., crosstalk terms). The wavelength assignment (WA) algorithm in [21] extends this method by initially considering the wavelengths that are most separated in terms of frequency. Then, in order to maximize the frequency separation, wavelengths are analyzed in an optimal order.

Considering a set of candidate routes, the wavelength selection phase can be performed either sequentially or in parallel. This is similar to the routing sub-problem. In the sequential approach, the first available wavelength that satisfies network layer and physical layer constraints is selected. This First-Fit (FF) selection approach has been considered in a large number of IA-RWA proposals [9, 13, 16, 18, 19, 24, 27, 30, 33, 38, 42–45]. Other IA-RWA algorithms try to look through all of the candidate wavelengths so as to find the Best-Fit (BF), i.e., the most appropriate wavelength [17, 27]. For example, based on the network state information given at the source node, as in the Least-Loaded algorithm [26], the wavelength with the lowest utilization in the network can be selected. Finally, a random wavelength selection can also be performed [15, 24]. In the random wavelength assignment scheme, a wavelength among the available ones is randomly selected. However, it is shown that the blocking rate of the random wavelength assignment is worse than that of the FF algorithm [10]. Nonetheless, since the random wavelength assignment tends to spread the wavelengths use across the links, the crosstalk effects may be decreased [15].

Some of the proposed WA algorithms make a decision based on strictly physical layer impairments. An example can be a PLI-aware algorithm, which aims at minimizing the FWM crosstalk effect [23]. Both FF and BF wavelength selection schemes are utilized in this algorithm. Two other algorithms are reported in [46] and while one of them focuses on the selection of the lightpath with the highest Q-factor margin, the other addresses the fairness issue and it also minimizes the impact of the new lightpath on the existing (i.e., already established) lightpaths.

There are some heuristics that intend to solve the IA-RWA problem jointly, in addition to separate R and WA solutions. The work in [47] presents an algorithm that is able to find an appropriate lightpath in one algorithmic step thanks to a layered representation of links and wavelengths in a single graph. The MC algorithm [48] is another example of an algorithm that solves the IA-RWA problem jointly. This algorithm runs an SP algorithm to find candidate routes for each wavelength. The link costs are constant and equal to the physical link lengths. The number of crosstalk components along the route is calculated for each candidate route. This algorithm selects the route among all candidate routes and the wavelength with the minimum crosstalk intensity. Finally, the Best-OSNR algorithm in [19] jointly assigns a route and wavelength to a given request in order to maximize the optical signal-to-noise ratio (OSNR).

12.2.1.2 Metaheuristics

Besides the heuristic-based algorithms, there is also a class of IA-RWA algorithms that exploit metaheuristic methods. Metaheuristics allow the convergence to an optimum solution through successive iterations. They are very attractive as far as they do not involve complex mathematical formulations.

One of the metaheuristics which is applied to solve the IA-RWA problem is the Ant Colony Optimization (ACO) [33, 45, 49]. ACO approach is characterized by ant-like mobile agents that cooperate and stochastically explore the network. The agents iteratively build solutions based on their own information and on the traces (called pheromones) left by other agents in the network nodes. The proposed ACO algorithm calculates the path on a hop-by-hop basis. The next hop is calculated based on pheromone values of the node, which accounts for the optical performance monitoring of the links. This algorithm is able to calculate a multi-constrained path in a distributed manner, considering the restrictions resulting from ASE noise and optical power budget.

Another metaheuristic IA-RWA algorithm makes use of a Genetic Algorithm (GA) as reported in [50]. A GA operates on a set of solutions called population. Appropriately selected solutions from one population are used to form, through a number of operations, a new population in each iteration of the algorithm. This new population is expected to be a better one. The proposed IA-RWA algorithm attempts to compute a lightpath in such a way that the average blocking rate and the usage of optical devices (e.g., wavelength converters and amplifiers) is minimized. Both PMD and ASE noise are considered as physical impairments.

The Tabu-Search (TS) metaheuristic has also been utilized in [38] to solve the problem of survivable lightpath provisioning. TS is a neighborhood search technique, which tries to avoid local minimum by accepting worse solutions and by using the solutions' search history. The proposed TS algorithm operates on a set of *k*-SP routes, where *k* is dynamically modified according to the direction of improvement. This adaptive approach improves the efficiency of the proposed algorithm. Both PMD and ASE noise physical impairments are considered in [38].

Finally, the authors in [26] propose a Predictive Algorithm (PA) to solve the IA-RWA problem. The main idea of this algorithm is to apply the branch prediction concept, which is originally used in computer architecture problems. The algorithm selects the lightpath based on the history of the previous connection requests. In order to compute the lightpath, it does not need any update message with global network information. Therefore, it can be used in distributed routing. The impact of physical impairments is considered in this algorithm by limiting the transmission distance.

12.2.1.3 Optimization Methods

IA-RWA algorithms based on network optimization theory are the last class of investigated proposals. Network optimization methods are usually suitable for planning (i.e., offline) of network resources, as well as for operation (i.e., online) and centralized lightpath provisioning. Most of the proposed solutions in the investigated literature target transparent optical networks.

A link-path formulation to solve an Integer Linear Programming (ILP) problem of RWA in a transparent network is proposed in [34]. A set of *k* paths is pre-calculated using an SP algorithm. It uses either a single physical impairment [34] or a Q-penalty [36] as the link cost. The ILP formulation considers the existence of sparse wavelength-conversion capable nodes in the network. More specific problems considering PLI constraints into the optimization problem were studied as well. A Mixed-ILP (MILP) formulation for the IA-RWA problem of multicast connections is proposed in [43], which considers optical power constraints. The authors in [49] propose algorithms for the logical topology design and traffic grooming problem in WDM networks with router interfaces and optical layer constraints. The proposed approach is based on a linear program, which is NP-complete. Authors in [49] also propose heuristic algorithms which use a graphical modeling tool. An ILP formulation for the problem of traffic grooming in optical virtual private networks with a bit error rate (BER) constraint is also presented in [51].

The regenerator placement problem in translucent networks with constraints on OSNR is solved in [52] using an ILP formulation. A solution to a similar problem, considering BER constraints is reported in [53], which exploits a Dynamic Programming (DP) technique. Moreover, the survivable lightpath provisioning problem in a translucent network is solved in [38] using an ILP formulation. PMD and ASE noise are considered in this work.

The implementation of an LP solver in a Path Computation Element (PCE) is reported in [39]. The objective function minimized the maximum link bandwidth utilization. The outcome of this objective function is a set of routes which satisfy the required constraint in terms of bandwidth and optical signal quality.

12.3 Classification of IA-RWA Algorithms

An IA-RWA algorithm has to consider not only the physical layer impairments but also the wavelength availability along the selected lightpath (route and wavelength). During network planning the entire set of connection requests is known in advance. The planning mode (offline) RWA problem for setting up these connection requests is named the Permanent Lightpath Establishment (PLD) problem. In a dynamic traffic scenario, the connections are requested in some random order, and the lightpaths have to be set up as needed (referred to as SLD or ALD). In the static (offline) case, there is enough time between the planning and provisioning (i.e., actual lightpath establishment) processes. Therefore, any additional equipment required can be deployed. In this context, the main goal is to accommodate the whole demand set (i.e., minimum blocking rate). In the dynamic traffic scenario, there is little time between planning and provisioning. It is assumed that the demand set must be accommodated using whatever equipment is already deployed in the network. Thus, the IA-RWA proposal must take into account any constraints posed by the current network state, which may force a demand to be routed over a suboptimal path. One big challenge for IA-RWA algorithms is the quality-of-transmission (QoT)-awareness, in the sense that they must ensure (during admission control) that all lightpaths in the network meet a QoT (e.g., BER) threshold, without disrupting previously established lightpaths.

There are few IA-RWA algorithms that consider the impact of the existing connections in their IA-RWA decisions. Some proposals address this problem by considering the crosstalk due to the already established connections in the network [21, 48, 54]. The HQ (Higher Q) and MmQ (Maximize Minimum Q) algorithms in [54] minimize the effect of new crosstalk when establishing a lightpath. The Minimum Crosstalk (MC) [48] wavelength assignment algorithm selects the wavelength with the minimum crosstalk intensity caused by already established lightpaths. In [24], a different approach is considered; during the lightpath selection, the BER of the selected and affected lightpaths are considered in the lightpath establishment. Few works address the problem of selecting the lightpath considering the effect of the selected lightpath on the possible future demands. In [55] and [18], the Dispersion Optimised Impairment Constraint-based (DOIC) IA-RWA algorithm assigns the wavelength with the lowest residual dispersion. This is performed to increase the availability of the wavelengths for the upcoming demands.

Due to its complexity, the re-routing feature of the IA-RWA algorithms is rarely considered in the proposed algorithms. Re-routing refers to the re-computation and re-establishment of already established connections to alleviate the impact of the new connection on the existing ones. In [56], re-routing is utilized to perform the restoration of Label Switched Paths (LSPs) based on a threshold of the OSNR values.

Considering the recent literature, three main approaches can be classified, when the PLIs are introduced in the RWA algorithms: (1) Compute the route and the wavelength without considering the impact of PLIs and finally verify the QoT of the selected lightpath considering the physical layer impairments, (2) Considering the PLIs in the routing and/or wavelength assignment decisions, and (3) Considering the PLIs in the route and/or wavelength assignment decision and finally verify the QoT of the candidate lightpath. These cases and their various combinations are shown in Fig. 12.3.

In case A-1, the route and the wavelength are selected without considering the PLI constraints, but after the verification step, only the wavelength assignment decision can be modified. In case A-2, the route is computed without considering the physical impairments, but there are re-routing attempts if the PLI constraints are not met for the candidate route(s); finally there is a final phase of wavelength assignment (without consideration for PLIs). In case A-3, the route and the wavelength are computed using plain RWA algorithms that are unaware of the PLI constraints (selecting both route and wavelength in one step or selecting the route and then the wavelength) and there is a final step of checking the PLI constraints in order to possibly change the RWA decision.

Some proposals have followed the A-2 approach (e.g., [13, 29, 57]). A combination of IA-RWA algorithm and regenerator placement for dynamic traffic is proposed in [13]. This algorithm computes paths between any combinations of nodes with a constraint of a minimum Q (Quality) value. The authors in [57]

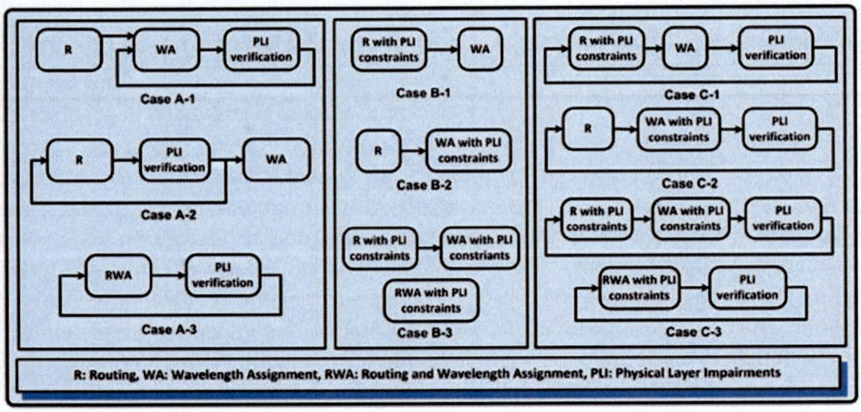

Fig. 12.3 Various IA-RWA approaches (From [14], Figure 4. Reproduced by permission of © 2009 Elsevier)

propose a modified Bellman–Ford algorithm to compute the path with the minimum number of hops with a certain (monetary) cost limit. At the destination, the physical layer requirements with different levels of agreement are checked. Also, for static traffic, the proposed algorithm in [29] computes *k*-shortest paths considering the costs of the links associated with impairments and OEO devices; the final path is chosen in a way that satisfies a minimum Q value. The path establishment process is also considered in [29] and two methods (sequential and parallel) are compared. In the sequential method, the source node sends out a single PATH message containing optical properties which are checked at the destination; if the source node receives back a Path-Error message, another path from the list of candidate paths will be tried out. In the parallel method, the source node sends out *k* PATH messages and the destination node makes the path selection.

The A-3 approach is adopted in [9, 17, 26, 47, 58]. The Lightpath Establishment and Regenerator Placement (LERP) algorithm is proposed in [9] as an IA-RWA and regenerator placement algorithm. LERP initially computes *k* alternate paths, then the wavelength is assigned using the FF or Random wavelength assignment, and finally there is the QoT verification step. In [59] the proposed IA-RWA algorithm is simply the SP combined with FF wavelength assignment, then the quality of the candidate lightpath in term of BER is verified. The A* algorithm is utilized for the RWA problem in [47]; it computes *k*-SPs, and the path which satisfies a minimum Q value and having the smallest cost in terms of fiber and regenerator utilization is chosen. A variation of the A-3 approach is reported in [16]. In this work the FF wavelength assignment is used and then the SP for the selected wavelength is computed. Afterwards, the final phase verifies the level of OSNR and the pulse broadening. In [58], the RWA is solved using the SP algorithm for each available wavelength and then the lightpath (route and wavelength) with the highest Q value is selected. In [26], the lightpath selection is done using a predictive approach and considering the possible inaccuracy in the wavelength availability information; then a verification of the Maximum Transmission Distance (MTD) for the selected wavelength is performed. Finally in [17], the lightpath is computed in two phases. First, a lightpath computation step is performed (Best Path selects the SP among all the available wavelengths or FF selects the SP for the first available wavelength) and then a lightpath verification step takes place based on a BER threshold.

In case A-1, the physical impairments are only verified in the wavelength assignment process and the potential improvement by selecting other routes to get better QoT values is ignored. On the contrary, in case A-2, different routes are considered to meet the required QoT, while eventually the wavelength is assigned without considering the impact of physical impairments. The last case (A-3) does not verify the impact of physical impairments neither in routing nor in the wavelength assignment processes. The QoT is verified only after finding a candidate lightpath and if the solution is not satisfactory the whole process is repeated.

The approaches in group B (cases B-1, B-2, and B-3) generally address the RWA problem considering the physical layer information: in case B-1 the route is computed using the PLI constraints; in case B-2 these constraints are considered in the wavelength assignment step; finally, in case B-3 the PLI constraints are considered both in the route and wavelength selection steps. Some of the works in this category assign the weight of the links (i.e., link costs) based on the physical layer information.

Case B-1 is adopted and reported in [18, 38, 42]. In [18] the minimum cost path is computed, in which the cost is defined as the total lightpath length and the number of consecutive transparent nodes. In [38] and [42] a variation of the B-1 approach is presented, in which the wavelength is initially assigned using the FF algorithm and then the route is computed. In [42] the selected route is the one with the minimum noise figure for the candidate wavelength and in [38] the route is computed using a link cost which is associated with different PLI constraints. The Minimum Crosstalk (MC) wavelength assignment in [48] is in line with the B-2 approach. It selects the wavelength with the minimum crosstalk intensity.

In [19, 35, 54, 55], several B-3 variations are reported, where the PLI constraints are taken into account in both route and wavelength selection steps. The DOIC algorithm presented in [55] considers the PLI constraints related to CD and OSNR in the route decision process, and also the residual dispersion range in the wavelength assignment. In [35] the shortest cost path or the k link-disjoint shortest cost paths are computed. The link cost in this work is the Q value, which is obtained from real-time Q measurements. Depending on the network conditions, the final decision is taken according to the Q-factor value or the wavelength balancing efficiency. The Best OSNR RWA algorithm is proposed in [19]. It selects the lightpath with the maximum OSNR. In [54] two IA-RWA algorithms are proposed, the HQ (Highest Q) and the MmQ (Maximum minimum Q) algorithms. The HQ selects the lightpath with the highest Q value and MmQ selects the lightpath that introduces the minimum impact on the already established lightpaths. In general, different variations of case B (i.e., B-1, B-2, and B-3) consider the impact of physical impairments in the routing and/or wavelength assignment step, however these schemes do not try to verify the QoT of the solution or find the optimum one [54].

Case C combines the two previous approaches. The PLI constraints are considered in the routing (case C-1), or in the wavelength assignment (case C-2) or in both steps (case C-3). There is also a final phase of verification of the PLI constraints that enables the re-attempt process in the lightpath selection phase.

Examples of the C-1 approach are found in [20, 24, 28–30, 36, 37]. Two algorithms proposed in [28] compute *k*-shortest paths ($k = 3$) using the worst case of Q values as the link costs. Then, the Q value of all lightpaths are verified at the destination. The values are compared with a threshold. The Q threshold is an offline value in the case of static routing, or a dynamic Q value, based on current network state, for dynamic routing. Another proposed algorithm in [29] selects the route between source and destination in a hop by hop manner. In each node the QoT of the lightpath is verified. Furthermore, some PLI constraints are also checked at the destination node. In [36] *k*-shortest paths are computed considering the network and physical characteristics such as the link costs; finally there is a validation of the PLI constraints considering the Q-factor, among the previously computed paths. A similar approach is adopted in [30], where alternate paths are selected by pruning those links that are not fulfilling the dispersion and ASE requirements in the topology. The lightpath selected is the one with the least number of links. Moreover, other PLI constraints are checked at the destination. In [37], the Q-factor penalty is used as the link cost in the network in order to compute the *k*-shortest paths. Finally, the BER is computed and verified at the destination; the selected path is the one with the lowest BER value considering a BER threshold. The proposed IA-RWA algorithms in [20] utilize link costs related to the Q-factor (as $1/Q$ or $1/Q^2$) for computing the SP; then a PLI constraints verification phase considers the availability of the lightpath or otherwise the maximum reachable node. In [24] the authors propose a variation of case C-1, in which the FF wavelength assignment (unaware of PLI constraints) is initially adopted, and then the SP for that wavelength is computed considering the noise variance of the PLI as the link cost. Finally, lightpaths which cause a BER value higher than a given threshold for the new lightpath or for other already established lightpaths are eliminated.

Authors in [21] propose the C-2 approach, where the SP is computed without considering the PLI constraints and the wavelength assignment uses a wavelength order, taking into account the PLI constraints. Finally, there is a QoT verification step for the lightpaths in terms of minimum Q value. The proposal in [22] is very similar to the aforementioned approach, however, a BER threshold is considered at the destination node to validate the QoT of the lightpath.

An instance of case C-3, where the PLI constraints are considered in both routing and wavelength assignment steps, is reported in [31]. In this work, *k*-shortest routes are computed considering a Q-penalty value as the link cost, and the selected wavelength is the one that maximizes the Q value. Finally, the QoT is verified for the chosen lightpaths.

Different categories in case C (i.e., C-1, C-2, and C-3) not only try to consider the impact of physical impairments, but also verify the QoT of the solutions and also explore the solution space in order to find the optimum one. Obviously, the cost of this scheme is its running time complexity.

Finally, specific wavelength (WA) assignment algorithms considering the PLI constraints are also reported in few works (e.g., [21, 31, 48, 55]). The reader should also note that different combination of the subcases presented in Fig. 12.3 may also be found in the literature.

The classical performance evaluation of the proposed IA-RWA algorithms in the literature has been to calculate the percentage of blocked connections versus the traffic load for dynamic (and static) traffic and to calculate the amount of necessary resources (fibers, wavelengths, regenerators, etc.) to serve the whole demand set for the static case. In a dynamic scenario, the network is already planned and the goal is to route the maximum number of connections. Thus, the percentage of blocked connections (or blocking rate) is used to evaluate the performance of different IA-RWA algorithms. On the other hand, for the static case (network planning), two scenarios can be identified: (1) If the resources of the network are fixed, i.e., number of fibers, wavelengths, regenerators, etc., then the objective is to route the maximum number of connection requests and (2) if the IA-RWA algorithms with static traffic are utilized in the planning phase of the network, then the objective will be to optimize the performance of the network (e.g., minimize the number of required wavelengths, minimize the number of regenerators in case of translucent networks, etc.). Moreover, authors report results of their proposed algorithms in terms of running time.

12.4 Wavelength Conversion

As previously mentioned, the wavelength-continuity constraint is imposed on the RWA problem in addition to the physical layer impairments in transparent optical networks. This constraint implies that for a given lightpath the same wavelength should be used on all links (along the lightpath) from source to destination. This requirement may affect both the network performance and the complexity of the RWA algorithms since the setup of a new lightpath is conditioned on the availability of the same wavelength in a number of links. The wavelength-continuity constraint can be relaxed in the nodes that are equipped to perform wavelength conversion, thus improving the connection blocking probability. In practice, wavelength conversion can be realized in switching nodes either by means of a dedicated all-optical device or with the assistance of an OEO conversion utilizing an opaque node. When an OEO conversion takes place an input wavelength is converted into an electronic signal and then is converted back into the optical domain on another wavelength. Since all-optical wavelength converters are still immature technology and very expensive, the OEO wavelength conversion becomes a viable alternative. In addition to potential wavelength conversion and increasing the optical reach of the signal, OEO can provide other functions as well. For instance, in a network with sub-rate traffic, it is essential to bundle multiple connections together to better utilize the capacity of a wavelength. This bundling process is most effective when the traffic can be groomed at various nodes in the network. The grooming process is typically performed in the electrical domain. The impact of physical impairments and features of the electrical layer on constrained routing is investigated in [20].

Most of the IA-RWA algorithms do not consider the wavelength conversion capability. Some of the algorithms that consider wavelength conversion and sparse regenerators in translucent networks are reported in [9, 30, 38, 47]. The term *sparse* in this context refers to the wavelength conversion, which is only available in selected nodes in the network. In the case that the node provides wavelength converters shared between different input and output ports the number of the simultaneous conversions may be restricted. The usual assumption is that full wavelength conversion, i.e., from any input to any output wavelength is available. On the contrary, wavelength converters with a limited conversion range allow an incoming wavelength to be switched only to a small subset of outgoing wavelengths [60]. The problem of the limited conversion with consideration for PLIs has not been addressed extensively in the literature.

An interesting problem arises in the translucent network scenario and it concerns the optimization of placement of wavelength conversion-capable regenerators. Here the objective is to minimize the connection blocking probability resulting from both physical and network layer constraints. Such a problem was addressed in [52] and [61] by using some traffic-prediction-based heuristics. The regenerator placement is also addressed in [9, 13, 28, 47]. The allocation of regenerators is proposed in [18, 30, 40]. Specifically, in [30] and [40], algorithms are presented that minimize the number of used regenerators along the selected lightpath.

12.5 Resilience and Protection

In a transparent (and to some extents translucent) optical network, the impact of a failure propagates (transparently) through the network and therefore failure localization is very challenging. The large amount of information transported in optical networks makes rapid fault localization and isolation a crucial requirement for providing guaranteed quality of service and bounded unavailability times. The identification and isolation of failures in transparent optical networks is complex due to three factors: (1) fault propagation, (2) lack of control (electronic) information, and (3) large processing effort. Some of the main fault localization challenges in transparent optical networks include the selection of performance parameters to cover the full range of faults while ensuring cost effectiveness and preserving transparency, the placement of monitoring equipment to reduce the number of redundant alarms and to lower capital expenses, and the design of fast localization algorithms.

There are very few works in the context of IA-RWA which have addressed the issue of resilience and protection. Authors in [40] propose an approach, which in addition to PLIs and traffic condition, also considers the path reliability in the framework of a constraint-based path selection algorithm. In [58] the effect of PLIs on dedicated path protection schemes is analyzed, investigating the performance of dark and lit backup (protection) paths. It is concluded that lit backup path scenarios introduce significant penalties in terms of blocking probability and vulnerability to failures. In the context of all-optical networks with various path protection schemes, the authors of [48] have proposed algorithms that exhibit low blocking probability without high computational complexity. The authors conclude that considering the blocking rate and required processing time, their dark backup path algorithm performs better than the case of lit backup paths; however, lit backup path eliminates the need for signaling and facilitates faster network recovery. The authors in [47] have proposed Suurballe's disjoint-path algorithm in the layered network graph, in order to find the shortest cycle passing through source and destination using disjoint nodes. This cycle is then divided into primary and protection paths. If no disjoint paths exist, the cycle having the minimum number of common vertices is selected. The work in [38] addresses the issue of survivability in optical mesh networks considering optical layer protection and realistic optical signal quality constraints. Three variations of resource sharing scenarios, including wavelength-link sharing, regenerator (i.e., OEO) sharing between protection lightpaths, and regenerator sharing between working and protections paths are investigated. In addition to an ILP-based solution, the authors have also proposed a local optimization heuristic approach and a tabu search heuristic to solve this problem.

12.6 Impairment-Aware Control Plane Extensions

A control plane (CP) is recognized as a necessary requirement for fast and flexible resource provisioning, easy network operation, and high reliability and scalability of optical networks. The standardization process for such a control plane is currently being done independently by two different bodies: the Automatically Switched Optical Network (ASON) concept [62] developed by the International Telecommunications Union (ITU) and the Generalized Multi Protocol Label Switching (GMPLS) suite of protocols [63] developed by the Internet Engineering Task Force (IETF).

The main benefit of the ASON approach is the definition of the architecture, the requirements, and the functionalities of the control plane independent of a particular choice of control protocol. Therefore, a variety of such protocols can be used ranging from the Asynchronous Transfer Mode (ATM) family to Multiprotocol Label Switching (MPLS) and Generalized-MPLS (GMPLS). On the other hand, GMPLS focuses on the implementation of the control plane, involving routing (Open Shortest Path First with Traffic Engineering extensions (OSPF-TE)), signaling (Resource Reservation Protocol with Traffic Engineering extensions (RSVP-TE)), and resource management (Link Management Protocol (LMP)) protocols.

The integration of IA-RWA algorithms in control planes [64–66] follows two different approaches: (1) Centralized and (2) Distributed. In the centralized approach, a single element stores the complete information about network topology, resource availability, and PLI parameters in a central repository. This element is therefore in charge of collecting and updating all these information and also responsible for computing the optimal routes satisfying the specific set of lightpath requirements such as QoT, resilience, etc. The central element could be either the Network Management System (NMS) or a Path Computation Element (PCE) [65].

In the distributed approach, each node is responsible to compute, set up, and maintain lightpaths using a common and distributed control plane (e.g., GMPLS) [64]. The nodes can collect the information about the status of the resource availability using a routing protocol, execute an RWA solver, and establish the lightpaths utilizing a signaling protocol. To include the PLI constraints in the RWA problem, some extensions are essential to be added to the current signaling and/or routing protocols.

Although both the GMPLS and ASON approaches for CP in optical networks are relatively mature and key standards are in place, they do not include any information related to the PLIs and therefore are unaware of the quality of optical signals. Some recent works deal with the problem of encompassing the PLI constraints into the GMPLS CP functionalities. Three different models have been proposed: (1) the Path Computation Element (PCE) model, (2) the Signaling model, and (3) the Routing model (see Fig. 12.4).

12.6.1 PCE-Based Approach

The PCE-based centralized approach (Fig. 12.4a) was first proposed in [39] and is based on the computation of the PLI-RWA lightpaths in a centralized way utilizing a PCE element. PCE is defined in [67] as an entity that is capable of computing a lightpath (route and wavelength) considering PLIs and/or other computational constraints. It aims at performing complex centralized route computation on behalf of the control module of the nodes.

The PCE stores two types of databases in this integration framework. The first is the Traffic Engineering Database (TED), located at each node and also includes a centralized TED, which is updated from the localized TEDs through a standard, distributed routing protocol (e.g., OSPF-TE). The second is the Physical Parameter Database (PPD) obtained by the Network Management System (NMS) or through a performance monitoring system. The PPD maintains up-to-date information about any possible PLI concerning any network link. Whenever a new connection request arrives at a node, it sends a query to the PCE. The PCE computes the

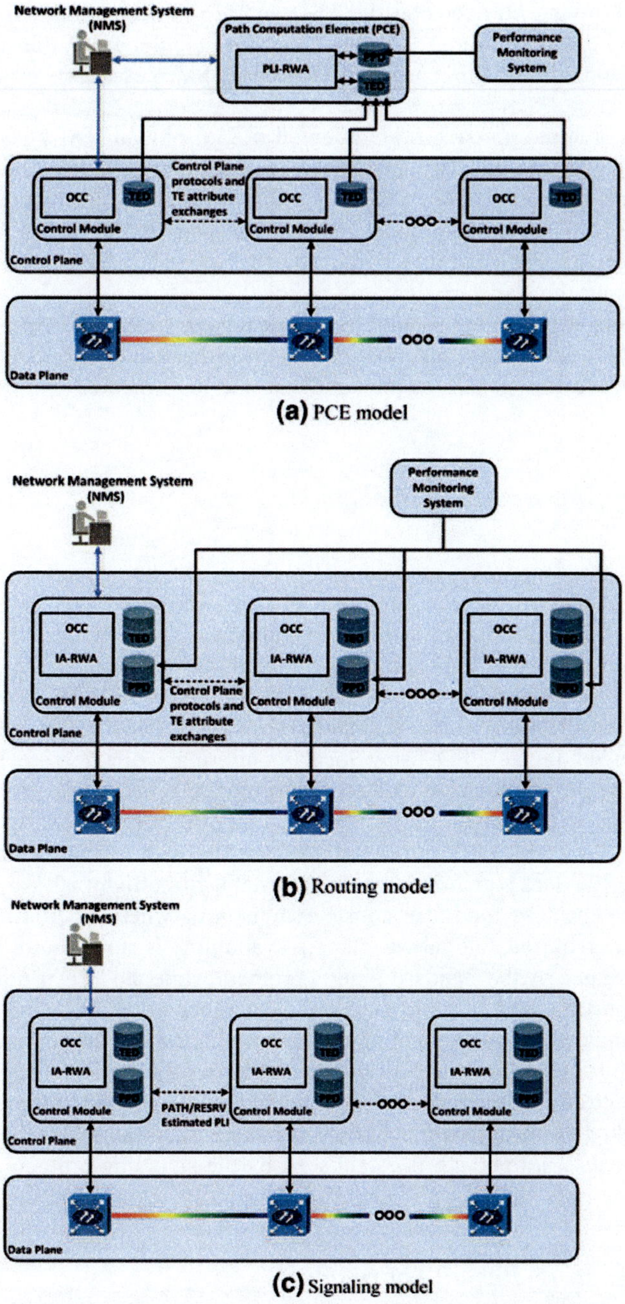

Fig. 12.4 Physical layer impairment-aware control plane extensions in integration schemes [optical cross-connect controller (OCC)] (From [14], Figure 5. Reproduced by permission of © 2009 Elsevier)

required path taking into account the TED and PPD information and sends back the computed explicit route to the source control module. Then, the source node, using the standard signaling protocol (PATH/RESV messages), establishes the lightpath.

Slightly different approaches are adopted in [56, 68]. In [68] PCE interworks only with the NMS. In this case, the NMS is in charge of collecting the connection requests, sending them to the PCE along with all required information (e.g., PLI performance, topology, and logical link status), and finding the lightpath. The NMS notifies the nodes of the computed routes, and the nodes establish the lightpaths using the standard signaling protocol.

The authors in [56] also propose an Optical Performance Monitoring (OPM) manager, which is directly integrated in the NMS instead of a separate PCE, but the behavior and functionalities are similar to the PCE model.

12.6.2 Routing-Based Approach

The routing-based approach, as shown in Fig. 12.4b was reported in [62] and consists of extending the routing protocol (e.g., the OSPF-TE in GMPLS) to include the PLI constraints into the RWA problem. Each node is in charge of storing updated TED and PPD databases on the resource utilization and on the PLI performance concerning any link in the network [63]. As in the case of the TE attributes, local PLI (i.e., PLI performance of local node and of the attached links) can be included in the PPD using local monitoring while remote PLI can be obtained by exploiting the extended routing protocol. Whenever a new connection request arrives at a node, an online IA-RWA algorithm computes the route considering the TED and PPD information. Once the path is computed, the node activates the standard signaling protocol to establish the lightpath.

A different routing model is considered in [44], where no routing protocol extensions are required. Global wavelength availability is stored in the TED of the nodes and updated by the standard routing protocol while the PPD is not necessary. Only PLI constraints with static nature (i.e., number of active wavelengths) are considered and their mathematical models are preloaded into the control module of the nodes. Any incoming connection request triggers the IA-RWA algorithm that computes a set of candidate routes according to the TED information and checks their feasibility by means of mathematical models. If at least one feasible computed route exists, a lightpath is established by the signaling protocol.

12.6.3 Signaling-Based Approach

The signaling-based approach (Fig. 12.4c) has been proposed in [32] and consists of extending the signaling protocol (e.g., RSVP-TE in GMPLS) to encompass the

PLI constraints. In this context no modifications are introduced in the routing protocol [44, 69]. Whenever a connection request arrives at a node, the route is computed according to the TED information and a setup request message is launched in the network. This message collects estimated PLI performance information of any traversed link between the source and destination nodes. In fact, each node must store updated local PPD and mathematical models to calculate the PLI performance. If the accumulated PLI performance on the receiver interface at the destination node is compliant with an acceptable signal quality, a positive response message is sent back to the source node and the lightpath is established. If the accumulated PLI performance is not satisfactory, an error message is sent back to the source node and another attempt can be triggered following a different route.

Some improvements to this approach are described in [29] and [70] to speed up the lightpath establishment process. In [29], four different approaches are proposed and compared: (1) in K-seq, the source node computes k different routes and sequentially attempts to establish a feasible lightpath; (2) in K-par, the source node sends k setup messages simultaneously and the destination node can choose one according to some criteria; (3) in HbH, the route is computed hop-by-hop, which means that each node considers only the information about the adjacent links; (4) in FF, the setup message is flooded to the entire network. The Lightpath Provisioning with Signaling Feedback (LPSF) concept is defined in [70]. It exploits the error message delivered to the source node adding some feedback information about the PLI performance of the rejected route. The source node can therefore store this information in the local PPD and compute additional feasible routes.

12.6.4 Comparison of Control Schemes

The centralized approach (e.g., PCE) is able to provide optimal path computation in terms of both network utilization and optical signal quality. It also does not require any modification or extension to the current signaling and routing protocols. Also, PCE has a global view of the network and, if there is no inconsistency in the databases, the setup procedure does not require any re-attempt, which can speed up the service provisioning. Nonetheless, it suffers from scalability problems and in case of failure fast restoration cannot be achieved. There are also some advantages when considering the PCE model for multi-domain scenarios, where an abstraction of the entire routing area may optimize the inter-domain paths [39, 63].

Maintaining accurate routing information about all network nodes under dynamic traffic is extremely difficult. Therefore, the routing model is a less advantageous solution, since in addition to the TED information it requires the global dissemination of the PLI data. It may have some benefits in case of static traffic, or less volatile traffic conditions, but in such a case the PCE model may outperform the routing one.

The signaling model seems to be the easiest and fastest way to encompass the PLI performance into the RWA problem. On the other hand, providing optimal

Table 12.1 Requirements, advantages and disadvantages of the control models (From [14], Table 8. Reproduced by permission of © 2009 Elsevier)

Model	Approach	Requirements	Pros	Cons
PCE	Centralized	PCE element with high reliability; Global TED and PLID databases	Global network view; Optimal path computation, signal quality and network resource utilization; No changes in control protocols; Useful for multi-domain scenario	Low flexibility and scalability; Vulnerability to database failure; Slow recovery; Depends on OPM; Intensive computation
Routing	Distributed	Global PLID database; Some extensions to disseminate efficiently the PLI performance	Distributed approach; Optimal path computation; Fast setup delay	Slow convergence; Intensive computation; Depends on OPM
Signaling	Distributed	Local PLID database; Mathematical models for PLI estimation; Some extensions to the signaling protocol	Distributed approach; Minor changes in signaling protocol; No dissemination overhead	High setup delay; Signaling overhead; No optimal resource utilization; No optimal signal quality

resource utilization and signal quality is not feasible in this approach. It may introduce high setup delay due to the re-attempts of failed lightpath establishment processes (Table 12.1).

12.7 Impairment-Aware Multicasting

Advances in optical WDM networking have made bandwidth-intensive multicast applications such as interactive distance learning, video-conferencing, distributed games, movie broadcasts from studios, etc., widely popular. Multicasting has been investigated in the research community since the early days of optical networking, but has only recently received considerable attention from the service providers, mainly because now many emerging applications can potentially utilize the optical multicasting feature. Wavelength-routed networks are now widely deployed in order to accommodate the bandwidth requirements of next-generation networks and they provide the flexibility to serve multicast or groupcast connections.

In optical networks, a light-tree that spans the source and the destination set is used to serve the multicast request. Figure 12.5 shows an example of a multicast tree that connects the source node *s* with the two destinations d_1 and d_2. In transparent optical networks, optical splitters can be used to split the incoming

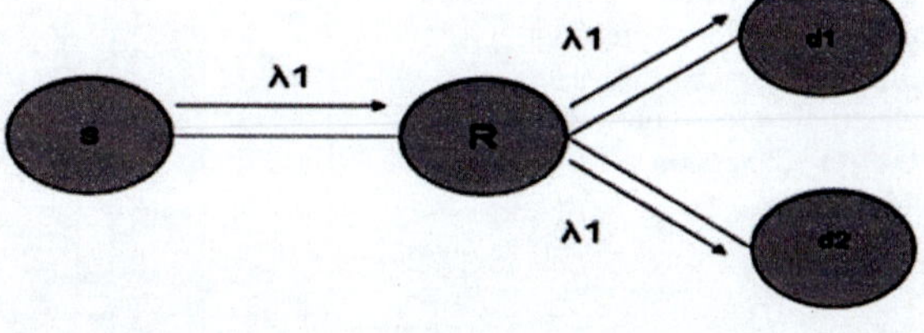

Fig. 12.5 Example of a multicast tree

signal into multiple output ports thus enabling a source node to establish connections with multiple destinations. In Fig. 12.5, the signal on the incoming wavelength λ_1 splits into the two output ports of the branching node *R*. In these networks, optical amplifiers are required to overcome the splitting losses as well as the attenuation losses. However, these amplifiers introduce Amplifier Spontaneous Emission (ASE) noise to the optical signal. Therefore, the proper engineering design of the multicast-capable WDM network is also important as it is affected by the physical impairments.

Constructing a cost-effective light-tree along with the wavelength assignment problem for these light-trees is proven to be an NP-complete problem [71]. Several ILP formulations have been proposed for designing optimum light-trees [72, 73] and several heuristics have been proposed for solving the multicast routing and wavelength assignment (MC-RWA) problem [74, 75]. However, algorithms that find the minimum cost multicast tree and assign a wavelength to the light-tree, are not enough to guarantee the viability of the multicast session, as it may violate the physical layer constraints. Recent work on the problem of routing and wavelength assignment (RWA) for provisioning multicast connections in transparent optical networks has included the effect of physical impairments to investigate whether a multicast connection should be admitted to the network. Initially, some works only considered the power loss due to the passive splitters as a constraint [43, 76–78], and more recently some works have investigated impairment-aware multicast RWA in a more general sense.

In Sect. 12.7.1 different multicast-capable architectures and engineering designs proposed in the literature are presented while Sect. 12.7.2 describes the MC-RWA problem and the different multicast routing algorithms proposed in the literature.

12.7.1 Multicast-Capable Architectures

To support multicasting in WDM optical networks, multicast-capable cross-connects (MC-OXCs) are needed at the network nodes, which can replicate the signal from one input port to multiple output ports. Two approaches exist to design nodes capable of supporting multicast traffic. In Fig. 12.6 the opaque approach is shown, where the optical signal is converted into the electronic domain, switched using an electronic cross-connect, duplicated to all the desired outputs, and then converted

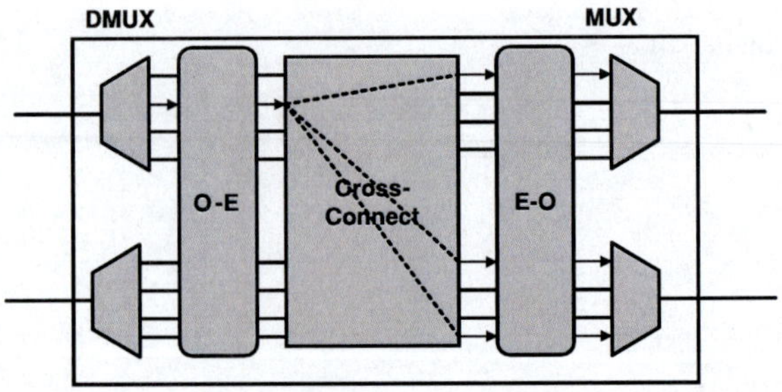

Fig. 12.6 Switch architecture for supporting multicast connections using electronic cross-connects and O-E-O conversion

back into the optical domain. The opaque approach is currently implemented widely due to the existence of mature technology to design high-bandwidth multi-channel *(N x N)* non-blocking electronic cross-connect fabrics. In the transparent approach, multicast connections can be supported more efficiently in the optical domain by utilizing optical splitters to split an incoming signal into two or more outputs rather than duplicating the data in the electronic domain.

Figure 12.7 shows a MC-OXC with splitting capabilities. The splitter equally splits the optical signal into *n* parts, where *n* is the number of outgoing (as well as incoming) ports. Then an optical switch routes the replicated optical signals to the destined output ports. Post-amplifiers are also required to boost the output signal power at the output of the node. This power consideration affects the design of multicast wavelength-routed networks [79–81], as well as the multicast routing algorithms [76, 79, 82]. The splitting capability of the MC-OXC is an important factor in supporting multicast connections in wavelength-routed WDM networks and a number of issues related to it are summarized in [83].

Also, as it is predicted that the cost associated with OXCs and MC-OXCs devices will still be expensive in the near future, in the design of an optical network, the number of nodes equipped with MC-OXCs must be taken into account. In [84–86] it is shown that an optical network with sparsely placed multicast capable nodes can achieve performance close to that of a network with all multicast-capable nodes and their optimal sparse placement allocation was studied in [86–88]. However, in networks where only some nodes are equipped with optical splitters, the typical multicast routing algorithms cannot be applied directly, as the physical network must be taken into consideration during the construction of the light-trees. Work in [82, 85, 89–98] is focused on finding multicast routing algorithms in networks with sparse splitters.

The wavelength conversion capability is also important in WDM optical networks as previously discussed in Sect. 12.4 for point-to-point connections. Similarly, for multicast connections, wavelength converters may also be required when the wavelength continuity constraint cannot be satisfied across each light-tree, to convert the optical signal from one wavelength into another. As all-optical wavelength converters are still expensive and difficult to implement when compared to other elements in WDM networks, their optimal sparse placement is studied in [99–101], and the MC-RWA problem that minimizes the amount of

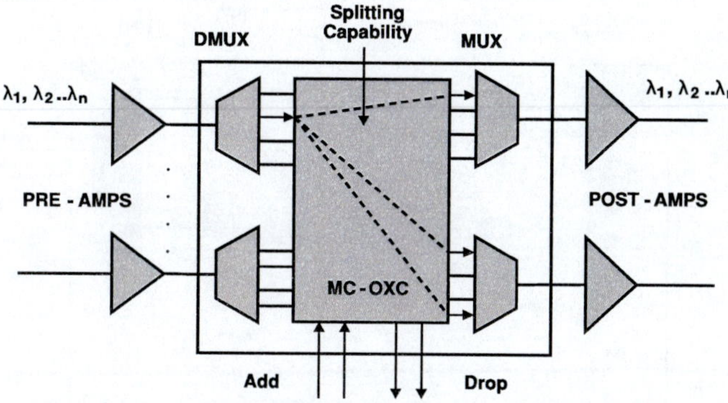

Fig. 12.7 Multicast-capable optical cross-connect (MC-OXC) node architecture

wavelength converters required to support the multicast traffic is investigated in [102–105].

Apart from the power constraints that optical splitters introduce, the physical layer constraints such as amplified spontaneous emission (ASE) noise introduced from the optical amplifiers, crosstalk, fiber nonlinearities, distortion due to optical filter concatenation, and polarization mode dispersion (PMD) amongst others must be taken into account in the design and engineering of transparent optical networks. Recent work on the problem of routing and wavelength assignment (RWA) for provisioning multicast connections in transparent optical networks has included physical impairments on the provisioning of multicast connections [79, 106, 107]. In these works, different node and engineering designs are proposed and the physical performance of the network is examined using a Q-budgeting approach described in detail in [79]. The effects of different node designs and proper system engineering designs on the performance of several multicast algorithms provide a good insight into the interaction of physical and network layers in a multicast-capable optical network.

Figure 12.8 illustrates a generic node architecture consisting of pre- and post-amplifiers (EDFAs), MUX/DMUXs, passive optical splitters, semiconductor optical amplifiers (SOAs), variable optical attenuators (VOAs), and optical switches. The gain of each post-amplifier compensates for the component losses in the node (Table 12.2 shows some best-case indicative numbers forcomponent losses) while the gain of each pre-amplifier compensates the loss of each preceding fiber span. Each node's EDFA is assigned a typical noise figure which depends on its gain (some typical values are shown in Table 12.3). VOAs are used to attenuate the input power to the post-amplifiers and in an actual system they would be used to control the effects of Polarization Depended Gain/Loss (PDG/PDL). SOAs are used as gates to block the power at outputs where the signal is not destined for and all gates are controlled together in an intelligent manner to avoid clashing at the same output port and/or same wavelength of the switch. A node architecture

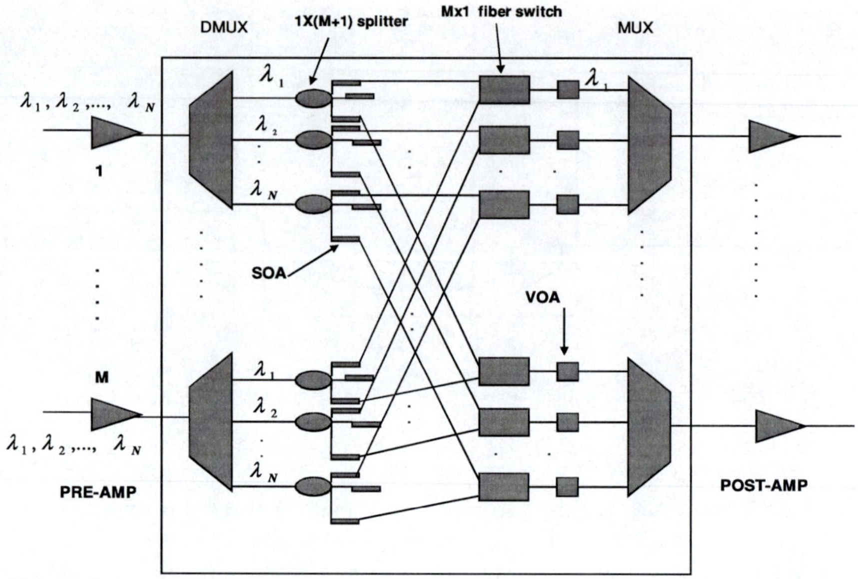

Fig. 12.8 Generic node architecture with passive splitters (From [107], Figure 1. Reproduced by permission of © 2010 The Institute of Electrical and Electronics Engineers.)

Table 12.2 Component losses in the node

Component	Losses in dB
MUX/DMUX	3
Splitter	10*log(Fanout)
Gate	0.6
VOA	0.5
Switch	1

Table 12.3 Noise figure of EDFAs

Gain in dB	NF
$G < 13$	7
$13 < G < 15$	6.7
$15 < G < 17$	6.5
$17 < G < 20$	6
$G > 20$	5.5

utilizing active optical splitters is also studied in [107] and for this design no gates are required, as the active splitters are assumed to split the power only as many times as is needed for the signal to be forwarded to the destined outputs. Performance results for these architectures showed that there is no particular advantage of using active instead of passive splitters as VOAs are used in both cases to attenuate the total power to a predetermined value that is calculated based on the worst-case scenario, defined in [79, 106, 107] as the maximum insertion loss the signal encounters passing through the maximum degree node.

12.7.2 Multicast Routing and Wavelength Assignment

In general, a multicast request is accepted into the network if a light-tree that spans the source and the destination set can be found, and if a wavelength assignment is possible for the entire light-tree. This is the well-known multicast routing and wavelength assignment problem (MC-RWA) that is proven to be NP-complete [71]. In its static case, where a set of multicast requests is given a priori, the MC-RWA problem is usually decomposed into two subproblems namely the routing problem and the wavelength assignment problem, and each problem is solved separately. First, a multicast routing algorithm is used to compute the routes of the given multicast requests, and then a wavelength assignment strategy is followed. The usual objectives of the MC-RWA problem in static cases are to maximize the multicast requests accepted in the system, given a number of available wavelengths, or to minimize the total number of wavelengths utilized while accommodating all the multicastrequests. In dynamic systems, the usual objective is to maximize the number of requests accepted into the network, given a fixed number of wavelengths, and therefore to minimize the blocking probability.

The multicast routing problem corresponds to finding the minimum cost tree or equivalently to finding the Steiner tree [71] which is a set of lightpaths connecting the source node with the destination nodes. The Steiner tree problem is proven to be NP-complete and therefore several heuristics exist that compute the multicast tree [108–110]. An alternate, simpler approach of finding the multicast tree is the shortest paths approach. A multicast tree consisting of shortest paths is found by computing all shortest paths from the source to every individual destination using Dijkstra's or Bellman Ford's algorithms [111], and then all unicast shortest paths are merged together to build the multicast tree. The Shortest Paths Tree (SPT) algorithm as described in [110] and compared to the Steiner tree heuristic, is inefficient in terms of recourses, as it requires a larger number of links to build the multicast tree. An analytical study of the tradeoffs between shortest path-based trees and Steiner-based trees can be found in [112]. An Optimized Shortest Paths Tree (OSPT) was proposed in [113] aiming at reducing the cost of the SPT tree. The difference between SPT and OSPT is that in the OSPT algorithm, each time a shortest path is found, the link costs on the network are updated to zero if they are used for the construction of that path. In that way, the same links can be reused for the construction of the next shortest path reducing the cost of the entire tree. Subsequently, when the multicast tree is calculated, a number of heuristics can be utilized for wavelength assignment such as *first fit, least used, most used,* and *random* as described in [114, 115].

12.7.2.1 Impairment-Aware Multicast Routing and Wavelength Assignment Algorithms

Most of the work related to the dynamic MC-RWA problem in all-optical networks is mainly focused on maximizing the multicast requests accepted in the network, by taking into account only the wavelength-continuity constraint. However, the constraints in the MC-RWA problem for all-optical networks include apart from the wavelength-continuity constraint, the power constraint introduced by the splitting losses and attenuation, and the physical constraints introduced by the physical layer transmission effects, such as amplified spontaneous emission (ASE) noise, PMD, crosstalk, fiber nonlinearities, distortion due to optical filter concatenation, and polarization-dependent gain/loss (PDG/PDL) amongst others.

Work in [76, 82, 116, 117] investigated the MC-RWA problem in WDM all-optical networks taking only power budget considerations into account. There, the objective of the MC-RWA problem was to minimize the session blocking probability and at the same time to ensure that the signal power at the destination nodes is detectable at the receivers (larger than the receiver sensitivity). The authors in [116, 117] formulated this problem as a Mixed-Integer Linear Program (MILP) optimization problem and also proposed a heuristic algorithm, called the *power aware multicasting* (PAM) algorithm, as the MILP was not scalable for large network sizes due to its high complexity, demonstrating that the PAM algorithm produces near-optimal solutions. A heuristic algorithm was also proposed in [82] starting by constructing the initial tree without taking into account the power considerations and then modifying the initial tree by replacing a set of adjacent splitters with a single splitting node, named *centralized splitting node*, in order to reduce the total amount of power loss. Another heuristic algorithm was proposed in [76], and is based on the *balanced light-tree* (BLT) approach. Initially, the BLT algorithm finds the multicast tree without taking into consideration the power budget constraints and then the balancing procedure takes place to minimize the power losses by decreasing the splitting losses of the light-tree (decreasing the split ratio of the maximum degree node after a delete/add operation). The balancing part is repeated until two successive iterations fail to reduce the maximum split ratio.

On the other hand, the authors in [79, 106, 107, 118] investigated the MC-RWA problem in metropolitan all-optical networks considering both the power budget constraints as well as other physical layer transmission effects (utilizing the Q-factor to account forthese effects [79]). The physical layer system modeling used for the calculation of the Q-factor is described in detail in [79] and the different node and engineering designs proposed and compared are described in [106, 107]. The objective of this impairment-aware MC-RWA problem is to minimize the session blocking probability and at the same time to ensure that the Q-factor at the destination nodes in every multicast tree accepted into the system is above a predetermined threshold. Two multicast routing algorithms that take the Q-factor into account during the construction of the light-trees are proposed in [79, 118] that aim at maximizing the multicast connections that can be admitted to the network. The *balanced light-tree-Q* (BLT-Q) algorithm is an extension of the BLT algorithm [76] that only took power budget constraints into consideration. Initially, the BLT-Q algorithm builds the light-tree without taking the Q-factor into consideration. Then the Q-factor for each destination node is computed and the balancing procedure begins. During the balancing procedure the algorithm identifies the minimum Q-factor node u and the maximum Q-factor node v, and attempts to delete node *u* from the tree, and add it back to the tree by connecting it to some node *y* in the path from the source node to node *v*. The balancing part is repeated until two successive iterations fail to increase the minimum Q-factor. The *BLT-*$Q_{tolerance}$ algorithm is a modification of the BLT-Q algorithm as the balancing procedure is performed only at those destination nodes where the Q-factor is below the predetermined Q-threshold. Thus, if after a number of iterations the minimum Q-value for all destination nodes is above the predetermined Q-threshold or if two successive iterations fail to increase the minimum Q-factor then the balancing algorithm terminates. BLT_Q aims at maximizing the average Q-factor of the light-tree, while BLT-$Q_{tolerance}$ aims at increasing the Q-factor only where it is needed in order to preserve the cost of the light-tree as low as possible. Performance results in [79, 106, 107, 118] showed that the existing multicast routing algorithms that take into consideration only the wavelength continuity and the power budget constraints, cannot be applied directly when the Q-factor is taken into account in the MC-RWA algorithm. Hence, multicast routing algorithms as those described in [79, 106, 107, 118] that take the physical impairments into consideration during the construction of the multicast tree must be considered.

12.8 Conclusions and Future Issues

Routing and wavelength assignment (RWA) is the core component in planning and operation of optical networks. Considering the physical layer impairments in the RWA decisions adds a new dimension to this problem. Heuristics, metaheuristics, and optimization techniques are proposed as algorithmic approaches to solve the IA-RWA problem. The general approach to address the IA-RWA problem can be divided into two main categories. The first trend utilizes traditional RWA algorithms and after selecting the lightpath the physical constraints are verified; in this approach the IA-RWA algorithm is not deliberately designed for routing with PLI constraints. The second approach is to use some metrics, which are related to the PLI constraints, as cost of the links in order to compute the shortest path(s). In general, QoT support in physical impairment-aware optical networks can have two meanings. It can refer to the predefined level of signal quality, as measured at the destination receiver, which allows for flawless network operation or it can refer to the support of differentiated lightpath requests according to a set of routing constraints (e.g., max. transmission quality degradation, max. delay, or reliability) [35, 40, 42, 46].

Assuming good algorithms are in place, a small number of wavelength conversions (either via OEO or all-optical conversion) is needed to approximate the performance of opaque network architectures. Clearly, regenerator and/or monitoring equipment placement are important factors in the design of the network. By using a proper regenerator or monitoring equipment placement strategy in some nodes of the network, it is possible to obtain performance (in terms of blocking probability) similar to that of an opaque network,with much lower cost.

There are also some proposals that address the resilience and protection issues in the IA-RWA algorithms. In addition to simulation studies, experimental and some analytical models are available to evaluate the performance of the proposed algorithms. None of the surveyed works considers the inaccuracy of the physical impairment information (analytically computed or measured) into their IA-RWA algorithms. The proposed adaptive IA-RWA algorithms simply change their decisions assuming that the physical information is completely accurate.

The physical impairments and IA-RWA algorithms can be incorporated in the control plane using one of the PCE-, routing-, or signaling-based models. The PCE model is able to provide optimal path computation in terms of both network utilization and optical signal quality. It also does not require any modification or extension of the current signaling and routing protocols. PCE has a global view of the network, which can speed up the service provisioning. Nonetheless, it suffers from scalability problems. The routing model seems to be a less advantageous solution, since in addition to the TED information it requires the global dissemination of the PLI performance data. The signaling model on the other hand seems to be the easiest and fastest way to encompass the PLI performance into the RWA problem. However, it is not able to provide optimal resource utilization and signal quality. It may also require high set-up delay due to the re-attempts of failed lightpath establishment processes and obtain possible sub-optimal route decisions due to impairment-unaware route computation.

Finally, impairment-aware MC-RWA algorithms are becoming critical in the provisioning of multicast services in optical networks by including not only power budget constraints when calculating the required light-tree, but also additional physical layer constraints. It is shown that different multicast-capable nodes, different network engineering designs, as well as different IA-MC-RWA algorithms are crucial in the design of such networks that are capable of supporting future multicast applications.

Although the IA-RWA algorithms presented in the literature focus mainly on the optical circuit-switching (or wavelength-routed) networks, there are still issues specific to the optical burst/packet switching networks (OBS/OPS) that have to be addressed. To support short connection holding times it was proposed to incorporate the information about physical impairments into a set-up control message, by means of a data vector that is processed at consecutive nodes [119]. Another problem concerns OBS networks, where the transmission offset time may be affected by optical signal dispersion effects [120]. Clearly, although the IA-RWA problem has been recently addressed by a number of researchers, there exist a number of open issues still to be considered.

References

1. Berthold J, Saleh AAM, Blair L, Simmons JM (2008) Optical networking: past, present, and future. IEEE/OSA J Lightwave Tech 26(9):1104–1118
2. Simmons JM (2005) On determining the optimal optical reach for a long-haul network. IEEE/OSA J Lightwave Tech 23(3):1039–1048

3. Sygletos S, Tomkos I, Leuthold J (2008) Technological challenges on the road toward transparent networking. OSA J Opt Netw 7(4):321–350
4. Striegler A, Meissner M, Cvecek K, Sponsel K, Leuchs G, Schmauss B (2005) NOLM-based RZ-DPSK signal regeneration. IEEE Photonics Tech Lett 17(3):639–641
5. Huang YK, Glesk LX, Baby I, Bin Li V, Prucnal PR (2005) Simultaneous all-optical 3R regeneration of multiple WDM channels. In: Proceedings of IEEE/LEOS annual meeting, paper MM2, Sydney, Australia
6. Shen G, Tucker RS (2007) Translucent optical networks: the way forward. IEEE Commun Mag 45(2):48–54
7. Saleh AAM (2003) Defining all-optical networking and assessing its benefits in metro, regional and backbone networks. In: Proceedings of IEEE/OSA optical fiber communication conference (OFC), paper WQ1, Atlanta, GA
8. Chiu AL, Guangzhi L, Dah-Min H (2006) New problems on wavelength assignment in ULH networks. In: Proceedings of IEEE/OSA optical fiber communication conference (OFC/NFOEC), paper NThH2, Anaheim, CA
9. Ali Ezzahdi M, Al Zahr S, Koubaa M, Puech N, Gagnaire M (2006) LERP: a quality of transmission dependent heuristic for routing and wavelength assignment in hybrid WDM networks. In: Proceedings on international conference computer communication and networks (ICCCN), Arlington, VA, pp 125–136
10. Zang H, Jue JP, Mukherjee B (2000) A review of routing and wavelength assignment approaches for wavelength-routed optical WDM networks. SPIE Opt Netw Mag 1(1): 47–59
11. Ramaswami R, Sivarajan KN (1995) Routing and wavelength assignment in all-optical networks. IEEE/ACM Trans Netw 5(3):489–500
12. Rouskas GN (2001) Routing and wavelength assignment in optical WDM networks. In: Proakis J (ed) Wiley encyclopedia of telecommunications. Wiley, New York
13. Pachnicke S, Paschenda T, Krummrich PM (2008) Physical impairment based regenerator placement and routing in translucent optical networks. In: Proceedings of IEEE/OSA optical fiber communication conference (OFC), San Diego, CA
14. Azodolmolky S, Klinkowski M, Marin E, Careglio D, Pareta JP, Tomkos I (2009) A survey on physical layer impairments aware routing and wavelength assignment algorithms in optical networks. Elsevier Comput Netw 53(10):926–944
15. Ramamurthy B, Datta D, Feng H, Heritage JP, Mukherjee B (1999) Impact of transmission impairments on the teletraffic performance of wavelength-routed optical networks. IEEE/OSA J Lightwave Tech 17(10):1713–1723
16. Pereira HA, Chaves DAR, Bastos-Filho CJA, Martins-Filho JF (2007) Impact of physical layer impairments in all-optical networks. In: Proceedings of IEEE international microwave and optoelectronics conference (IMOC), Salvador, Brazil, pp 536–541
17. Huang Y, Heritage JP, Mukherjee B (2005) Connection provisioning with transmission impairment consideration in optical WDM networks with high-speed channels. IEEE/OSA J Lightwave Tech 23(3):982–993
18. Carpenter TJ, Menendez RC, Shallcross DF, Gannett JW, Jackel J, Von Lehmen AC (2004) Cost-conscious impairment-aware routing. In: Proceedings of IEEE/OSA optical fiber communication conference (OFC) 1, San Diego, CA
19. Cardillo R, Curri V, Mellia M (2006) Considering transmission impairments in configuring wavelength routed optical networks. In: Proceedings of IEEE/OSA optical fiber communication conference (OFC), Anaheim, CA
20. Zsigmond S, Németh G, Cinkler T (2007) Mutual impact of physical impairments and grooming in multilayer networks. In: Proceedings of international conference on optical networking design and modeling (ONDM), Athens, Greece, pp 38–47
21. He J, Brandt-Pearce M, Pointurier Y, Subramaniam S (2007) QoT-Aware routing in impairment-constrained optical networks. In: Proceedings of IEEE global communications conference (GLOBECOM), Washington, DC, pp 2269–2274

22. He J, Brandt-Pearce M, Subramaniam S (2007) QoS-aware wavelength assignment with ber and latency guarantees for crosstalk limited networks. In: Proceedings of IEEE international communication conference (ICC), Glasgow, Scotland
23. Marsden A, Maruta A, Kitayama K (2008) Routing and wavelength assignment encompassing FWM in WDM lightpath networks. In: Proceedings of international conference on optical networking design and modeling (ONDM), Vilanova i la Geltrú, Spain
24. He J, Brandt-Pearce M, Pointurier Y, Subramaniam S (2007) Adaptive wavelength assignment using wavelength spectrum separation for distributed optical networks. In: Proceedings of IEEE international communication conference (ICC), Glasgow, Scotland, pp 24–28
25. Farahmand M, Awduche D, Tibuleac S, Atlas D (2002) Characterization and representation of impairments for routing and path control in all-optical networks. In: Proceedings IEEE/OSA national fiber optics engineers conference (NFOEC), Dallas, TX
26. Marin E, Sánchez S, Masip X, Solé J, Maier G, Erangoli W, Santoni S, Quagliotti M (2007) Applying prediction concepts to routing on semi transparent optical transport networks. In: Proceedings of IEEE international conference on transparent optical networks (ICTON), Rome, Italy, pp 32–36
27. Cardillo R, Curri V, Mellia M (2005) Considering transmission impairments in wavelength routed networks. In: Proceedings of international conference on optical networking design and modeling (ONDM), Milan, Italy, pp 421–429
28. Pachnicke S, Paschenda T, Krummrich P (2008) Assessment of a constraint based routing algorithm for translucent 10 Gbits/s DWDM networks considering fiber nonlinearities. OSA J Opt Netw 7(4):365–377
29. Salvadori E, Ye Y, Zanardi A, Woesner H, Carcagni M, Galimberti G, Martinelli G, Tanzi A, La Fauci D (2007) Signaling-based architectures for impairment-aware lightpath set-up in GMPLS networks. In: Proceedings of IEEE global communications conference (GLOBECOM), Washington, DC, pp 2263–2268
30. Yang X, Ramamurthy B (2005) Dynamic routing in translucent WDM optical networks: the intradomain case. IEEE/OSA J Lightwave Tech 23(3):955–971
31. Markidis G, Sygletos S, Tzanakaki A, Tomkos I (2007) Impairment aware based routing and wavelength assignment in transparent long haul networks. In: Proceedings of international conference on optical networking design and modeling (ONDM), Athens, Greece, pp 48–57
32. Cugini F, Andriolli N, Valcarenghi L, Castoldi P (2004) A novel signaling approach to encompass physical impairments in GMPLS networks. In: Proceedings of IEEE global communications conference (GLOBECOM), Dallas, TX, pp 369–373
33. Pavani GS, Zuliani LG, Waldman H, Magalhães MF (2008) Distributed approaches for impairment-aware routing and wavelength assignment algorithms in GMPLS networks. Comput Netw 52(10):1905–1915
34. Tomkos I, Vogiatzis D, Mas C, Zacharopoulos I, Tzanakaki A, Varvarigos E (2004) Performance engineering of metropolitan area optical networks through impairment constraint routing. IEEE Commun Mag 42(8):S40–S47
35. Deng T, Subramaniam S (2005) Adaptive QoS routing in dynamic wavelength-routed optical networks. In: Proceedings of international conference on broadband communication networks and systems (Broadnets), vol 1, Boston, MA, pp 184–193
36. Kulkarni P, Tzanakaki A, Mas C, Tomkos I (2005) Benefits of Q-factor based routing in WDM Metro Networks. In: Proceedings of European conference on optical communications (ECOC), vol 4. Glasgow, Scotland, pp 981–982
37. Markidis G, Sygletos S, Tzanakaki A, Tomkos I (2007) Impairment-constraint-based routing in ultralong-haul optical networks with 2R regeneration. IEEE Photonics Tech Lett 19(6):420–422
38. Yang X, Shen L, Ramamurthy B (2005) Survivable lightpath provisioning in WDM mesh networks under shared path protection and signal quality constraints. IEEE/OSA J Lightwave Tech 23(4):1556–1567

39. Cugini F, Paolucci F, Valcarenghi L, Castoldi P (2007) Implementing a path computation element (PCE) to encompass physical impairments in transparent networks. In: Proceedings of IEEE/OSA optical fiber communication conference (OFC/NFOEC), Anaheim, CA, pp 1–3
40. Jukan A, Franzl G (2004) Path selection methods with multiple constraints in service-guaranteed WDM networks. IEEE/ACM Trans Netw 12(1):59–72
41. He J, Brandt-Pearce M (2006) RWA using wavelength ordering for crosstalk limited networks. In: Proceedings of IEEE/OSA optical fiber communication conference (OFC/NFOEC), Anaheim, CA
42. Martins-Filho JF, Bastos-Filho CJA, Arantes EAJ, Oliveira SC, Coelho, de Oliveira JPG, Dante RG, Fontana E, Nunes FD (2003) Novel routing algorithm for transparent optical networks based on noise figure and amplifier saturation. In: Proceedings of IEEE international microwave and optoelectronics conference (IMOC), vol 2. Iguaza Falls, Brazil, pp 919–923
43. Hamad AM, Kamal AE (2005) Routing and wavelength assignment with power aware multicasting in WDM networks. In: Proceedings of international conference on broadband communication networks and systems (Broadnets), vol 1. Boston, MA, pp 31–40
44. Salvadori E, Ye Y, Zanardi A, Woesner H, Carcagnì M, Galimberti G, Martinelli G, Tanzi A, La Fauci D (2007) A study of connection management approaches for an impairment-aware optical control plane. In: Proceedings of international conference on optical networking design and modeling (ONDM), Athens, Greece, pp 229–238
45. Pavani GS, Waldman H (2006) Adaptive routing and wavelength assignment with power constraints using ant colony optimization. In: Proceedings of international telecommunication symposium, Fortaleza, Brazil, pp 637–642
46. Pointurier Y, Brandt-Pearce M, Deng T, Subramaniam S (2006) Fair QoS-aware adaptive routing and wavelength assignment in all-optical networks. In: Proceedings of IEEE international communication conference (ICC), Istanbul, Turkey, pp 2433–2438
47. Morea A, Brogard N, Leplingard F, Antona JC, Zami T, Lavigne B, Bayart D (2008) QoT function and A* Routing an optimized combination for connection search in translucent networks. OSA J Opt Netw 7(1):42–61
48. Zhai Y, Pointurier Y, Subramaniam S, Brandt-Pearce M (2007) QoS-aware RWA algorithms for path-protected DWDM Networks. In: Proceedings of IEEE/OSA optical fiber communication conference (OFC/NFOEC), Anaheim, CA
49. Lee K-I, Shayman M (2005) Optical network design with optical constraints in IP/WDM networks. IEICE Trans Commun E88(B5):1898–1905
50. Lima MAC, Cesar AC, Araujo AFR (2003) Optical network optimization with transmission impairments based on genetic algorithm. In: Proceedings of IEEE international microwave and optoelectronics conference (IMOC), vol 1. Foz do Iguaqu, Brazil, pp 361–365
51. Szodenyi A, Zsigmond S, Megyer B, Cinkler T (2005) Design of traffic grooming optical virtual private networks obeying physical limitations. In: Proceedings of IFIP/IEEE international conference on wireless and optical communication networks (WOCN), Dubai, pp 221–225
52. Zang H, Huang R, Pan J (2000) Methodologies on designing a hybrid shared-mesh-protected WDM network with sparse wavelength conversion and regeneration. Proc SPIE 4910(1):188–196
53. Kim S, Seo S (2001) Regenerator placement algorithms for connection establishment in all-optical networks. IEE Proc Commun 148(1):25–30
54. Pointurier Y, Brandt-Pearce M (2006) Fair routing and wavelength assignment in all optical networks. In: Proceedings of IEEE/OSA optical fiber communication conference (OFC/NFOEC), Anaheim, CA
55. Zulkifli N, Okonkwo C, Guild K (2006) Dispersion optimized impairment constraint based routing and wavelength assignment algorithms for all-optical networks. In: Proceedings of IEEE international conference on transparent optical networks (ICTON), vol 3. Nottingham, UK, pp 177–180

56. Lee JH, Tsuritani T, Guo H, Okamoto S, Yoshikane N, Otani T (2008) Field trial of GMPLS-controlled all-optical networking assisted with optical performance monitors. In: Proceedings of IEEE/OSA optical fiber communication conference (OFC/NFOEC), San Diego, CA
57. Jukan A, Franzl G (2002) Constraint-based path selection methods for on-demand provisioning in WDM networks. In: Proceedings of IEEE international conference on computer communication (INFOCOM), New York, NY
58. Zhai Y, Pointurier Y, Subramaniam S, Brandt-Pearce M (2007) Performance of dedicated path protection in transmission impaired DWDM networks. In: Proceedings of IEEE international communication conference (ICC), Glasgow, Scotland, pp 2342–2347
59. Mukherjee B, Huang Y, Heritage JP (2004) Impairment-aware routing in wavelength-routed optical networks. In: Proceedings of IEEE lasers and electro-optics society annual meeting (IEEE/LEOS), vol 1(1). Puerto Rico, pp 428–429
60. Turkcu O, Subramaniam S (2007) Blocking in reconfigurable optical networks. In: Proceedings of IEEE international conference on computer communication (INFOCOM), Anchorage, AK, pp 188–196
61. Shen G, Grover WD, Cheng TH, Bose SK (2002) Sparse placement of electronic switching nodes for low-blocking in translucent optical networks. OSA J Opt Networking 1(12):424–441
62. Architecture for the Automatically Switched Optical Network (ASON) (2001) ITU-T Rec. G8090/Y.1304
63. Generalized Multi-Protocol Label Switching (GMPLS) Architecture (2004) IETF RFC 3945
64. Strand J, Chiu AL, Tkach R (2001) Issues for routing in the optical layer. IEEE Commun Mag 39(2):81–87
65. Martinez R, Pinart C, Cugini F, Andriolli N, Valcarenghi L, Castoldi P, Wosinska L, Comellas J, Junyent G (2006) Challenges and requirements for introducing impairment-awareness into the management and control planes of ASON-GMPLS WDM networks. IEEE Commun Mag 44(12):76–85
66. Castoldi P, Cugini F, Valcarenghi L, Sambo N, Le Rouzic E, Poirrier MJ, Adriolli N, Paolucci F, Giorgetti A (2007) Centralized vs. distributed approaches for encompassing physical impairments in transparent optical networks. In: Proceedings of international conference on optical networking design and modeling (ONDM), vol 4534. Athens, Greece, pp 68–77
67. A Path Computation Element (PCE)-Based Architecture (2006) IETF RFC 4655
68. Tsuritani T, Miyazawa M, Kashihara S, Otani T (2006) Optical path computation element interworking with network management system for transparent mesh networks. In: Proceedings of IEEE/OSA optical fiber communication conference (OFC/NFOEC), vol 8. San Diego, CA, pp 1–10
69. Cugini F, Andriolli N, Valcarenghi L, Castoldi P (2005) Physical impairment aware signaling for dynamic lightpath set up. In: Proceedings of European conference on optical communications (ECOC), vol 4. Glasgow, Scotland, pp 979–980
70. Sambo N, Giorgetti A, Andriolli N, Cugini F, Valcarenghi L, Castoldi P (2006) GMPLS signaling feedback for encompassing physical impairments in transparent optical networks. In: Proceedings of IEEE global communications conference (GLOBECOM), San Francisco, CA, pp 1–5
71. Stern T, Ellinas G, Bala K (2008) Multiwavelength optical networks: architectures, design and control, 2nd edn. Cambridge University Press, Cambridge, UK
72. Kamal A, Ul-Mustafa R (2003) Multicast traffic grooming in WDM networks. In: Proceedings of SPIE/IEEE Opticom, Dallas, TX
73. Yang D-N, Liao W (2003) Design of light-tree based logical topologies for multicast streams in wavelength routed optical networks. In: Proceedings of IEEE international conference on computer communication (INFOCOM), San Francisco, CA
74. Sun Y, Gu J, Tsang DHK (2001) Multicast routing in all-optical wavelength-routed networks. SPIE Opt Netw Mag 2(4):101–109

75. Sahin G, Azizoglu M (1998) Multicast routing and wavelength assignment in wide-area networks. In: Proceedings of SPIE conference on all-opt networks, Boston, MA
76. Xin Y, Rouskas GN (2004) Multicast routing under optical layer constraints. In Proceedings of IEEE international conference on computer communication (INFOCOM), Hong Kong, pp 2731–2742
77. Hamad AM, Kamal AE (2004) Optimal power-aware design of all-optical multicasting in wavelength routed networks. In: Proceedings of IEEE international communication conference (ICC), Paris, France, pp 1796–1800
78. Wu KD, Wu J-C, Yang C-S (2001) Multicast routing with power consideration in sparse splitting WDM networks. In: Proceedings of IEEE international communication conference (ICC), Helsinki, Finland, pp 513–517
79. Ellinas G, Antoniades N, Panayiotou T, Hadjiantonis A, Levine AM (2009) Multicast routing algorithms based on Q-factor physical layer constraints in metro networks. IEEE Photonics Tech Lett 21(6):365–367
80. Ali M, Deogun JS (2000) Power-efficient design of multicast wavelength-routed networks. IEEE J Select Areas Commun 18(10):1852–1862
81. Ali M, Deogun JS (2000) Cost-effective implementation of multicasting in wavelength-routed networks. IEEE/OSA J Lightwave Tech 18(12):1628–1638
82. Wu JWK, Yang C (2001) Multicast routing with power consideration in sparse splitting WDM networks. In: Proceedings of IEEE international communication conference (ICC), Helsinki, Finland, pp 513–517
83. He J, Chan SHG, Tsang DHK (2002) Multicasting in WDM networks. IEEE Commun Mag 4(1):2–20
84. Malli R, Zhang X, Qiao C (1998) Benefit of multicasting in all-optical networks. Proc SPIE 35319(12):209–220
85. Zhang X, Wei J, Qiao C (2000) Constrained multicast routing in WDM networks with sparse light splitting. In: Proceedings of IEEE international conference computer communication (INFOCOM), Tel-Aviv, Israel, pp 1781–1790
86. Ali M (2002) Optimization of splitting node placement in wavelength-routed optical networks. IEEE J Select Areas Commun 20(8):1571–1579
87. Ali M, Deogun JS (2000) Allocation of splitting nodes in wavelength-routed networks. Photonic Netw Commun 2(3):247–265
88. Hamad AM, Kamal AE (2004) Optimal power-aware design of all-optical multicasting in wavelength routed networks. In: Proceedings of IEEE international communication conference (ICC), vol 3. Paris, France, pp 1796–1800
89. Sreenath N, Satheesh K, Mohan G, Murthy CSR (2000) Virtual source-based multicast routing in WDM optical networks. In: Proceedings of IEEE international conference networks (ICON), Singapore, pp 385–389
90. Sreenath N, Krishna Mohan Reddy N, Mohan G, Siva Ram Murthy C (2001) Virtual source-based multicast routing in WDM networks with sparse light splitting. In: Proceedings of IEEE workshop on high perform switching and routing, Dallas, TX, pp 141–145
91. Tseng W–Y, Kuo S–Y (2001) All-optical multicasting on wavelength-routed WDM networks with partial replication. In: Proceedings of international conference information networking (ICOIN), Beppu City, Japan, pp 813–818
92. Yan S, Ali M, Deogun J (2001) Route optimization of multicast sessions in sparse light-splitting optical networks. In: Proceedings of IEEE global communications conference (GLOBECOM), San Antonio, TX, pp 2134–2138
93. Sun Y, Gu J, Tsang DHK (2001) Multicast routing in all-optical wavelength-routed networks. SPIE Opt Netw Mag 2(4):101–109
94. Liang W, Liu Y (2007) Online broadcasting and multicasting in WDM networks with shared light splitter bank. In: Proceedings of international conference on broadband communication networks and systems (Broadnets), Raleigh, NC, pp 504–513

95. Hsieh C–Y, Liao W (2003) All optical multicast routing in sparse-splitting optical networks. In: Proceedings of IEEE international conference on local computer networks (LCN), Bonn/ Königswinter, Germany, pp 162–167
96. De T, Sen S (2006) Multicast routing and wavelength assignment in sparse splitting all optical networks. In: Proceedings of international conference on wireless and optical communication networks, Bangalore, India
97. Hsieh C–Y, Liao W (2007) All-optical multicast routing in sparse splitting WDM networks. IEEE J Select Areas Commun 25(6):51–62
98. Zhou F, Molnar M, Cousin B (2008) Avoidance of multicast incapable branching nodes for multicast routing in WDM networks. In: Proceedings of IEEE international conference on local computer networks (LCN), Montreal, Canada, pp 336–344
99. Cao Y, Yu O (2005) Optimal placement of light splitters and wavelength converters for multicast in WDM networks. In: Proceedings of international conference on communication, circuits and systems (ICCCAS), vol 1. HongKong, pp 585–589
100. Yong K-M, Poo G-S, Cheng T-H (2005) Optimal placement of multicast and wavelength converting nodes in multicast optical virtual private network. In: Proceedings of IEEE international conference on local computer networks (LCN), Bonn, Germany, pp 505–506
101. Yan F, Hu W, Sun W, Guo W, Jin Y, He H (2009) Allocation of wavelength selective and convertible cross connects in optical multicast networks. In: Proceedings of communication photonics conference and exhibition (ACP), Shanghai, China, pp 1–7
102. Chen B, Wang J (2002) Efficient routing and wavelength assignment for multicast in WDM networks. IEEE J Select Areas Commun 20(1):97–109
103. Zhou Y, Poo G-S (2006) Multicast wavelength assignment for sparse wavelength conversion in WDM networks. In: Proceedings of IEEE international conference on Computer Communication (INFOCOM), Barcelona, Spain, pp 1–10
104. Huang W, Tang L, Rago M, Sivasankaran A, Tacca M, Fumagalli A (2010) Routing and wavelength assignment computed jointly for a given set of multicast trees reduces the total wavelength conversion. In: Proceedings of IEEE international conference on transparent optical networks (ICTON), Munich, Germany, pp 1–5
105. Yong K-M, Cheng T-H, Poo G-S (2009) Dynamic multicast routing and wavelength assignment with minimal conversions in delay-constrained WDM networks. In: Proceedings of international conference on computer communication and networks (ICCCN), San Francisco, CA, pp 1–6
106. Panayiotou T, Ellinas G, Antoniades N, Levine AM (2010) Designing and engineering metropolitan area transparent optical networks for the provisioning of multicast sessions. In: Proceedings of IEEE/OSA optical fiber communication conference (OFC/NFOEC), San Diego, CA
107. Panayiotou T, Ellinas G, Antoniades N, Hadjiantonis A (2010) Node architecture design and network engineering impact on optical multicasting based on physical layer constraints. In: Proceedings of IEEE international conference on transparent optical networks (ICTON), Munich, Germany, pp 1–4
108. Singhal NK, Sahasrabuddhe LH, Mukherjee B (2003) Provisioning of survivable multicast sessions against single links failures in optical WDM mesh networks. IEEE/OSA J Lightwave Tech 21(11):2587–2594
109. Sun Y, Gu J, Tsang DHK (2001) Multicast routing in all-optical wavelength-routed networks. SPIE Opt Netw Mag 2(4):101–109
110. Sahasrabuddhe LH, Mukherjee B (2000) Multicast routing algorithms and protocols: a tutorial. IEEE Netw 14(1):90–102
111. Bertsekas D, Gallager R (1992) Data networks, 2nd edn. Prentice-Hall, Englewood Cliffs
112. Barath-Kumar K, Jaffe J (1983) Routing to multiple destinations in computer networks. IEEE Trans Commun 31(3):343–351
113. Khalil A, Hadjiantonis A, Ellinas G, Ali MA (2005) Pre-planned multicast protection approaches in WDM mesh networks. In: Proceedings of European conference on optical communications (ECOC), vol 1. Glasgow, Scotland, pp 25–26

114. Choi JS, Golmie N, Lapeyrere F, Mouveaux F, Su D (2000) A functional classification of routing and wavelength assignment schemes in DWDM networks: static case. In: Proceedings of international conference on optical communication and networks (ICON), Singapore
115. Sun Y, Gu J, Tsang DHK (1999) Routing and wavelength assignment in all optical networks with multicast traffic. In: Proceedings of international tele-traffic congress (ITC), Edinburgh, Scotland
116. Hamad AM, Kamal AE (2005) Routing and wavelength assignment with power aware multicasting in WDM networks. In: Proceedings of international conference on broadband networks and systems (Broadnets), Boston, MA
117. Hamad AM, Kamal AE (2010) Power-aware connection provisioning for all-optical multicast traffic in WDM networks. IEEE/OSA J Opt Commun Netw 2(7):481–495
118. Ellinas G, Panayiotou T, Antoniades N, Hadjiantonis A, Levine AM (2008) Multicasting with physical-layer constraints in metropolitan area networks. In: Proceedings of IEEE international conference on transparent optical networks (ICTON), vol 3. Athens, Greece, pp 75–78
119. Guild K (2006) Ultra-fast optical routing including impairment constraints (UFORIC). Photon in switching conference, Heraklion, Greece
120. Willner A (2003) Optical burst switching (OBS): issues in the physical layer. In: Proceedings of IEEE international workshop on optical burst switching, co-located with IEEE global communications conference (GLOBECOM), Dallas, TX

Chapter 13
Provisioning and Survivability in Multi-Domain Optical Networks

Nasir Ghani, Min Peng and Ammar Rayes

Abstract Modern wireline infrastructures have morphed into a complex interconnection of distributed decentralized domains, as delineated by ownership, location, and technology type. These developments have prompted the need for advanced control plane solutions to achieve services/applications provisioning across multiple-possibly heterogeneous-domains with guaranteed quality of service (QoS) and reliability. As a result, multiple efforts are underway to address these concerns, including standardization initiatives as well as research studies in key topic areas, i.e., multi-domain routing, distributed path computation, survivability, and others. Along these lines, this chapter presents a detailed survey of some of these contributions, with a particular emphasis on those in the multi-domain optical networking area.

N. Ghani (✉)
Department of Electrical and Computer Engineering
University of New Mexico
Albuquerque, NM 87131, USA
e-mail: nghani@ece.unm.edu

M. Peng
Wuhan University, Wuhan
People's Republic of China
e-mail: pengm@whu.edu.cn

A. Rayes
Cisco Systems, Inc.
170 West Tasman Drive
San Jose, CA 95134, USA
e-mail: rayes@cisco.com

N. (Neo) Antoniades et al. (eds.), *WDM Systems and Networks*, Optical Networks,
DOI: 10.1007/978-1-4614-1093-5_13,

13.1 Introduction

The Internet has been built-up as a decentralized set of network domains, termed as *autonomous systems* (AS), with each managed by its own authority and operating under its own routing policy. Today there are tens of thousands of such domains (organized in a well-established hierarchy) that provide connectivity to billions of users. Now a major facilitator of this growth has been the development of scalable reachability exchange mechanisms for disseminating end-point *Internet Protocol* (IP) addresses between domains. Most notably, the *border gateway protocol* (BGP) has emerged as the de-facto inter-domain routing solution for inter-AS communication. Overall, this setup has worked very well for interconnecting hosts across "best-effort" packet-switched IP domains.

Nevertheless, continual advances in a host of "next-generation" technologies have drastically altered the networking space, both at the IP and underlying "non-IP" layers [1, 2]. Foremost, developments in state-of-the-art electronic devices have enabled full *quality of service* (QoS) support for packet flows and enabled new control plane paradigms such as *multi-protocol label switching* (MPLS). Indeed, many of these advances have benefited from earlier developments in cell-based *asynchronous transfer mode* (ATM) technologies. As these technologies and standards have matured, many related capabilities have also permeated into Layer 2, where new "carrier-grade" Ethernet technologies now also offer hardened services support. Meanwhile the further development of reconfigurable optical *dense wavelength division multiplexing* (DWDM) devices at the Layer 1 fiber-optic level has also ushered in a whole new set of circuit-switched paradigms. For example, related commercial systems can implement flexible wavelength connectivity over arbitrary fiber-plants (e.g., rings and meshes) and support extended reach with high-channel counts, i.e., terabits/fiber yields. Channelized Layer 1.5 SONET/SDH technologies have also been developed and offer very flexible mapping solutions for IP, Ethernet, and even storage networking protocols onto optical wavelength circuits. Finally, a revamped *generalized MPLS* (GMPLS) framework has been standardized to help streamline and automate services provisioning at the dynamic optical and SONET/SDH layers.

As these networking technologies have been introduced, modern wireline infrastructures have morphed into a complex interconnection of distributed, decentralized domains, see Fig. 13.1. Namely, each domain here can be treated as a standalone entity running its own management and provisioning paradigms, as delineated by ownership (multi-carrier), location, technology type, and others [1–3]. Concurrent to these developments, end-user services have also expanded, both in terms of bandwidth scale and distance reach. Moreover, many end-users are also demanding increasingly flexible service durations on the order of hours-days—a definitive change from traditional long-standing "telco" leased line demands. Inevitably, these trends have prompted the need for advanced control plane solutions to achieve services/applications provisioning across multiple—possibly heterogeneous—network domains with guaranteed QoS and reliability.

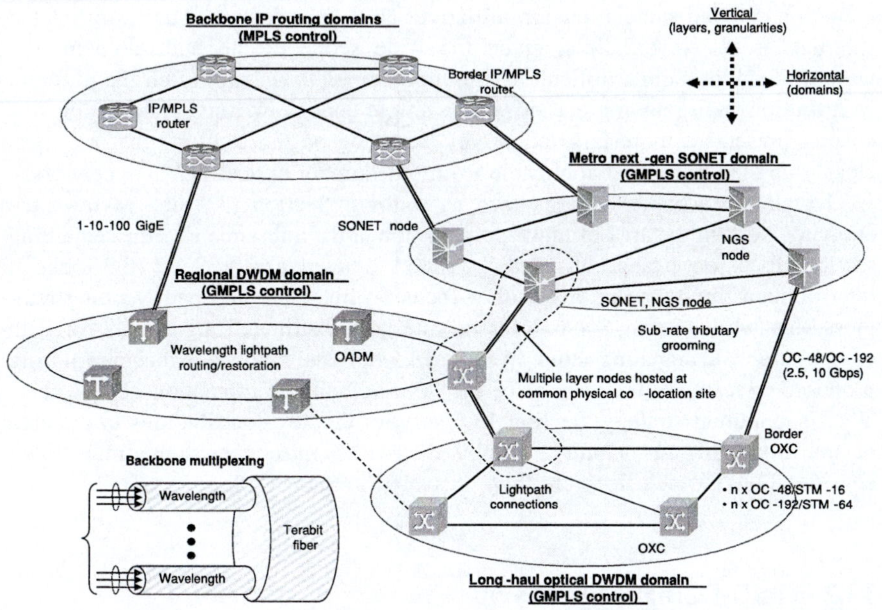

Fig. 13.1 Multi-layer backbone network comprising of multiple domains

Some key examples here include high-definition video conferencing, *Ethernet private line* (EPL) services, *virtual private networks* (VPN) among others.

Now a wide range of QoS *traffic engineering* (TE) and survivability schemes have already been developed for dynamic provisioning in IP and DWDM networks, see [4–6]. However, most of these schemes assume single-domain settings with full network visibility, i.e., topology, resources, diversity, etc. Clearly, such assumptions will not hold in general distributed multi-domain settings. For example, existing inter-domain reachability exchange protocols (such as BGP) can only provide address aggregation and not additional link-state exchange. As a result, these solutions will be inadequate for provisioning deterministic services, i.e., constraint-based QoS-routing based, path diversity, and other requirements [1, 7]. Moreover, scalability limitations will also prevent a single entity from maintaining detailed up-to-date state for large numbers of domains—even at the intra-carrier level. Finally, additional privacy and policy concerns will further complicate inter-domain visibility, particularly in competitive inter-carrier environments [2, 3]. In particular, the latter rules govern access to resources and services based upon administrative criteria, e.g., business relationships and competitive rules.

In light of the above, carriers and operators require a comprehensive set of protocols and algorithms to provision services across next-generation multi-domain infrastructures. Specifically, these solutions must provide key capabilities in terms of inter-domain routing, signaling, path computation, and survivability. As a result, multiple efforts are underway to address these pressing concerns, including detailed standardization initiatives as well as research studies in related subtopic areas. Nevertheless, given the wide scope of the multi-domain networking field, it is quite difficult to present a complete account of all the different contributions and ongoing activities in a single concise writing. As a result, this chapter presents a more targeted survey, focusing on those efforts that are most closely aligned with (and applicable to) multi-domain optical DWDM networks.

Overall, this chapter is organized as follows. Section 13.2 first reviews the evolving work in the area of multi-domain standards. Subsequently, the remaining sections focus on a range of research-related topics. Namely, Sect. 13.3 looks at inter-domain routing solutions, with a focus on link-state and path-vector strategies. Meanwhile, Sect. 13.4 focuses on alternate distributed algorithms for path computation and signaling setup. In Sect. 13.5 optional crankback mechanisms are proposed to resolve the concerns of the "per-domain" computation being generally a suboptimal strategy. Section 13.6 surveys the key contributions in the area of multi-domain survivability. Finally Sect. 13.7 introduces some multi-layer considerations. Conclusions are then presented in Sect. 13.8.

13.2 Multi-Domain Standards

Multi-domain network standardization has been ongoing at various levels over the past few decades. Most notably, the *Internet Engineering Task Force* (IETF) has pioneered a host of solutions for routing, and more lately signaling and path computation support across domains. The *International Telecommunications Union* (ITU-T) has also contributed some well-defined standards for transport domain architectures. These efforts have built upon earlier frameworks developed in the ATM Forum for advanced hierarchical routing [8]. In addition, the *Optical Internetworking Forum* (OIF) has also played a vital role in developing boundary-level protocols for client–carrier and carrier–carrier interfacing. Finally, the grid computing community has also outlined some proposals of its own. These efforts are summarized in Fig. 13.2 and now reviewed here.

13.2.1 International Telecommunications Union Standards

The ITU-T has matured a comprehensive framework for multi-domain transport networks under its *automatic switched transport network* (ASTN) standard (G.8080) [9]. This work has been conducted by the ITU's telecoms standardization sector, Study Group 15, and has evolved from the pre-cursor *automatic switched optical network* (ASON) effort [8, 10] to define services support over SONET/SDH and DWDM-based networks. Overall, ASTN defines the main components of a control plane architecture to support on-demand provisioning of *end-to-end* (E2E) call services in operator networks, i.e., set up, modify, and release. The key

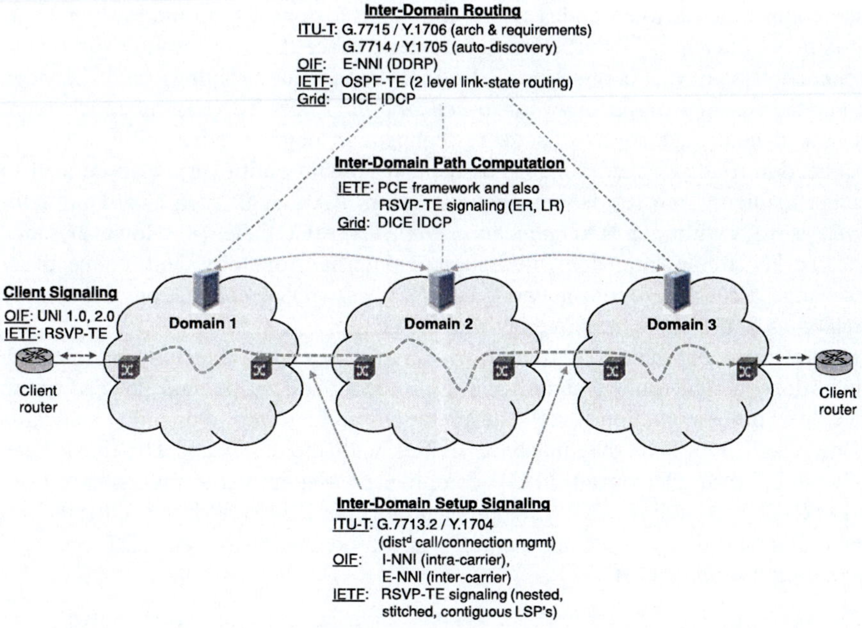

Fig. 13.2 Multi-domain (optical) networking standards overview

aim here is to define the interfaces that support the interactions between these components, i.e., based upon a client–server model, with a focus on those interactions that occur across a multi-vendor divide [9, 11]. To facilitate this, several recommendations have been generated for routing architecture/requirements, automated discovery, and distributed call/connection control (including signaling).

Overall, the base ASTN architecture uses a hierarchical setup in which the network consists of multiple routing layers, with each layer having several *routing area* (RA) domains [9] connected via *sub-network point pool* (SNPP) links. Here, the lowest level routing domains are comprised of physical nodes/links. In turn, these lower-layer domains are aggregated into single logical nodes to form the next layer, and so on and so forth in a recursive manner. Associated standards also define requirements for inter-area auto-discovery, auto-provisioning, and auto-restoration. Now at the inter-carrier level, domains are typically delineated by ownership. Meanwhile at the intra-carrier level, domains can be subdivided based upon a range of operator policies, e.g., administrative or managerial responsibilities, trust relationships, geographical regions, technology, addressing schemes, survivability techniques and distributions of control functionality.

Overall, the interconnection between and within domains is defined by appropriate reference points, which use abstract interface definitions for information exchange. For example, interconnection between the user and provider domains is done over a *user network interface* (UNI) reference point (i.e., call control, resource discovery, connection control, connection selection). Similarly, interconnection between nodes within a domain is done via an *internal-network-to-network interface* (I-NNI) reference point interface (i.e., for resource discovery, connection control, connection selection, connection routing) and between domains via an *external-network-to-network interface* (E-NNI) reference point (i.e., call control, resource discovery, connection control, connection selection, connection routing). Additionally, associated domain entities are also defined to setup, maintain, and release connections across RA's, e.g., such as *routing controllers* (RC) within an RA, *connection controllers* (CC), etc. [8]. However, since the ASTN framework does not support internal topology and connectivity exchange across the client network interface, associated connections are simply treated as SNPP links, i.e., overlay model [8].

Now given that the ASTN framework only focuses on architecture and interface definitions, actual routing and signaling protocols are not defined. Instead, these detailed "implementation-level" considerations are deferred to the other standards bodies, and it is here that liaison activities with the IETF and OIF have been crucial. For example, various ITU-T compliant routing implementations have been specified by extending GMPLS routing protocols [12]. In addition, other efforts have also addressed signaling support via the adaptation of the IETF and OIF signaling protocols [13].

13.2.2 Optical Internetworking Forum Standards

The OIF was formed by a series of industrial partners in the late 1990s to focus on the design of internetworking standards for optical network devices. To date, this forum has developed several interfacing control standards, including a client-to-carrier UNI [14, 15] and a carrier-to-carrier NNI [16, 17] (as reflecting the roles defined in the ITU-T standards). Namely, the UNI standard implements bandwidth signaling for client devices (i.e., such as IP/MPLS routers) and allows them to request/release "optical" connections from underlying carrier SONET/SDH or DWDM domains, i.e., akin to "optical dial-tone" standard. Since there is no trust relationship here, carrier topology information is not propagated to the client side, i.e., this follows the overlay model [8]. Currently, the latest UNI 2.0 standard [15] features much-improved capabilities for security, bandwidth modification, etc.

Meanwhile the NNI protocol has been designed to support more detailed *inter-domain* functions, i.e., including reachability and resource exchange as well as setup signaling. Herein, two different variants have been designed, as per intra- and inter-carrier requirements, i.e., I-NNI [18] and E-NNI [16], respectively. The former standard interfaces nodes within the same administrative area whereas the latter serves adjacent areas. As such, the E-NNI relegates all interfacing issues to domain boundaries, thereby removing restrictions on internal domain control and equipment interoperability. Now in terms of resource propagation, the NNI standard includes a link-state routing protocol, termed as the *domain-to-domain routing protocol* (DDRP) [8, 19, 17]. This setup essentially runs stacked instances of GMPLS link-state routing protocols in a hierarchical manner. However, since this solution imposes a fully hierarchical design, it represents a significant change from the current Internet routing model.

The overall OIF UNI-NNI combination facilitates rapid deployment of new services across multi-vendor "optical" domains. Moreover this framework works for both underlying optical DWDM and SONET/SDH domains, as per the different granularities supported in UNI/NNI signaling. Hence the NNI standard can support multi-layer interfacing at the circuit-switching level, and these capabilities have been demonstrated in several successful multi-vendor demonstrations. However, since there is no resource exchange across the UNI, higher-layer IP/MPLS networks must treat underlying "optical" links as tunneled links, i.e., overlay model [2, 8]. Clearly, this setup will be problematic for client layer routing since the number of Layer 3 adjacencies will grow per the square of border nodes, i.e., connection mesh problem. Hence clients may have to run data-plane inter-domain routing protocols, e.g., such as BGP, across UNI-provisioned connections.

13.2.3 Internet Engineering Task Force Standards

Over the years, the IETF has developed numerous control protocols. In terms of routing, most early efforts were focused on the design of intra-AS *interior gateway protocols* (IGP) for link-state propagation, i.e., as exemplified by the *open shortest path first* (OSPF) and *intermediate-system to intermediate-system* (IS–IS) protocols [8]. Nevertheless, OSPF also provides an added "second" level of routing between *area border routers* (ABR), thereby allowing carriers to achieve link-state exchange across multiple areas. Meanwhile at the inter-AS level, related IP *exterior gateway protocol* (EGP) standards were developed to provide basic connectivity support, e.g., BGP. However, given the massive scalability requirements of open public domain networks, these protocols only address (aggregated) end-point reachability exchange via the use of scalable path-vector routing techniques.

Now with the emergence of advanced "QoS-capable" IP/MPLS technologies in the late 1990s, the IETF slowly began to expand its protocols framework. First, a new *resource reservation* (RSVP) protocol was introduced to allow routers to reserve labels (and bandwidth) for user flows. This solution followed a soft-state design and used backwards reservation principles, as driven by the sink end-points. Additionally, RSVP also introduced a *loose route* (LR) capability in which sources nodes could specify partial (high-level) node sequences and then rely upon downstream nodes to perform *explicit route* (ER) expansion. Subsequently, with the growth in "circuit-switched" optical technologies, the IETF also began to work on a broader range of solutions to control lower "non-IP" networking layers, e.g., such as Ethernet (Layer 2), optical DWDM (Layer 1), and next-generation SONET/SDH (Layer 1.5). This expansion was driven by the realization that many of these layers required very similar control functionalities as those provided by "IP-layer" MPLS protocols [1]. Broadly termed as GMPLS [8], this solution suite included some targeted protocol extensions for existing routing and signaling designs, as well altogether new solutions for link management and path computation support. These are briefly reviewed here.

On the routing side, the IETF revamped its legacy IGP protocol suite by adding new features to support *constraint-based routing* (CBR) [2]. Namely, the base OSPF standard has been revised to a more capable OSPF-*traffic engineering* (OSPF-TE) offering via the addition of new *opaque link-state* (LSA) definitions for new TE and SONET/DWDM links [20–22]. As a result, OSPF-TE routing *TE databases* (TEDB) can store information fields for wavelengths, timeslots, *shared risk link groups* (SRLG), etc. [8]. Meanwhile, further signaling enhancements to RSVP, termed as RSVP-TE, have also been added to support DWDM lightpath and SONET/SDH timeslot circuit setup, e.g., bi-directional setup, link bundling and recovery signaling [23]. Finally, a new *link management protocol* (LMP) has also been developed for discovery and fault localization, see [8].

In all, the GMPLS routing and signaling protocols provide some key capabilities for multi-domain support. Foremost, OSPF-TE extends full hierarchical link-state propagation between optical domains with varying link granularities via its two-level routing design (see also Sect. 13.3.1). Moreover, this solution can also be adapted across "non-GMPLS" protocol domains by using software proxy devices at domain boundaries to translate proprietary control messages into standardized OSPF-TE messages [24]. Recently, opaque LSA extensions have also been defined for OSPF to provide more TE information for inter-AS links [25]. These additions are designed to minimize excessive domain-internal state propagation. Furthermore, the RSVP-TE signaling protocol is also fully conducive to multi-domain path setup (in limited visibility settings), owing to its LR capabilities. Namely, domains can iteratively compute partial "skeleton" routes and then use hard state RSVP-TE signaling to expand and reserve the full domain-level sequences (see Sect. 13.3.1). Finally, RSVP-TE has recently been augmented to support crankback signaling [26] via the inclusion of new crankback and error reporting *type-length-value* (TLV) fields. Carefully note that GMPLS framework does not detail extensions to inter-AS path-vector routing protocols such as BGP. The likely reasoning here is that link-state protocols are more germane for deterministic path provisioning needs (e.g., lightpaths) as they provide more detailed state information and also faster dissemination behaviors. Nevertheless, some draft proposals have outlined evolutions of BGP to support QoS routing, lightpath routing, and improved reliability (discussed further in Sects. 13.3.2 and 13.6.2).

Furthermore, the IETF has also standardized a framework for path computation in next-generation networks, with direct considerations for multi-domain needs. Consider that earlier, path computation was treated as a by-product of routing in legacy IP networks, i.e., individual nodes were expected to compute the "next-hop" to a destination address. However, with growing demands for TE-based provisioning (subject to multiple constraints), more complex and "optimized" path computation support was required. As a result, a *path computation element* (PCE) architecture [7, 27] was defined to formally decouple path computation from routing and signaling. Specifically, domains are designated with one or more PCE

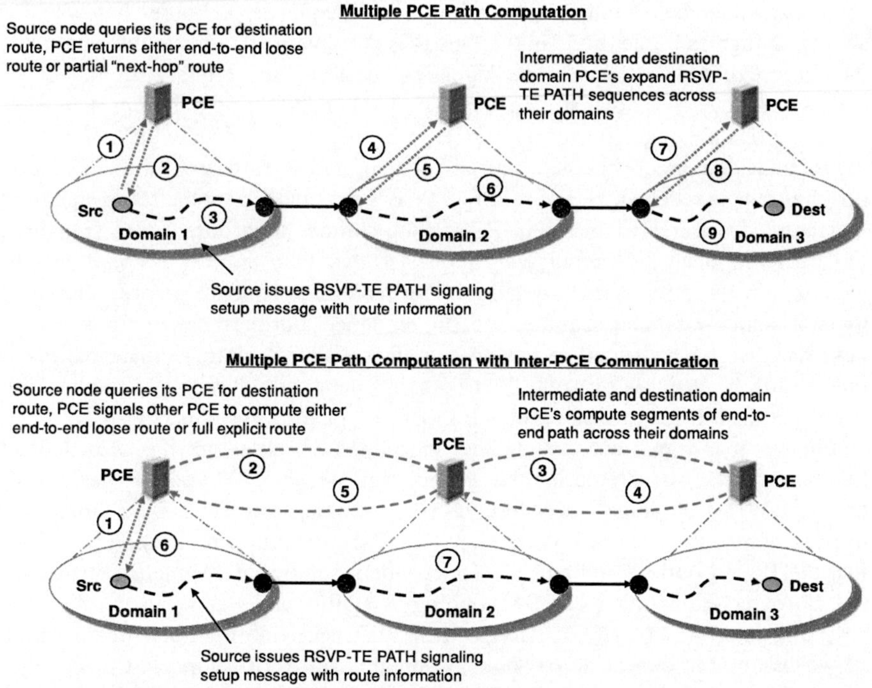

Fig. 13.3 Distributed multi-PCE (multi-domain) path computation scheme

entities with access to intra-/inter-domain TEDB and policy databases. Then these entities use this information to provide path computation services for their *path computation clients* (PCC) via a new *PCE protocol* (PCEP) [28]. Automated discovery is also provided for PCC nodes to locate a PCE. More recently, PCE extensions have also been outlined for optical lightpath routing and wavelength assignment RWA [29].

Now with regard to actual PCE-based computation strategies, two models have been proposed in [27], i.e., centralized and distributed. The former uses a given PCE to resolve all requests in an E2E manner—most realistic for intra-domain settings with full resource/policy views. Meanwhile the latter is more suited for decentralized multi-domain (multi-carrier) settings with partial visibility and also provides improved scalability and reliability. Distributed PCE computation can also support "domain-specific" constraints, allowing for heterogeneous-domain provisioning, i.e., bandwidth/delay in IP domains, wavelength/color continuity in DWDM networks, etc. Along these lines, two distributed sub-approaches are outlined to handle the contingencies of varying levels of "global" state, i.e., multi-PCE path computation with and without inter-PCE signaling [27], Fig. 13.3. Consider these strategies in the context of multi-domain provisioning.

In multi-PCE computation without inter-PCE signaling, Fig. 13.3, the source domain PCE basically computes a partial, i.e., LR, to the destination domain. This route can either be a complete "domain-level" sequence of border gateways or simply an ingress border node in the "next-hop" domain. This route information is then inserted into an appropriate signaling message, most likely an RSVP-TE *PATH* message, for downstream expansion. Namely, ingress border gateways receiving this message consult their PCE entities to resolve detailed ER sequences across their domains to the next partial LR hop, and so forth until the destination domain/node is reached. Hence multiple PCE's are invoked by the setup signaling messages. Conversely, in multi-PCE computation with inter-PCE signaling, Fig. 13.3, the head PCE recursively requests other PCE entities to compute sub-path segments using PCEP signaling. The resulting sequence returns either an explicit source–destination route or a LR sequence. Furthermore, the use of path keys has also been defined to preserve confidentiality on intra-domain path segments [30]. Note that a separate "per-domain" path computation scheme has also been specified in [31] using regular RSVP-TE signaling (see Sect. 13.4).

Finally, it is important to note that the GMPLS framework has also defined several models for integration between "higher-layer" IP and "lower layer" optical DWDM domains, i.e., overlay, peer, and integrated [2, 3, 8]. Namely, the overlay approach is based upon earlier IP-ATM integration models and assumes that the IP and optical domains run independent routing and signaling instances, i.e., IP routers do not have DWDM topology visibility and vice versa. As such, the signaling interfaces (e.g., OIF UNI or RSPV-TE) provide the only means for IP networks to interconnect across optical domains and form "tunneled link" adjacencies over provisioned lightpath connections. Clearly, this model is most applicable for edge client–carrier domain integration in inter-carrier settings where privacy and scalability concerns are paramount. By contrast, the peer model allows IP routers and optical nodes to operate as full peers via direct (routing and signaling) control message exchange. Hence in this "unified" approach, only a single routing protocol instance is required, e.g., as supported by OIF NNI or hierarchical OSPF-TE. Although this approach can facilitate the design of more effective inter-layer TE and protection schemes, it blurs the boundaries between the IP and optical domains/layers and entails higher control overheads at individual nodes (scalability and security concerns). As such, this model is more germane for IP-optical integration in more selective intra-carrier settings. Finally, the integrated/augmented model achieves a trade off between the above two strategies by allowing selected nodes to have more information about external domains, e.g., only interfacing border routers will run full peering with optical domains and not all routers.

13.2.4 Grid Computing Community

The grid computing community has long been interested in inter-domain provisioning as a means to support more "globalized" services paradigms [32]. Along these lines, the *Open Grid Forum* (OGF) has a working group that has defined a new *network description language* (NDL) [33] to specify network topologies. Namely, NDL characterizes multi-layer networking infrastructures using the *resource description framework* (RDF) standard for web data interchange, i.e., semantic web. This is done by building a comprehensive network ontology comprising of five "core" schemas, i.e., topology, layer, capability, domain, and physical. Owing to its high level of generality, NDL can be used to represent very complex network topologies and interconnections, and thereby feed subsequent path computation engines [34]. However, the NDL does not rely upon any "layer-specific" information retrieval/distribution model and hence can essentially import topological information from multiple sources, e.g., such as OSPF-TE link-state databases, centralized TL1 repositories and operator-specified *extensible markup language* (XML) files. Currently the *Global Lambda Integrated Facility* (GLIF) consortium has leveraged NDL to detail its individual domains and generate global GLIF inter-domain network maps.

In addition, some prominent "e-science" network operators have also formed their own multi-domain integration initiative, i.e., the *Dante, Internet2, CANARIE, and ESnet* (DICE) group. This body has recently proposed an *inter-domain controller protocol* (IDCP) [35] for dynamic services provisioning across multiple administrative domains. Specifically, IDCP defines messages for reserving connections and also gathering information on resources using web services formats. Here, detailed messages are passed between *inter-domain controllers* (IDC) to help compute and reserve multi-domain paths, loosely similar to PCE and also RSVP signaling. Detailed authentication capabilities are also supported.

Finally, a sample grid-based multi-domain services implementation is also presented in [36]. Namely, the authors build a "service plane" overlay to control diverse underlying control layers and interface different technology domains, akin to "thin layer" third party integration approach in [3]. This solution provides full authentication support between domains (i.e., trust establishment, policy enforcement mechanisms) and also defines new *dynamic resource allocation controller* (DRAC) entities to implement web services based control plane and inter-domain peering functionalities. Results from a real testbed are also presented, see [36].

13.3 Multi-Domain Routing and Path Computation

TE path computation typically mandates the availability of accurate link-level connectivity and resource state information [1]. In general, this is not a concern in single-domain (IP/MPLS or DWDM) settings, which readily support full TE databases, either via a centralized *network management system* (NMS) [2] or distributed link-state routing protocols (OSPF-TE, IS–IS). However, when extending to multi-domain settings, it is generally difficult to implement complete topology and resource information exchange between large numbers of domains. For example, in most inter-carrier settings, privacy and policy restrictions will likely supersede pure TE provisioning concerns and severely restrict inter-domain state exchange. Conversely, even in intra-carriers settings, full topology exchange may simply not scale beyond a few domains with moderate link and node counts.

To address these challenges, researchers have investigated a wide range of solutions based upon "hierarchical" inter-domain routing (and path computation) [1]. The main objective here is to provide peering exchange between domains and deliver a level of "global" visibility to selected gateways. Furthermore, in order to meet scalability requirements, most hierarchical routing protocols typically implement some form of *quantity* and *frequency* reduction [37]. Specifically, the former entails techniques such as topology (link-state) and/or path aggregation. Meanwhile, the latter pertains to routing update policies, e.g., such as periodic and threshold change-based. Now many initial studies have looked at hierarchical routing in legacy ATM and/or "best-effort" IP routing networks [1]. Nevertheless, recent efforts have extended these paradigms to broader "QoS-enabled" IP and optical DWDM settings, and these contributions are now detailed.

13.3.1 Topology Aggregation Approaches

Many hierarchical routing protocols implement routing quantity reduction by using node or link-state aggregation. For example, this concept was extensively used in older ATM networks running the *private network-to-network interface* (PNNI) routing protocol [29]. Specifically, this protocol defines a routing setup in which physical layer nodes are clustered into groups, and then further clustered into higher-level nodes and groups in a recursive manner, i.e., to build a multi-level hierarchy. Hence a node at a particular layer aggregates state from the nodes in its underlying layer and then shares this information with other nodes on its level, see [38] for a sample study. Note that the current ASTN framework also leverages this generalized hierarchical routing model [9].

Meanwhile other studies in multi-domain IP and ATM networking have also proposed the use of *topology abstraction* schemes for routing quantity reduction, Fig. 13.3. Here, graph transformation algorithms are used to condense domain-level topologies into smaller "virtualized" graphs consisting of fewer "abstract" (or virtual) nodes and links. These algorithms are typically run at designated computational entities within a domain and the reduced state is then propagated between domain gateways (along with regular inter-domain link-state updates) to provide aggregated "global" TE views [37]. Note that this approach fits well with the peer exchange model [8] and is most suited for core domains as opposed to edge (client) domains. To date, a wide range of abstraction schemes have been studied [37–42], including simple node, star, full-mesh, tree, etc. (Fig. 13.4). Namely, simple node abstraction reduces a domain to a single vertice, yielding little/no intra-domain visibility and thereby trading off blocking for scalability. Meanwhile, the star and mesh abstractions retain domain-level border nodes and use abstract links to summarize domain state, i.e., O(N) and O(N^2), respectively,

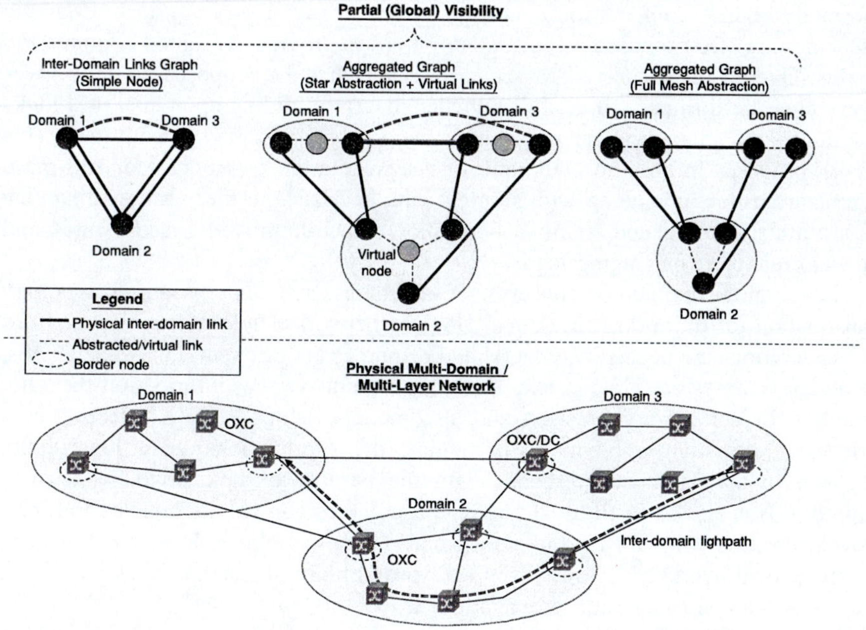

Fig. 13.4 Sample topology abstraction schemes: simple node, star, mesh

for N border nodes. For example, mesh abstraction computes abstract links to summarize available resources when traversing between domain border nodes.

Now initial topology abstraction studies have been done in the context of packet/cell-switched networks in which the key link attributes are bandwidth and delay. For example, [39] builds star and mesh abstractions for IP bandwidth-provisioning networks using various inter-domain "skeleton" route selection strategies, i.e., widest-shortest path (hop-count minimization), shortest-widest path (load-balancing) etc. Overall results here show that full-mesh abstraction gives the lowest blocking, but has higher overheads and requires shorter inter-domain routing update intervals to be effective. Additional results also show that shortest-widest path selection is more susceptible to increased routing update intervals. Meanwhile, [40] proposes topology aggregation for multiple QoS parameters in ATM networks (generally also extendible to IP networks). Namely, source-oriented techniques are introduced to selectively advertise abstract links to peering border nodes (to reduce redundancy and improve full-mesh scalability). In particular, two types are proposed here (for sparse and dense topologies) along with multi-QoS parameter summarizations, see [40]. Findings reveal that the proposed schemes are very competitive with full-mesh abstraction, but can be sensitive to larger update intervals. Finally, some authors have also applied information-theoretic approaches for topology abstraction. For example, [41] studies abstraction in directed graphs with additive link cost metrics (based upon residual capacities) and formally bounds compression distortion as per an asymmetry constant. Heuristic schemes are also proposed using spanning trees to provide very good compression gains. Overall results for ATM scenarios show that the proposed strategy gives very close performance to the full-knowledge case. Building upon this, [42] looks at multi-domain routing networks with bandwidth *and* delay sensitivities. This work proposes information-theoretic line segmentation techniques for full-mesh and star abstractions along with shortest path heuristics to meet bandwidth/delay constraints. The related findings here show much-improved success rates and lower crankback messaging loads.

Leveraging the above concepts, researchers have also proposed topology abstraction for multi-domain optical DWDM networks. Indeed, more specialized considerations are needed here as related resources are different, i.e., wavelengths, converters, timeslots, etc. For example, [43] presents various information-theoretic models for domain compression and also models lightpath RWA selection as a Bayesian decision problem. In particular, the formulations yield logarithmic growth in management overheads with the number of link wavelengths. The solution is analyzed for limited bus-type topologies and results indicate that network blocking can be traded off to a small degree to achieve large reductions in management overheads. However, practical routing implementation issues are not addressed for abstract state propagation. In another study, [44] proposes hierarchical routing for all-optical ASON networks. Specifically, aggregation algorithms are defined to first summarize aggregate wavelength and delay information on lightpath routes between border nodes. A routing update triggering policy is also proposed based upon the change in aggregate reserved wavelengths in a domain. Finally, "skeleton" LR lightpath computation algorithms are specified over the hierarchical topology, with specialized provisions for state inaccuracies (via the provision of bypass paths, as proposed for single-domain DWDM networks [45]). The results show good blocking gains (and sizeable routing overhead reductions) with performance approaching that of "flat" routing, i.e., full global state. However, this study does not address distributed signaling issues or handle conversion/regeneration at domain boundaries.

Building upon the above, [46] presents simple node and full-mesh abstraction schemes for all-optical multi-domain DWDM networks. These solutions introduce graph-theoretic RWA schemes (using *k*-shortest path and load-balancing heuristics) to compute LR domain sequences and also specify LR expansion signaling (RSVP-TE based) for intra-domain route resolution. In addition, relative change-based triggering policies are also applied for propagation of inter-domain link-state, both physical and abstract. The findings show lowest blocking rates (and signaling overheads) for full-mesh abstraction approach, albeit inter-domain routing loads are sizably higher, depending upon the number of border nodes [46]. The authors further extend their work in [47] to consider more realistic settings with wavelength conversion at border and interior nodes. Specifically, modified full-mesh abstractions are built by using the most congested subpaths between border nodes. Again, results show that full-mesh abstraction gives the lowest blocking and also signaling overheads. More importantly, findings show that augmenting sparse conversion networks with full-mesh topology abstraction gives better blocking performance than deploying full conversion networks with more basic simple node abstraction. Note that [37] also gauges the impact of delayed (periodic) update strategies on the performance of multi-domain topology abstraction in all-optical DWDM networks. Here the authors introduce a *total topology element* (TTE) metric and derive formulas to summarize the complexity of various abstractions, i.e., simple node, full-mesh, star, and hybrid. Simulations are then presented and show the lowest blocking with the full-mesh scheme, albeit it is also more sensitive to larger update intervals (as per [40]) and can give longer paths.

Meanwhile, [48] also studies simple node abstraction in multi-domain DWDM networks with wavelength convertible gateway nodes. Specifically, three schemes are proposed here, E2E, *concatenated shortest path routing* (CSR), and *hierarchical routing* (HIR). The E2E scheme is a baseline solution which assumes a "flat" global network with full visibility, i.e., unrealistic. Meanwhile, the HIR scheme computes shortest path LR routes on a global simple node abstracted topology whereas CSR relies upon "domain-to-domain" segment expansion along the fixed segment sequence (akin to "per-domain" scheme, detailed in Sect. 13.4). Additional domain gateway selection strategies are also proposed if there are multiple distinct border nodes within a domain, i.e., random, closest-distance, and least-loaded selection. Expectedly, results show higher blocking with the HIR and CSR strategies, with HIR doing slightly better (in conjunction with least-loaded gateway selection). However, both the HIR and CSR schemes only use fixed "hierarchical" topologies to compute LR sequences. As such they do not incorporate resource state (i.e., no inter-domain routing required) and will yield higher blocking versus load-balancing schemes [47].

Nevertheless, one of the major concerns with topology abstraction is the increased level of inter-domain routing state, particularly for full-mesh schemes. This reduces scalability and also increases control plane bandwidth requirements. As a result, some novel update triggering schemes have been studied to limit dissemination. For example, [49] proposes novel gradient-based triggering schemes which only generate updates for abstract links experiencing the maximum relative changes (for the most-used wavelengths). These algorithms are shown to give very sizeable reductions in full-mesh routing loads, by almost 50% at higher loads, with negligible increase in blocking. Meanwhile [50] outlines a partial link monitoring scheme which advertises updates for selected abstract links by monitoring a subset of intra-domain physical links. Namely, a load-based heuristic algorithm is used to rank and select paths for subsequent use in abstraction computation. In addition, detailed analytical probabilistic models are also developed to predict triggering rates (routing loads) for this revised scheme as well as for periodic and threshold-based policies. Overall findings show that the proposed scheme gives notably lower triggering rates than the threshold approach, with nearly the same blocking probabilities. In a related study, [51] also looks at routing update behaviors in multi-layer ASON networks and suggests a faster "parallel" update strategy. A generic bound on the convergence time is then presented.

Now from a real-world perspective, many of the above topology abstraction schemes can be implemented via the emerging protocols standards outlined in Sect. 13.2. For example, two-level hierarchical link-state routing in OSPF-TE (or the more generalized OIF inter-domain routing standard, Sect. 13.2.2) can readily distribute abstract link-state. Moreover, the use of topology aggregation has also been mentioned in ASTN and GMPLS standards as a means for improving scalability [37, 44]. Related "skeleton" path signaling and expansion can also be done using the RSVP-TE protocol or via the PCE framework (see end of Sect. 13.4). Now it is noted that most of the above schemes only focus on resource management issues. However, related policy concerns can also be factored into the associated algorithms, e.g., by leveraging work on hierarchical policy models for BGP to improve rule abstraction and efficiency [52].

Carefully note that many single domain DWDM networking studies now incorporate physical layer impairment into the RWA process, e.g., see [53] for a comprehensive survey as well as [54, 55] for distributed signaling and PCE-based algorithms, respectively. Clearly, such considerations will also be very important in "all-optical" multi-domain scenarios, owing to the extended hop counts and distances involved. Nevertheless, it is very plausible that carriers will deploy full opto-electronic regeneration at domain boundaries [47] in order to enforce bit-level user *service level agreements* (SLA) monitoring, especially in inter-carrier setting. Indeed, this may very well preclude the need for complicated impairment provisions at the multi-domain-level, i.e., routing, path computation and signaling.

13.3.2 Path-Vector Routing Approaches

As noted earlier, hierarchical routing (topology abstraction) poses some key scalability concerns in terms of messaging overheads and computational complexities [56]. Moreover, link-state propagation may be difficult to implement in generalized inter-carrier settings owing to privacy concerns and non-uniform control plane instances. As an alternative, path-vector routing solutions have also been developed for secure multi-domain control. An early example here is the *"optical" BGP* (O-BGP) solution [57, 58] which is designed to run between border (OXC) gateway nodes and maintain path and wavelength state to other domains. Here, regular BGP reachability updates are augmented to include the available wavelength sets along the path routes. Hence inter-domain gateway nodes simply maintain "next-hop" entries to other domains, specifying the next-hop nodes (border nodes) and available wavelengths. In addition, specialized signaling messages are also defined to actually signal inter-domain lightpath setups. Namely, [59] introduces a new "O-BGP" message type and details two-phase (discovery, setup) and four-phase (discovery, reservation, setup, confirmation) lightpath setup mechanisms. However, the overall O-BGP solution was not included in the GMPLS framework. Meanwhile, in a similar approach, [60] also proposes a novel *constraint-based optical path-vector routing protocol* (COPRP) solution for

DWDM networks. This solution extends the vanilla BGP approach by advertising multiple routes to each destination domain (as per different RWA constraints) and also supports wavelength conversion inside domains. In addition, a two-phase backward reservation protocol is defined to help minimize setup delay (akin to RSVP-TE). Finally, parallel signaling with local rerouting (upstream crankback) is also provided to lower blocking. However, although some average/worst-case timing analysis is presented, detailed performance evaluations are not conducted in [60].

Meanwhile, some other studies have analyzed the performance of path-vector RWA schemes. For example, [61, 62] propose a framework in which domains have one or more *inter-domain routing agents* (IDRA) which implement routing and path computation and also tie in with intra-domain routing databases, see also [56]. Using this, the IDRA entities not only advertise multiple paths to a single destination but also use adaptive filtering to predict the number of wavelengths on the inter-domain paths. Specifically, linear stochastic estimation is applied to estimate the *effective number of available wavelengths* (ENAW) on the paths, and these values are then refined using Kalman filtering prediction-correction (based upon updated wavelength information in the updates). Specialized load values are also computed for each path, thereby providing additional information to improve route selection and avoid overly lengthy paths with highest ENAW values. Inter-domain RWA signaling then uses this average wavelength state to select and expand E2E domain routes. Note that this solution can also provide policy control, e.g., such as restricting the numbers of wavelengths to certain destinations. Overall, simulation results for both small and large network topologies show very impressive blocking reductions (1–2 orders of magnitude) with the proposed ENAW-based scheme as compared to shortest path O-BGP. In addition, update message loads are also much lower. However, broader comparisons with hierarchical *link-state* routing schemes (Sect. 13.3.1) are not done. Another path-vector RWA scheme (with optional border gateway conversion) is also studied in [63]. This solution assumes that gateways propagate and maintain alternate "next-hop" routes to all destination domains and proposes a distributed scheme to concatenate domain sequences to build an E2E route. In addition to minimizing cost, the scheme also checks *bit-error rate* (BER) costs to limit impairments on all-optical segments. Overall results show good gains versus all-optical or opto-electronic schemes. However, associate path-vector route computation and dissemination concerns are not studied here.

Finally, various studies have also addressed QoS extensions for BGP path-vector routing. The basic idea here is to augment update messages and advertise multiple destination routes, as per specific QoS metrics (bandwidth, delay, cost, etc.). Nevertheless, most of this work has focused on QoS support in IP packet-forwarding networks and may not be directly applicable to MPLS/GMPLS networks, i.e., which require further signaling considerations for multi-domain *label switched path* (LSP) or lightpath setups. In light of this, these solutions are only briefly reviewed here. Foremost, some IETF drafts have proposed BGP signaling extensions to advertise QoS metrics, see [64]. Leveraging this, [65] proposes a

"QoS-planes" concept in which traffic is forwarded along different "slices" of the network as per a given metric and differentiated service class. This setup uses offline TE mechanisms to pre-optimize resource peerings with neighboring domains and then uses propagated QoS state for route selection. Results show notable improvements for delay and bandwidth guarantee support. Meanwhile, [66] uses histogram techniques to compute a compound bandwidth metric for a route, termed the *available bandwidth index* (ABI). This statistical average helps improve scalability by reducing the amount of updates generated during periods of resource fluctuation. This QoS state is then propagated with BGP routes and used for subsequent path selection. Nevertheless, advertising multiple routes in BGP can impose sizeable update loads, as noted in [67]. Hence, various scalability improvements have been proposed to address this. For example, [67] outlines some novel path reduction algorithms to decrease the number of updates generated for multiple (QoS metric) BGP route advertisements. Meanwhile [68] proposes another solution which allows AS domains to "pull" (i.e., request) additional routes via bilateral negotiations on a per-need basis.

13.4 Distributed PCE-Based Path Computation

More recently, researchers have also started to leverage the new IETF PCE framework (Sect. 13.2.3) to design novel inter-domain path computation solutions. Now one of the earlier solutions here has been the "per-domain" approach which implements joint signaling/computation using PCE and RSVP-TE standards [31]. Namely, ingress border nodes simply specify the "next-hop" domain toward the destination and perform local route expansion to the appropriate egress border node. This process is repeated at downstream domains, thereby allowing individual domains to compute their own traversing segments and iteratively construct concatenated E2E path sequences. As such, this approach can be classified under the base multi-PCE computation designation (without inter-PCE signaling, see Fig. 13.3, Sect. 13.2.3). However, "per-domain" computation is generally a suboptimal strategy as it may not find routes even if they exist [69]. Hence optional crankback mechanisms have been proposed to resolve these concerns, further discussed in Sect. 13.5.

Meanwhile, a host of studies have looked at more elaborate PCE-based computation strategies, albeit mostly for IP/MPLS networks. These schemes basically rely upon inter-PCE signaling, as shown in Fig. 13.3 (Sect. 13.2.3). A key example here is the *backward recursive path computation* (BRPC) solution [70–72] which assumes pre-computed E2E domain route sequences, as generated via prior TE computation/optimization or policy agreements. Using this information, the algorithm implements a recursive reverse search using PCEP signaling (Sect. 13.2.3) to construct a *virtual span path tree* (VSPT) from the destination to the source domains. This is essentially a "sweep" method which enumerates all potential ingress/egress domain node combinations along a pre-specified domain sequence and then allows the source PCE to choose the best one given a particular metric. Expectedly, the BRPC method is most effective when it can exploit a multiplicity of border nodes between two domains, i.e., three or more.

Overall, the performance of the BRPC scheme and its variants has been well-studied for IP/MPLS networks. For example [72] compares per-domain computation (with crankback) versus BRPC operation for post-fault survivability scenarios. Overall findings show notably better performance with the latter scheme, but this work is more appropriately discussed in Sect. 13.6.1. Meanwhile, [73] analyzes several PCE strategies, all of which assume pre-computed shortest inter-AS (domain) routes. Specifically, these include per-domain backward, per-domain ping-pong, and per-domain backward tree schemes. The per-domain backward method incrementally computes/reserves shortest domain-traversing path segments from the destination back to the source. Meanwhile, the ping-pong scheme uses the RSVP-based forward computation/reverse reservation method to choose the shortest paths. Finally, the per-domain backward tree scheme is essentially the same at the BRPC method (with possibly more selective a priori inter-AS route selection). Detailed results for a variety of networks (linear, mesh, etc.) show the best blocking performance with the tree-based scheme, albeit compute times are slightly larger. In general, the gains are more evident in the more complicated mesh topologies, and propagation delays show notable impacts (versus "idealized" instantaneous methods). Finally, [74] proposes a detailed PCE-based inter-domain TE QoS framework for the European Union EuQoS project. Here, TE LSP entities are termed as "EQ-links" and a specialized request/response *inter-AS path computation protocol* (IA-PCP) is defined for their setup. Overall, this solution builds upon the BRPC scheme by further addressing E2E domain sequence computation, e.g., by using inter-AS BGP path state. The timing overheads of the proposed scheme are also analyzed and shown to be of the same order as the BRPC method.

Recently, the BRPC scheme has also been adapted for multi-domain RWA in optical networks. For example, [75] focuses on E2E wavelength selection along pre-specified domain sequences and proposes both sequential and parallel resource reservation (signaling) strategies. Results show the best performance (i.e., lowest blocking, setup delays) with *first-fit* (FF) wavelength selection. In addition, [76] also outlines BRPC enhancements with new PCEP objects to track available wavelength statuses between PCE entities. Overall results show that this revised BRPC method gives good blocking reduction (5–20%) and less signaling failures versus the base "colorless" BRPC scheme. However, setup delays are notably larger [76]. Carefully note that the above schemes are solely designed for "all-optical" E2E domains. As mentioned earlier in Sect. 13.3.1, physical layer impairments may preclude such operation, particularly if inter-carrier SLA monitoring is required at domain boundaries. Hence this may simplify BRPC design by relegating wavelength selection to the intra-domain-level.

With the increased prominence of PCE standards, various other studies are also starting to emerge. In particular, several studies have addressed E2E domain sequence computation, something which is generally assumed as an input in most of the above studies. For example, [77] proposes new PCEP extensions to carry additional route information during the distributed inter-PCE computation phase. This requires domain PCE's to advertise extra routes to destinations by storing all BGP updates, and optionally also disseminating inter-domain link bandwidth state (i.e., pull method). Two related schemes are proposed here to leverage this information and then construct improved inter-domain sequences, i.e., route and bandwidth, respectively. Results show that both schemes can give lower blocking versus basic BGP (shortest domain sequence) selection, with the latter link-state bandwidth scheme doing the best. Building upon this, [78] proposes a more elaborate *"hierarchical" BGP* (HBGP) solution to improve domain sequence selection. Namely, two routing strategies are specified, *single route* (SR) and *multiple route* (MR), reflecting the number of routes stored to each destination prefix. To further address scalability concerns, only a single entity in each domain is required to participate in HBGP routing. Now given the multiplicity of HBGP routes, both random and maximum-bandwidth route selection policies are proposed, and these can be further combined with either per-domain or BRPC computation (for domain sequence expansion). Overall, bandwidth-based domain route selection is found to give the lowest inter-domain blocking when coupled with BRPC route expansion, see [78].

Dynamic domain-sequencing is also addressed in [79] by leveraging the PCEP framework proposed in [77]. Namely, a virtual topology global graph is built by extracting inter-domain link connectivity from BGP tables and then augmenting it with virtual intra-domain links (propagated between the PCE's only using PCEP). Overall findings show much lower inter-domain blocking when intra-domain links are assigned dynamic "load-based" costs. Finally, [80] also addresses E2E domain selection in multi-domain networks and proposes two "cooperative" schemes, model-based and adhoc. The former assumes that PCE's have access to (and exchange) topology and resource information in an aggregated form, be it link/path-vector state. This information is then used during inter-PCE signaling to build routes. Albeit detailed algorithms are not presented for the model-based scheme, some analysis is done to show the benefit of "domain-level" blocking probability estimates. Meanwhile the latter assumes that a PCE has little/no information on other domains and relies upon a form of inter-PCE "flooding" to search for routes, i.e., borrowing from mobile adhoc routing protocols such as *dynamic source routing* (DSR). It is stated that the model-based strategy is more suitable for larger networks whereas the adhoc one is more suited to smaller domain counts.

Carefully note that the PCE standards can also implement many of the hierarchical routing algorithms presented in Sect. 13.3.1 (even though these solutions were proposed before the emergence of this framework). For example, topology abstraction schemes can readily be adapted to multi-PCE settings by either using per-domain or PCE-based computation. Namely, for per-domain adaptation, source PCE's can compute the full LR skeleton route and then initiate RSVP-TE signaling to expand intra-domain routes. Conversely, if using PCE-based adaptation, source PCE's can again compute the skeleton path but then initiate full PCEP signaling (with downstream domains) to expand the E2E sequence.

13.5 Crankback Strategies

In general, hierarchical routing schemes entail notable overheads in terms of state dissemination and path computation complexities. Moreover, the dated nature of routing state is known to have deleterious impacts on performance [37, 49]. Hence some have proposed the use of signaling crankback "re-try" mechanisms to avoid resource-deficient (or failed) links and improve setup success rates. This approach is particularly amenable to basic "per-domain" computation schemes which operate with minimal global inter-domain visibility (Sect. 13.4), albeit it can be applied to hierarchical routing schemes as well.

Overall, crankback operates by having downstream nodes detect resource setup failures (or link failures) and send notifications to appropriate upstream nodes. These nodes can then re-try several alternate paths to the destination and explicitly avoid the problematic links. To support such operation, the RSVP-TE protocol has been augmented with added fields for crankback support, i.e., such as counter limits, failed node/link tracking, etc. [26]. In addition, several crankback methodologies have also been outlined—including E2E (source), boundary (i.e., domain ingress), and segment—all of which are applicable in multi-domain settings.

However, to date only a handful of studies have applied crankback to multi-domain settings (mostly for IP/MPLS settings but readily extendible to DWDM settings). For example, [72] combines crankback with "per-domain" computation (Sect. 13.4) to re-try multiple domain egress border nodes, and upon failure, cranks back to upstream (ingress) headend routers. However no limits are placed upon the number of crankback attempts (i.e., exhaustive search), and "next-hop" domain selection strategies are not specified (random). As a result, findings show higher request blocking rates and crankback delays, particularly when compared to PCE-based strategies utilizing pre-determined inter-domain routes such as BRPC. Meanwhile, [81] proposes crankback schemes to minimize E2E path delays in multi-domain IP/MPLS networks. Namely, this work tables two next-hop domain selection strategies, one which selects the next-hop as the "nearest" egress border node in the domain and another which uses specialized inter-domain *round-trip time* (RTT) estimates, i.e., pre-computed global state. Overall the latter heuristic is shown to yield slightly higher carried load and reduced re-tries, although it requires adoption of a specialized coordinates system, see [81].

The authors in [82, 83] also develop another crankback solution for multi-domain MPLS networks operating with minimal "global" state, i.e., *compute while switching* (CWS) scheme. This method first uses basic "per-domain" crankback, as in [72], to come up with an initial route, i.e., by probing egress nodes specified by interior and/or EGPs. Transmission is then started on this route and at the same time simultaneous crankback is initiated to search for shorter "near-optimal" (efficient) routes. If such route is found, data switchovers are done. As such, this scheme implements a form of distributed exhaustive search and findings show very high success rates, yielding inter-domain paths with minimum number of domains. However, associated signaling overheads (and delays for secondary setups) are not analyzed and the scheme requires non-standard extensions to RSVP-TE attributes. Additionally, implementing hitless switchovers at high speeds may pose added complexities for carriers.

Building upon the above, [84–86] have tabled a generalized enhanced intra/inter-domain crankback framework for "per-domain" settings where nodes have full domain link-state visibility (via OSPF-TE) but limited "next-hop" inter-domain visibility (via BGP). The solution is adapted for both IP/MPLS [84, 85] and DWDM networks [86] and introduces some key innovations. Foremost, full history tracking is done and dual counters are introduced to limit the number of intra- and inter-domain crankback attempts and prevent excessively long resource inefficient routes. Next, inter-AS path-vector routing state is processed to build "approximate" global topology views and pre-compute multi-entry "next-hop" domain tables for all destinations. Furthermore, [85] also tracks setup failure histories at border nodes to improve success rates. The results show that this improved scheme is notably better than the "per-domain" scheme of [72] and can closely match (sometimes exceed) the performance of full hierarchical routing with mesh abstraction (Sect. 13.3.1, [46]), i.e., particularly for higher interfering intra-domain loads. However, setup delays are notably higher, by 30–50% over hierarchical routing in certain cases. Note that these schemes have also extended for single/multi-failure recovery across domains, as detailed in Sect. 13.6.4.

Carefully note that most of the above crankback studies combine intra-domain path computation with the inter-domain signaling/expansion phase, i.e., assume RSVP signaling. Alternatively, these algorithms can also be implemented using the PCE framework, i.e., crankback information can be passed between PCE entities computing the routes. Overall, there remains significant latitude to develop further crankback schemes by leveraging existing work on single domain crankback. For example [87] uses path prediction methods to improve crankback setup gains, whereas [88] outlines a suggested vector scheme to minimize wavelength reservation blocking during the backward reservation phase. Finally, [89] also incorporates optical layer impairments into the crankback process and uses equal-cost link-disjoint or maximally link-disjoint path selection.

13.6 Multi-Domain Survivability

Survivability in multi-domain networks is a major concern for operators. In general, this is a rather complicated task since one must contend with partial topological (i.e., intra and inter-domain link diversity) state as well as distributed path computation environments. Although most existing survivability studies have focused on single-domain networks, various efforts have started to look at broader multi-domain concerns. Generally, these efforts can be classified along the traditional lines of pre-configured protection (dedicated, shared) and post-fault restoration [6]. However, within the multi-domain context, protection strategies can be further delineated as per the availability (or lack thereof) of global diversity state, i.e., “per-domain” protection or E2E path protection (link-disjoint or domain-disjoint), as shown Fig. 14.5. These contributions are now surveyed with a focus on the provisioning aspects. Namely, related fault detection/localization issues are not detailed as they are mostly relegated to lower link layer mechanisms.

13.6.1 Per-Domain Protection Strategies

Protection schemes provide proactive recovery by pre-computing and pre-reserving backup resources for individual working connections. These approaches are well-suited for single node/link fault occurrences as they can provide guaranteed recovery. To date, a full range of protection schemes have been developed for single domain networks, including link, segment (i.e., subpath), and E2E path protection. Now given the broad scope of multi-domain networks, all of these strategies can be leveraged here as well, in some shape or form [90].

Foremost, various efforts have proposed “per-domain” protection strategies [69, 91], loosely following the similarly named path computation strategy (Sect. 13.4). Also termed as “intra-domain protection” in [72], this approach is shown in Fig. 13.5 and handles all intra-domain faults using “domain-internal” methods—usually dedicated path protection. Meanwhile, inter-domain faults are relegated to more “localized” point-to-point interconnection strategies, see Fig. 13.5. In particular, the latter typically relies upon variations of older SONET/SDH dual/multi-homing setups between gateway border nodes. By and large, “per-domain” protection is well-suited to multi-carrier scenarios with tight policy control, as it ensures that both working and protection (sub) paths traverse the same sequence of domains. Moreover, this overall approach also offers some key saliencies. Foremost, “per-domain” protection allows operators to apply independent protection schemes within their respective domains and achieve very rapid recovery times over shorter segments, typically under 100 ms. In addition, this approach is relatively easier to implement and is also quite scalable, i.e., with little/no inter-domain routing requirements. Moreover, alarm propagation across domain boundaries is also minimized, a key plus in multi-operator settings. Finally, “per-domain” protection can provide recovery from multiple simultaneous failures, provided they occur in differing domains.

Along these lines, some initial efforts have looked at dedicated “per-domain” protection strategies. For example, [92] presents an initial “per-domain” solution for multi-domain MPLS networks using several inter-domain interconnection strategies. In addition, [93] outlines a broader solution for LSP recovery by using full-mesh and 1:1 domain inter-connection schemes. This scheme assumes the availability of condensed “global” topological state to first compute E2E “skeleton” primary/backup path sequences, i.e., including all ingress/egress border nodes. Subsequently, two *sequential* signaling strategies are proposed. Namely, the first scheme expands the E2E primary working route and then uses the stored path information to avoid overlaps on subsequent backup route expansion. As this is

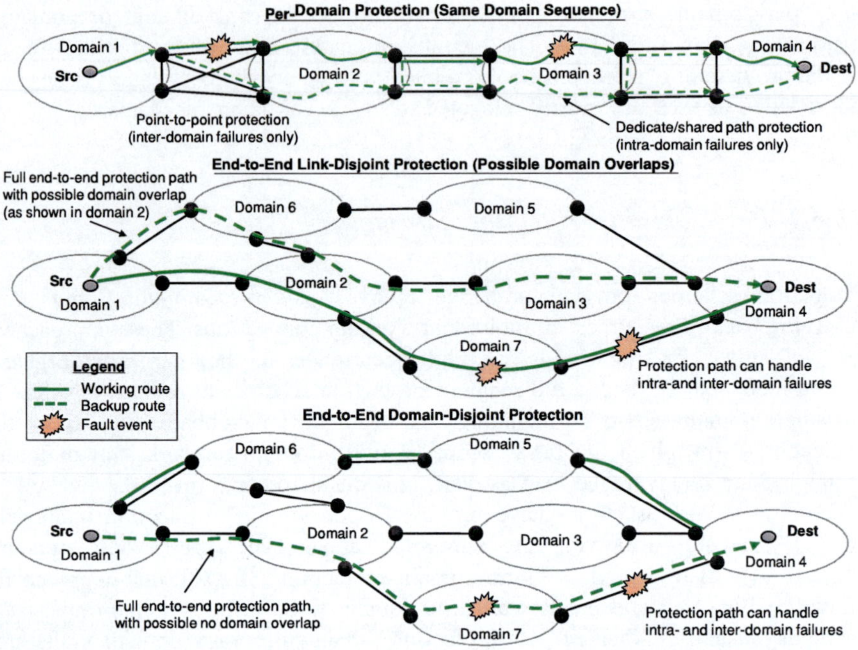

Fig. 13.5 Multi-domain protection schemes (intra-domain routes not shown)

vulnerable to *intra-domain* trap topologies, the second strategy jointly expands intra-domain working/protection routes, e.g., via Surballe's algorithm. Overall results show that the joint scheme gives improved setup success and lower path costs.

Meanwhile, [69] presents a more comprehensive look at "per-domain" diverse path pair routing and addresses both computation and setup signaling issues in distributed multi-PCE settings. Specifically, the authors outline two broad computation strategies based upon *sequential* and *simultaneous* (parallel) setup. Namely, sequential schemes first signal and resolve complete E2E primary paths (without any considerations for backup paths) and then use recorded route information to setup link-diverse diverse backup paths. Along these lines, two variants are proposed, i.e., signaling-based (RSVP-TE) route exclusion and PCE-based route exclusion. In particular, PCE-based route exclusion uses the BRPC method to compute a protection VSPT (by tracking routes via PCEP messages). However, as noted earlier, sequential schemes are susceptible to *inter-domain* trap topologies and can also yield sub-optimal route sequences, i.e., cannot guarantee diverse path pair discovery even if it exists. To resolve this, the authors propose a simultaneous setup scheme (using inter-domain PCE communication) which performs joint primary/backup pair computation and better optimizes TE constraints. Nevertheless, the associated overheads and complexities with "purely" simultaneous PCE computation can be quite high. As a result, a novel PCE-based route exclusion scheme is also proposed to compute multiple diverse paths between domain boundary nodes (and also avoid hops that can block setup). This scheme essentially mimics simultaneous setup behaviors in a sequential fashion. Overall results show that the latter PCE-based route exclusion scheme gives the lowest blocking. This improvement is most evident when the number of boundary nodes in a domain is small, i.e., two, as this leads to higher backup path blocking.

In a similar theme, [94] augments the "working-mode" BRPC scheme (Sect. 13.4) to achieve dedicated 1 + 1 inter-domain LSP protection setup. Here it is assumed that E2E domain sequences are pre-specified, and hence the focus is only on intra-domain path computation and border node selection. Specifically, PCEP message extensions are defined to indicate the availability of multiple disjoint trees between egress/ingress border nodes, thereby preventing suboptimal choices and increasing path success rates. However, detailed performance results are not presented for any of the proposed schemes.

Finally, it is well understood that dedicated protection strategies imply relatively higher resource overheads. Hence some researchers have also implemented resource sharing for "per-domain" protection, i.e., by leveraging existing work for single domain networks [6]. These schemes basically allow sharing of resources between backup routes which have disjoint working paths. For example, [95] proposes a solution for interfacing IP and DWDM networks supporting two-connected (or more) domains, where each domain has at least two border nodes connecting to another domain. Here, inter-domain paths are pre-computed using ILP formulations to prioritize ingress/egress gateways, and the results are stored by a PCE, assuming long-standing demands. Subsequently, RSVP-TE signaling is used to identify primary and backup domain ingress/egress border node pairs and expand dedicated or shared *intra-domain* protection routes. For dedicated protection, these local paths are computed using Surballe's disjoint paths algorithm, whereas for shared protection, optimal ILP formulations are used. The results here show good efficiency gains with ILP-based shared protection, albeit protection paths are longer. Meanwhile, [96] also develops another shared "per-domain" shared (intra-domain) protection scheme for DWDM networks with full wavelength conversion. The scheme assumes redundant inter-domain fiber links as well as virtual inter-domain graph topologies to compute loose working routes (via same algorithm as [97]). Conversely, shared intra-domain routes are computed in a load-balancing manner and results show notably faster recovery times versus [97], albeit utilizations are higher as well. Note that further renditions can also make use of shared link protection between gateways.

13.6.2 End-to-End Path and SubPath Protection Strategies

Although "per-domain" protection offers some key benefits (including rapid recovery), it forces both primary and backup paths to traverse the same E2E domain sequences [69]. This mandates some prior knowledge of domain-level connectivity, as provided by pre-computation and prior agreements (not discussed in detail in most studies). Hence "per-domain" schemes impose stringent connectivity/diversity requirements at domain boundaries, and may still give reduced efficiency and increased vulnerability to multi-failure events. It is here that alternate strategies have been developed to increase "domain diversity" [91] between E2E primary/backup routes and thereby improve resiliency, see Fig. 13.5. These approaches are particularly suited to hierarchical link-state routing setups (Sect. 13.3.1) and are now detailed further.

Various studies have looked at the *joint* computation of inter-domain protection routes. For example, earlier work in [90] proposes three different E2E path protection strategies for multi-segment networks, where a "segment" is essentially a domain. The first scheme assumes full global knowledge and is really an idealization, i.e., *basic E2E* (BE2E). Meanwhile, the second scheme computes "domain-disjoint" working/protection routes and requires a "top-layer" skeleton network view, i.e., *disjoint segment protection* (DSP). Finally, the third scheme is similar to "per-domain" strategies [92] (without inter-domain link protection) and assumes that the backup path follows the same domain sequence and uses intra-domain recovery, i.e., *concatenated segment protection* (CSP). Expectedly, results show the lowest blocking and hop counts with the global knowledge BE2E scheme. Meanwhile the more restrictive CSP approach gives the worst performance, whereas the DSP achieves a medium between the two, i.e., partial global state. However, details on the disjoint pair path computation algorithms are not presented nor the impacts of dated inter-domain routing state analyzed.

In a similar theme, [98] studies dedicated *diverse* inter-AS path selection in multi-domain PCE-enabled networks (Fig. 13.5) supporting bandwidth/delay and also policy constraints. Here a two-step heuristic is developed to first compute feasible routes to a destination domain by using a breadth-first search, i.e., modified constrained k-shortest path algorithms. These routes are then searched for a subset of least-cost diverse routes using an ILP formulation. The findings show that the scheme can work in large networks and is very competitive with an idealized minimum cost ILP formulation. However, mandating domain-disjoint backup paths [98] can be overly restrictive and lead to high blocking. Even though this requirement can be relaxed by allowing primary/backup paths to traverse common domains, further provisions are required to ensure intra-domain link-disjointness (and avoid trap topologies) along intra-domain routes, as noted in [2] (see also Fig. 13.5). Along these lines, [99] proposes a novel hierarchical routing solution (Sect. 13.3.1) that uses full-mesh topology abstraction to extract and propagate critical *path-level* diversity state, based upon Surballe's algorithm. A distributed algorithm is then presented to compute link-disjoint inter-domain path pairs and it is formally shown that this abstracted state can enable simultaneous computation of link-disjoint path pairs. However, associated dimensionalities are quite high here, on the order of $O(N^4)$, for N border nodes, and this will limit scalability.

In general, shared protection can provide very good improvements in backup resource efficiencies. Hence related schemes for domain-diverse protection have also been studied [97, 99–101]. For example, in [97] the authors propose novel virtual edge (full-mesh) topology abstractions to aggregate specialized information along intra-domain routes, e.g., residual capacity, allocated backup capacity and maximal backup bandwidth. Using this information, "two-step" *shared path protection* (SSP) schemes are built, in which the first step computes skeleton LR paths and the second step expands intra-domain routes. Moreover two different SSP schemes are proposed, *sequential* and *joint* working/backup path computation. Results show improved blocking reduction with the latter. Further renditions of this work in [100] also propose backup re-optimization strategies—a first for multi-domain networks. The objective here is to intermittently re-route existing backup paths to free up resources, e.g., after fixed periods, or excessive blocking intervals. In particular, three schemes are proposed using ILP modeling to minimize backup capacity, i.e., global rerouting (across entire network), local rerouting (serially per-domain), and local rerouting with least effort (for smallest numbers of backup paths). Although the global scheme is projected to give the best results, it is not tested due to its computational complexity. Meanwhile the localized schemes are shown to give good gains, i.e., 20–40% blocking reductions, with minimum overheads when used with blocking-based triggering.

In addition, [101] also studies shared protection (termed as "restoration") in multi-domain networks running hierarchical routing with full-mesh topology abstraction. Two schemes are proposed here, one for computing "link-disjoint" backup paths and the other for computing "domain-disjoint" backup paths. However, since both of these schemes perform sequential computation (working and then protection), they are vulnerable to inter-domain trap topologies. Now in terms of topology abstraction, each virtual link is derived by computing the shortest path between the border nodes and recording its bottleneck bandwidth. In addition, a global transit traffic matrix is also generated to track link dependencies for (shared) backup bandwidth. Finally, for the link-disjoint scheme, a binary matrix is introduced to record link diversity between (the shortest paths of) all virtual links, i.e., resulting in $O(N^4)$ overheads for N border nodes. Overall findings for an NSFNET-based multi-domain network show slightly higher link utilizations and backup resource consumption with the domain-disjoint scheme, albeit blocking is comparable between the two approaches.

Subpath protection has also been proposed as a means to improve recovery times and protect against border node failures in multi-domain settings. This approach strikes a balance between purely localized "per-domain" protection strategies (Sect. 13.6.1) and E2E path protection strategies. Now a range of subpath protection schemes have been studied for single-domain networks, see [102, 103]. However, these algorithms are not readily extendible to multi-domain settings owing to the limitations of partial (abstracted) state. To address these concerns, [104] proposes an overlapped segment (subpath) scheme, using the shared protection framework introduced in [97]. Specifically, a working path is first computed on the same virtual topology and then segmented into subpaths as bounded by a maximum length. Here, two different segmentation schemes are proposed, one using a greedy method the other using a dynamic programing approach. Finally, minimum cost shared backup subpaths are computed for each segment, i.e., where the cost metrics are the same as in [97]. In addition, a "blocking-go-back" option is also specified to allow re-computation re-tries during intra-domain expansion failures, i.e., essentially a form of crankback. A range of results show much-improved sharing efficiency (versus dedicated protection) as well-added benefits from intra-domain retries. However, blocking rates can increase with shorter segment sizes as well. Some further variations of this work are also presented in [105] along with some discussions on real-world inter-domain routing issues. Namely, the authors propose sending updates after each request (i.e., resource change) and also note related scalability concerns with this approach. Although optional timer-based updates are also outlined, the degradation in performance resulting from dated state (and/or associated scalability issues of the various update policies) are not analyzed.

In addition, [106] also tables shared segment protection for multi-domain DWDM networks. This algorithm first computes a LR working path on a virtual (global) topology and then divides it into several subpath segments spanning "two domains" each, i.e., overlapped to provide domain boundary protection. The algorithm then computes two intra-domain and one inter-domain shared protection subpaths to protect against local and external faults. Here, all path routing is done using shortest path routing with the same link cost metrics as in [97, 105]. Overall, the findings show that the proposed scheme gives notably faster recovery (over 50% lower) but consumes more resources and yields higher blocking.

Carefully note that increased diversity between primary/backup domains may still not prevent failures for cases where operators lease underlying fibers along a common route [95]. As a result, it will be further necessary to incorporate added SRLG information into the path computation algorithms. Although this issue has been considered in detail for single domains [107], multi-domain extensions require further study, as also noted in [69].

Finally, some multi-domain path-vector recovery schemes have also been investigated. The key goal here is to reduce the lengthy recovery timescales associated with "post-fault" re-computation in "vanilla" BGP, i.e., re-convergence times in the minutes range [56]. Akin to QoS extensions for BGP (Sect. 13.3.2), the common approach here is to augment message updates to disseminate alternate "diverse" route information and activate them upon failures, i.e., essentially a form of E2E protection. For example, *reliable BGP* (R-BGP) [108] requires domains to advertize additional "failover" paths to themselves which are most disjoint (and policy-compliant) from their primary paths. Results show notable improvements in recovery/reconvergence times with as well as reduced domain disconnectivity. Meanwhile, [109] runs separate BGP processes to compute "complimentary", i.e., AS node-disjoint, policy-compliant routes to each destination. Here the gateway nodes directly solicit (i.e., pull) alternate routes, thereby limiting advertisements to a need-only basis and improving scalability. Bilateral negotiation is also added to allow intermediate domains more control over their transit traffic. Furthermore, [110] also presents path splicing mechanisms to improve inter-domain recovery with BGP. Finally, [111] leverages multi-path routing and runs separate BGP "routing planes" to compute diverse destination routes. This solution implements optimized pre-selection of backup egress points and is shown to achieve rapid recovery against intra- and inter-AS faults (even with a small number of routing planes). Nevertheless, all of these schemes are designed for packet-forwarding domains and will require further adaptations for optical DWDM path-vector routing scenarios.

13.6.3 p-*Cycle Protection Strategies*

The concept of *protection cycles* (*p*-cycles) was originally proposed for optical network survivability in [112]. The basic idea here is to route spare capacity "cycle" overlays across generalized network topologies in order to achieve rapid "ring-like" recovery speeds. Namely, a *p*-cycle provides a restoration path for every span which has both end nodes on the cycle. In addition, protection is also provided for straddling spans, i.e., those with end-points on cycle but not on the cycle. To date a wide range of studies have been done to improve the basic *p*-cycle concept, see survey in [113]. More importantly, some recent efforts have also extended this solution for multi-domain optical networks. Below we consider some of the details.

The first application of *p*-cycles for multi-domain survivability was presented in [114]. Here authors focus on providing resilience for inter-domain links and the solution resolves the survivability problem at two levels, intra-domain and inter-domain. Namely, intra-domain protection uses "domain-internal" mechanisms to recover faults whereas inter-domain protection uses *p*-cycles to recover inter-domain link failures. The basic idea here is to compute *p*-cycles over a skeleton virtual topology (with domains represented as nodes, i.e., simple node) and then determine the requisite border nodes to traverse. In addition, pre-configured intra-domain connections are also linked to the *p*-cycle to protect straddling links [115]. Detailed procedures are then presented to improve sharing of backup bandwidth on inter-domain links. Overall, the findings indicate that *p*-cycle schemes can not only give good availability but also consume more resources than traditional protection strategies. This work is also extended in [116] by developing a more elaborate ILP-based *origin–destination* (OD) cycle framework.

In general, the work in [114, 115] uses the entire multi-domain virtual topology to compute the *p*-cycles. As such, this can result in overly lengthy routes with unacceptably high impairments. Moreover, computational complexities can also grow to prohibitive levels here. To address these concerns, the authors in [116] apply domain partitioning strategies (based upon spectral clustering) to segment domains into smaller "sub-multi-domain" networks. Localized *p*-cycles protection is then applied within these reduced entities and further provisions are also introduced to protect inter-domain links connecting the partitions. Results show much-improved computational scalability, by almost two orders of magnitude, with a very small increase in redundancy overheads, i.e., 2–4% range.

Carefully note that multi-domain *p*-cycle protection is a static, pre-computation-type approach. As such, it will require the availability of detailed domain-internal topological state as well as off-line computation to provision the requisite *p*-cycle overlay routes. This contrasts with the earlier-discussed dynamic "on-line" protection (Sects. 13.6.1, 13.6.2) and restoration strategies (Sect. 13.6.4) for multi-domain survivability.

13.6.4 Post-Fault Restoration Strategies

Overall, multi-domain protection strategies pose some notable complexities in terms of path computation and routing overheads (as discussed in Sect. 13.3.1). As a result, many carriers are interested in *post-fault* restoration schemes which can dynamically re-compute failed routes using crankback. Even though these algorithms offer more latent recovery (100's ms to seconds range) and no specific guarantees [1, 8], they are quite suitable for mid-tiered data services and intra-carrier settings (with homogeneous signaling paradigms). Moreover, from a practical perspective, restoration strategies impose minimal routing requirements and can readily be implemented using existing RSVP-TE signaling or PCE standards. Perhaps most important of all, these reactive strategies consume much less resources than counterpart protection schemes, and also provide a viable means of recovery against random correlated multi-failure attacks.

Nevertheless, there are very few studies on multi-domain restoration recovery. For example, the work in [72] compares crankback recovery versus more advanced PCE recovery using the BRPC scheme (from Sect. 13.4) for single link failures in multi-domain IP/MPLS networks. Here, a very basic "per-domain" scheme is used to crankback to upstream ingress border nodes and re-initiate setup on failed routes, i.e., a form of intermediate restoration. The results show that this approach gives notably lower performance versus BPRC, both in terms of higher recovery times and lower restoration success rates. However, this overall scheme lacks some of the key crankback enhancements proposed in [84, 85], i.e., such as improved "next-hop" domain selection and non-exhaustive search behaviors.

To resolve some of the above challenges, [117] adapts the joint intra-/inter-domain working-mode crankback scheme of [85] (Sect. 13.5) for both intermediate and E2E restoration. The former initiates crankback re-tries from the working ingress border node immediately upstream of the failed node/link, i.e., to preserve parts of the existing working route. Meanwhile the latter re-attempts setup from the source node. Also, both of these schemes track failed links using RSVP-TE exclude route fields and akin to [84], assume minimal inter-domain knowledge for "next-hop" domain selection (via inter-AS BGP protocol databases). Overall simulation results for two test-case topologies show the best gains with the E2E scheme, which gives over 80% restoration success rates even under high-load settings (without any increase in blocking for working-mode connections). In addition, E2E restoration gives lower hop count utilization and recovery times.

Note that the authors in [118] have also applied this scheme to study *multi-failure* recovery in multi-domain networks, i.e., as resulting from large-scale *weapons of mass destruction* (WMD) attacks. However, most of this work is focused on IP bandwidth routing networks only. As such there remains significant scope to develop further crankback solutions for DWDM networks.

13.7 Multi-Layer Considerations

In general, most operational backbone networks are not homogenous in terms of their technology usage. Hence multi-domain provisioning will require further interfacing between domains operating at different network technology layers, i.e., *multi-layer/multi-domain integration* [1, 3]. Indeed, there is a clear grooming dimension to this problem, owing to the differing link granularities that may be encountered along E2E routes, as noted in [119] (Chap. 15). For example, multiple packet LSP flows (or SONET circuits) originating in MPLS (TDM) domains can be groomed over a single coarse DWDM lightpath "tunneled link connection" traversing one/more DWDM transport domains. Now even though regular grooming has been widely studied [119], many of these proposed strategies cannot be re-applied within distributed multi-domain environments as they assume global visibility across all layers (flat networks). As a result, only a handful of studies have looked at multi-layer integration and this is an area which requires further work. Below we consider a brief overview.

The authors in [120] study grooming across SONET-DWDM network domains and propose threshold-based mechanisms for lightpath addition/deletion. Here, both centralized and distributed schemes are proposed, with the latter only using traversing path state at a node, i.e., no routing. The overall results for ring and mesh networks show good cost reductions (in terms of port costs) for mid-range threshold values and smaller request granularities. However, this work only focuses on lightpath addition/deletion and not its application in broader inter-domain routing setups. Meanwhile, [121] addresses path provisioning in multi-domain networks with multiple granularities. Namely, a *multi-segment* multi-layer graph model is defined where gateway nodes have boundary grooming ports and weight assignments are chosen to minimize hop counts or mux/demux points. The graph model is then evaluated for various path selection schemes mirroring those in [48], e.g., centralized (full knowledge, unrealistic), domain-by-domain (local knowledge), and hierarchical source routing (partial inter-domain knowledge). The latter approach only disseminates domain segment information for specified granularity levels, although details on state compression are not given. Results for two-granularity SONET networks show lower blocking with the hierarchical approach (increased global state) versus the domain-by-domain approach. In addition, [122] also studies grooming in multi-domain waveband switching networks in order to reduce switching port costs. Here a multi-layer *waveband grooming with layered auxiliary graph* (WGLAG) model is proposed using full-mesh abstraction, along with LR schemes to compute E2E lightpath route sequences. Results show reduced per-port costs and blocking with the grooming model, albeit critical inter-domain routing dissemination concerns are not addressed.

Meanwhile [123] studies Ethernet-DWDM domain integration and outlines an "integrated" routing and signaling framework to handle two request granularities, i.e., subrate *Ethernet virtual circuits* (EVC) and DWDM lightpaths. In contrast to conventional provisioning, signaling messages for sublambda EVC requests are only processed at the source lightpath OXC nodes and end-points are allowed to own the E2E tunneled link. Tunneled links (lightpath segments) are taken down using an idle (i.e., no usage policy) and ownership rules are also defined to avoid deadlock conditions, see [123]. However, the scheme assumes a single underlying DWDM domain, thus simplifying dynamic routing design. Overall simulation results for various routing update strategies (timer, threshold-based) show minimal increase in setup latencies and signaling loads. In addition, the highest resource contention is also seen at lower loads with infrequent updates.

Meanwhile, the work in [124] looks at more realistic distributed multi-layer/ multi-domain networks with grooming capabilities. Namely, a new *GMPLS exchange point* (GXP) solution is proposed for resource visibility and exchange between carrier and client-network domains. In particular, resource visibility is used to define usage policies for transmission, multiplexing, and switching at a layer. The GXP then collects routing and policy information and acts as a proxy, by transforming physical network topologies to generate *service visibility graph* (SVG) views for customers. These entities are then used for path computation, and two strategies are proposed, i.e., layer-by-layer (which does TE computation one plane at a time) to more complex combined routing. In addition, various tunneled link release policies are also proposed, i.e., no release, scheduled and idle release. Overall results show lower blocking when there is better visibility between client/ provider domains, although resource partitioning may be required at higher loads to ensure QoS support. In addition, adaptive tunneled link release mechanisms also tend to give better results.

Meanwhile the work in [125] presents a detailed "distributed grooming" framework for multi-layer IP (or Ethernet) and DWDM domain integration based upon the peer model [8] (Sect. 13.2.3). Namely, an integrated hierarchical link-state routing scheme is introduced to disseminate dynamic information between domains at both layers, i.e., physical, abstract, *and* tunneled inter-domain link-state. Commensurate inter-domain routing/triggering policies are then defined along with tunneled link setup/takedown signaling procedures. Namely, such underlying "link" connections are taken down when capacity usages drop to zero, i.e., akin to idle release in [124]. Using this, distributed grooming LR algorithms are proposed using a layer-by-layer load-balancing TE approach [125]. Detailed findings show much lower blocking with the use of full-mesh abstraction and increased tunneled link counts. However, associated inter-domain routing overheads are very significant here owing to tunneled link advertisements.

Finally, [126] also addresses inter-layer TE in PCE-based settings, with a particular focus on defining frameworks to manage "tunneled" LSP connections. Namely, two different models are presented, one based upon higher-layer signaling and another using a *virtual network topology manager* (VNTM) entity. The former scheme defines additional signaling functions for higher-layer border nodes (that source/sink tunneled links) to setup/takedown originating lower-layer LSP segments. Meanwhile the latter approach introduces a new VNTM entity which cooperates with a domain PCE to setup and maintain a set of "virtual" underlying links. However, this work only details these frameworks and does not present any analysis results for different inter-layer TE mechanisms.

Finally, [127] also studies protection in multi-domain IP-over-DWDM networks. Here it is assumed that each IP domain is connected via at least two different links to at least two different OXC nodes. A range of recovery schemes are then considered at the optical layer, including one which sets up working and backup lightpaths per-domain-pair and the other which only sets ups working lightpaths. Dynamic post-fault recovery (restoration) of lightpaths is also studied. Overall results show that resource sharing at the optical layer is very beneficial in reducing capacity overheads. However, dynamic restoration is found to give higher port costs and may also impact convergence at the higher-layer IP routing level.

13.8 Conclusions

This chapter presents a survey of the major provisioning and survivability solutions for multi-domain networking, a very challenging and evolving area. First, the latest standards activities and contributions are surveyed across the key forums such as the ITU-T, IETF, and OIF. Subsequently, the latest research work in related multi-domain subtopic areas is presented, with a focus on optical DWDM networks. Namely, this includes work on hierarchical routing strategies—using link-state and path-vector approaches—as well as other distributed path computation strategies based upon the IETF PCE framework. In addition, a full range of survivability schemes are also discussed, including "per-domain" protection, E2E protection, crankback restoration, and multi-domain p-cycle strategies. Finally, some broader contributions in multi-layer operation are also overviewed.

Acknowledgments We are extremely grateful to the editors, Profs. George Ellinas and Neo Antoniades, for their patience and encouragement during the preparation of this manuscript. We would also like to thank the series editor, Prof. Biswanath Mukherjee, for his guidance and vision in building this invaluable series. Finally, we are very thankful to Mr. Feng (Stephen) Xu and Mostafa Esmaeili for their assistance with the diagrams and proof-reading. Overall, this work has been supported in part by the Department of Energy Office of Science (under award DE-SC0001229), the National Science Foundation (under award CNS-0806637), and the Defense Threat Reduction Agency (Basic Research Program). We are indeed very grateful to these agencies for their generous support.

References

1. Ghani N et al (2008) Control plane design in multidomain/multilayer optical networks. IEEE Commun Mag 46(6):78–87
2. Chamania M, Jukan A (2009) A survey of inter-domain peering and provisioning solutions for the next generation optical networks. IEEE Commun Surv Tutor 11(1):33–51
3. Alanqar W, Jukan A (2004) Extending end-to-end optical service provisioning and restoration in carrier networks: opportunities, issues, and challenges. IEEE Commun Mag 42(1):52–60
4. Medhi D (2002) Quality of service (QoS) routing computation with path caching: a framework and network performance. IEEE Commun Mag 40(12):106–113
5. Zang H, Jue J, Mukherjee B (2000) A review of routing and wavelength assignment approaches for wavelength- routed optical WDM networks. Opt Netw Mag 1(1):47–60
6. Cholda P, Mykkeltveit A, Helvik BE, Wittner OJ, Jajszczyk A (2007) A survey of resilience differentiation frameworks in communication networks. IEEE Commun Surv Tutor 9(4):32–55
7. Farrell A (2006) A framework for inter-domain traffic engineering. IETF RFC 4726
8. Bernstein G, Rajagopalan B, Saha D (2003) Optical network control-architecture, protocols and standards. Addison-Wesley Longman, Boston
9. Tomic S et al (2004) ASON and GMPLS overview and comparison. Photonic Netw Commun 7(2):111–130
10. Architecture for the Automatic Switched Optical Networks (ASON) (2001) ITU-T Rec. G.8080/Y.1304
11. Requirements for Automatic Switched Transport Networks (2001) ITU-T Rec. G.807/ Y.1302
12. Ceccarelli D, Caviglia D, Zhang F, Li D, Belotti S, Grandi P, Rao R, Pithewan K, Drake J (2010) Traffic engineering extensions to OSPF for generalized MPLS (GMPLS) control of evolving G.709 OTN networks. IETF Draft draft-ceccarelli-ccamp-gmpls-ospf-g709-03
13. Papadimitriou D (2006) Generalized multi-protocol label switching (GMPLS) signaling extensions for G.709 optical transport networks control. IETF RFC 4328
14. User Network Interface (UNI) 1.0 Signaling Specification (2001) OIF-UNI-01.0
15. User Network Interface (UNI) 2.0 Signaling Specification (2008) OIF-UNI-02.0-Common
16. OIF E-NNI Signaling Specification (2009) OIF-E-NNI-Sig-02.0
17. External Network–Network Interface (E-NNI) OSPF-Based Routing 1.0 (Intra-Carrier) Implementation Agreement. (2007) OIF-ENNI-OSPF-01.0
18. Intra-Carrier E-NNI Signaling Specification (2004) OIF-E-NNI-Sig-01.0
19. Bernstein G, Cheng D, Pendarakis D (2002) Domain to domain routing using GMPLS, OSPF extension V1.1. Optical internetworking forum OIF2002.23.06
20. Apostolopoulos G, Kama S, Williams D, Guerin R, Orda A, Przygienda T (1999) QoS routing mechanisms and OSPF extensions. IETF RFC 2676
21. Katz D (2003) Traffic engineering (TE) extensions to OSPF version 2. IETF RFC 3630
22. Kompella K, Rekhter Y (2005) OSPF extensions in support of generalized multi-protocol label switching (GMPLS). IETF RFC 4203
23. Papadimitriou D, Farrel A (2007) Generalized MPLS (GMPLS) RSVP-TE signaling extensions in support of calls. IETF RFC 4974
24. Lehman T, Yang X, Guok CP, Rao NSV, Lake A, Vollbrecht J, Ghani N (2007) Control plane architecture and design considerations for multi-service, multi-layer, multi-domain hybrid networks. In: Proceedings of IEEE international conference on computer communications (INFOCOM), high speed networking (HSN) workshop, Anchorage, AK
25. Chen M, Zhang R, Duan X (2009) OSPF extensions in support of inter-autonomous system (AS) MPLS and GMPLS traffic engineering. IETF RFC 5392
26. Farrel A, Satyanarayana A, Iwata A, Fujita N, Ash G (2007) Crankback signaling extensions for MPLS and GMPLS RSVP-TE. IETF RFC 4920

27. Farrel A, Vasseur J-P, Ash J (2006) A path computation element (PCE)-based architecture. IETF RFC 4655
28. Vasseur J, Le Roux J (2009) Path computation element (PCE) communication protocol (PCEP). IETF RFC 5440
29. Bernstein G, Yee Y (2009) Extending GMPLS/PCE for use in wavelength switched optical networks. In: Proceedings of IEEE/OSA optical fiber communication conference (OFC), San Diego, CA
30. Farrel A, Bradford R, Vasseur JP (2009) Resource reservation protocol (RSVP) extensions for path key support. IETF RFC 5553
31. Vasseur J, Ayyangar A, Zhang R (2008) Per-domain path computation method for establishing inter-domain traffic engineering (TE) label switched paths (LSPs). IETF RFC 5152
32. Jukan A, Karmous-Edwards G (2007) Optical control plane for the grid community. IEEE Commun Surv 9(3):33–40
33. Open Grid Forum (OGF) working group, http://www.ogf.org
34. Network Description Language (NDL) website, http://www.science.uva.nl/research/sne/ndl
35. DICE (Dante-Internet2-CANARIE-ESnet) collaboration, see website at http://www.controlplane.net/
36. Gommans L et al (2006) Applications drive secure lightpath creation across heterogeneous domains. IEEE Commun Mag 44(3):100–106
37. Maier G, Busca C, Pattavina A (2008) Topology-information periodic updates in multi-domain ASON networks with topology aggregation. Fiber Integr Opt 27(4):265–277
38. Iliadis I (2000) Optimal PNNI complex node representations for restrictive costs and minimal path computation time. IEEE/ACM Trans Netw 8(4):493–506
39. Hao F, Zegura E (2000) On scalable QoS routing: performance evaluation of topology aggregation. In: Proceedings of IEEE international conference on computer communications (INFOCOM), Tel Aviv, Israel, pp 147–156
40. Kormaz T, Krunz M (2000) Source-oriented topology aggregation with multiple QoS parameters in hierarchical networks. IEEE/ACM Trans Model Comp Simul 10(4):295–325
41. Awerbuch B, Shavitt Y (2001) Topology aggregation for directed graphs. IEEE/ACM Trans Netw 9(2):82–90
42. Liu K, Nahrstedt K, Chen S (2004) Routing with topology abstraction in delay-bandwidth sensitive networks. IEEE/ACM Trans Netw 12(1):17–29
43. Liu G, Ji C, Chan VWS (2005) On the scalability of network management information for inter-domain light-path assessment. IEEE/ACM Trans Netw 13(1):160–172
44. Sanchez-Lopez S, Masip-Bruin X, Marín-Tordera E, Solé-Pareta J, Domingo-Pascual J (2005) A hierarchical routing approach for GMPLS-based control plane for ASON. In: Proceedings of IEEE international conference on communications (ICC), Seoul, Korea
45. Zhou J, Yuan X (2002) A study of dynamic routing and wavelength assignment with imprecise network state information. In: Proceedings of international conference parallel processing (ICPP), Workshop on optical networks, Vancouver, Canada, pp 1–3
46. Liu Q, Kok MA, Ghani N, Gumaste A (2006) Hierarchical routing in multi-domain optical networks. Comput Commun 30(1):122–131
47. Liu Q, Ghani N, Rao NSV, Gumaste A, Garcia ML (2007) Distributed inter-domain lightpath provisioning in the presence of wavelength conversion. Comput Commun 30(18):3362–3375
48. Zhu Y, Jukan A, Ammar M (2003) Multi-segment wavelength routing in large-scale optical networks. In: Proceedings of IEEE international conference on communications (ICC), Anchorage, AK
49. Liu Q, Xie C, Frangieh T, Ghani N, Gumaste A, Rao NSV (2009) Routing scalability in multi-domain DWDM networks. Photonic Netw Commun 17(1):63–74
50. Yu Y, Jin Y, Sun W, Guo W, Hu W (2009) On the efficiency of inter-domain state advertising in multi-domain networks. In: Proceedings of IEEE global communications conference (GLOBECOM), Honolulu, Hawaii, pp 1–3

51. Zhang N, Bao H, Liu A (2007) Analysis on routing convergence in optical network. In: Proceedings of IEEE international conference on network and parallel computing workshops (IFIP), Liaoning, China
52. Vanbever L, Quoitin B, Bonaventure O (2009) A hierarchical model for BGP routing policies. In: Proceedings of ACM SIGCOMM workshop on programmable routers for extensible services of tomorrow, Barcelona, Spain
53. Azodolmolky S, Klinkowski M, Marin E, Careglio D, Solé-Pareta J, Tomkos I (2009) A survey on physical layer impairments aware routing and wavelength assignment algorithms in optical networks. Comput Netw 53(7):926–944
54. Martínez R, Pinart C, Cugini F, Andriolli N, Valcarenghi L, Castoldi P, Wosinska L, Comellas J, Junyent G (2006) Challenges and requirements for introducing impairment-awareness into the management and control planes of ASON/GMPLS WDM networks. IEEE Commun Mag 44(12):76–85
55. Salvadori E, Ye Y, Zanardi A, Woesner H, Carcagni M, Galimberti G, Martinelli G, Tanzi A, La Fauci D (2007) Signaling-based architectures for impairment-aware lightpath set-up in GMPLS networks. In: Proceedings of IEEE global communications Conference (GLOBECOM), Washington, DC
56. Yannuzzi M, Yannuzzi M, Sánchez-López S, Masip-Bruin X, Solé-Pareta J, Domingo-Pascual J (2005) A combined intra-domain and inter-domain QOS routing model for optical networks. In: Proceedings of international conference on optical networking design and modeling (ONDM), Milan, Italy
57. St Arnaud B, Weir M, Coulter J (2001) BGP optical switches and lightpath route arbiter. Opt Netw 2(2):73–81
58. Blanchet M, Parent F, St Arnaud B (2001) Optical BGP (OBGP): InterAS Lightpath Provisioning. IETF Draft draft-parent-obgp-01.txt
59. Francisco M, Francisco MJ, Pezoulas L, Huang C, Lambadaris I (2002) End-to-end signaling and routing in optical IP networks. In: Proceedings of IEEE international conference on communications (ICC), New York, NY
60. Yu O (2004) Intercarrier interdomain control plane for global optical networks. In: Proceedings of IEEE international conference on communications (ICC), New York, NY
61. Yannuzzi M, Masip-Bruin X, Sanchez-Lopez S, Tordera EM, Sole-Pareta J, Domingo-Pascual J (2006) Interdomain RWA based on stochastic estimation methods and adaptive filtering for optical networks. In: Proceedings of IEEE global communications conference (GLOBECOM). San Francisco, CA
62. Yannuzzi M, Masip-Bruin X, Fabrego G, Sanchez-Lopez S, Sprintson A, Orda A (2008) Toward a new route control model for multidomain optical networks. IEEE Commun Mag 46(6):104–111
63. Yang X, Ramamurthy B (2003) Inter-domain dynamic routing in multi-layer optical transport networks. In: Proceedings of IEEE global communications conference (GLOBECOM), San Francisco, CA
64. Boucadair M (2005) QoS-enhanced border gateway protocol. IETF Draft draft-boucadair-qos-bgp-spec-01.txt
65. Griffin D, Spencer J, Griem J, Boucadair M, Morand P, Howarth M, Wang N, Pavlou G, Asgari A, Georgatsos P (2007) Interdomain routing through QoS-class planes. IEEE Commun Mag 45(2):88–95
66. Xiao L, Lui K-S, Wang J, Nahrstedt K (2002) QoS extension to BGP. In: Proceedings IEEE international conference on network protocols (ICNP), Paris, France
67. Zhang T, Cui Y, Zhao Y, Fu L, Korkmaz T (2006) Scalable BGP QoS extension with multiple metrics. In: Proceedings of IEEE international conference on networking and services (ICNS), Silicon Valley, CA
68. Xu W, Rexford J (2006) MIRO: multi-path interdomain routing. In: Proceedings of ACM SIGCOMM, Pisa, Italy
69. Takeda T, Oki E, Shiomoto K (2008) Diverse path setup schemes in multi-domain optical networks. In: Proceedings of IEEE BROADNETS, London, England

70. Vasseur JP, Zhang R, Bitar N, Le Roux JL (2009) A backward recursive PCE-based computation (BRPC) procedure to compute shortest constrained inter-domain traffic engineered label switched paths. IETF RFC 5441
71. Dasgupta S, de Oliveira J, Vasseur JP (2009) Performance analysis of inter-domain path computation methodologies. IETF RFC 5468
72. Dasgupta S, De Oliveira J, Vasseur J (2007) Path-computation-element-based architecture for interdomain MPLS/GMPLS traffic engineering: overview and performance. IEEE Netw 21(4):38–45
73. Geleji G, Perros HG, Xin Y, Beyene T (2008) A performance analysis of inter-domain QoS routing schemes based on path computation elements. In: Proceedings of IEEE high capacity optical networks (HONET), Penang, China
74. Dugeon O, Mingozzi E, Stea G, Bisti L (2008) Provisioning QoS in inter-domain traffic engineering. Ann Telecom Springer 63(11):545–557
75. Zhao Y, Zhang J, Ji Y, Gu W (2010) Routing and wavelength assignment problem in PCE-based wavelength-switched optical networks. IEEE/OSA J Opt Comm Netw 2(2): 196–205
76. Casellas R, Martinez R, Muñoz R, Gunreben S (2009) Enhanced backwards recursive path computation for multi-area wavelength switched optical networks under wavelength continuity constraint. In: Proceedings of IEEE/OSA Optical Fiber Communication Conference (OFC), San Diego, CA
77. Cugini F, Paolucci F, Valcarenghi L, Castoldi P, Welin A (2009) PCE communication protocol for resource advertisement in multi-domain BGP-based networks. In: Proceedings of IEEE/OSA optical fiber communication conference (OFC), San Diego, CA
78. Buzzi L, Bardellini MC, Siracusa D, Maier G, Paolucci F, Cugini F, Valcarenghi L, Castoldi P (2006) Hierarchical border gateway protocol (HBGP) for PCE-based multi-domain traffic engineering. In Proceedings of IEEE international conference on communications (ICC), Istanbul, Turkey
79. Wan X, Chen Y, Zhang H, Zheng X (2009) Dynamic domain-sequencing for inter-domain path computation in WDM networks. In: Proceedings of IEEE Asia communications and photonics conference and exhibition (ACP), Shanghai, China
80. Torab P, Jabbari B, Xu Q, Gong S, Yang X, Lehman T, Tracy C, Sobieski J (2006) On cooperative inter-domain path computation. In: Proceedings of IEEE international symposium on computers and communications (ISCC), Pula-Cagliari, Italy
81. Pelssner C, Bonaventure O (2006) Path selection techniques to establish constrained interdomain MPLS LSPs. In: Proceedings of IFIP international network conference, Coimbra, Portugal
82. Aslam F, Uzmi ZA, Farrel A, Pioro M (2007) Inter-domain path computation using improved crankback signaling in label switched networks. In: Proceedings of IEEE international conference on communications (ICC), Glasgow, Scotland
83. Aslam F, Uzmi ZA, Farrel A (2007) Interdomain path computation: challenges and solutions for label switched networks. IEEE Commun Mag 45(10):94–101
84. Xu F, Esmaeili M, Xie C, Ghani N, Peng M, Liu Q (2010) Enhanced crankback signaling for multi-domain traffic engineering. In: Proceedings of IEEE international conference on communications (ICC), Cape Town, South Africa
85. Xu F, Peng M, Esmaeili M, Ghani N (2009) Advanced crankback provisioning for multi-domain networks. In: Proceedings of IEEE high perform switching routing (HPSR), Dallas, TX
86. Esmaeili M, Xu F, Ghani N, Xie C, Peng M, Liu Q (2010) Enhanced crankback for lightpath setup in multi-domain optical networks. IEEE Comm Lett 14(5):480–482
87. Shariff A, Woodward M (2006) A distributed algorithm for unicast QoS-routing using path feasibility prediction. In: Proceedings of IEEE high perform switching routing (HPSR), Poznan, Poland

88. Giorgetti A, Sambo N, Cerutti I, Andriolli N, Castoldi P (2010) Suggested vector scheme with crankback mechanism in GMPLS-controlled optical networks. In: Proceedings of international conference on optical networking design and modeling (ONDM), Kyoto, Japan
89. Cugini F, Sambo N, Andriolli N, Giorgetti A, Valcarenghi L, Castoldi P, Le Rouzic E, Poirrier J (2008) Enhancing GMPLS signaling protocol for encompassing quality of transmission (QoT) in all-optical networks. IEEE/OSA J Lightwave Technol 26(19):3318–3328
90. Seetharaman S, Jukan A, Ammar M End-to-end dedicated protection in multi-segment optical networks. http://citeseerx.ist.psu.edu/viewdoc/download?doi=10.1.1.131.3016&rep=rep1&type=pdf
91. Takeda T, Farrel A, Ikejiri Y, Vasseur JP (2008) Analysis of inter-domain label switched path (LSP) recovery. IETF RFC 5298
92. Huang C, Messier D (2004) A fast and scalable inter-domain MPLS protection mechanism. J Comm Netw 6(1):1–8
93. Ricciato F, Monaco U, Ali D (2005) Distributed schemes for diverse path computation in multidomain MPLS networks. IEEE Commun Mag 43(6):138–146
94. Paolucci F, Cugini F, Valcarenghi L, Castoldi P (2008) Enhancing backward recursive PCE-based computation (BRPC) for inter-domain protected LSP provisioning. In: Proceedings of IEEE/OSA optical fiber communication conference (OFC), San Diego, CA
95. Staessens D, Colle D, Lievens U, Pickavet M, Demeester P, Colitti W, Nowe A, Steenhaut K, Romeral R (2008) Enabling high availability over multiple optical networks. IEEE Commun Mag 46(6):120–126
96. Guo L (2007) LSSP: a novel local segment-shared protection for multi-domain optical mesh networks. Comput Comm 8(8):1794–1801
97. Troung D, Thiongane B (2006) Dynamic routing for shared path protection in multidomain optical mesh networks. OSA J Opt Netw 5(1):58–74
98. Secci S, Rougier J, Pattavina A (2008) On the selection of optimal diverse AS-paths for inter-domain IP/(G)MPLS tunnel provisioning. In: Proceedings of telecommunication network workshop on QoS in multiservice IP networks, Venice, Italy
99. Sprintson A, Yannuzzi M, Orda A, Masip-Bruin X (2007) Reliable routing with QoS guarantees for multi-domain IP/MPLS networks. In: Proceedings of IEEE international conference on computer communications (INFOCOM), Anchorage, AK
100. Jaumard B, Troung D (2007) Backup path re-optimizations for shared path protection in multi-domain networks. In: Proceedings of IEEE global communications conference (GLOBECOM), San Francisco, CA
101. Gao Z, Nasser H (2008) End-to-end restoration algorithms in multi-domain mesh networks. In: Proceedings of IEEE international symposium on computers and communications (ISCC), Marrakech, Morocco
102. Ho P, Tapolcai J, Cinkler T (2004) Segment shared protection in mesh communications networks with bandwidth guaranteed tunnels. IEEE/ACM Trans Netw 12(6):1105–1118
103. Ou C, Zang H, Singhal NK, Zhu K, Sahasrabuddhe LH, MacDonald RA, Mukherjee B (2004) Subpath protection for scalability and fast recovery in optical WDM mesh networks. IEEE J Sel Areas Commun 22(9):1859–1875
104. Troung D, Jaumard B (2007) Using topology aggregation for efficient segment shared protection solutions in multi-domain networks. IEEE J Sel Areas Commun 25(9):96–107
105. Troung D, Jaumard B (2008) Recent progress in dynamic routing for shared protection in multidomain networks. IEEE Commun Mag 46(6):112–119
106. Xie X, Sun W, Hu W, Wang J (2008) A shared subpath protection strategy in multi-domain optical networks. In: Proceedings of IEEE global communications conference (GLOBECOM), New Orleans, LA
107. Xu D, Xiong Y, Qiao C, Li G (2004) Failure protection in layered networks with shared risk link groups. IEEE Netw 18(3):36–41
108. Kushman N, Kandula S, Katabi D, Maggs B (2007) R-BGP: staying connected in a connected world. In: Proceedings of USENIX symposium networked systems design and implementation (NSDI), Cambridge, UK

109. Liao Y, Gao L, Guerin R, Zhang Z-L (2008) Reliable interdomain routing through multiple complementary routing processes. In: Proceedings of ACM international conference emerging networking experiments and technology (CoNEXT), Madrid, Spain
110. Motiwala M, Elmore M, Feamster N, Vempala S (2008) Path splicing. In: Proceedings of ACM SIGCOMM, Seattle, WA
111. Wang N, Guo Y, Ho K-H, Howarth M, Pavlou G (2009) Fast network failure recovery using multiple BGP routing planes. In: Proceedings of IEEE global communications conference (GLOBECOM), Honululu, Hawaii
112. Grover W, Stamatelakis D (1998) Cycle-oriented distributed pre-configuration: ring-like speed with mesh-like capacity for self-planning network restoration. In: Proceedings of IEEE international conference on communications (ICC), Atlanta, GA
113. Grover W, Doucette J, Kodian A, Leung D, Sack A, Clouqueur M, Shen G (2006) Design of survivable networks based on *p*-cycles. In: Handbook of optimization in telecommunication, Springer, Germany, pp 391–434
114. Farkas A, Szigeti J, Cinkler T (2005) *p*-Cycles based protection schemes for multi-domain networks. In: Proceedings of IEEE design of reliable communication networks (DRCN), Ischia, Italy
115. Szigeti J, Romeral R, Cinkler T, Larrabeiti D (2009) *p*-cycle protection in multi-domain optical networks. Photonic Net Comm 17(1):35–47
116. Drid H, Cousin B, Molnár M, Ghani N (2010) Graph partitioning for survivability in multi-domain optical networks. In: Proceedings of IEEE international conference on communications (ICC), Cape Town, South Africa
117. Xu F, Esmaeili M, Peng M, Ghani N (2010) Multi-domain restoration with crankback IP/MPLS networks. Opt Switch Netw 8(1):68–78
118. Xu F, Peng M, Esmaeili M, Rahnamay-Naeini M, Khan S, Ghani N, Hayat M (2010) Post-fault restoration in multi-domain networks with multiple failures. In: Proceedings of IEEE military communications conference (MILCOM), San Jose, CA
119. Dutta R, Kamal A, Rouskas G (ed) (2008) Traffic grooming for optical networks: foundations, techniques, and frontiers. Springer, New York
120. Widjaja I, Saniee I, Qian L, Elwalid A, Ellson J, Cheng L (2003) A new approach for automatic grooming of SONET circuits to optical express links. In: Proceedings of IEEE international conference on communications (ICC), Anchorage, AK
121. Zhu Y, Jukan A, Ammar M, Alanqar W (2004) End-to-end service provisioning in multi-granularity multi-domain optical networks. In: Proceedings of IEEE international conference on communications (ICC), Paris, France
122. Guo L, Cao J, Wang X, Jiang D, Yang T (2010) Waveband grooming based on layered auxiliary graph in multi-domain optical networks. Opt Fiber Technol 16(3):162–171
123. Hadjiantonis A, Ali MA, Chamas H, Bjorkman W, Elby S, Khalil A, Ellinas G (2007) Evolution to a converged layer 1, 2 in a global-scale, native ethernet over WDM-based optical networking architecture. IEEE J Sel Areas Commun 25(5):1048–1058
124. Tomic S, Jukan A (2007) Performance analysis of infrastructure service provision with GMPLS-based traffic engineering. IEEE J Sel Areas Commun 25(5):881–894
125. Liu Q, Frangieh T, Xu F, Xie C, Ghani N, Gumaste A, Lehman T, Gouk C, Klasky S (2009) Distributed grooming in multi-domain IP/MPLS-DWDM networks. In: Proceedings of IEEE global communications conference (GLOBECOM), Oahu, Hawaii
126. Oki E, Inoue I, Shiomoto K (2007) Path computation element (PCE)-based traffic engineering in MPLS and GMPLS networks. In: Proceedings of IEEE Sarnoff Symposium, Princeton, NJ
127. Staessens D, Depre L, Colle D, Lievens I, Pickavet M, Demeester P (2006) A quantitative comparison of some resilience mechanisms in a multidomain IP-over-optical network environment. In: Proceedings of IEEE international conference on communications (ICC), Istanbul, Turkey

Chapter 14
Future Directions in WDM Systems and Networks

Georgios Ellinas, Neophytos (Neo) Antoniades and Ioannis Roudas

Abstract The field of optical communication systems and networks will most likely continue to flourish for years to come, mainly due to the continuous creation of new applications that demand more and more bandwidth in conjunction with more advanced functionalities required for the optical systems and networks at the transport, as well as the control layer. This chapter highlights some of the main emerging trends for WDM systems and networks.

14.1 Emerging Trends: What the Future Holds

The main focus of this book is to describe some of the most important current trends in WDM systems and networks design and engineering, including work on different physical effects causing system degradation in the WDM transport layer, advanced modulation formats and electronic compensation techniques,

G. Ellinas (✉)
Department of Electrical and Computer Engineering
University of Cyprus
1678 Nicosia, Cyprus
e-mail: gellinas@ucy.ac.cy

N. (Neo) Antoniades
Department of Engineering Science and Physics
College of Staten Island/The City University of New York
Staten Island, NY 10314, USA
e-mail: neo.antoniades@csi.cuny.edu

I. Roudas
Department of Electrical Engineering
University of Patras
26504 Rio, Greece
e-mail: roudas@ece.upatras.gr

N. (Neo) Antoniades et al. (eds.), *WDM Systems and Networks*, Optical Networks,
DOI: 10.1007/978-1-4614-1093-5_14, © Springer Science+Business Media, LLC 2012

state-of-the-art in device and network element level modeling, and WDM system/ network simulation techniques as well as commercial simulation tools. WDM systems described in this book included long-haul and ultra-long-haul optical communications systems, metro, access, and short range (in-building) systems and networks, as well as optical interconnects. Interactions between the physical (transport) and logical (control) layers are also investigated, with a focus on impairment-aware routing and protection and the book also addresses provisioning and survivability in multi-domain optical networks.

The preceding chapters in this book described in detail the current research efforts and technological advancements and commercial developments in the design and engineering of WDM systems and networks. However, from the description of the work in these chapters, it is clear that there are several challenging problems and future directions for these systems and networks. Since this book cannot address the large number of questions that arose, as well as all the new emerging trends, some of the main ones are noted in this chapter as topics for future exploration.

One of the important directions in which the technology is moving includes complex modulation and electronic compensation techniques and coherent systems to improve performance and spectral efficiency of optical communication systems in long-haul, metro, and access networks. Other emerging trends include advances in access networks such as converged optical-wireless networks and technologies, and next-generation WDM-PON architectures. Furthermore, advances in new types of fibers including multimode and plastic optical fibers (POFs) are in the forefront, to be used for high-speed short-reach communications, and new applications for avionics and in-home/in-building communication systems. New directions with optical interconnects, optical transport systems supporting 100/400 Gbps (or even 1 Tbps) bit rates, data center interconnections, green optical networking, as well as multiformat/multirate systems are also included in this discussion.

14.1.1 Complex Modulation and Electronic Compensation Techniques: Higher Data Rate Optical Transport Systems

In the past few decades, scientists and engineers in the optical communications field have developed WDM and optical amplifier technologies to increase the optical communication systems bandwidth and capacity. However, in recent years people have started to seek complex solutions beyond simple intensity modulation and direct detection (IMDD) architectures. As described in Chap. 10, it is widely expected that the combination of advanced modulation formats, such as M-ary QAM and OFDM, and coherent and self-coherent detections using high-speed digital signal processing (DSP) and digital compensation technologies can provide improvements in spectral efficiency and overcome the transmission impairments of optical communication systems. Such techniques are expected to become a part of the next generation of optical communication systems, which can reduce the signaling rate and overcome the fiber-induced distortion and the bandwidth limitations of high-speed optics and electronics (allowing high-speed transmission at 100 Gbps and above). As a result, steadily increasing resources are allocated to the study of these topics [1–18]. This ongoing research is essential for the understanding of the impact of linear and nonlinear transmission effects on future high-speed transmission systems using these formats, and of the techniques to improve their performance.

Another future research direction includes the study of the performance of OFDM/QAM techniques over multi-mode silica fiber (MMF) and POF transmission for short reach transmission (i.e., in-building networks, avionics/car/naval vessel communications, etc.). POF is already been used in high-end vehicles and is starting to be considered in the avionics arena where the majority of airplane data links are short and low speed. However, a few relatively long runs up to 100 m and multiple in-line connectors present a challenge for the POF optical power budget. The main drawback of MMF and POF are high attenuation and a reduced transmission bandwidth due to modal dispersion. A recent capacity experiment in avionics tested MMF segments of various core radii (50, 62.5, and 100 μm) and showed data rate transmission capabilities above 5 Gbps [19]. Several research efforts exist in defining/developing various launching conditions, transmitter technologies, and modulation techniques such as subcarrier multiplexing to increase the usable bandwidth of MMF [20]. This is a result of the observation that unlike in single-mode fiber (SMF) where the fiber frequency response falls steadily toward zero, in MMF the frequency response contains high-frequency components [20]. By modulating an NRZ data stream onto a high-frequency subcarrier using some digital modulation technique (QPSK, OFDM, QAM, etc.) it is possible to increase the available transmission bandwidth of MMF. Complementary to proposed works using MMF and POF, the use of SMF for next-generation avionic platforms is also another area of future research.

Finally, development of multi-format/multi-rate optical systems and networks is another area of future research activity. In these systems and networks there is increased flexibility but at the same time additional considerations that need to be taken into account during network operation. For example, in the physical layer one must consider not only the transmission reach for each modulation format/rate individually, but also the interactions (interference) among the different connections that use different modulation formats/rates. From the network (control plane) perspective, new routing algorithms need to be developed that will account for the cross-rate interference effects, enabling the transmission of connections with acceptable quality of transmission (QoT).

As capacity demand in the backbone network keeps increasing, optical transport systems with higher and higher bit rates are developed. Even though a few short years ago the focus was on 40 and then on 100 Gbps systems, current development and research efforts focus on DWDM systems with 400 Gbps (and even 1 Tbps) data rates. Clearly, there are a number of issues, especially in the physical layer, that need to be resolved prior to the widespread deployment of networks utilizing these high data rates, including increased optical signal-to-noise ratio (OSNR), impact of nonlinearities and others. As outlined in Chap. 10 of this book and in the previous discussion in this section, coherent optical communications in conjunction with the usage of forward error correction (FEC) and DSP at the transmitter and receiver are among the main enablers for the push toward higher and higher data rate transport systems. A number of other approaches that include mode division multiplexing as well as multicore fibers are also topics of current and future research. Especially the latter is the logical expansion of bandwidth in space since the frequency (WDM) aspect is moving toward full utilization.

14.1.2 Next-Generation PON-Based Access Networks

Current access technologies represent a significant bottleneck in bandwidth and service quality between a high-speed residential/enterprise network and a largely overbuilt core backbone network. Backbone networks are provisioned for operation under worst-case scenarios of link failures, and thus backbone links are lightly loaded most of the time. In addition, high capacity routers and ultra-high capacity fiber links have created a true broadband architecture. However, large backbones are not the whole of the equation; distribution of that connectivity to individual enterprises and homes is just as critical for meeting the huge demand for more bandwidth. Unfortunately, the cost of deploying true broadband access networks with current technologies remains prohibitive. This in turn makes it difficult to support end-to-end quality of service (QoS) for a wide variety of applications, particularly non-elastic applications such as voice, video, and multimedia that cannot tolerate variable or excessive delay or data loss.

Passive optical network (PON), as described in detail in Chap. 9, is a technology viewed by many as an attractive solution to the last mile problem as PONs can provide reliable yet integrated data, voice, and video services to end-users at bandwidths far exceeding current access technologies. A PON minimizes the number of optical transceivers, central office terminations, and fiber deployment compared to point-to-point and curb-switched fiber solutions. By using passive components and eliminating regenerators and active equipment normally used in fiber networks, PONs reduce the installation and maintenance costs of fiber as well as connector termination space.

It is widely accepted that multi-channel wavelength-division multiplexed PON (WDM-PON) architectures will emerge as the most promising next-generation access solution that can provide evolutionary upgrade to current time-division multiplexed PONs (TDM-PONs) and meet the growing end-user demands [21–25]. Traditional WDM-PON systems allocate a separate pair of dedicated upstream and downstream wavelength channels to each subscriber, enabling the delivery of a symmetric 100 Mb/s or more of dedicated bandwidth per subscriber/ONU in each direction [21, 22]. In contrast to traditional TDM-PON architectures, WDM-PONs have a unique ability to support any future bandwidth upgrade without altering the physical infrastructure. In addition to their operational simplicity, WDM-PONs provide dedicated point-to-point optical connectivity to each subscriber with bit rate and protocol transparencies, guaranteed QoS, and increased security.

Despite these advantages, traditional tree-based WDM-PON architectures suffer from centralized dynamic bandwidth allocation (DBA), inability to efficiently utilize limited available network resources and to cope with the dynamic and bursty traffic patterns of the emerging integrated triple pay services, inability to support a truly shared Local Area Network (LAN) capability among end users, and lack of simple and cost-effective protection and/or restoration capabilities. Research efforts are currently underway in trying to address the above limitations and devise, model, and analyze WDM-PON architectures, survivability and bandwidth allocation schemes for QoS applications for these architectures [26]. It is anticipated that once the bandwidth and quality barrier of today's access networks is removed, new and unforeseen applications will emerge and attain widespread popularity. Furthermore, as previously mentioned in Chap. 10 new modulation techniques can also be used with the WDM-PON-based access networks to more efficiently deliver broadband services to the access users. Specifically, the use of OFDM in WDM-PON-based access architectures is an area of current and future research interest, as it can allow for the usage of existing low-bandwidth optical components while increasing the data rate of the system.

14.1.3 Converged Optical-Wireless Access Networks

Concurrent with the upsurge of PON-based wireline broadband access solutions, the growing demand for advanced data-centric mobile multimedia services including multimedia messaging, mobile video, mobile music, mobile TV, and IPTV, has accelerated the development and deployment of new wireless broadband access technologies. These emerging technologies, including High-Speed Packet Access (HSPA), fourth-generation (4G) mobile WiMAX, and cellular Long-Term Evolution (LTE) are capable of delivering speeds comparable to or better than current fixed-line broadband access systems—up to 15–200 Mb/s peak air throughput per user. A recent research study suggests that LTE is well positioned to lead the world's mobile infrastructure markets after 2011. A typical LTE architecture consists of two types of network elements supporting the user and control planes. The first is the new enhanced ("Evolved NodeB (eNB)") per 3GPP standards. The second is the new Access GateWay (AGW), which is equivalent to the 3G Radio Network Controller (RNC).

Currently, the wired and wireless services are separately provided by two independent physical networks. Wired networks based on fiber-to-the-home (FTTH) access technologies provide huge bandwidth to users but are not flexible enough to allow roaming connections. On the other hand, wireless networks offer mobility to users, but do not possess abundant bandwidth to meet the ultimate demand for various services and applications. By leveraging the advantages of both of these access technologies combined on an integrated architecture platform, next-generation Fiber-Wireless (FiWi) networks will enable the support of a wide range of emerging fixed-mobile applications and services independent of the access infrastructure. Several research efforts have recently emerged to address the integration of optical and wireless technologies into the envisioned fixed-mobile platform by combining the practically unlimited capacity of fiber-based PON infrastructures with the ubiquity and mobility of wireless networks [15, 16, 27–30]. Most of these research efforts, however, have mainly focused on investigating RoF transmission characteristics and modulation techniques (as outlined in Chap. 8), as well as characterizing the performance gain of the wireless segment (mostly WiFi) when it is collocated with the fiber-based PON segment of the hybrid architecture (hybrid optical-wireless mesh network (WMN) architectures, and hybrid broadband access architectures that attempt to overlay PON and WiFi/WiMAX technologies [28–30]).

However, the overlay architecture essentially consists of the collocation of the two access networks. In this case, the performance gain of the wireless system is mainly due to its utilization of the unlimited capacity of the fiber access infrastructure as a mobile backhaul transport network. To realize the full potential of combining wireline PON and 4G mobile broadband access technologies, an integrated fixed-mobile access infrastructure has to be developed in terms of both hardware and software functionalities that can lead to a multiservice, all-packet-based converged fixed-mobile access networking transport infrastructure that conforms with both PON and 4G mobile access standards [31]. This integrated architecture can then efficiently backhaul and support a mix of advanced wired and wireless multimedia traffic and services along with the diverse QoS, capacity/rate, and reliability requirements set by these services.

14.1.4 Optical Interconnects

As the bandwidth requirements keep increasing, optoelectronic data communication has been migrating from fiber-optic long-distance communication in the 80s to local area networks in the 90s to chip-to-chip optical interconnects for high-performance computing and data transport technology (as described in Chap. 6) [32, 33] now and into the future. As a result, it has become feasible to transfer data at ultra-high data rates at most system hierarchy levels.

Optical interconnects are appealing for chip-to-chip high data rate performance applications, mainly because metal mediums (commonly copper) are becoming the limiting factor of achievable bandwidth. Apart from offering much greater bandwidth capabilities, the optical medium also reduces or eliminates problems like crosstalk, electromagnetic absorption, wave reflection, and impedance matching. However, at the small scale of microelectronics it is difficult to create optical devices and particularly integrate them together. The continuing challenge over the past few years has been to develop technologies and to integrate together the main components of an optical interconnect (laser, waveguide, and photodetector).

There are several research efforts in progress in these areas, as well as in the area of intra-chip connections [34–40]. We foresee that these efforts will intensify, mainly in the intra-chip domain, as these interconnections are becoming more and more complicated due to the extremely small scales.

Current and future research in optical interconnects is also tightly coupled to the advancement of next-generation data centers that house an enormous amount of servers (nodes). These efforts include both the interconnection of different physical data centers, as well as the optical interconnection of different nodes within a single data center [41]. The latter, for example, provides for the efficient creation of high-performance cloud infrastructures with enhanced (distributed) computing capabilities. Research efforts in this specific area of data center interconnections also includes the design of large port count, low-latency, and low-power optical switches that can be used for a variety of applications, with different traffic pattern characteristics, supported by these data centers [42, 43], as well as the design of efficient, fault tolerant topologies and routing approaches for inter-node or inter-data center communication [44–46].

14.1.5 Green Optical Networking

There has been much research activity lately related to energy consumption in traditional telecommunication networks that operate in the electronic domain, mainly due to the large amount of energy consumed by these networks and the rapid increase that is anticipated in the coming years. The optical communications research community has also recently began to tackle this issue, clearly identifying that the energy bottleneck will be an important factor in future optical communication networks. Recent studies have focused on the power consumption of different optical and electronic switching systems and technologies [47–50], as well as on various networking techniques for reducing the number of network resources used in order to reduce the energy consumption. A number of approaches to achieve this include (a) utilizing different grooming algorithms that limit the power consumption as much as possible (trying to route low-rate demands through higher-rate lightpaths in order to use a smaller number of network interfaces) [51], (b) switching-off elements (placing them in sleep-mode when they are not used) or placing entire parts of the network in sleep-mode (taking advantage of time variations in traffic, network over-provisioning, and redundant elements in survivable networks) [52], (c) increasing the use of optical bypass utilizing all-optical wavelength selective cross-connects (WSXCs) or reconfigurable optical add drop multiplexers (ROADMs), and (d) utilizing intelligent routing algorithms that reduce the average number of hops per connection, thus reducing the energy consumption passing through the intermediate nodes. Future research efforts in this area will include the more power-efficient engineering and design of the optical network elements, sub-systems, and systems, the more power-efficient engineering and design of the optical network (e.g., in terms of the numbers of optical amplifiers utilized), and the design of power-efficient networking techniques and algorithms in the same manner as the approaches outlined in (a)–(d) above.

14.1.6 Spectrum-Sliced Elastic Optical Path Networks

Another area of recent research activity that is a paradigm shift to the work that has been conducted since the 1990s on optical systems and networks is the area of *spectrum-efficient elastic optical transport networks* (OTNs) [53]. The motivation behind this work is the need to increase the capacity utilization of the OTN and to eliminate the inefficiencies and capacity over-provisioning that is observed in today's systems that utilize fixed ITU-T grid and wavelength spacing schemes. The concept of "SLICE" is introduced in such networks, which provides an arbitrary contiguous concatenation of optical spectrum, thus allowing the allocation of tailored bandwidth. Having this elastic approach we are now able to better match traffic demands to the allocated SLICEs, thus increasing the capacity utilization of the network. Thus, in this case, instead of implementing RWA on the fixed ITU-T grid, routing and slice assignment (RSA) is implemented over the entire available spectrum.

There are a number of technological approaches to flexible bandwidth networking including orthogonal frequency division multiplexing (OFDM), coherent wavelength division multiplexing (CoWDM), or dynamic optical arbitrary waveform generation/measurement (OAWG/OAWM). The first two multicarrier-based approaches distribute the data on many low-speed subcarriers, while the third approach uses parallel synthesis and coherent combination of many lower bandwidth spectral slices to create the data waveforms [54]. In contrast to the first two approaches, this technique is better in the sense that for the generated waveforms there is no restriction placed on the subcarrier bandwidth and modulation format (bounded only by the bandwidth of the transmitter).

In these types of networks there are a host of problems that are of interest and are currently topics of current and future research. On the components and systems level these include the design of different components on how to best generate/detect these slices (flexible bandwidth transmitters/receivers), different spectrum-efficient transmission techniques, as well as various node architectures that incorporate the "elastic" concept (e.g., flexible bandwidth wavelength selective cross-connects) [53–55]. On the network level there are a multitude of research issues including, among others, efficient RSA algorithms and techniques [56–58], subwavelength grooming techniques, techniques to better match traffic demands to available resources, and finally control plane extensions to incorporate this new networking paradigm.

Clearly, there is a number of open research issues for WDM systems and networks in a multitude of directions. Even though the field has matured considerably in the last few decades there are still questions to be answered and we look forward to the exciting times ahead.

References

1. Kaminow I, Li T, Willner AE (2008) Optical fiber telecommunications VA&VB. Academic Press, New York
2. Tsukamoto S, Katoh K, Kikuchi K (2006) Coherent demodulation of optical multilevel phase-shift keying signal using homodyne detection and digital signal processing. IEEE Photon Technol Lett 18(10):1131–1133
3. Griffin R, Carter A (2002) Optical differential quadrature phase-shift key (ODQPSK) for high capacity optical transmission. In: Proceedings of the IEEE/OSA optical fiber communication conference (OFC), paper WX6, Anaheim, CA
4. Leven A, Kaneda N, Koc UV, Chen YK (2007) Coherent receiver for practical optical communication systems. In: Proceedings of IEEE/OSA optical fiber communication conference (OFC/NFOEC), paper OThK4, Anaheim, CA
5. Sun H, Wu K-T, Roberts K (2008) Real-time measurements of a 40 Gb/s coherent system. Opt Express 16(2):873–879
6. Savory SJ (2008) Digital signal processing options in long haul transmission. In: Proceedings of IEEE/OSA optical fiber communication conference (OFC), paper OTuO3, Anaheim, CA
7. Ip E, Kahn JM (2007) Digital equalization of chromatic dispersion and polarization mode dispersion. IEEE/OSA J Lightwave Technol 25(8):2033–2043
8. Savory SJ, Stewart AD, Wood S, Gavioli G, Taylor MG, Killey RI, Bayvel P (2006) Digital equalization of 40 Gbit/s per wavelength transmission over 2480 km of standard fibre without optical dispersion compensation. In: Proceedings of European conference on optical communications (ECOC), paper Th2.5.5, Cannes, France
9. Renaudier J, Charlet G, Salsi M, Pardo OB, Mardoyan H, Tran P, Bigo S (2008) Linear fiber impairments mitigation of 40-Gbit/s polarization-multiplexed QPSK by digital processing in a coherent receiver. IEEE/OSA J Lightwave Technol 26(1):36–42
10. Shieh W, Bao H, Tang Y (2008) Coherent optical OFDM theory and design. Opt Express 16(2):841–859
11. Gasulla I, Capmany J (2007) High-frequency radio over fiber QPSK transmission through a 5 km multimode fiber link. In: Proceedings of European conference on optical communications (ECOC), Berlin, Germany, p 2
12. Tang JM, Lane PM, Shore KA (2006) Transmission performance of adaptively modulated optical OFDM signals in multimode fiber links. IEEE Photon Technol Lett 18(1):205–207
13. Winzer PJ, Raybon G (2007) 100G Ethernet—a review of serial transport options. In: Proceeding of IEEE LEOS summer topical meetings, paper MA2.2, Portland, OR
14. Fludger CRS, Duthel T, Schulien C (2007) Towards robust 100G Ethernet transmission. In: Proceedings of IEEE LEOS summer topical meetings, paper ME3.1, Portland, OR
15. Madamopoulos N, Pathak B, Antoniades N, Ummy MA (2009) PON ring architectures for truly shared LAN capability and dynamic bandwidth allocation for fiber wireless (FiWi) applications. In: Proceedings of SPIE Photon West (OPTO), San Jose, CA
16. Chang G-K, Jia Z, Yu J, Wang T, Ellinas G (2008) Super broadband optical wireless access technologies. In: Proceedings of IEEE/OSA optical fiber communication conference (OFC/NFOEC), paper OThD1, San Diego, CA
17. Lethien C, Loyez C, Vilcot J-P, Clavier L, Bocquet M, Rolland P (2009) Indoor coverage improvement of MB-OFDM UWB signals with radio over POF system. Opt Commun 282(24):4706–4715
18. Cioffi JM (1991) A multicarrier primer. ANSI T1E1.4 Committee Contribution, 91–157
19. Harres DN (2004) Military aircraft fiber optics networks: status & direction. IEEE Aerospace and Electronic Systems Magazine 19(6):16–22
20. Tyler EJ, Kourtessis P, Webster M, Rochart E, Quinlan T, Dudley SEM, Walker SD, Penty RV, White IH (2003) Toward terabit-per-second capacities over multimode fiber links using SCM/WDM techniques. IEEE/OSA J Lightwave Technol 21(12):3237–3243

21. Lee C-H, Sorin W, Kim B (2006) Fiber to the home using a PON infrastructure. IEEE/OSA J Lightwave Technol 24(12):4568–4583
22. Kazovsky L, Shaw W, Gutierrez D, Cheng N, Wong S (2007) Next-generation optical access networks. IEEE/OSA J Lightwave Technol 25(11):3428–3442
23. An F-T, Kim KS, Gutierrez D, Yam S, Hu E, Shrikhande K, Kazovsky L (2004) SUCCESS: a next-generation hybrid WDM/TDM optical access network architecture. IEEE/OSA J Lightwave Technol 22(11):2557–2569
24. Martinez JJ, Gregorio JIG, Lucia AL, Velasco AV, Aguado JC, Binue MAL (2008) Novel WDM-PON architecture based on a spectrally efficient IM-FSK scheme using DMLs and RSOAs. IEEE/OSA J Lightwave Technol 26(3):350–356
25. Song H, Mukherjee B, Park Y, Yang S (2006) Shared-wavelength WDM-PON access network for supporting downstream traffic with QoS. In: Proceedings of IEEE/OSA optical fiber communication conference (OFC/NFOEC), paper OThK2, Anaheim, CA
26. Erkan H, Hossain ASMD, Dorsinville R, Ali MA, Hadjiantonis A, Ellinas G, Khalil A (2008) A novel ring-based WDM-PON access architecture for efficient utilization of network resources. In: Proceedings of IEEE international conference on communication (ICC), Beijing, China
27. Jia Z, Yu J, Chowdhury A, Ellinas G, Chang G-K (2007) Simultaneous generation of independent wired and wireless services using a single modulator in millimeter-wave-band radio-over-fiber systems. IEEE Photon Technol Lett 19(20):1691–1693
28. Sarkar S, Dixit S, Mukherjee B (2007) Hybrid wireless-optical broadband-access network (WOBAN): a review of relevant challenges. IEEE/OSA J Lightwave Technol 25(11):3329–3340
29. Shaw W-T, Wong S-W, Cheng N, Balasubramanian K, Zhu X, Maier M, Kazovsky LG (2007) Hybrid architecture and integrated routing in a scalable optical-wireless access network. IEEE/OSA J Lightwave Technol 25(11):3443–3451
30. Shen G, Tucker RS, Chae C-J (2007) Fixed mobile convergence architectures for broadband access: integration of EPON and WiMAX. IEEE Commun Mag 45(8):44–50
31. Ali MA, Ellinas G, Erkan H, Hadjiantonis A, Dorsinville R (2010) On the vision of complete fixed-mobile convergence. IEEE/OSA J Lightwave Technol 28(16):2343–2357
32. Chen RT, Lin L, Choi C, Liu YJ, Bihari B, Wu L, Tang S, Wickman R, Picor B, Hibbs-Brenner MK, Bristow J, Liu YS (2000) Fully embedded board-level guided-wave optoelectronic interconnects. Proceedings of the IEEE 88(6):780–793
33. Tang S, Li T, Li F, Wu L, Dubinovsky M, Wickman R, Chen RT (1996) 1-GHz clock signal distribution for multi-processors super-computers. In: Proceedings of the international conference in massively parallel proccessing using optical interconnections (MPPOI), Maui, Hawaii, pp 186–191
34. Estevez C, Guidotti D, Chang G-K (2009) A novel lightwave device integration and coupling process for optical interconnects. In: Proceedings of electronic components and technology conference, San Diego, CA, pp 1859–1864
35. Berger C, Offrein BJ, Schmatz M (2006) Challenges for the introduction of board-level optical interconnect technology into product development roadmaps. Proc of the SPIE 6124(1):61240J-1–61240J-12
36. Oda M, Sakai J, Takahashi H, Kouta H (2007) Chip-to-chip optical interconnection for next-generation high-performance systems. In: Proceedings of IEEE/LEOS annual meeting, Lake Buena Vista, FL, pp 638–639
37. Glebov AL, Lee MG, Yokouchi K (2007) Integration technologies for pluggable backplane optical interconnect systems. Opt Eng 46(1):15403-1–15403-10
38. Schares L, Kash JA, Doany FE, Schow CL, Schuster C, Kuchta DM, Pepeljugoski PK, Trewhella JM, Baks CW, John RA, Shan L, Kwark YH, Budd RA, Chiniwalla P, Libsch FR, Rosner J, Tsang CK, Patel CS, Schaub JD, Dangel RD, Horst F, Offrein BJ, Kucharski D, Guckenberger D, Hegde S, Nyikal H, Lin C-K, Tandon A, Trott GR, Nystrom M, Bour DP,

Tan MRT, Dolfi DW (2006) Terabus: terabit/second-class card-level optical interconnect technologies. IEEE J Sel Top Quant Electron 12(5):1032–1044
39. Dangel R, Berger C, Beyeler R, Dellmann L, Horst F, Lamprecht T, Meier N, Offrein BJ (2007) Prospects of a polymer-waveguide-based board-level optical interconnect technology. In: Proceedings of IEEE workshop on signal propagation on interconnects, Genova, Italy, pp 131–134
40. Fan S-H, Guidotti D, Estevez C, Chang G-K, Chang Y-J, Lu DD (2008) Short-reach flexible optical interconnection using embedded edge-emitting lasers and edge-viewing detectors. Proc of the SPIE 6899:689905-1–689905-11
41. Glick M (2008) Optical interconnects in next generation data centers: an end to end view. In: Proceedings of IEEE symposium on high performance interconnects, Stanford, CA, pp 178–181
42. Proietti R, Ye X, Yin Y, Potter A, Yu R, Kurumida J, Akella V, Yoo SJB (2010) 40 Gb/s 8×8 low-latency optical switch for data centers. In: Proceedings of IEEE/OSA optical fiber communication conference (OFC/NFOEC), paper OMV4, Los Angeles, CA
43. Bergman K (2010) Optically interconnected high performance data centers. In: Proceedings of European conference on optical communications (ECOC), paper Mo.2.D.1, Torino, Italy
44. Al-Fares M, Loukissas A, Vahdat A (2008) A scalable, commodity data center network architecture. In: Proceedings of the ACM special interest group on data communication (SIGCOMM), Seattle, WA, pp 63–74
45. Greenberg A, Lahiri P, Maltz DA, Patel P, Sengupta S (2008) Towards a next generation data center architecture: scalability and commoditization. In: Proceedings of the ACM PRESTO, Seattle, WA, pp 57–62
46. Greenberg A, Hamilton JR, Jain N, Kandula S, Kim C, Lahiri P, Maltz DA, Patel P, Sengupta S (2009) Vl2: a scalable and flexible data center network. In: Proceedings of the ACM special interest group on data communication (SIGCOMM), Barcelona, Spain, pp 51–62
47. Tucker R (2006) The role of optics and electronics in high-capacity routers. IEEE/OSA J Lightwave Technol 24(12):4655–4673
48. Tucker R, Parthiban R, Baliga J, Hinton K, Ayre R, Sorin W (2009) Evolution of WDM optical IP networks: a cost and energy perspective. IEEE/OSA J Lightwave Technol 27(3):243–252
49. Tucker R (2010) Green optical communications—part I: energy limitations in transport. IEEE J Select Top Quant Electron. doi:10.1109/JSTQE.2010.2051216:1-14
50. Tucker R (2010) Green optical communications—part II: energy limitations in network. IEEE J Select Top Quant Electron. doi:10.1109/JSTQE.2010.2051217:1-16
51. Hasan M, Farahmand F, Patel A, Jue J (2010) Traffic grooming in green optical networks. In: Proceedings of IEEE international conference on communication (ICC), Capetown, South Africa
52. Chiaraviglio L, Mellia M, Neri F (2009) Energy-aware backbone networks: a case study. In: Proceedings of IEEE international conference on communication (ICC), Dresden, Germany
53. Jinno M, Takara H, Kozicki B, Tsukishima Y, Sone Y, Matsuoka S (2009) Spectrum-efficient and scalable elastic optical path network: architecture, benefits, and enabling technologies. IEEE Commun Mag 47(11):66–73
54. Yoo SJB (2011) Flexible bandwidth terabit coherent optical communication networks by optical arbitrary waveform generation and measurement. IEEE Photonics Society Newsletter pp 5–12
55. Fontaine NK, Scott RP, Yoo SJB (2011) Dynamic optical arbitrary waveform generation and detection in InP photonic integrated circuits for Tb/s optical communications. Optics Commun 284(15):3693–3705
56. Christodoulopoulos K, Tomkos I, Varvarigos EA (2011) Elastic bandwidth allocation in flexible OFDM-based optical networks. IEEE/OSA J Lightwave Technol 29(9):1354–1366

57. Jinno M, Kozicki B, Takara H, Watanabe A, Sone Y, Tanaka T, Hirano A (2010) Distance-adaptive spectrum resource allocation in spectrum-sliced elastic optical path network. IEEE Commun Mag 48(8):138–145
58. Takara H, Kozicki B, Sone Y, Tanaka T, Watanabe A, Hirano A, Yonenaga K, Jinno M (2010) Distance-adaptive super-wavelength routing in elastic optical path network (SLICE) with Optical OFDM. In: Proceedings of European conference on optical communications (ECOC), Torino, Italy